ADVANCED APPLICABLE
ENGINEERING MATHEMATICS

ADVANCED APPLICABLE ENGINEERING MATHEMATICS

Volume 2

S. VAIDYANATHAN

Professor (Retired)
Anna University
Chennai

IKON BOOKS
Publishers and Distributors

CBS Publishers & Distributors Pvt Ltd

New Delhi • Bengaluru • Chennai • Kochi • Pune
Hyderabad • Kolkata • Mumbai • Nagpur • Patna

CBS Publishers & Distributors Pvt Ltd
4819//XI, 24 Ansari Road, Daryaganj
New Delhi-110 002 (India)

IKON BOOKS, Publishers and Distributors
B-37, First Floor, Office No. 105
Street No. 14, Madhu Vihar
I.P. Extention, Delhi - 110 092
Telephone: 91-11-43618579
Email: ikonbooks@gmail.com
         cs@ikonbooks.com
         Website: www.ikonbooks.com

ADVANCED APPLICABLE
ENGINEERING MATHEMATICS
VOLUME 2

First Edition: 2013

ISBN: 978-81-239-2263-8

CBS Publishers & Distributors Pvt Ltd
Ph: 23289259, 23266861, 23266867
Website: www.cbspd.com

Fax: 011-23243014
e-mail: delhi@cbspd.com;
cbspubs@airtelmail.in

Corporate Office: 204 FIE, Industrial Area, Patparganj, Delhi 110 092
Ph: 4934 4934   Fax: 4934 4935   e-mail: publishing@cbspd.com.
                                               publicity@cbspd.com

Branches

- **Bengaluru:** Seema House 2975, 17th Cross, K.R. Road,
  Banasankari 2nd Stage, Bengaluru 560 070, Karnataka
  Ph: +91-80-26771678/79   Fax: +91-80-26771680   e-mail: bangalore@cbspd.com
- **Chennai:** 20, West Park Road, Shenoy Nagar, Chennai 600 030, Tamil Nadu
  Ph: +91-44-26260666, 26208620   Fax: +91-44-42032115   e-mail: chennai@cbspd.com
- **Kochi:** 36/14 Kalluvilakam, Lissie Hospital Road, Kochi 682 018, Kerala
  Ph: +91-484-4059061-65   Fax: +91-484-4059065   e-mail: kochi@cbspd.com
- **Pune:** Bhuruk Prestige, Sr. No. 52/12/2 + 1 + 3/2 Narhe, Haveli
  (Near Katraj-Dehu Road Bypass), Pune 411 041, Maharashtra
  Ph: +91-20-64704058, 64704059, 32342277 Fax: +91-20-24300160   e-mail: pune@cbspd.com

Representatives

• **Hyderabad** 0-9885175004	• **Kolkata** 0-9831437309	• **Mumbai** 0-9833017933	
• **Nagpur** 0-9021734563	• **Patna** 0-9334159340		

ANNA UNIVERSITY
CHENNAI - 600 025 | INDIA

DR. P MANNAR JAWAHAR
PhD., ME, DMIT, B.Sc., MISTE, MISME,
MISNDT, MSAE(USA), FIE, Chartered Engineer

VICE CHANCELLOR

FOREWORD

The pursuit of gaining knowledge by one's own merited distinction to spread sweetness and light into the younger minds through the language of Mathematics in an environment for science and engineering academic institutions of higher learning is found to be reflected in this excellent treatise on **"Advanced Applicable Engineering Mathematics"** for postgraduate students who study in various Universities in India.

The author S. Vaidyanathan is to be complemented for his passion to spread the love of learning among the teaching and students for understanding the concepts enshrined in mathematics.

The compilation is a reflection of his sincere efforts as an academician in various capacities for more than three decades under the Directorate of Collegiate Education, Directorate of Technical Education and later at Anna University, Chennai functioning under the Government of Tamil Nadu.

The material in this book presented in two volumes covers and extensive array of topics for postgraduate students. The first volume covers topics on Convergence, Analytic Functions, Transform Calculus, Matrices, Vectors, Tensors, Calculus of Variation, Simulation, Fuzzy Logic, Graph Theory, Special Functions, Probability, Statistics, Stochastic Processes and Queuing Theory.

The Second volume deals with Linear and Non-Linear Optimization, Dynamic Programming, Multi Variate Analysis Theory of Estimation, Testing of Hypothesis, Design of Experiments, Statistical Quality Control and Numeral Analysis and Numerical Methods.

The Book is well planned and the treatments lucid. Each chapter has a large number of examples well chosen and many are from the University papers and it richly deserves for students of the Universities in India.

29/12/11

(Dr. P. Mannar Jawahar)

PREFACE

Mathematics is a science of space and numbers having many branches of specialization. Advanced applicable Mathematics deals with progress suitable for the application in modern science and engineering dealing with natural and physical phenomena. Numerical methods deal with the study of extensive numerical complex computations of algorithms. Numerical analysis is a way to do higher mathematical problems on a computer. It is a technique widely used by scientists and engineers to solve their problems. A numerical analysis solution is always numerical.

Analytical methods usually give a result in terms of mathematical function that can be evaluated for specific instances. Numerical results can be plotted to show some of the behaviors of the solution. Another important distinction is that the result from numerical analysis is an approximation, but results can be made as accurate as desired. Even though there are limitations to achieve high accuracy, innumerable operations must be carried out but computers do them so rapidly without even making mistakes. The analysis of computer errors and other sources of errors in numerical methods critically play an important part in achieving the intended results.

The material in this book is divided into three parts – part I has 11 chapters, part II has 8 chapters and part III has 6 chapters. This book starts with the **Chapter One** dealing with fundamental background for insight into the Applicable Mathematics pertaining to Number systems, Convergence, Analytic functions, Taylor Series, Contour Integrals, Residues, Mappings, Schwarz –Christoffel Transformation, Simulation, Fuzzy Logic and Graph Theory. **The Second Chapter** is concerned with Ordinary Differential Equations. **Chapter Three** contains the notion of vector algebra Victor calculus and Tensor Analysis. **Chapter Four** deals with Advanced Matrix Theory with special emphasis on matrix norms, Jordan Canonical From, Singular Value Decomposition, Least square approximations and the Q–R Algorithm.

Chapter Five deals with classification of Partial Differential Equations, Solutions to Initial and Boundary value Problems, Laplace Transform methods for solving Parabolic, Hyperbolic and Elliptic Partial Differential Equations, General Properties of Harmonic Functions, Dirichlet and Neumann Problems, Potential Theory, Fourier Transforms and Convolution Theorem: **Chapter Seven** deals with Transform calculus, Laplace Transforms of derivatives and integrals, Fourier Analysis and Fourier Transforms, both Continuous and discrete and Z-Transforms. The **Eighth Chapter** consists of the various Special Functions. **Chapter Nine** discusses Probability, Random variables, Distribution Functions continuous and discrete probability functions, Bivariate Multiple and Partial Correlation, Characteristic Functions. **Chapter Ten** is concerned with Stochastic Processes, Markov Process and Markov Chain Random Processes, Auto and Cross Correlation, Power Spectral Density and Gaussian Processes. **Chapter Eleven** is devoted to Queuing Theory, Markovian, queuing models, Single and Multiple Servers, Non-Poisson Queues, various types of queueing systems, Distribution of waiting time and Reliability.

Part two begins with **Chapter Twelve** on Linear Programming and deals with Graphical and Simplex Methods, Transportation and Assignment Problems, Dual Simplex Method and Integer Programming. **Chapter Thirteen** deals with Non-linear Optimization, Quadratic Forms, Gradient, Directional derivatives, unconstrained optimization methods like steepest ascent and descent methods, Newton-Raphson Univariate search, Fibonaci search, Hooke's and Jeeves' search methods, Lagrangian method. **Chapter Fourteen** gives emphasis on Dynamic Programming, Principle of optimality. Recursive computation, Shortest path, Cargo loading and Capital budgeting problems, Dimensionality in Dynamic Programming. **Chapters from Fifteen to Eighteen** give a brief account on Separable Programming, Quadratic Programming, Geometric Programming and Stochastic Programming respectively. **Chapter Nineteen** is on Genetic Algorithms, understanding genetics, evolution, biodiversity, robustness in search, Genetic Algorithms Vs Traditional Methods, Simulated Annealing search, Heuristic for discrete optimization. In part three, **Chapters from Twenty to Twenty Four** give prominence to statistical approach to Multivariate Analysis Estimation Theory, Testing of Hypothesis, Design of Experiments and Statistical Quality Control which are essential for Scientists and Engineers in their decision making in every practical situation while solving their problems. The **Last Twenty fifth Chapter** gives prominence to Numerical Methods and Applications in all aspects. It devotes to the following aspects: Determination of the roots of an equation, Direct and Iterative methods, Solution to system of linear equations, Interpolation, Splines, Numerical solution of Ordinary differential equations, Finite Element methods, Shooting method and Stability of Non-linear systems.

The book has evolved in its present form out of the resource material prepared by the author on a period of three decades in teaching Applied Mathematics, Statistical Techniques, Linear and non-Linear optimization techniques to the Post Graduate students of the various

Engineering Branches in some of the established Government Engineering Colleges in Tamil Nadu and after 1970 in the College of Engineering, Guindy and a finally at Anna University in 1978, based on the syllabus which is revised from time to time.

A reasonable number of examples have been worked out and exercises in each section generally based in that section have been included. The author has included University questions set by him from time to time as an examiner to many of the well established Universities in Tamil Nadu offering Post Graduate Programmes in Science and Engineering courses of studies.

Many of the teaching members in Mathematics and Engineering departments have been a perennial source of inspiration and encouragement to the author all through his cordial association with the teaching members of the various faculties. The author has utilized the facilities offered by the Main library and Departmental libraries in the Anna University main campus. The content of the book is based on the latest revised syllabus in Mathematics for more than fifty branches of engineering specializations offered by Anna University Chennai for the Post-graduate students in the engineering departments. The author earnestly hopes that this book will prove to be much of greater utility to the students as well as the teaching members handling the subject.

Suggestions for improvements will be highly appreciated and duly acknowledged.

In writing a text book, it is difficult to make full acknowledgement to the individual authors and publishers of the references in the Bibliography. However the author conveys grateful acknowledgements collectively to the authors and publishers of the books in the Bibliography in referring to the contents of these books.

February, 2010   S.VAIDYANATHAN

Author

CONTENTS

CHAPTER 1

NON LINEAR OPTIMIZATION

1.1 NON LINEAR OPTIMIZATION

1.1.1 Preliminaries

A **convex set** is a collection of points such that, for each pair of points in the collection, the entire line segment joining these two points is also in the collection.

An **extreme point** of a convex set is appoint in the set that does not lie on any line segment which joins two other points in the set.

The simplex multipliers are precisely the dual variables, dual prices or shadow prices.

Economic interpretation of the dual variables

Primal

Min $Z = C^T X = \displaystyle\sum_{j=1}^{n} c_j x_j$

S.T $AX \leq b$, $\displaystyle\sum_{j=1}^{n} a_{ij} x_j \leq b_i$,

$i = 1, \dots m$

$X \geq 0$, $x_j \geq 0$, $j = 1, \dots n$

$Z = W$

$\displaystyle\sum_{j=1}^{n} c_j x_j = \sum_{i=1}^{m} b_i y_i = \sum_{i=1}^{m} \pi_i b_i$

Dual

Max $W = b$

S.T $\displaystyle\sum_{j=1}^{m} a_{ij} y_j \geq c_j$,

$j = 1, \dots n$

y_i unrestricted, $i = 1, \dots m$

$y_i = \pi_i$, $i = 1, 2, \ldots m$, the simplex multipliers are precisely the dual variables or dual prices or shadow prices.

The dual variables represent the worth per unit of resource i. The y_i or π_i are referred to as dual prices or shadow prices.

1.1.2 *A linear programming*

Consists in the determination of the values of the decision variables which maximize (or minimize) the value of the linear objective function subject to linear side constraints.

The **three** components of a typical linear programming problem are:

- An objective function – measures the effectiveness of the system as a mathematical function of its decision variables.
- A set of side constraints – accounts for the physical limitations of the system. And
- Non negative decision variables – these are unknowns to be determined from the solution of the problem.

1.1.3 *The assumptions in the linear programming formulation are*

• Linearity

All the nonnegative decision variables that occur in the formulation of the LPP are of degree one. Hence the objective function and the constraints are linear.

• Additivity

The additivity assumption requires that given any activity levels $(x_1, x_2, \ldots x_n)$, the total usage of each resource and the resulting total measure of effectiveness equal the sum of the corresponding quantities generated by each activity conducted by itself.

• Divisibility

Sometimes the decision variables would have physical significance only if they have integer values. However, the solution obtained by linear programming often is not **integer**. Therefore, the divisibility assumption is that activity units can be divided into **Fractional Levels**, so that **Non-Integer** values for the decision variables are permissible.

• Deterministic

All parameters of L.P. Model – the technological coefficients a_{ij}, the resources b_i and cost per unit or profit per unit of the items produced are known constants.

1.1.4 *Definitions*

Solution – Any specification of values for the decision variables $(x_1, x_2, \ldots x_n)$ is called a solution, regardless of whether it is a desirable or even an allowable choice.

A **Feasible Solution** – is a solution for which all the constraints are satisfied.

Basic Soluution – Given a system of m simultaneous linear equations in n unknowns, $Ax = b$, $(m < n)$, $X^T \in R^n$, where A is an $m \times n$ matrix of rank m. Let B be any $m \times m$ submatrix formed by m linearly independent columns of A. Then a solution obtained by suffering $(n - m)$ variables not associated with the columns of B, equal to zero and solving the system, is called a basic solution to the given system of equations.

The m variables which may be all different from zero are called basic variables.

Degenerate Solution – A basic solution to the system $AX = b$ is called degenerate if one or more of the basic variables vanish.

Basic Feasible Solution – A feasible solution to a LPP which is also a basic solution to the problem is called a basic feasible solution with LPP.

Improved Basic Feasible Solution – Let X_B and $\hat{X}_B$ be two basic feasible solutions to the standard LPP, then $\hat{X}_B$ is said to be an improved basic feasible solution as compared to X_B, if

$\hat{C}_B \hat{X}_B \geq C_B X_B$, $\hat{C}_B$ is the constituted costs components to $\hat{X}_B$.

Optimum Basic Feasible Solution – A basic feasible solution X_B to LPP Max $Z = CX$, S.T. $AX = b$, $x \geq 0$ is called an optimum feasible solution, if

$Z_0 = C_B X_B \geq Z^*$, where Z^* is the value of the objective function for any feasible solution.

A **Corner Point Feasible Solution** – is a feasible solution that does not lie an any line segment connecting two other feasible solutions.

Optimality – guarantees that only better solutions will be encountered.

Feasibility – ensures that of the starting solution is basic feasible, only basic feasible solution will be obtained during computation.

We introduce an **Artificial Variable** to obtain an initial basic feasible solution.

The artificial variables used in linear programming for computational device. They have a high penalty cost and will not appear in the final solution.

Two difficulties of the big-M method when the problem is to be solved on a digital computer.

- M is to be assigned larger value than any other cost coefficients and
- if M is not chosen large enough, incorrect answers may be obtained.

Detection of degeneracy during the solution stage of LPP – The same min θ_0 (with the usual notation) may occur in two or more rows at any stage of iteration so that by introducing one vector into the current basis 2 or more vectors may have to be removed from the basis, making the solution to be come degenerate

1.1.5 Complementary Slackness Conditions

The two results regarding the optimal primal and dual solution:

(i) If at the optimum a primal variable x_j has $z_j - c_j > 0$, then x_j must be non basic and hence at zero level.

(ii) If at the optimum a dual variable y_i has a positive value, the ith primal condition

$\sum_{j=1}^{n} a_{ij} x_j \le b_i$ must be satisfied in equation form because its associated slack must be zero.

$z_j - c_j$ represents the difference between the left and right sides of the dual constraint and hence must represent the dual surplus variable.

If we assume that v_j and s_i are the surplus and slack variables for the jth dual and ith primal constraints, then

$$v_j = z_j - c_j = \sum_{i=1}^{m} a_{ij} y_i - c_j$$

$$s_i = \sum_{j=1}^{n} a_{ij} x_j$$

or where $v_j > 0$, $x_j = 0$
and where $y_i > 0$, $s_i = 0$

We can express both the above conditions in a compact form. For the optimal primal and dual solutions, we have

$v_j x_j = y_i s_i = 0$ for all i and j.

i.e., $x_j \left(\sum_{i=1}^{m} a_{ij} y_i - c_j \right) = 0$, $j = 1, 2, \dots n$

and $y_i \left(b_i - \sum_{j=1}^{n} a_{ij} x_j \right) = 0$, $i = 1, 2, \dots m$

These two are called **Complementary Slackness Conditions**.

1.1.6 Two Computational Advantages of the Revised Simplex Method

In the revised simplex method, computations at each iteration are obtained from the current inverse B^{-1} and the original data of the problem. Thus the adverse effect of machine round off error can be minimized by controlling the round off error in computing B^{-1}.

Additionally, the revised simplex algorithm may lead to fewer computations than in the regular tableau algorithm, depending on the relationship between the number of constraints and the number of variables.

1.1.7 *The linear programming problem*

It consists in determining the values of the decision variables, which minimize (or maximize) the value of the linear objective function subject to linear constraints.

The problem of maximizing or minimizing a given function.

$$Z = f(X) \qquad \qquad ...(1)$$

subject to given constraints

$$g_i(X) \leq, \ = \ or \geq b_i \qquad \qquad ...(2)$$

is called the general constrained optimization problem. The function Z appearing in (1) is called the objective function. In (2), the number of independent equality constraints must be less than n, the number of variables, otherwise the problem is over specified.

Inequalities $\leq$ and $\geq$ types can always be converted into equations by introducing **slack** and **surplus** variables respectively.

Strict inequality constraints have been omitted from (2). This is not a serious limitation in practice, since any constraint of $<$ or $>$ type can be replace by one of $\leq$, $=$ or $\geq$ type by means of some simple manipulations.

A special case is the general unconstrained optimization problem in which there are no constraints, and the problem is merely to find values of x_j which maximize $f(X)$.

When ever constraint in (2) is an equation, we have **classical optimization problem**:

Maximize $Z = f(X)$

Subject to $g_i(X) = b_i$ \qquad \qquad ...(3)

In this problem, the functions f and g_i are assumed to possess continuous first order – partial derivatives with respect to all the variables. Functions with this property are said to belong the class C_1. Necessary conditions for a maximum can be found by classical analytic method of Lagrange multipliers; if we assume further that the functions f and g_i possess continuous second – order partial derivatives with respect to all the variables; i.e. if $f, g_i \in C_2$, then sufficient conditions for a maximum can also be found.

If both $f(X)$ and all $g_i(X)$ are linear functions of the x_j, we have a **Linear Programming Problem**.

Linear Programming is still one of the most widely used optimization techniques, since it has hundreds of useful applications and extremely large problems can now be solved on electronic computer by means of simplex method.

1.1.8 Non Linear Programming

The non linear programming problem can be stated mathematically in the following general form :

Maximize (Minimize) $f(x_1, x_2, \dots x_n)$ subject to $g_i(x_1, x_2, \dots x_j \dots x_n) \leq b_i$, $i = 1, 2, \dots m$ and $x_j \geq 0$ for $j = 1, 2, \dots n$. ...(4)

or maximize (or minimize)

$f(x_j)$ for $i = 1, 2, \dots n$ subject to $g_i(x_j) \leq b_i$ for $i = 1, 2, \dots m$ and $x_j \geq 0$, where $f(x_j)$ and $g_i(x_j)$ are real valued non linear functions of the n positive decision variables. ...(5)

1.1.9 Convex and Concave Functions

Finding the global optimum of the objective function, a fundamental difficulty in the theory and practice of optimization is that most optimization techniques find only **Local Optima** – dynamic programming is a notable exception to this general statement. An obvious way out of this difficulty is to find local optimum and then by comparison the global optimum. This procedure is often time – consuming and at times impracticable. However, there is a special class of functions, namely convex and concave functions for which local and global optima are closely related.

Definitions

A function $f(X)$ is said to be convex over a convex set C in E^n if, for any two points X_1, $X_2 \in C$ and for all $\lambda \in [0, 1]$,

$$f[\lambda X_2 + (1-\lambda)X_1] \leq \lambda f(X_2) + (1-\lambda) f(X_1) \qquad \dots(6)$$

A function $f(X)$ is said to be concave over a convex set C in E^n if, for any two points $X_1, X_2 \in C$ and for all $\lambda \in [0, 1]$,

$$f[\lambda X_2 + (1-\lambda)X_1] \geq \lambda f(X_2) + (1-\lambda) f(X_1) \qquad \dots(7)$$

Convex Set

A set C is said to be convex in E^n, if for any two points $X_1, X_2 \in C$ and for all $\lambda \in [0, 1]$, the entire line segment

$\lambda X_2 + (1 - \lambda)X_1$ also belongs to C.

The set of all feasible solutions to a linear programming problem is a convex set.

The following elementary results are of considerable practical importance.

(i) If $f(x)$ is convex, then $-f(x)$ is concave and vice versa.

(ii) The linear function $Z = C^T X$ is both convex and concave through E^n.

(iii) A convex (concave) function has the property that its value at an interpolated point is less than (greater than) or equal to the value that would be obtained by linear interpolation.

(iv) The sum of finite number of convex (concave) functions is itself a convex (concave) functions.

1.1.10 The following results can be established easily

(v) The positive semidefinite quadratic form $Z = X^T A X$ is convex function throughout E^n.

(vi) A negative semi-definite quadratic form is a concave function throughout E^n.

(vii) The positive definite quadratic form $Z = X^T A X$ is strictly convex function throughout E^n; for in this case $X^T A X > 0$ for $x \neq 0$. Hence for $\lambda \in (0, 1)$, the strict inequality. $\lambda^2 X^T A X < \lambda \, X^T A X$ for all $x \neq 0$

(viii) A negative definite quadratic form is a strictly concave function throughout E^n.

1.1.11 The following theorems can be established

Theorem 1: Let $f(X) \in C_1$ throughout the interior of the convex set X in E^n, and suppose that $f(X)$ is convex over X, then

$$(X_2 - X_1)^T \nabla f(X_2) - f(X_1)$$

for all interior points $X_1 \in X$ and all points $X_2 \in X$.

Theorem 2: Suppose that $f(X)$ is a convex function for $X \geq 0$ and let V be the non-empty set

$$V = \{X : f(X) \leq b, X \geq 0\}$$

Then V is a convex set

Theorem 3: Suppose that $f(X)$ is convex function for $X \geq 0$ and that $f(X) \in C_1$. Then the tangent hyper plane to the hyper surface $f(X) = b$ at the point $x_0 \geq 0$ is a supporting hyperplane at X_0 to the convex set.

$$K = \{X : f(X) \leq b, X \geq 0\}$$

Theorem 4: Let $f(X) \in C_1$ throughout the interior of the convex set X in E_n and suppose that $f(X)$ is concave over X. Then

$$(X_2 - X_1)^T \nabla f(X) \geq f(X_2) - f(X_1)$$

For all interior points $X_1 \in X$ and all points $X_2 \in X$.

Theorem 5: Suppose that $f(X)$ is a concave function for all $X \geq 0$ and let V be the non-empty set

$$V = \{X : f(X) \leq b, X \geq 0\}$$

Then V is convex set.

Theorem 6: Let $f(X)$ be convex function over the closed convex set X in E_n. Then any local minimum of $f(X)$ is also the global minimum $gf(X)$ over X.

Theorem 7: Let $f(X)$ be a convex function over the closed convex set X in E^n. Then the set of points at which $f(X)$ takes on its global minimum is convex set.

1.2 Uni Modal Multimodal Functions

Suppose that a function $\phi(X)$ of real variable X attains its maximum value in the interval $[0, L]$ at $X = X^*$, is strictly increasing for $X < X^*$ and is strictly decreasing for $X > X^*$. Then $\phi(X)$ is said to be unimodal in $[0, L]$. Note that $\phi(X)$ may or may not be continuous at $X = X^*$ **Unimodal Functions.** If over a given region, a function increases (decreases) to a certain point and then decreases (increases) monotonically, the function is said to be **Unimodal, A unimodal** function has only one peak (valley). **Multi Modal:** Functions with two or more peaks are called "Multi Modal".

1.3 Gradient – Steepest Ascent and Steepest Descent Directions

1.3.1 *Directional Derivative*

P is point in space and a direction at P, given by the unit vector $\vec{b}$. C is any ray from P in the direction $\vec{b}$. Q is a point on C at distance s from P. If the limit

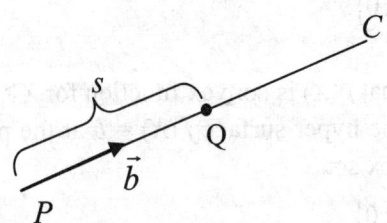

$$\frac{\partial f}{\partial s} = \lim_{s \to 0} \frac{f(Q) - f(P)}{s}, \text{ exists,} \qquad \text{...(8)}$$

it is called the directional derivative of f at P in the direction of $\vec{b}$.

In the **Cartesian** frame of reference, if P has a position vector $\vec{a}$ then the ray C can be represented in the form

$$\vec{r}(s) = x(s)i + y(s)j + z(s)k = \vec{a} + s\vec{b}, (s \geq 0) \qquad \text{...(9)}$$

and $\dfrac{\partial f}{\partial s}$ is the derivative of the function, $f[x(s), y(s), z(s)]$ with respect to the arc length s of

C. Assuming that f has continuous first partial derivatives, we have

$$\frac{\partial f}{\partial s} = \frac{\partial f}{\partial x}\frac{dx}{ds} + \frac{\partial f}{\partial y}\frac{dy}{ds} + \frac{\partial f}{\partial z}\frac{dz}{ds} \qquad \qquad \text{...(10)}$$

From (9),

$$\frac{\partial \vec{r}}{\partial s} = \frac{dx}{ds}i + \frac{dy}{ds}j + \frac{dz}{ds}k = \vec{b}$$

Introducing the vector

$$\text{grad } f = \frac{\partial f}{\partial x}i + \frac{\partial f}{\partial y}j + \frac{\partial z}{\partial z}k \qquad \qquad \text{...(11)}$$

And write (10) in the form of an inner product (dot product)

$$\frac{\partial f}{\partial s} = \vec{b}.\text{grad } f \quad \left(|\vec{b}|=1\right) \qquad \qquad \text{...(12)}$$

The vector grad f is called the gradient of the scalar function f. By introducing the differential operator

$$\nabla = \frac{\partial}{\partial x}i + \frac{\partial}{\partial y}j + \frac{\partial}{\partial z}k \quad \text{(del)}$$

we may write

$$\text{grad } f = \nabla f = \frac{\partial f}{\partial x}i + \frac{\partial f}{\partial y}j + \frac{\partial z}{\partial z}k$$

The notation ∇f for the gradient is frequently used in literature.

If in particular $\vec{b}$ has the direction of the positive x axis, then $\vec{b} = i$ and

$$\frac{\partial f}{\partial s} = \vec{b}.\text{grad } f = \frac{\partial f}{\partial x}i.i = \frac{\partial f}{\partial x}$$

Similarly,

$$\frac{\partial f}{\partial s} = \frac{\partial f}{\partial y}j.j = \frac{\partial f}{\partial y}$$

and

$$\frac{\partial f}{\partial s} = \frac{\partial f}{\partial z}k.k = \frac{\partial f}{\partial z}$$

1.3.2 Gradient as normal vector to surfaces

Consider a differentiable scalar function $f(x, y, z)$ in space. Suppose that for each constant c, the equation

$$f(x, y, z) = c = \text{const} \qquad \qquad ...(13)$$

represents a surface S in space. Then by letting c assume all values we obtain a family of surfaces, which are called the **Level Surfaces** of the function f.

A curve C in space may be represented in the form

$$\vec{r}(t) = x(t)i + y(t)j + z(t)k \qquad \qquad ...(14)$$

If we now require that C lies on S, then the function $x(t), y(t), z(t)$ in (14) must be such that

$$f[x(t), y(t), z(t)] = c.$$

By differentiating this with respect to t and using the chain rule, we obtain

$$\frac{\partial f}{\partial x}\frac{dx}{dt} + \frac{\partial f}{\partial y}\frac{dy}{dt} + \frac{\partial f}{\partial z}\frac{dz}{dt} = (\text{grad } f).\frac{d\vec{r}}{dt} = 0$$

when the vector

$$\frac{d\vec{r}}{dt} = \frac{\partial x}{\partial t}i + \frac{\partial y}{\partial t}j + \frac{\partial z}{\partial t}k$$

is tangent to C.

By considering curves of S passing through a point P of S in various directions, these tangents will lie in the same plane which touches S at P. This plane is called the **Tangent Plane** of S at P. The straight line through P and perpendicular to the tangent plane is called the normal of S at P.

1.3.3 *Gradient Methods for Unconstrained Optimization*

Gradient methods for finding a local maximum or minimum of an unconstrained function $f(X)$ are based ultimately on the simple fact that $f(X)$ increases or decreases in the directo

n $\vec{d}$ ascending as the directional derivative $[\nabla f(X)]^T.\vec{d}$ is positive or negative.

We consider functions that are twice continuously differentiable. The general idea is to generate successive points, starting from a given initial point, in the direction of the fastest increase (maximization) of the function. The technique is known as the **Gradient Method** because the gradient of the function at a point is indicative of the fastest rate of increase.

1.3.4 *Steepest Ascent Method*

Termination of the gradient method occurs at the point where the gradient vector becomes null. This is only a necessary condition for optimality. It is emphasized that optimality cannot be verified unless it is known a priori that $f(X)$ is concave or convex.

Suppose that $f(X)$ is maximized. Let X^0 be the initial point from which the procedure starts and define $\nabla f(X^k)$ as the gradient of f at kth point X^k. The idea of this method is to determine a particular path p along which $\dfrac{df}{dp}$ is maximized at a given point. This result is achieved if successive points X^k and X^{k+1} are selected such that

$$X^{k+1} = X^k + r^k \nabla f(X^k) \qquad \qquad ...(15)$$

where r^k is a parameter called the optimal step size.

The parameter r^k is determined such that X^{k+1} results in the largest improvement in f. In other words, if the function $h(r)$ is defined such that

$$h(r) = f\left[X^k + r^k \nabla f(X^k)\right] \qquad \qquad ...(16)$$

r^k is the value of r maximizing $h(r)$. Since $h(r)$ is a single variable function, the direct search method may be used to find the optimum provided that $h(r)$ is strictly unimodel.

The proposed procedure terminates when two successive trial points X^k and X^{k+1} are approximately equal. This is equivalent to having

$$r^k \nabla f(X^k) \cong 0$$

Under the assumption that $r^k \neq 0$ which will always be true unless X^0 happens to be the optimum of $f(X)$, that in equivalent to the necessary condition $\nabla f(X^k) = 0$

1.3.5 *Method of Steepest Descent*

This method is also iterative, proceeding from an initial approximation say X^1 for the minimizing point to successive points $X^2, X^3, ...$, until some stopping condition is satisfied.

Given the current point X^k the point X^{k+1} is obtained by a linear search in the direction $\vec{d}^k$, where

$$\vec{d}^k = -\nabla f(X^k) = -g(X^k) = -g^k \qquad \qquad ...(17)$$

i.e. $\vec{d}^k$ is the negative gradient vector at the point X^k. It is well known that this direction is direction from X^k in which the initial rate of decrease of $f(X)$ is greatest.

Thus the sequence $\{X^k\}$ is defined by

$$X^{k+1} = X^k + (\lambda^k) * \vec{d}^k$$
$$= X^k + (\lambda^k) * g^k \qquad \qquad ...(18)$$

where X^1 is given $(\lambda^k)*$ is determined by the linear search so that X^{k+1} minimizes $f(X)$ in the direction $-g^k$.

The search E^n for a minimum of $f(x_1, x_2, ... x_n)$ is therefore reduces to a sequence of linear searches.

It is essential that the sequence $\{X^*\}$ should converge to desired optimal point.

1.4 UNCONSTRAINED NON LINEAR ALGORITHM

Classical optimization uses calculus to determine the points of maxima and minima for unconstrained and constrained problems. These are not suitable for computation. Hence there is necessity to develop working algorithms.

In practice the rate of convergence of the sequence $\{X_k\}$ is important. A major disadvantage of the steepest decent (or ascent) method is that no account is taken of the second derivatives of $f(X)$ and yet the curvature of the function – which determines its behavior near the minimum (or maximum) depends on these derivatives. This means that the rate of convergence of the sequence $\{X^*\}$ may be very slow. The Newton – Raphson method over comes this disadvantage.

1.4.1 The Newton – Raphson Method

For analytic purposes, the simplest function (of any member of variables) with a strong minimum is quadratic with a positive definite Hessian matrix. The basic idea of the Newton – Raphson method is to approximate the given function $f(X)$ in each iteration by such a quadratic function and then to move the current point to the turning point of quadratic.

Assuming that $f(X) \in C_2$ in a suitable neighbour of the current point X_k, we evaluate $f(X)$ and its first and second derivatives at $X = X_k$, and find quadratic function $y_k(X)$ which matches these values at $X = X_k$. This gives

$$y_k(X) = \frac{1}{2}(X - X_k)^T G_k(X - X_k) + (X - X_k)g_k + f(X_k) \qquad \text{...(19)}$$

Where G_k is the Hessian matrix and g_k is the gradient vector $f(X)$, both evaluated at $X = X_k$.

Suppose that $y_k(X)$ takes its minimum value at $X = X_m$. Then $\nabla y_k(X_m) = 0$, i.e., $G_k(X_m - X_k) + g_k = 0$, which yields

$$X_m = X_k - G_k^{-1}g_k \text{ (if } G_k \text{ is non-singular)} \qquad \text{...(20)}$$

The Newton – Raphson method uses X_m as the next current point, giving the iterative formula

$$X_{k+1} = X_k - G_k^{-1}g_k, \quad k = 1, 2, \dots \qquad \text{...(21)}$$

A variant of equation (21) that is often used and which is usually superior is

$$X_{k+1} = X_k - \lambda_k^* G_k^{-1}g_k, \quad k = 1, 2, \dots \qquad \text{...(21)}$$

Where λ_k^* is determined by a linear search from X_k in the direction $-G_k^{-1}g_k$.

Greater numerical accuracy is attained if the variables X_j are scaled so that the Hessian matrices G_k are approximately equal to the unit matrix – that is true of all algorithms which make use of Hessian matrices or their inverses, Normally, the variables are scaled only once, at the beginning of the calculation.

The convergence of both forms (21) and (22) of the Newton – Raphson method is rapid when X_k is near the optimal point X^*. Whichever form is used both $|X_{k+1} - X_k|$ and $|g_k|$ should be tested in a convergence criterion. The success of the method depends on the direction $-G_k^{-1}g_k$ being a direction descent. Thus we require

$$(-G_k^{-1}g_k)^T\, g_k\ < 0$$

i.e., $\quad g_k^T G_k^{-1} g_k\ > 0$ ...(23)

which is satisfied at all points for which $g_k \neq 0$ if G_k is positive definite.

1.4.2 *Taylor Series Expansion*

A second order Taylor expansion of $f(X)$ about the point X_0 in some convex region R in E^n in terms of the gradient operator and the Hessian is given by

$$f(X) =\ f(X_0) + (X - X_0)\nabla f^T(X_0) + \frac{1}{2}(X - X_0)^T H\big(X_0 + \theta(X - X_0)\big)(X - X_0)$$

where $0 \leq \theta \leq 1$

1.5 Necessary and Sufficient Conditions for Optimum

Let $f(X)$ be an n variable function. We assume that the first and second partial derivatives of $f(X)$ are continuous at every X.

A necessary condition for X_0 to be an extreme point of $f(X)$, is that $\nabla f(X_0) = 0$

Proof

By Taylor's theorem for $0 < \theta < 1$,

$$f(X_0 + \mathbf{h}) - f(X_0) =\ \nabla f(X_0)\mathbf{h} + \frac{1}{2}\mathbf{h}^T H \mathbf{h}\big|_{X_0 \to \theta\mathbf{h}}$$

where $\mathbf{h} = (h_1, h_2, \dots h_n)$ and $|h_j|$ is sufficiently small for all j.

For sufficiently small $|h_j|$, the remainder term $\frac{1}{2}\mathbf{h}^T H \mathbf{h}$ is of the order h_j^2, hence

$$f(X_0 + \mathbf{h}) - f(X_0) =\ \nabla f(X_0)\mathbf{h} + O(h^2) \approx \nabla f(X_0)\mathbf{h}$$

It can be shown by contradiction that $\nabla f(X_0)$ must vanish at a minimum point X_0. Otherwise if it does not, then for a specific j the following condition will hold.

$$\frac{\partial f}{\partial x_j} < 0 \text{ or } \frac{\partial f}{\partial x_j} > 0$$

By selecting h_j with appropriate sign, it is always possible to have

$$h_j \frac{\partial f}{\partial x_j}(X_0) < 0$$

Setting all other h_j equal to zero, Taylor's expansion yields

$$f(X_0 + \mathbf{h}) < f(X_0)$$

This result contradicts the assumption that X_0 is a minimum point. Consequently $\nabla f(X_0)$ must equal zero. A similar proof can be established for the maximization case.

Because the necessary condition is also satisfied for inflection and saddle points, the points obtained from the solution of

$$\nabla f(X_0) = 0$$

are referred to as **stationary** points.

A sufficient condition for a stationary point X_0 to be an extreme point is that the Hessian H evaluated at X_0 satisfy the following conditions.

- H is positive definite if X_0 is a minimum point.
- H is negative definite if X_0 is a maximum point.

Proof

By Taylor's theorem, for $0 < \theta < 1$,

$$f(X_0 + \mathbf{h}) - f(X_0) = \nabla f(X_0)\mathbf{h} + \frac{1}{2}\mathbf{h}^T H\mathbf{h}\,|_{X_0 \to \theta\mathbf{h}}$$

Given X_0 is a stationary point, then $\nabla f(X_0) = 0$. Thus

$$f(X_0 + \mathbf{h}) - f(X_0) = \frac{1}{2}\mathbf{h}^T H\mathbf{h}\,|_{X_0 \to \theta\mathbf{h}}$$

If X_0 is a minimum point, then $f(X_0 + \mathbf{h}) = f(X_0), \mathbf{h} \neq 0$

Thus for X_0 to be a minimum point, it must be true that

$$\frac{1}{2}\mathbf{h}^T H\mathbf{h}\,|_{X_0 \to \theta\mathbf{h}} > 0$$

Given that second partial derivative is continuous, the expression $\frac{1}{2}\mathbf{h}^T H\mathbf{h}$ must have

the same sign both at X_0 and $X_0 + \theta\mathbf{h}$. Because $\mathbf{h}^T H\mathbf{h}\,|_{X_0 \to \theta\mathbf{h}}$ defines a quadratic form this expression and hence $\mathbf{h}^T H \mathbf{h}|_{X_0 \to \theta\mathbf{h}}$ is positive if and only $H|_{X_0}$ is positive definite. This means a sufficient condition for the stationary point X_0 to be a minimum is that Hessian matrix, H evaluate at the same point is positive – definite.

A similar proof for the maximization case shows that the corresponding Hessian matrix must be negative – definite.

1.6 Jacobian Method for Equality Constraints

Constrained derivatives (Jacobian) method.

Minimize $\quad z = f(X)$

Subject to $g(X) = 0$ where

$$X = (x_1, x_2, \ldots x_n) \text{ and}$$
$$g = (g_1, g_2, \ldots g_m)^T$$

The functions $f(X)$ and $g_i(X)$, $i = 1, 2, \ldots, m$ are twice continuously differentiable.

The idea of using the constrained derivatives is to develop a closed – form expression for the first partial derivatives of $f(X)$ at all points that satisfy the constraints $g(X) = 0$. The corresponding stationary points are identified as the points at which these partial derivatives vanish.

By Tayler's theorem for $X + \Delta X$ in the feasible neighbourhood of X, we have

$$f(X + \Delta X) - f(X) = \nabla f(X)\Delta X + O(\Delta x_j^2)$$

and $g(X + \Delta X) - g(X) = \nabla g(X)\Delta X + O(\Delta x_j^2)$

As $\Delta x_j \to 0$, the above equations reduce to

$$\delta f(X) = \nabla f(X)\delta X$$

and $\quad \delta g(X) = \nabla g(X)\delta X$

For feasibility, we must have $g(X) = 0$, $\delta g(X) = 0$ and it follows that

$\delta f(X) - \nabla f(X)\delta X = 0$

$\nabla g(X)\delta X = 0$

This gives $(m + 1)$ equations in $(n + 1)$ unknowns $\delta f(X)$ and δX. Note that $\delta f(X)$ is a dependent variable and hence is determined once δX is known. This means that we have m equations in n unknowns.

If $m > n$, at least $(m - n)$ equations are redundant. Eliminating redundancy the system reduces to $m \le n$. If $m = n$, the solution is $\delta X = 0$ and X has no feasible neighbourhood, which means that the solution space consists of one point only. The remaining case, when $m < n$ requires further elabaration.

Define $\quad X = (Y, Z)$

such that

$$Y = (y_1, y_2, \dots y_n), Z = (z_1, z_2, \dots z_{n-m})$$

The vectors Y and Z are called the dependent and independent variables, respectively. Rewriting the gradient vectors of f and g in terms of Y and Z.

We get,

$$\nabla f(Y, Z) = (\nabla_y f, \nabla_z g)$$

and $\nabla g(Y, Z) = (\nabla_y g, \nabla_z g) = (J, C)$

Define

$$J = (\nabla_y g_1, \nabla_y g_2, \dots \nabla_y g_m)^T$$

$$C = (\nabla_z g_1, \nabla_z g_2, \dots \nabla_z g_m)^T$$

$J_{m \times m}$ is called **Jacobian Matrix** and $C_{m \times (n-m)}$, the control matrix. The Jacobian matrix J is assumed to be non singular. This is always possible because the given in equations are independent by definition. The components of the vector Y must thus be selected such that the matrix J is non-singular.

The original set of equations in $\delta f(X)$ and δX may be written as

$$\delta f(Y, Z) = \nabla_y f \delta y + \nabla_z f \delta z$$

and $\qquad J \delta y = - C \delta Z$

Because J is nonsingular, its inverse J^{-1} exists. Hence

$$\delta y = - J^{-1} C \delta Z$$

Substituting for δy in the equation for $\delta f(X)$ gives δf as a function of δz, that is

$$\delta f(Y, Z) = (\nabla_z f - \nabla_y f J^{-1} C) \delta z$$

From this equation, the constrained derivative with respect to the independent vector Z is given by

$$\nabla_c f = \frac{\delta_c f(Y, Z)}{\delta_c Z}$$

$$= \nabla_z f - \nabla_y f J^{-1} C,$$

Where $\nabla_c f$ is the constrained gradient vector of f with respect to Z. Thus $\nabla_c f(Y, Z)$ must be null at the stationary points.

1.7 Univariate Search Method

In this method, each time we minimize with respect to one variable. For example consider the problem.

Minimize $f(X) = f(x_1, x_2) = x_1^2 - x_1 x_2 + 3x_2^2$

Let the starting point be (given) taken as $X_1 = (1, 1)$

Step 1: Find the step size say λ_1 along x_1 by minimizing $f(1 + \lambda_1, 1)$

$\therefore \quad f(1 + \lambda_1, 1) = (1 + \lambda_1)^2 - (1 + \lambda_1) + 3$

For minimum

$$\frac{df}{d\lambda_1} = 2(1 + \lambda_1) - 1 = 0 \Rightarrow 2\lambda_1 = -1 \text{ or } \lambda_1 = -\frac{1}{2}$$

$$\frac{d^2 f}{d\lambda_2^2} = 2 > 0 \text{ for } \textbf{Minimum} \text{ is satisfied}$$

$\therefore$ The new point is $\left(1, -\frac{1}{2}, 1\right) = \left(\frac{1}{2}, 1\right) = X_2$ (say)

Step 2: Now find the step size λ_2 where X_2 by minimizing

$$f\left(\frac{1}{2}, 1 + \lambda_2\right) = \frac{1}{4} - \frac{1}{2}(1 + \lambda_2) + 3(1 + \lambda_2)^2$$

For minimum $\dfrac{df}{d\lambda^2} = -\dfrac{1}{2} + 6(1 + \lambda_1) = 0$ or $\lambda_2 = -\dfrac{11}{12}$

$$\frac{d^2 f}{d\lambda_2^2} = 6 > 0 \text{ for Minimum.}$$

$\therefore$ The new point is $\left(\dfrac{1}{2}, 1 + \lambda_2\right)$ is $\left(\dfrac{1}{2}, 1 - \dfrac{11}{12}\right) = \left(\dfrac{1}{2}, \dfrac{1}{12}\right) = X_3$

Repeat step 1 and 2, with starting point as $X_3 = \left(\dfrac{1}{2}, \dfrac{1}{12}\right)$

The procedure is repeated till the quantities $|\lambda_2|$ are less than some prefixed tolerance.

1.8 Fibonacci Sequence and Search

1.8.1 *The Fibonacci Sequence {F_n} is defined by the Recurrence Relation*

$F_0 = F_1 = 1, F_n = F_{n-1} + F_{n-2}, n \geq 2$ ...(8.1)

$F_n - F_{n-1} - F_{n-2} = 0$

$E^2 F_{n-2} - EF_{n-2} - F_{n-2} = 0$

or $(E^2 - E - 1)F_{n-2} = 0$

$m^2 - m - 1 = 0$ $(E^2 - E - 1)F_n = 0$

$$m = \frac{1 \pm \sqrt{1+4}}{2} = \frac{1 \pm \sqrt{5}}{2}$$

$$\therefore \qquad m_1 = \frac{\sqrt{5}+1}{2}, m_2 = \frac{-(\sqrt{5}-1)}{2}$$

$$\therefore \qquad F_n = A\left(\frac{\sqrt{5}+1}{2}\right)^n + B(-1)^n \left(\frac{\sqrt{5}-1}{2}\right)^n \qquad \qquad ...(8.2)$$

$n = 0$ $F_0 = A + B = 1$

$n = 1$ $F_1 = A\left(\frac{\sqrt{5}+1}{2}\right) + B(-1)\left(\frac{\sqrt{5}-1}{2}\right) = 1$

$$A\frac{\left(\sqrt{5}+1\right)}{2} - B\frac{\left(\sqrt{5}-1\right)}{2} = 1$$

$$A\left(\frac{\sqrt{5}+1}{2}\right) - (1-A)\left(\frac{\sqrt{5}-1}{2}\right) = 1$$

$$A\left(\frac{\sqrt{5}+1}{2} + \frac{\sqrt{5}-1}{2}\right) = 1 + \frac{\sqrt{5}-1}{2}$$

$$A\sqrt{5} = \frac{\sqrt{5}+1}{2}$$

$$A = \frac{\sqrt{5}+1}{2\sqrt{5}} = \frac{1}{2}\frac{\sqrt{5}}{5}\left(\sqrt{5}+1\right)$$

$$B = 1 - A = 1 - \frac{\sqrt{5}}{10}\left(\sqrt{5}+1\right)$$

$$= \frac{10 - 5 - \sqrt{5}}{10} = \frac{5 - \sqrt{5}}{10} = \frac{\sqrt{5}\left(5 - \sqrt{5}\right)}{10}$$

$$\therefore A = \frac{\sqrt{5}\left(\sqrt{5}+1\right)}{10} \quad B = \frac{\sqrt{5}}{10}\left(\sqrt{5}-1\right)$$

$$\therefore F_n = \frac{\sqrt{5}}{10} \frac{\left(\sqrt{5}-1\right)^{n+1}}{2^n} + (-1)^n \frac{\sqrt{5}\left(\sqrt{5}-1\right)^{n+1}}{10 \quad 2^n} \qquad \ldots(8.3)$$

1.8.2 Fibonacci Search

The function $\phi(X)$ of the real variable X in the internal $[0, L]$ is Unimodal with X^* as the maximum value. It is strictly increasing for $X < X^*$ and strictly decreasing $X > X^*$, the function $\phi(X)$ may or may not be continuous at $X = X^*$.

X^* is to be found by means of function evaluations and comparisons only.

The Fibonacci sequence $\{F_n\}$ is defined by

$F_0 = F_1 = 1, F_n = F_{n-1} + F_{n-2}, n \geq 2$

...(8.4)

The Fibonacci search is based on the following theorem.

Theorem: Let $\phi(X)$ be a unimodal function defined on a closed interval $[0, L_n]$ where L_n is such that X^* can be located with an error of less than unity by making at most n evaluations of $\phi(X)$. Let $F_n = l.u.b. \ L_n$.

Then the sequence $\{F_n\}$ is Fibonacci sequence.

Proof

By induction

If $n = 0$ or 1 i.e. we have no value of $\phi(X)$ or only one value of $\phi(X)$, then we have no information about X^* and hence $F_0 = F_1 = 1$ (the 1 being the given length of the final interval of uncertainty). If $n \geq 2$, we evaluate $\phi(X)$ at $X = X_1$ and $X = X_2$, where $X_1, X_2 \in (0, L_n)$ and $X_1 < X_2$. These two values of X are to be determined subsequently.

If $n = 2$, $L_2 = 2 - \epsilon$, where $0 < \epsilon < 1$. We can take $X_1 = 1 - \frac{1}{2}\epsilon$ (the mid point of the interval) and $X_2 = 1$.

Choice of X_1, X_2 when $n = 2$

The choice of X_1 and X_2 locates X^* with an error of less than unity in the interval $(0, 1)$ if $\phi(X_1) > \phi(X_2)$ and in the interval $\left(1 - \frac{1}{2}\epsilon, 2 - \epsilon\right)$ if $\phi(X_1) < \phi(X_2)$.

Hence

$$l.u.b\ L_2 \geq 2$$

On the other hand,

$$l.u.b\ L_2 < 2 + \delta$$

for any $\delta > 0$, for if $L_2 > 2$, it is impossible to find two overlapping segments each of length unity which together cover the interval $[0, L_2]$. Therefore

$$F_2 = 2 = F_1 + F_2,$$

which begins the induction.

Now assume that

$$F_k = F_{k-1} + F_{k-2}, (k = 2, 3, \dots n-1) \qquad \qquad \dots (8.5)$$

We have to show that above equation is also true for $k = n$ and we do this by proving

(i) $F_n \leq F_{n-1} + F_{n-2}$,

(ii) $F_n \geq F_{n-1} + F_{n-2}$

Suppose the evaluate $\phi(X)$ at $X = X_1$ and $X = X_2$, where $X_1, X_2 \in (0, L_n)$ and $X_1 < X_2$. If $\phi(X_1) > \phi(X_2)$, then the remaining interval of uncertainty is $[0, X_2)$ and we already know $\phi(X_1)$, where $X_1 \in (0, X_2)$. Hence $X_1 < F_{n-2}$, since X^* may lie in the interval $[0, X_1]$ and we are allowed only $(n-2)$ further evaluations. Also $X_2 < F_{n-1}$, since X^* must be located with an error of less than unity in $[0, X_2]$ by evaluating $\phi(X)$ at $X = X_1$ and at $(n-2)$ other points at most.

Similarly if $\phi(X_1) < \phi(X_2)$, then the remaining interval of uncertainty is $(X_2, L_n]$, and

$$L_n - X_1 < F_{n-1}$$

Hence

$$L_n < F_{n-1} + X_1$$
$$< F_{n-1} + F_{n-2}$$

It follows that

$$F_n = l.u.b\ L_n \leq F_{n-1} + F_{n-2} \qquad \qquad \dots (8.6)$$

Hence (i) is proved.

To prove (ii), we can choose

$$L_n = \left(1 - \frac{1}{2}\in\right)(F_{n-1} + F_{n-2})$$

with

$$X_1 = \left(1 - \frac{1}{2}\in\right)F_{n-2},\ X_2 = \left(1 - \frac{1}{2}\in\right)F_{n-1}\ \text{where}\ 0 < \in < 1 \qquad \dots (8.7)$$

The choice of the factor $\left(1 - \frac{1}{2}\in\right)$ in the equation (8.7) is consistent with $L_2 = 2 - \in$, as used previously. Equation (8.7) shows that is L_n is taken to be any fixed number less than $(F_{n-1} + F_{n-2})$, then it is always be increased, since $\in$ may be made arbitrarily small. It follows that

$$F_n = l.u.b \geq F_{n-1} + F_{n-2} \qquad \qquad \dots (8.8)$$

These inequalities (8.6) and (8.8) give

$$F_n = F_{n-1} + F_{n-2} \qquad \qquad ...(8.9)$$

Thus the equation (8.5) is true for $k = n$ and hence for all $k \geq 2$. But $F_0 = F_1 = 1$ and so $\{F_n\}$ is the Fibonacci sequence (8.4).

The following method for maximizing $f(X)$ on the interval $0 \leq x \leq 1$ is suggested by the proof of theorem. For any $\in$, $0 < \in < 1$, suppose that $\phi(X) \equiv f\left(\dfrac{lX}{L_n}\right)$ is defined on the interval $0 \leq X \leq L_n$ where $L_n = F_n$. This formula comes from equation (8.7) and (8.9) using X_1 and X_2 from equation (8.7), compare the values $\phi(X_1)$ and $\phi(X_2)$, leaving an interval of uncertainty of length $L_{n-1} = \left(1 - \dfrac{1}{2}\in\right) F_{n-1}$, together with the value of $\phi(X)$ at one of the two optimal points of evaluation for this smaller interval. One further evaluation and comparison leaves an interval uncertainty of length $L_{n-2} = \left(1 - \dfrac{1}{2}\in\right) F_{n-2}$, proceeding in this way we have after $(n - k + 1)$ evaluations $L_k = \left(1 - \dfrac{1}{2}\in\right).F_k$, $(2 \leq k \leq n)$. In particular, after $(n - 1)$

evaluations, we have $L_2 = \left(1 - \dfrac{1}{2}\in\right) F_2, 2 - \in$.

The nth and the final evaluation is at one of the points X_1, X_2 indicated in figure 1 and hence the final interval of uncertainty is less that unity. (Note the final error in X^* is not greater than max $\left(1 - \dfrac{1}{2}\in, \in\right)$. Set $\in = 0$. X_1 and X_2 coincide.

1.8.3 Unconstrained Algorithm

This section presents the Direct Search Algorithm

Director search method

This method applies only to unimodal single – variable functions. The optimization of single variable functions plays a key role in the development of the more general multi variate algorithms.

The idea of direct search methods is to identify the interval of uncertainty that includes the optimum solution point. By successive iteration, the interval of uncertainty narrows till the desired level of accuracy is obtained.

There are two algorithms:

- Dichotomous Procedure/Method
- Golden Search Method

We consider only the Golden search method.

General step i:

Let $I_{i-1} = (x_L, x_R)$ be the current interval of uncertainty.

At iteration 0, $x_L = a$ and $x_R = b$ – the original interval (a, b).

Next define, x_1 and x_2 such that

$$x_L < x_1 < x_2 < x_R$$

$x_L = a$ ●————————————————————————● $x_R = b$

The next interval of uncertainity, I_i is determined as follows :

(i) If $f(x_1) > f(x_2)$, then $x_L < x^* < x_2$. Set $x_R = x_2$ and $I_i = (x_L, x_2)$

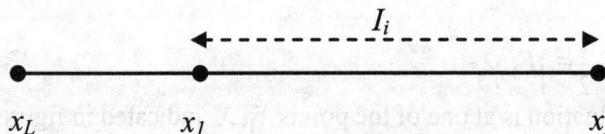

x_L ●————————●————————————●————————————● x_2

(ii) If $f(x_1) < f(x_2)$, then $x_1 < x^* < x_R$. Set $x_L = x_1$ and $I_i = (x_1, x_R)$

$$I_i$$

x_L ●————————●————————————————————————————● x_R
x_1

(iii) If $f(x_1) < f(x_2)$, then $x_1 < x^* < x_2$. Set $x_L = x_1$ and $x_R = x_2$ and $I_i = (x_1, x_2)$

The manner in which x_1 and x_2 are determined guarantees that $I_i < I_{i-1}$. The algorithm terminates at iteration k if $I_k \leq \Delta$, where Δ is a **user specified** level of accuracy.

Define for $0 < \alpha < 1$,

$$x_1 = x_R - \alpha(x_R - x_L)$$
$$x_2 = x_L + \alpha(x_R - x_L)$$

Then the interval of uncertainty I_i at iteration i equals (x_L, x_2) or (x_1, x_R).

Consider the case: $I_i = (x_L, x_2)$, which means that x_1 is included in I_i. In iteration $i + 1$, we select x_2 equal to x_1 in iteration i, which leads to the following equation:

x_2 (iteration $i + 1$) $= x_1$ (iteration i)

Substitution yields,

$$x_L + \alpha[x_2 \text{ (iteration } i) - x_L] = x_R - \alpha(x_R - x_L)$$
$$x_L + \alpha [x_L + \alpha(x_R - x_L) - x_L] = x_R - \alpha(x_R - x_L)$$
$$\alpha^2(x_R - x_L) + \alpha(x_R - x_L) - (x_R - x_L) = 0$$

This yields $\alpha^2 + \alpha - 1 = 0$

The roots are

$$\alpha = \frac{\left(-1 \pm \sqrt{5}\right)}{2}. \text{ Since } 0 \leq \alpha \leq 1,$$

We select the positive root, namely

$$\alpha = \frac{\left(\sqrt{5} - 1\right)}{2} \approx 0.618$$

The design of the golden section computations guarantees on α – reduction in successive intervals of uncertainty; that is

$$I_{i+1} = \alpha I_i.$$

This golden section method converges more rabidly to the desired level of accuracy. Further, each iteration requires half the computations because the method always recycles one set of computations from the immediately proceeding iteration.

1.8.4 An Alternate Approach to Golden Section Search

Golden section search is based on the fact that

$$r = \lim_{n \to \infty} \frac{F_n}{F_{n+1}} = \frac{\sqrt{5} - 1}{2} \doteq 0.618034$$

where F_n and F_{n+1} are successive terms of the Fibonacci sequence $\{F_n\}$, $n \geq 2$. In this search technique, the interval of uncertainty is reduced by the constant factor r at each step; thus Golden Section search is a limiting form of Fibonacci search. In general, it can be used only on functions of a continuous variable. Note that r satisfies,

$$r^2 + r - 1 = 0$$

Let $\phi(X)$ be a unimodal function of a continuous variable X defined on the closed interval $[0, L_n]$. The points of evaluation in Golden Section Search may be found by dividing equations

$$X_1 = \left(1 - \frac{1}{2}\in\right)F_{n-2}, X_2 = \left(1 - \frac{1}{2}\in\right)F_{n-1},$$

by equation

$$L_n = \left(1 - \frac{1}{2}\in\right)(F_{n-1} + F_{n-2}) \text{ and letting } n \to \infty. \text{ This gives}$$

$$X_1 = \frac{F_{n-2}}{F_{n-1} + F_{n-2}} \cdot L_n$$

$$= \frac{F_{n-2}}{F_{n-1}} \cdot \frac{F_{n-1}}{F_{n-1} + F_{n-2}} \cdot L_n$$

$$= \left(\frac{F_{n-2}}{F_{n-1}} \right) \left(\frac{F_{n-1}}{F_n} \right) L_n$$

Letting $n \to \infty$

$$= \lim_{n \to \infty} \left(\frac{F_{n-2}}{F_{n-1}} \right) \lim_{n \to \infty} \left(\frac{F_{n-1}}{F_n} \right) L_n$$

$$= r^2 L_n$$

Also $\quad X_2 = \dfrac{F_{n-1}}{F_{n-1} + F_{n-2}} \cdot L_n = \dfrac{F_{n-1}}{F_n} \cdot L_n$

$$= \lim_{n \to \infty} \left(\frac{F_{n-1}}{F_n} \right) L_n = r \cdot Ln$$

Hence $\quad X_1 = r^2 L_n, X_2 = rLn$ ...(1)

The Golden section search to maximize the unimodal function $\phi(X)$ on the interval $[0, L_n]$ therefore proceeds as follows:

(i) Evaluate $\phi(X_1)$ and $\phi(X_2)$ where X_1 and X_2 are given by (1)

(ii) If $\phi(X_1) > \phi(X_2)$, discard interval $(X_2, L_n]$. The remaining interval is a length rLn and $X_1 = r(rL_n) = r^2 L_n$ is one of the points of evaluation on it. The other point of evaluation is

$X = r^2(rL_n) = r^3 L_n$

On the other hand, if $\phi(X_2) > \phi(X_1)$ discard the initial $[0, X_1)$. The remaining interval is again of length rL_n and X_2 is one of the points of evaluation on it, since

$L_n - X_1 = (1 - r^2)L_n = rL_n$

and $X_2 - X_1 = rL_n - r^2 L_n = r(1 - r)L_n$

$= r \cdot r^2 L_n = r^2 (rL_n)$

The other point of evaluation is

$X = X_1 + r(rL_n)$

$= r^2 L_n + r^2 L_n = 2r^2 L_n$

(iii) Repeat steps (i) and (ii) for the successive remaining intervals of lengths $rL_n, r^2 L_n, r^3 L_n$ ..., until the desired accuracy in X^* is attained.

We observe here that Golden Section is simple, though less efficient, than Fibonacci search. Also, in particular, when optimizing the function $f(X)$ there is no need to scale the variable x.

1.8.5 Hook's and Jeeves' Search (method)

Hook's and Jeeves' method which dates from 1961 is one of the most widely used direct search methods. It attempts in a simple though ingenious way to find the most profitable search directions.

Consider the problem minimizing $f(X)$.

Choose an initial base point b_1 and step lengths h_j for the respective variables x_j. For greater numerical accuracy, it is advisable to choose the h_j so as to equalize, as for as possible, the quantities

$$\left| f(\mathbf{b}_1 + h_j e_j) - f(\mathbf{b}_1) \right|;$$

these are the magnitudes of the changes in $f(\mathbf{b}_1)$ due to a change of one step length in each variable in turn. After $f(\mathbf{b}_1)$ has been evaluated, the method proceeds by a sequence **exploratory**, and **pattern moves**. If an exploratory move leads to decrease in the value of $f(X)$ it is called a success, otherwise it is called a **failure**. A pattern move is not tested for success or failure.

Exploratory moves

The purpose of are exploratory move is to acquire information about $f(X)$ in the neighbourhood of the current base point. The procedure for an exploratory move about the point $\mathbf{b}_1$ is as follows:

E(i): Evaluate $f(\mathbf{b}_1 + h_1 e_1)$. If the move from $\mathbf{b}_1$ to $\mathbf{b}_1 + h_1 e_1$ is a success, replace the base point $\mathbf{b}_1$ by $\mathbf{b}_1 + h_1 e_1$. If it is a failure evaluate $f(\mathbf{b}_1 + h_1 e_1)$. If this move is a success replace $\mathbf{b}_1$ by $\mathbf{b}_1 + h_1 e_1$. If it is another failure retain the original base point $\mathbf{b}_1$.

E(ii): Repeat E(i) for variable x_2 by considering variations $\pm h_2 e_2$ from the point which result from E(i). Apply this procedure to each variable in turn, finally arriving at a new base point $\mathbf{b}_2$ after $(2n + 1)$ function evaluations at most, including $f(\mathbf{b}_1)$.

E(iii) If $\mathbf{b}_2 = \mathbf{b}_1$, halve each of the step length hj and return to E(i). The calculations terminate when the step lengths have been reduced to some prescribed level. If $\mathbf{b}_2 \neq \mathbf{b}_1$, make a pattern move from $\mathbf{b}_2$.

Pattern moves and subsequence moves

A pattern or leapfrog, move attempts to speed up the search by using information already acquired about $f(X)$. It is invariably followed by a sequence or exploratory moves, with a view to finding the improved direction of search in which to make another pattern move. We denote by $p_1, p_2, \dots$ the points reached by successive pattern moves. It seems sensible to move from $\mathbf{b}_2$ in the direction $(\mathbf{b}_2 - \mathbf{b}_1)$, since a move in this direction has already led to a decrease in the value of $f(X)$. The procedure for a pattern move from $\mathbf{b}_1$ is therefore as follows:

P(i) move from $\mathbf{b}_2$ to $p_1 = 2\mathbf{b}_2 - \mathbf{b}_1$ and continue with a new sequence of exploratory moves about p_1.

P(ii) If the lowest function value obtained during the pattern and exploratory moves of P(i) is less than f(b2), then a new base point b3 has been reached. In this case, return to P(i) with all suffixes increased by unity. Otherwise, abandon the pattern move from $\mathbf{b}_2$ and continue with a new sequence of exploratory moves about $\mathbf{b}_2$.

Curve fitting problems and large systems of ill – conditioned linear equations can be solved economically by this method, by reformulating them as optimization problems.

1.8.6 The Gradient Projection Method

This method is based on a general non linear programming algorithm due to **Rosen**. The current point is assumed to lie on one or more of the constraint boundaries and the basic idea of the method is to search in a direction in which $f(X)$ decreases but which is "tangential" to the boundaries of the active constraints. Specifically, let the constraints be

$$g_i(X) \geq 0$$

The gradient vectors $\nabla g_1, \nabla g_2, \dots \nabla g_l, (l \leq m)$ evaluated at the current point of the active constraint functions $g_1(X), g_2(X), \dots g_l(X)$ generate a subspace S of E^n. It is assumed for simplicity that the vectors $\nabla g_1, \nabla g_2, \dots \nabla g_l$ are linearly independent. If they are linearly dependent, the method can still be used by selecting a linearly independent subset of them; this case causes no difficulty, in Rosen's algorithm. The direction of search is given by projection of $-\nabla f(X)$ on $O(S)$, the orthogonal complement of S. we shall now determine this direction.

If $\mathbf{s}$ is any direction in S then, for some constants μ_k.

$$\mathbf{s} = \sum_{k=1}^{l} \mu_k \nabla g_k = N\mu \text{ say,} \qquad \qquad \dots(8.6.1)$$

where $\mu = [\mu_1, \mu_2, \dots \mu_l]$. It is now assumed without loss of generality, that are unit vectors (the values of ∇g_k are suitably adjusted). Then the matrix N_1 of order $(n \times l)$ and rank l is the matrix of unit column vectors $\nabla g_k = [g_{1k}, g_{2k}, \dots g_{nk}]$, i.e.

$$N = \begin{bmatrix} g_{11} \cdots g_{1l} \\ \cdots\cdots\cdots \\ g_{n1} \cdots g_{nl} \end{bmatrix}$$

where $\sum_{j=1}^{n} g_{jk}^2 = 1, k = 1, 2, \dots 1.$

Let $\mathbf{g} = \nabla f(X)$. Then we can write and

$$\mathbf{g} = \mathbf{r} + \mathbf{s}, \text{ where } \mathbf{r} \in O(S), \mathbf{s} \in S - r \qquad \text{...(8.6.2)}$$

and $-\mathbf{r}$ is the required direction of search. From equations (8.6.1) and (8.6.2)

$$\mathbf{g} = r + N\mu \qquad \text{...(8.6.3)}$$

Since $\mathbf{r} \in O(S)$, we have

$$(\nabla g)^T \mathbf{r} = 0 \qquad \text{...(8.6.4)}$$

Hence from equations (8.6.3) and (8.6.4)

$$N^T\mathbf{g} = N^T N\mu$$

And the matrix $N^T N$ is non-singular because the columns of N are linearly independent. Thus

$$\mu = (N^T N)^{-1} N^T\mathbf{g} \qquad \text{...(8.6.5)}$$

Now, from equations (8.6.1) and (8.6.5)

$$\mathbf{s} = N(N^T N)^{-1} N^T\mathbf{g} \qquad \text{...(8.6.6)}$$

and combining equations (8.6.2) and (8.6.6), we find

$$\mathbf{r} = \mathbf{g} - N(N^T N)^{-1} N^T\mathbf{g}$$
$$= \mathbf{P}\mathbf{g},$$

Where $\mathbf{P} = I - N(N^T N)^{-1} N^T$

Thus the direction of search is given by

$$-\mathbf{r} = -\mathbf{P}\mathbf{g}$$

The matrix $\mathbf{P}$ is called a **projection matrix**: it projects the direction on which it operates into the orthogonal complement of the directions represented by the columns of N. In practice, P is used whenever the current point is on or very close to one or more of the constraint boundaries.

Since $-\mathbf{r}^T \nabla f(X) = -\mathbf{r}^T\mathbf{g} = -\mathbf{r}^T(\mathbf{r} + \mathbf{s})$

$$= -\mathbf{r}^T\mathbf{r} < 0,$$

the objective function $f(X)$ decreases the direction of search $-\mathbf{r}$. The current point is moved in the direction $-\mathbf{r}$ until either $f(X)$ is a minimum in this direction or a new constraint boundary is reached. It sometimes happens that the current point becomes non-feasible during the course of the calculations – this case is considered below.

An important case arises when every constraint is linear, for the gradient projection method then moves the current point along the constraint boundary in the direction in which $f(X)$ decreases. In this case, the current constraint boundary may be regarded as the intersection of a number of hyperplanes. In each iteration, this intersection consists of a set of hyperplanes which only differs from the previous set by one: either a hyperplane is dropped from the set or one is added.

Rosen's algorithm takes advantage of this by providing two recursion relation to compute $(N^T N)^{-1}$ one for discarding and the other for adding a hyperplane.

When the current point becomes non – feasible, it is necessary to compute a return direction. At a non – feasible point, suppose that the absolute the values of the constraint functions for the violated constraints are $w_1, w_2, ... w_r$. The return direction $s \in S$ is determined as follows:

Let $\qquad w_k = (\nabla \mathbf{g}_k)^{-1} \mathbf{s} = |\mathbf{s}| \cos \theta_k, k = 1, 2, ... l,$ $\qquad$ ...(8.6.8)

where the unit vectors $\nabla \mathbf{g}_k$ are as defined earlier and θ_k is the angle between $\nabla \mathbf{g}_k$ and $\mathbf{s}$. It is clear from equation (8.6.8) that is a particular w_k say w_r increases relative to the remaining w_k then the direction of $\mathbf{s}$ approaches that of $\nabla \mathbf{g}_k$. Equations (8.6.8) can be written

$\mathbf{w} = N^T \mathbf{s}$, where $\mathbf{w} = [w_1, w_2, ... w_l]$ using equation (8.6.1), we obtain $w = N^T N \mu$, and hence $\mu = (N^T N)^{-1} w$ ..(8.6.9)

From (8.6.1) and (8.6.9), the return direction is $\mathbf{s} = (N^T N)^{-1} \mathbf{w}$.

1.9 Equality Constraints – Lagrangian Method

The classical optimization problem is

maximize $\qquad z = f(X)$

subject to $g_i(X) = b_i$, where $f, g_i \in C_1$ $\qquad$ ...(9.1)

It is assumed that the constraints are independent.

The basic idea in using Lagrange multipliers is to convert the constrained problem (9.1) into an unconstrained problem.

We say that $f(X)$ has a **constrained local maximum** at $X = X^*$, if a number $\in > 0$ exists such $f(X) \leq f(X^*)$ for all X satisfying both the inequality $|X - X^*| < \in$ and the constraints in (9.1). An important special case of constrained local maximum we say that $f(X)$ has a **constrained global maximum** at $X = X^*$ if $f(X) \leq f(X^*)$ for all X satisfying the constrain (9.1). For completeness, we may add that for the un constrained optimization problem we say simply that $f(X)$ has a **local maximum** or a **global maximum** at $X = X^*$ if the above definitions still hold when reference to the constraints in (9.1) is omitted.

Any maximum at $X = X^*$ is said to be **strong maximum** if there exists a neighbourhood of X^* in which no other maximum occurs. Otherwise, the maximum is said to be a **weak maximum**.

Necessary conditions for a constrained local maximum.

Assume now that $f(X)$ has a constrained local maximum at $X = X^*$, Then at $X = X^*$, we must have

$$df = \sum_j \frac{\partial f}{\partial x_j} dx_j = 0 \qquad \text{...(9.2)}$$

and
$$dg_i = \sum_j \frac{\partial g_i}{\partial x_j} dx_j = 0 \qquad \qquad ...(9.3)$$

Multiply equations (9.3) by constraints $-\lambda_i$ respectively, and add all the resulting equations to equation (9.2). We can then write

$$dF = \sum_j \frac{\partial f}{\partial x_j} dx_j = 0 \qquad \qquad ...(9.4)$$

where the **Lagrangian function** F is defined

$$F(\mathbf{X}, \boldsymbol{\lambda}) = f(\mathbf{X}) + \sum_j \lambda_i [b_i - g_i(X)] \qquad ...(9.5)$$

and the **Lagrange multipliers** λ_i are the components of the $m -$ vector $\boldsymbol{\lambda}$. Because of the m independent constraints in (9.1), only $(n - m)$ of the variables x_j are independent, Let these be $x_{m+1}, x_{m+2}, ... x_n$. Now choose the λ_i, which so for are arbitrary, in such a way that

$$\frac{\partial F}{\partial x_i} = 0 \qquad \qquad ...(9.6)$$

Then equation (9.4) becomes

$$\frac{\partial F}{\partial x_{m+1}} dx_{m+1} + ... \frac{\partial F}{\partial x_n} dx_n = 0 \qquad \qquad ...(9.7)$$

But the variables $x_{m+1}, x_{m+2}, ... x_n$ are independent and, by allowing each of these in turn to vary, we obtain, from equation (9.7),

$$\frac{\partial F}{\partial x_{m+1}} = ... = \frac{\partial F}{\partial x_n} = 0 \qquad \qquad ...(9.8)$$

Thus, from equations (9.6) and (9.8) we have

$$\frac{\partial F}{\partial x_j} \equiv \frac{\partial f}{\partial x_j} - \sum_i \lambda_i \frac{\partial g_i}{\partial x_j} = 0 \,...(9.9)$$

which are n necessary conditions for the existence of a constrained local maximum of $f(X)$.

Conditions (9.9) also apply when $f(X)$ is to be minimized subject to the constraints of the problem (9.1).

Hence any point satisfying the equation (9.8) is **stationary point** or **critical point** of $f(X)$.

A stationary point at which $f(X)$ takes on a maximum or minimum value is called an **extreme value** or **turning value**.

Note:
(i) Not all maxima or minima occur at stationary points.
(ii) Not all stationary points correspond to maxima or minima.

An optimal point X^* is any point at which $f(X)$ takes on a maximum or minimum value: an asterisk (*) will always denote an optimal value.

Assuming the existence of the Lagrange multipliers in equation (9.9), the implicit function theorem guarantees that these equations together with the constraint equations (9.1), can be solved for the $(m + n)$ unknowns λ_i, x_j to give $\lambda = \lambda^*$, $X = X^*$

The necessary conditions can be expressed in symmetrical form

$$\frac{\partial F}{\partial x_j} = 0, \frac{\partial F}{\partial \lambda_i} = 0 \qquad \qquad ...(9.10)$$

Sufficient conditions for a constrained local maximum.

Sufficient conditions involve second or higher – order derivatives of $f(X)$ as in the unconstrained optimization problem.

We assume that $f, g_i \in C_2$. A sufficient condition for the existence of a constrained local maximum of the function f is that $d^2f < 0$ at the optimal point, subject to $dg_i = 0$ and $d^2g_i = 0$. Hence taking the differential of the left hand side of equation (9.4),

We find

$$d(df) = d^2 f = \mathbf{dX}^T H_f(\mathbf{X}^*)d\mathbf{X} + [\nabla f(\mathbf{X}^*)]d^2\mathbf{X} < 0 \qquad ...(9.11)$$

where

$$H_f(\mathbf{X}) \equiv \left\{ \frac{\partial^2 f}{\partial x_j \partial x_k} \right\}, j, k = 1, 2, ... n$$

Is the Hessian matrix of $f(X)$ in (9.11) the vector $\mathbf{dX} \equiv [dx_1, dx_2, ... dx_n]$ must satisfy the vector matrix form of equation (9.3) namely,

$$dg_i = [\nabla g_i(\mathbf{X}^*)]^T d\mathbf{X} = 0 \qquad \qquad ...(9.12)$$

and must also satisfy

$$d^2g_i = \mathbf{dX}^T Hg(X^*)\mathbf{dX} + [\nabla g_i(\mathbf{X}^*)]^T d^2\mathbf{X} \qquad ...(9.13)$$

1.10 EXAMPLES

1.10.1 *Steepest Ascent & Descent Methods*

EXAM. 1:

Minimize $f(x) = x1^2 - x1x2 + 3x2^2$ given $x0 = (1, 2)$

$$X^{k+1} = X^k - r(\nabla f(x^k))$$

Iteration I:

$$\Rightarrow \qquad X^1 = x^0 - r(\nabla f(x^0))$$

$$\nabla f(x) = [2x1 - x2, -x1 + 6x2]$$

at $X^0 = (1, 2)$

$$\nabla f(x^0) = [0, 11]$$

$$X^1 = (1, 2) - r(0, 11)$$

$$X^1 = (1, 2 - 11r)$$

$$h(r) = f(1, 2 - 11r) = 1 - (1)(2 - 11r) + 3(2 - 11r)^2$$

$$= 1 - 2 + 11r + 3(4 + 121r^2 - 44r)$$

$$= 1 - 2 + 11r + 12 + 363\, r^2 - 132r$$

$$h(r) = 11 - 121r + 363r^2$$

$$\frac{dh(r)}{dr} = 0 \Rightarrow -121 + 726r = 0$$

$\Rightarrow \qquad 726r = 121r = \dfrac{121}{726} = \dfrac{1}{6}$

$$X^1 = (1, 2 - 11r), \text{ sub } r = \frac{1}{6}$$

$$X^1 = (1, 2 - 11r) = \left(1, \frac{1}{6}\right)$$

$\therefore \qquad X^1 = \left(1, \dfrac{1}{6}\right)$

$$\frac{d^2 h(r)}{dr^2} = 726 > 0$$

Iteration II:

$$X^2 = x^1 - r(\nabla f(x^1))$$

$$\nabla f(x) = [2x1 - x2, -x1 + 6x2]$$

at $X^1 = \left(1, \dfrac{1}{6}\right)$

$$\nabla\left(f(X^1)\right) = \left[2 - \frac{1}{6}, -1 + 1\right]$$

$$\nabla f(X^1) = \left[\frac{11}{6}, 0\right]$$

$$X^2 = \left(1, \frac{1}{6}\right) - r\left(\frac{11}{6}, 0\right)$$

$$X^2 = \left(1, -\frac{11r}{6}, \frac{1}{6}\right)$$

$$h(r) = x1^2 - x1x2 + 3x2^2$$

$$\text{sub } x_1 = 1 - \frac{11r}{6}, \; x_2 = \frac{1}{6}$$

$$\Rightarrow \left(1 - \frac{11r}{6}\right)^2 - \left(1 - \frac{11r}{6}\right)\frac{1}{6} + 3\left(\frac{1}{6}\right)^2$$

$$\Rightarrow \left(\frac{6 - 11r}{6}\right)^2 - \left(\frac{6 - 11r}{36}\right) + \frac{3}{36}$$

$$\Rightarrow \frac{36 + 121r^2 - 132r - 6 + 11r + 3}{36}$$

$$\Rightarrow \frac{121r^2 - 121r + 33}{33}$$

$$\frac{dh(r)}{dr} = 0 \Rightarrow 242r - 121 = 0 \Rightarrow r = \frac{121}{242} = \frac{1}{2}$$

$$\frac{d^2 h(r)}{dr^2} = 242 > 0$$

$$X^2 = \left(\frac{6 - 11r}{6}, \frac{1}{6}\right)$$

$$\text{sub } r = \frac{1}{2}$$

$$X^2 = \left[\frac{6 - \dfrac{11}{2}}{6}, \frac{1}{6}\right]$$

$$\therefore \quad X^2 = \left(\frac{1}{12}, \frac{1}{6}\right)$$

EXAM. 2: Solve by both univariate search and steepest descent method.

Min. $f(x) = 16x1^2 - 3x1 + 16x2^2 - 3x2$ with $X^0 = (0, 0)$

Min $f(x) = -3x1 - 3x2 + 16x1^2 + 16x2^2$

$$X^{k+1} = X^k - r(\nabla f(x^k))$$

Iteration I:

$$X^1 = x^0 - r(\nabla f(x^0))$$

$$\nabla f(x) = [-3 + 32x1, -3 + 32x2]$$

at $X^0 = (0, 0)$

$$\nabla f(x^0) = [-3, -3]$$

$$X^1 = (0, 0) - r(-3, -3) \Rightarrow X^1 = (3r, 3r)$$

$$h(r) = f(X^1) = (-3x1, -3x2 + 16x1^2 + 16x2^2)(3r, 3r)$$

$$h(r) = -3(3r) - 3(3r) + 16(9r^2) + 16(9r^2)$$

$$h(r) = -18r + 288r^2$$

$$\frac{dh(r)}{dr} = -18 + 576r = 0 \Rightarrow r = \frac{18}{576} = \frac{1}{32}$$

$$X^1 = (3r, 3r) \Rightarrow \left(\frac{3}{32}, \frac{3}{32}\right)$$

Iteration II:

$$X^2 = X^1 - r\left(\nabla f(x^1)\right)$$

$$\nabla f(x) = [-3 + 32x1, -3 + 32x2]$$

at $X^1 = \left(\frac{3}{32}, \frac{3}{32}\right)$

$$\nabla f(X^1) = -3 + 32 \cdot \frac{3}{32}, -3 + 32 \cdot \frac{3}{32}$$

$$\nabla f(x^1) = (0, 0)$$

$$X^2 = \left(\frac{3}{32}, \frac{3}{32}\right) - r(0, 0)$$

$$X^2 = \left(\frac{3}{32}, \frac{3}{32}\right)$$

The iteration stops and the optimal value is $X^{(1)} = X^{(2)} = \left(\frac{3}{32}, \frac{3}{32}\right)$

EXAM. 3: Solve by steepest descent method. Min $f(x) = x1^2 - x1x2 + 3x2 + x1 - 1$ with $X^0 = (1, 2)$ as the initial point.

$$\text{Min } f(x) = x1 + 3x2 + x1^2 - x1x2 - 1$$

$$X^{k+1} = X^k - r\left(\nabla f(x^k)\right)$$

Iteration I:

$$X^1 = x^0 - r\left(\nabla f(x^0)\right)$$

$$\nabla f(x) = [1 + 2x1 - x2, 3 - x1]$$

$$\text{sub } X^0 = (1, 2)$$

$$\nabla f(X^0) = (1, 2)$$

$$X^1 = (1, 2) - r(1, 2)$$

$$X^1 = (1 - r, 2 - 2r)$$

$$h(r) = f(X^1) = [x1 + 3x2 - 1 + x1^2 - x1x2](1 - r, 2 - 2r)$$

$$\text{sub } x1 = (1 - r), x2 = (2 - 2r)$$

$$h(r) = (1 - r) + 3(2 - 2r) - 1 + (1 - r)^2 - (1 - r)(2 - 2r)$$

$$h(r) = (1 - r + 6 - 6r - 1) + (1 + r2 - 2r) - (2 - 2r - 2r + 2r^2)$$

$$h(r) = (1 - r + 6 - 6r - 1) + 1 + r2 - 2r - 2 + 2r + 2r - 2r^2)$$

$\Rightarrow\qquad h(r) = (5 - 5r - r^2)$

$$\frac{dh(r)}{dr} = 0 \Rightarrow -2r - 5 = 0 \Rightarrow r = -\frac{5}{2}$$

$$X^1 = (1 - r, 2 - 2r) \Rightarrow X^1 = \left(1 + \frac{5}{2}, 2 + 5\right) = \left(\frac{7}{2}, 7\right)$$

$\therefore$ Condition fails for minimization condition.

EXAM. 4: Solve by steepest descent method.

$$\text{Min } f(x) = x1^2 + x2^2 - 12x1 \text{ with } X^0 = (1, 1)$$

$$\text{Min } f(x) = -12x1 + x1^2 + x2^2$$

Iteration I:

$$X^1 = x^0 - r\left(\nabla f(x^0)\right)$$

$$\nabla f(x) = [-12 + 2x1, 2x2]$$

at $\qquad X^0 = (1, 1)$

$$\nabla f(X^0) = [-10, 2]$$

$$X^1 = (1, 1) - r(-10, 2)$$

$$X^1 = (1 + 10r, 1 - 2r)$$

$$h(r) = -12(1 + 10r) + (1 + 10r)^2 + (1 - 2r)^2$$

$$= (-12 - 120r) + (1 + 20r + 100r^2) + (1 + 4r^2 - 4r)$$

$$h(r) = -10 - 104r + 104r^2$$

$$\frac{dh(r)}{dr} = 0 \Rightarrow -104 + 208r = 0 \Rightarrow r = \frac{1}{2}$$

$$\boxed{\begin{aligned} \nabla f(x) &= \left(-12 + 2x_1, 2x_2\right) \\ \text{at } X^1 &= \left(6, 0\right) \\ \nabla f(X^1) &= (0, 0) \end{aligned}}$$

$$\therefore \qquad X^1 = (1 + 5, 1 - 1) \Rightarrow X^1 = (6, 0)$$

$$\frac{d^2 h(r)}{dr} = 208 > 0$$

$$X^2 = (6, 0) - r(0, 0)$$
$$X^2 = (6, 0)$$
$$h(r) = -12\,(6) + (6)2 + (0)2 = -72 + 36 = -36$$

$$\frac{dh(r)}{dr} = 0 \text{ and}$$

$$\frac{d^2 h(r)}{dr} = 0$$

∴ conditions fails for steepest descent.

∴ $X^1 = X^2 = (6, 0)$ makes $f(X)$ a minimum.

EXAM. 5: Solve by the steepest ascent method the problem.

Max. $f(x) = 4x1 + 6x2 - 2x1^2 - 2x1x2 - 2x2^2$ with initial point at $X^0 = (1, 1)$

Ans.: $X^1 = \left(\dfrac{1}{2}, 1\right), X^2 = \left(\dfrac{1}{2}, \dfrac{5}{4}\right), X^3 = \left(\dfrac{3}{8}, \dfrac{5}{4}\right), X^4 = \left(\dfrac{3}{8}, \dfrac{21}{16}\right),$

$$X^{k+1} = X^k + r\left(\nabla f(x^k)\right)$$

Iteration I :

$$X^1 = x^0 + r\nabla f(x^0)$$

$$\nabla f(x) = [4 - 4x1 - 2x2,\ 6 - 2x1 - 4x2]$$

at $\qquad X^0 = (1, 1)$

$$\nabla f(X^0) = (-2, 0)$$
$$X^1 = X^0 + r(-2, 0)$$
$$X^1 = (1, 1) + r(-2, 0)$$
$$X^1 = (1 - 2r, 1)$$
$$f(1 - 2r, 1) = 4(1 - 2r) + 6 - 2(1 - 2r)2 - 2(1 - 2r) - 2(1)$$
$$= (4 - 8r) + 6 - 2(1 - 4r + 4r^2) - (2 - 4r) - 2$$
$$= -8r^2 + 4r + 4$$
$$h(r) = 4\,(-2r^2 + r + 1)$$

$$\frac{dh(r)}{dr} = 0 \Rightarrow -4r + 1 = 0 \Rightarrow r = \frac{1}{4}$$

$$\frac{d^2h(r)}{dr} = -4 < 0$$

$$\therefore \quad X^{-1} = \left(1 - \frac{1}{2}, 1\right) \Rightarrow X^1 = \left(\frac{1}{2}, 1\right)$$

Iteration II:

$$X^2 = X^1 + r\nabla f(X^1)$$

$$X^2 = \left(\frac{1}{2}, 1\right) + r(0, 1)$$

$$\boxed{\begin{array}{l} \nabla f(X) = [4 - 4x_1 - 2x_2, 6 - 2x_1 - 4x_2] \\ \text{sub } X_1 = \frac{1}{2}, X_2 = 1 \\ \nabla f(X^1) = (0, 1) \end{array}}$$

$$X^2 = \left(\frac{1}{2}, 1\right) + (0, r)$$

$$X^2 = \left(\frac{1}{2}, 1 + r\right)$$

$$h(r) = 4x1 + 6x2 - 2x1^2 - 2x1x2 - 2x2^2$$

$$h(r) = 4\left(\frac{1}{2}\right) + 6(1 + r) - 2\left(\frac{1}{4}\right) - 2\left(\frac{1}{2} + \frac{1}{2} - r\right) - 2(1 + 2r + r^2)$$

$$h(r) = -2r^2 + r + \frac{9}{2} \Rightarrow \frac{-4r^2 + 2r + 9}{2}$$

$$\frac{dh(r)}{dr} = 0 \Rightarrow -8r + 2 = 0 \Rightarrow r = \frac{1}{4}$$

$$\frac{d^2h(r)}{dr} = -8 < 0$$

$$\therefore \quad X^2 = \left(\frac{1}{2}, 1 + \frac{1}{4}\right) = \left(\frac{1}{2}, \frac{5}{4}\right)$$

$$\therefore \quad X^2 = \left(\frac{1}{2}, \frac{5}{4}\right)$$

Iteration III:

$$X^3 = X^2 + r\nabla f(X^2)$$

$$= \left(\frac{1}{2}, \frac{5}{4}\right) + r \cdot \left(-\frac{1}{2}, 0\right)$$

> $\nabla f(X) = \left[4 - 4x_1 - 2x_2, 6 - 2x_1 - 4x_2\right]$
>
> at $X^2 = \left(\frac{1}{2}, \frac{5}{4}\right)$
>
> $\nabla f(X^2) = \left(-\frac{1}{2}, 0\right)$

$$X^2 = \left(\frac{1-r}{2}, \frac{5}{4}\right)$$

$$h(r) = 4x1 + 6x2 - 2x1^2 - 2x1x2 - x2^2$$

$$= 4\frac{(1-r)}{2} + 6\left(\frac{5}{4}\right) - 2\frac{(1-r)^2}{2} - 2\left(\frac{1-r}{2}\right)\left(\frac{5}{4}\right) - 2\left(\frac{25}{4}\right)$$

$$= 2(1-r) + \frac{15}{2} - \frac{1}{2}(1 + r^2 - 2r)\frac{5}{4}(1-r) - \frac{25}{2}$$

$$= 2(1-r) + \frac{15}{2} - \frac{1}{2}(1 + r^2 - 2r)\frac{5}{4}(1-r) - \frac{25}{2}$$

$$= 2 - 2r + \frac{15}{2} - \frac{1}{2}r^2 + r - \frac{5}{4} + \frac{5}{4}r - \frac{25}{2}$$

$$= \frac{(8 - 8r + 30 - 2 - 2r^2 + 4r - 5 + 5r - 30)}{4}$$

$$h(r) = \frac{-2r^2 + r + 1}{4}$$

$$\frac{dh(r)}{dr} = 0 \Rightarrow -4r + 1 = 0 \Rightarrow r = \frac{1}{4}$$

$$\frac{d^2h(r)}{dr} = -4 < 0$$

$$X^3 = \left(\frac{1-r}{2}, \frac{5}{4}\right) = \left(\frac{3}{8}, \frac{5}{4}\right)$$

$$\therefore \qquad X^3 = \left(\frac{3}{8}, \frac{5}{4}\right)$$

Iteration IV:

$$X^4 = X^3 + r\nabla f(X^3)$$

$$= \left(\frac{3}{8}, \frac{5}{4}\right) + r\left(0, \frac{1}{4}\right)$$

$$\Rightarrow \left(\frac{3}{8}, \frac{5}{4} + \frac{r}{4}\right)$$

$$X^4 = \left(\frac{3}{8}, \frac{5+r}{4}\right)$$

$$h(r) = 4x1 + 6x2 - 2x1^2 - 2x1x2 - 2x2^2$$

$$= 4\left(\frac{3}{8}\right) + 6\left(\frac{5+r}{4}\right) - 2\left(\frac{3}{8}\right)^2 - 2\left(\frac{3}{8}\right)\left(\frac{5+r}{4}\right) - 2\left(\frac{5+r}{4}\right)^2$$

$$= \frac{3}{2} + \frac{1}{2}(15+3r)\frac{-9}{32}\frac{-1}{16}(15+3r)\frac{-1}{8}(25+r^2+10r)$$

$$= \frac{3}{2} + \frac{(15+3r)}{2}\frac{-9}{32} - \frac{(15+3r)}{16}\frac{-1}{8}(+r^2+10r+25)$$

$$h(r) = \frac{(-4r^2 + 2r + 149)}{32}$$

$$\frac{dh(r)}{dr} = 0 \Rightarrow -8r + 2 = 0 \Rightarrow 8r = 2 \Rightarrow r = \frac{1}{4}$$

$$\frac{d^2h(r)}{dr} = -8 < 0$$

$$X^4 = \left(\frac{3}{8}, \frac{5+r}{4}\right) \Rightarrow \left(\frac{3}{8}, \frac{5+\frac{1}{4}}{4}\right)$$

$$\therefore \qquad X^4 = \left(\frac{3}{8}, \frac{21}{16}\right)$$

Proceed further $X^5 = (1/3, 4/3)$

EXAM. 6: Solve by steepest ascent method the following problem.

Max $f(x) = -x1 + x2 - x1^2 + x1x2$ with $X^0 = (0, 0)$

$$X^1 = x^0 + r\nabla f(x^0)$$

$$\nabla f(x) = [-1 - 2x1 + x2, 1 + x1]$$

at $\qquad X^0 = (0,0)$

$$\nabla f(X^0) = [-1, 1]$$
$$X^1 = (0,0) + r(-1, 1)$$
$$X^1 = (-r, r)$$
$$h(r) = (-x1 + x2 - x1^2 + x1x2)(-r, r)$$
$$h(r) = r + r - r^2 - r^2$$
$$h(r) = -2r^2 + 2r \Rightarrow -2(r^2 - r)$$

$$\frac{dh(r)}{dr} = 0 \Rightarrow -4r + 2 = 0 \Rightarrow 4r = 2 \Rightarrow r = \frac{1}{2}$$

$$X^1 = \left(\frac{-1}{2}, \frac{1}{2}\right)$$

$$\frac{d^2h(r)}{dr} = -4 < 0$$

Proceed further

EXAM. 7: Solve the following by steepest ascent method.

Max $f(x) = -x1^2 + x1x2 - 3x2^2$ with $X^0 = (1, 2)$

Ans. $X^1 = \left(1, \frac{1}{6}\right), X^2 = \left(\frac{1}{12}, \frac{1}{6}\right), X^3 = \left(\frac{1}{12}, \frac{1}{72}\right), X^4 = \left(\frac{1}{144}, \frac{1}{72}\right)$

Iteration I:

$$X_1 = x^0 + \nabla f(x^0)$$
$$\nabla f(x) = [-2x1 + x2, x1 - 6x2]$$

at $\qquad X^0 = (1, 2)$

$$\nabla f(x^0) = [0, -11]$$
$$X^1 = (1, 2) + r(0, -11)$$
$$X^1 = [1, 2 - 11r]$$
$$h(r) = -1 + (2 - 11r) - 3(4 + 121r^2 - 44r)$$
$$h(r) = -1 + 2 - 11r - 12 - 363r^2 + 132r$$
$$h(r) = -11 + 121r - 363r^2$$

$$\frac{dh(r)}{dr} = 0 \Rightarrow 121 - 726r = 0 \Rightarrow r = \frac{121}{726} \Rightarrow r = \frac{1}{6}$$

$$X^{-1} = \left[1, 2 - \frac{11}{6}\right]$$

$\therefore \qquad X^1 = \left[1, \frac{1}{6}\right]$

Iteration II:

$$X^2 = x^1 + r\nabla f(x^1)$$

$$X^2 = \left[1, \frac{1}{6}\right] + r\left[\frac{-11}{6}, 0\right]$$

$$X^2 = \left[1\frac{-11r}{6}, \frac{1}{6}\right]$$

$$\nabla f(X) = \left[-2x_1 + x_2, x_1 - 6x_2\right]$$
$$\text{at } X^1 = \left(1, \frac{1}{6}\right)$$
$$\nabla f(X^1) = \left[-2 + \frac{1}{6}, 1 - 1\right]$$
$$\nabla f(X^1) = \left[\frac{-11}{6}, 0\right]$$

$$h(r) = -x1^2 + x1x2 - 3x2^2$$

$$= -\frac{(6-11r)^2}{6} + \frac{(6-11r)}{6}\left(\frac{1}{6}\right) - 3\left(\frac{1}{36}\right)$$

$$= \left(\frac{1}{36}\right)\left[-36 - 121r^2 + 132r + 6 - 11r - 3\right]$$

$$h(r) = \left(\frac{1}{36}\right)\left[-33 - 121r^2 + 121r\right]$$

$$\frac{dh(r)}{dr} = 0 \Rightarrow -242r + 121 = 0 \Rightarrow r = \frac{121}{242} \Rightarrow r = \frac{1}{2}$$

$$X^2 = \left[\frac{6 - \frac{11}{2}}{6}, \frac{1}{6}\right] = \left[\frac{1}{12}, \frac{1}{6}\right]$$

$$\therefore \quad X^2 = \left[\frac{1}{12}, \frac{1}{6}\right]$$

Iteration III:

$$X^3 = x^2 + r\nabla f(x^2)$$

$$X^3 = \left[\frac{1}{12}, \frac{1}{6}\right] + r\left[0, \frac{-11}{12}\right]$$

$$X^3 = \left[\frac{1}{12}, \frac{1}{6}\frac{-11r}{12}\right]$$

$$X^3 = \left[\frac{1}{12}, \frac{2-11r}{12}\right]$$

$$\nabla f(X) = \left[-2x_1 + x_2, x_1 - 6x_2\right]$$
$$\text{at } X^2 = \left[\frac{1}{12}, \frac{1}{6}\right]$$
$$\nabla f(X^2) = \left[\frac{-1}{6} + \frac{1}{6}, \frac{1}{12} - 1\right]$$
$$\nabla f(X^2) = \left[0, \frac{-11}{12}\right]$$

$$h(r) = -\left(\frac{1}{12}\right)^2 + \left(\frac{1}{12}\right)\left(\frac{2-11r}{12}\right) - 3\left(\frac{2-11r}{12}\right)^2$$

$$= \frac{1}{144}\left[-1+2-11r-3(4+121r^2-44r)\right]$$

$$= \frac{1}{144}\left[-11r+1-12-363r^2+132r\right]$$

$$h(r) = \frac{1}{144}\left[-11+121r-363r^2\right]$$

$$\frac{dh(r)}{dr} = 0 \Rightarrow 121-726=0 \Rightarrow r=\frac{121}{726} \Rightarrow r=\frac{1}{6}$$

$$X^3 = \left[\frac{1}{12}, \frac{2-\frac{11}{6}}{12}\right]$$

$$X^3 = \left[\frac{1}{12}, \frac{1}{72}\right]$$

Iteration IV:

$$X^4 = x^3 + r\nabla f(x^3)$$

$$X^4 = \left[\frac{1}{12}, \frac{1}{72}\right] + r\left[\frac{-11}{72}, 0\right]$$

$$X^4 = \left[\frac{6-11r}{72}, \frac{1}{72}\right]$$

$$\boxed{\begin{aligned} \nabla f(X) &= \left[-2x_1+x_2, x_1-6x_2\right] \\ \text{at } X^3 &= \left(\frac{1}{12}, \frac{1}{72}\right) \\ \nabla f(X^3) &= \left[\frac{-1}{6}+\frac{1}{72}, \frac{1}{12}-\frac{1}{12}\right] \\ &= \left[\frac{-12+1}{72}, 0\right] = \left[\frac{-11}{72}, 0\right] \end{aligned}}$$

$$h(r) = f(X^4) = \left[\frac{6-11r}{72}\right]^2 + \left[\frac{6-11r}{72}\right]\left[\frac{1}{72}\right] - 3\left[\frac{1}{72}\right]^2$$

$$h(r) = \frac{1}{72^2}\left[-(36+121r^2-132r)+(6-11r)-3\right]$$

$$= \frac{1}{72^2}\left[-36-121r^2+132r+6-11r-3\right]$$

$$h(r) = \frac{1}{72^2}\left[-33+11r-121r^2\right]$$

$$\frac{dh(r)}{dr} = 0 \Rightarrow 11 - 242r = 0 \Rightarrow r = \frac{11}{242} \Rightarrow r = \frac{1}{22}$$

$$\frac{d^2h(r)}{dr^2} = -242 < 0$$

$$\therefore \qquad X^4 = \left[\frac{6-\dfrac{11}{2}}{72}, \frac{1}{72}\right]$$

$$X^4 = \left[\frac{11}{144}, \frac{1}{72}\right]$$

$$X^4 = \left[\frac{1}{144}, \frac{1}{432}\right]$$

EXAM. 8: Solve by steepest ascent method.

Max. $f(x) = -2x1^2 + 6x2 + 4x1 - 2x1x2 - 2x2^2$ with $X^0 = \left(\dfrac{3}{8}, \dfrac{5}{4}\right)$

$$X^1 = x^0 + r\nabla f(x^0) f(x) = 4x1 + 6x2 - 2x1^2 - 2x1x2 - 2x2^2$$
$$\nabla f(x) = [4 - 4x1 - 2x2, \; 6 - 2x1 - 4x2]$$

at $\qquad X^0 = \left(\dfrac{3}{8}, \dfrac{5}{4}\right)$

$$\nabla f(X^0) = \left[4 - \frac{3}{2} - \frac{5}{2}, \; 6 - \frac{3}{4} - 5\right]$$

$$\nabla f(X^0) = \left[0, \frac{1}{4}\right]$$

$$X^1 = \left[\frac{3}{8}, \frac{5}{4}\right] + r\left[0, \frac{1}{4}\right]$$

$$X^1 = \left[\frac{3}{8}, \frac{5+r}{4}\right]$$

$$h(r) = 4x1 + 6x2 - 2x1^2 - 2x1x2 - 2x2^2$$

$$h(r) = 4\left(\frac{3}{8}\right) + 6\left(\frac{5+r}{4}\right) - 2\left(\frac{9}{64}\right) - 2\left(\frac{3}{8}\right)\left(\frac{5+r}{4}\right) - 2\left(\frac{5+r}{4}\right)^2$$

$$= \frac{3}{2} + \frac{3}{2}(5+r) - \frac{9}{32} - \frac{3}{16}(5+r) - \frac{1}{8}(25 + r^2 + 10r)$$

$$= \left[\frac{3}{2} + \frac{15}{2} - \frac{9}{32} - \frac{15}{16} - \frac{25}{8}\right] + \left[\frac{3}{2}r - \frac{3}{16}r - \frac{10}{8}r\right] - \frac{1}{8}r^2$$

$$= \left[\frac{48 + 240 - 9 - 30 - 100}{32}\right] + \left[\frac{24r - 3r - 20r}{16}\right] - \frac{r^2}{8}$$

$$h(r) = \frac{149}{32} + \frac{1}{16}r - \frac{r^2}{8}$$

$$\frac{dh(r)}{dr} = 0 \Rightarrow r = \frac{1}{16} - \frac{2r}{8} = 0 \Rightarrow \frac{r}{4} = \frac{1}{16} \Rightarrow r = \frac{1}{4}$$

$$\therefore \quad X^1 = \left[\frac{3}{8}, \frac{5+r}{4}\right] = \left[\frac{3}{8}, \frac{5 + \frac{1}{4}}{4}\right]$$

$$\therefore \quad X^1 = \left[\frac{3}{8}, \frac{21}{16}\right]$$

$$\frac{d^2 h(r)}{dr^2} = \frac{-1}{4} < 0$$

Iteration I:

$$X^2 = x^1 + r\nabla f(x^1)$$

$$\therefore \quad X^2 = \left[\frac{3}{8}, \frac{21}{16}\right] + r[0, 0]$$

$$\therefore \quad X^2 = \left[\frac{3}{8}, \frac{21}{16}\right]$$

Hence the optimal solution is

$$\nabla f(X) = [4 - 4x_1 - 2x_2, 6 - 2x_1 - 4x_2]$$
$$\text{at } X^1 = \left[\frac{3}{8}, \frac{21}{16}\right]$$
$$\nabla f(X^1) = \left[4 - \frac{3}{2} - \frac{21}{8}, 6 - \frac{3}{4} - \frac{21}{4}\right]$$
$$\nabla f(X^1) = \left[\frac{32 - 12 - 21}{8}, \frac{24 - 3 - 21}{4}\right]$$
$$\nabla f(X^1) = [0, 0]$$

$$X^* = X^1 X^2 = \left[\frac{3}{8}, \frac{21}{16}\right]$$

EXAM. 9: Minimize $f(x) = 3x1^2 - 2x1x2 + x2^2 + x1$ taking $(1, 1)$ as the initial point by the steepest descent method.

$$f(x) = 3x1^2 - 2x1x2 + x2^2 + x1$$
$$X^0 = (1, 1)$$

$$\nabla f(x) = (6x1 - 2x2 + 1, -2x1 + 2x2)$$
$$\nabla f(x) = (6 - 2 + 1, -2 + 2) = (5, 0)$$

Iteration I:

$$X^{(1)} = X^0 - r\nabla f(x^0) = (1, 1) - r(5, 0) = (1 - 5r, 1)$$
$$h(r) = f(1 - 5r, 1) = 3(1 - 5r)^2 - 2(1 - 5r)(1) + 1^2 + (1 - 5r)$$
$$= 75r^2 - 25r + 3$$
$$h'(r) = 150r - 25 = 0 \Rightarrow r = 0.166$$
$$X^{(1)} = [1 - 5 \times 0.166, 1] = [0.1666, 1]$$

Iteration II:

$$X^{(1)} = (0.1666, 1)$$
$$\nabla f(X^{(1)}) = (0, 1.666)$$
$$X^{(2)} = (0.1666, 1) - r(0, 1.666)$$
$$= (0.1666, 1 - 1.666r)$$
$$h(r) = 3(0.1666)^2 - 2(0.1666)(1 - 1.666r) + (1 - 1.666r)^2 + 0.1666 + 1$$
$$+ 2.778r^2 - 3.332r$$
$$= 2.778r^2 - 2.777r + 0.9164$$
$$h'(r) = 5.556r - 2.777 = 0$$

$$\Rightarrow \qquad r = \frac{2.777}{5.556} = 0.49$$

$$X^{(2)} = (-0.5, 1.49)$$

Iteration III:

$$\nabla f(X^{(2)}) = 6(-0.5) - 2(1.49) + 1, -2(-0.5) + 2(1.49)$$
$$= (-4.98, 3.98)$$

$$X^{(3)} = X^{(2)} - r\nabla f(X^{(2)}) =$$
$$= (-0.5, 1.49) - r(-4.98, 3.98)$$
$$= (-0.5 + 4.98r, 1.49 - 3.98r)$$
$$R(r) = 3(-0.5 + 4.98)^2 - 2(-0.5 + 4.98r)(1.49 - 3.98r) + (1.49 - 3.98r)^2$$
$$+ (-0.5 + 4.98r)$$
$$= r^2[74.4 + 39.6 + 15.8] + r[-14.94 - 14.89 - 3.98 - 11.86 + 4.98]$$
$$+ [0.75 + 1.49 + 2.2201 - 0.5]$$
$$= 129.8r^2 - 40.69r + 3.960$$
$$h'(r) = 259.6r - 40.69 = 0 \Rightarrow r = 0.156$$
$$X^3 = (0.28, 0.866)$$

EXAM. 10: Find the minimum of $f(x) = x1 - x2 + 2x1^2 + 2x1x2 + x2^2$ by the steepest descent method with $X^0(0, 0)$.

$$f(x) = x1 - x2 + 2x1^2 + 2x1x2 + x2^2$$

$$X^0 = (0, 0)$$
$$\nabla f(X) = (1 + 4x1 + 2x2, -1 + 2x1 + 2x2)$$

Iteration I:

$$X^1 = X^0 - r\nabla f(X^0)$$
$$= (0, 0) - r(1, -1) = (-r, r)$$
$$h(r) = f(-r, r) = -r - r + 2r^2 - 2r^2 + r^2 = r^2 - 2r$$
$$h'(r) = 2r - 2 = 0 \Rightarrow r = 1 \text{ and } h''(r) > 0$$

Now $\qquad X^1 = (-1, 1)$

Iteration II:

$$X^1 = (-1, 1)$$
$$\nabla f(X^1) = (1 - 4 + 2, -1 - 2 + 2) = (-1, -1)$$
$$X^2 = X^1 - r\nabla f(X^1) = (-1, 1) - r(-1, 1) = (-1 + r, 1 + r)$$
$$h(r) = 5r^2 - 2r - 1$$
$$h'(r) = 10r - 2 = 0 \Rightarrow r = 0.2$$
$$X^2 = (-1 + r, 1 + r) = (0.8, 1.2)$$

Iteration III:

$$X^2 = (0.8, 1.2)$$
$$\nabla f(X^2) = [1 + 4 \times 0.8 + 2(1.2), -1 + 2(0.8) + 2(1.2)] = [6.6, 3]$$
$$X^3 = X^2 - r[6.6, 3]$$
$$= (0.8, 1.2) - (6.6r, 3r)$$
$$= (0.8 - 6.6r, 1.2 - 3r)$$
$$R(r) = 0.8 - 6.6r - 1.2 + 3r + 2(0.8 - 6.6r)^2 + 2(0.8 - 6.6r)(1.2 - 3r)$$
$$+ (1.2 - 3r)^2$$
$$= r^2[87.12 + 39.6 + 9] + r[-3.6 - 21.12 - 15.84 - 4.8 - 7.2] + [-0.4$$
$$+ 1.28$$
$$+ 1.92 + 1.44]$$
$$= 135.72r^2 - 52.56r + 4.24$$
$$h'(r) = 271.44r - 52.56 = 0$$

$$r = \frac{52.56}{271.44} = 0.193$$

$$X^3 = (-0.477, 0.6190981).$$

EXAM. 11: Minimize $f(x) = 5x1^2 + 2x2^2 - 2x1x2 - 4x1 - 4x2 + 4$ using the method of steepest descent with $X^0 = (0, 0)$

$$f(x) = 5x1^2 + 2x2^2 - 2x1x2 - 4x1 - 4x2 + 4$$
$$\nabla f(x) = (10x1 - 2x2 - 4, 4x2 - 2x1 + 4)$$

Iteration I:

$$\nabla f(X^0) = (-4, -4)$$
$$X^1 = (0, 0) - r(-4, -4)$$
$$X^1 = (4r, 4r)$$
$$h(r) = f(4r, 4r)$$
$$h(r) = 80r^2 - 32r + 4$$
$$\frac{dh}{dr} = 160r - 32 \Rightarrow r = \frac{1}{5}$$
$$X^{-1} = \frac{4}{5}, \frac{4}{5}$$

Iteration II:

$$\nabla f(X^1) = \left(\frac{12}{5}, \frac{-12}{5}\right)$$
$$X^2 = \left(\frac{4}{5}, \frac{4}{5}\right) - r\left(\frac{12}{5}, -\frac{12}{5}\right)$$
$$X^2 = \left(\frac{4 - 12r}{5}, \frac{4 + 12r}{5}\right)$$
$$h(r) = f\left(\frac{4 - 12r}{5}, \frac{4 + 12r}{5}\right)$$
$$h(r) = \frac{1296}{25}r^2 - \frac{288}{25}r + \frac{20}{25}$$
$$\frac{dh}{dr} = \frac{2592}{25}r - \frac{288}{25} = 0 \Rightarrow r = \frac{1}{9}$$
$$X^2 = \left(\frac{8}{15}, \frac{16}{15}\right)$$

Iteration III:

$$\nabla f(X^2) = \left(\frac{-4}{5}, \frac{-4}{5}\right)$$
$$X^3 = \left(\frac{8}{15}, \frac{16}{15}\right) + r\left(\frac{4}{5}, \frac{4}{5}\right) = \left(\frac{8 + 12r}{15}, \frac{16 + 12r}{15}\right)$$
$$X^3 = \left(\frac{8 + 12r}{15}, \frac{16 + 12r}{15}\right)$$
$$h(r) = f\left(\frac{8 + 12r}{15}, \frac{16 + 12r}{15}\right)$$
$$h(r) = \frac{720}{225}r^2 - \frac{288}{225}r + \frac{36}{225}$$

$$\frac{dh}{dr} = \frac{1440}{225}r - \frac{288}{225} = 0 \Rightarrow r = \frac{1}{5}$$

$$X^3 = \left(\frac{52}{75}, \frac{92}{75}\right)$$

Iteration IV:

$$\nabla f(X^3) = \left(\frac{12}{25}, \frac{-12}{25}\right)$$

$$X^4 = \left(\frac{52}{75}, \frac{92}{75}\right) - r\left(\frac{12}{25}, \frac{-12}{25}\right)$$

$$X^4 = \left(\frac{52 - 36r}{75}, \frac{92 - 12r}{75}\right)$$

$$h(r) = f\left(\frac{52 - 36r}{75}, \frac{92 - 12r}{75}\right)$$

$$h(r) = \frac{6480}{(78)^2}r^2 + \frac{180}{(75)^2}$$

$$\frac{dh}{dr} = 0 \Rightarrow r = 9$$

$$\therefore \qquad X^4 = \left(\frac{52}{75}, \frac{92}{75}\right)$$

1.10.2 Hook's and Jeeves' Method

EXAM. 1: Solve the problem by Hooke's and Jeeves' method.

Minimize $f(x_1, x_2) = x_1^2 + x_2^2 - 2x_1 - 2x_2$ starting with $x^{(0)} = (0, 0)$ and step length $h_1 = h_2 = 1$ and stepping length is $h_1 = h_2 < 1/2$.
SOLUTION

$F(x_1, x_2) = x_1^2 + x_2^2 - 2x_1 - 2x_2$
$\delta f/\delta x1 = 2x_1 - 2 = 0; \quad \delta f/\delta x2 = 2x2 - 2 = 0$
$x_1 = 1, \quad x2 = 1$
$\delta^2 f/\delta x_1^2 = 2; \quad d^2 f/dx_1 dx_2 = 0; \quad \delta^2 f/\delta x_2^2 = 2$
$H = ; \quad + \qquad + |H| = 4 > 0$

$\begin{vmatrix} 2 & 0 \\ 0 & 2 \end{vmatrix}$

$+ \qquad +$

therefore $f(x_1, x2)$ is minimum at $x^* = (1, 1)$, with $f(x^*) = -2$

Here $b_1 = (0, 0)$ $f(b_1) = 0$

E (i) $E(b_1) = f(1, 0) = -1$ F

Explorative $f(-1, 0) = 3$ F

Move $f(1, 1) = -2$ S

               $f(1, 1) = 2$ F

               $f(1, 1) = -1$ F

               $f(1, 1) = 3$ F

               $f(1, 1) = -2$ ignored

$E(b_1)$: $E(1, 1)$ is a success and hence the new base point is $b_2 = (1, 1)$, $f(b_2) = f(1, 1) = -2$

Start the pattern move at $b_2 = (1, 1)$

P(i) $p_1 = 2b_2 - b_1 = (2, 2) - (0,0) = (2, 2)$

   $f(p_1) = f(2, 2) = 0$

Here $f(p_1) = f(2, 2) = 0$ is more than $f(b_2) = f(1, 1) = -2$

P(ii) Then the pattern move is abandoned from $b_2 = (1, 1)$ with $f(b_2) = -2$.

E (i) $f(2, 1) = $ $4 + 1 - 4 - 2 = -1$ F

     $f(1, 2) = $ $1 + 4 - 2 - 4 = -1$ F

     $f(2, 2) = $ $4 + 4 - 4 - 4 = 0$ F

     $f(2, 0) = $ $4 - 4 = 0$ F

There is no possibility of either to continue with pattern or exploratory move from $b_2 = (1, 1)$, with $f(b_2) = -2$.

The optional point is reached and is given by $x^* = b = (1, 1)$ with $f(x^*) = f(b_1) = -2$.

GRAPHICAL REPRESENTATION

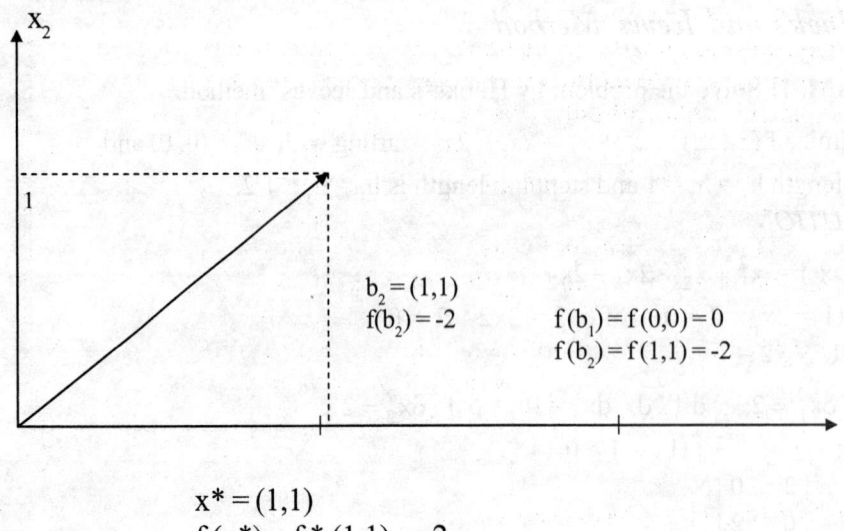

$b_2 = (1,1)$
$f(b_2) = -2$ $f(b_1) = f(0,0) = 0$
                 $f(b_2) = f(1,1) = -2$

$x^* = (1,1)$
$f(x^*) = f^*(1,1) = -2$

EXAM. 2: Use Hooke's and Jeeves' search method to the problem.

Minimize $f(x_1, x_2) = x_1^2 - 2x_1x_2 + 2x_2^2 - 2x_1 - 5x_2$ starting with $x(0) = (1, 1)$, with $h_1 = h_2 = 1$; proceed till $h_1 = h_2 < 1/2$.

SOLUTION

Explorative move: E(i)

$b_1 = (1, 1) \ f(1, 1) = 1 - 2 + 2 - 2 - 5 = -6$

E(i)

$E(b_1):$ $f(2, 1) = -7$ S

            $f(0, 1) = -7$ S

            $f(1, 2) = -7$ S

            $f(1, 0) = -7$ S

            $f(2, 2) = -7$ S

$E(2, 2) = -10$ is a success and

hence the new base point is $b_2 = (2, 2)$, with $f(b_2) = -10$.

Pattern move at $b_2 = (2, 2)$

$P(b_2): p_1 = 2b_2 - b_1 = (4, 4) - (1, 1) = (3, 3)$

and $f(p_1) = f(3, 3) = -12$

Explorative move at $p_1 = (3, 3)$

$E(p_1)$ $f(4, 3) = -13$ S

           $f(3, 4) = -9$ F

           $f(4, 4) = -12$ F

$E(4, 3)$ is a success with (-13) and hence

the new base point is $b_3 = (4, 3)$ with $f(4, 3) = -13$

Pattern move at $b_3 = (4, 3)$ with $f(4, 3) = -13$

$P(b_3)$

$P_2 = 2b_3 - b_2 = (8, 6) - (2, 2) = (6, 4)$

$f(b_3) = f(4, 3) = -13$

Explorative move at $p_2 = (6, 4)$, $f(p_2) = 36 - 48 + 32 - 12 - 20 = 68 - 80 = -12$

E(i) $f(6, 5) = 36 - 60 + 50 - 12 - 25 = 86 - 97 = -11$

          $f(5, 4) = 25 - 40 + 32 - 10 - 20 = 57 - 70 = -13$

          $f(5, 3) = 25 - 30 + 18 - 10 - 15 = 43 - 55 = -12$

Explorative move at $(6, 4)$ has values less than $f(b_3)$ and $f(b_2)$.

Hence continue with explorative move about $b_3 = (4, 3)$.

E(i) $f(4, 3) = -13$

          $f(4, 4) = -12$ F

          $f(3, 3) = -12$ F

          $f(3, 4) = -9$ F

          $f(5, 4) = -13$ S

$$f(5, 3) = -12 \quad F$$

$b_4 = (5, 4) \ f(b_4) = f(5, 4) = -13$

Start the pattern move at $b_4 = (5, 4)$, $f(5, 4) = -13$

$p_3 = 2b_4 - b_3 = (10, 8) - (4, 3) = (6, 5)$

$E(p3): \qquad f(p_3) \qquad = -13$

$\qquad\qquad f(5, 5) = -10$

$\qquad\qquad f(5, 4) = -13 \quad S$

$\qquad\qquad f(6, 4) = -12$

$\qquad\qquad f(6, 5) = -13$

$b_5 = (5, 4)$ is a success $f(b5) = f(5, 4) = -13$

Since $b_5 = b_4 = (5, 4)$

$F(b_5) = f(5, 4) = -13$

Explorative move at b_5

$E(iii)$

$f(4\frac{1}{2}, 4) = -12.75 \ F$

$f(4\frac{1}{2}, 4\frac{1}{2}) = -11.25 \ F$

$f(5, 4\frac{1}{2}) = -12 \ F$

$f(4\frac{1}{2}, 3\frac{1}{2}) = -13.25 \ S$

Further exploration and pattern moves are not necessary as $h_1 = h_2 = 1/2$.

The optimal point is at

$x^* = b_6 = (4.5, 3.5)$ with $f(x^*) = -13.25$

Graphical representation of Hooke's and Jeeves' search

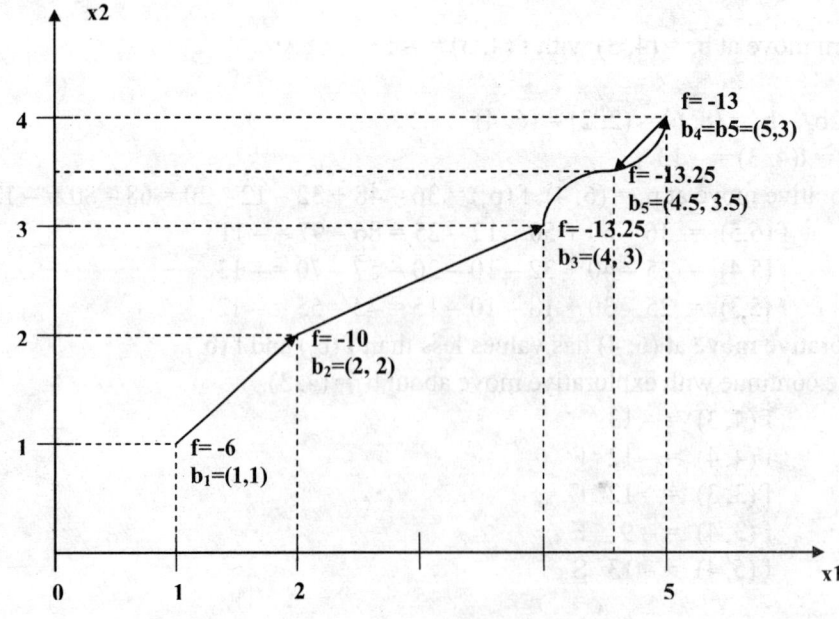

$$x(o) = \quad b1 = (1, 1) \qquad f(1, 1) = -6$$
$$b_2 = (2, 2) \qquad f(2, 2) = -10$$
$$b_3 = (4, 3) \qquad f(4, 3) = -13$$
$$b_4 = (5, 4) \qquad f(5, 4) = -13$$
$$b_5 = (5, 4) \qquad f(5, 4) = -13$$
$$b_6 = (4.5, 3.5) \quad f(4.5, 3.5) = -13.25$$
$$x^* = (4.5, 3.5); \ f(x^*) = -13.25$$

EXAM. 3: Solve the following problem by Hooke's and Jeeves' search method.

Maximize $f(x_1, x_2) = x_1 + x_2 - \dfrac{1}{2}x_1^2 + x_1 x_2 - x_2^2$

Starting with $x(o) = (1/2, 1/2)$ with step length $h_1 = h_2 = 1/2$; proceed till $h_1 = h_2 < 1/4$

SOLUTION

Minimize $f(x_1, x_2) = -x_1 - x_2 + 1/2 x_1^2 - x_1 x_2 + x_2^2$

$\delta f / \delta x_1 = -1 + x_1 - x_2 = 0 \quad \delta f / \delta x_2 = -1 - x_1 + 2x_2 = 0$

$x_1 - x_2 = 1$

$-x_1 + 2x_2 = 1$
_____

Add $\quad x_2 = 2$

$x_1 = 1 + x_2 = 3$

$x = (3, 2)$

$f(x^*) = f(3, 2) = -3 - 2 + \dfrac{9}{2} - 6 + 4$

$= -11 + 8\frac{1}{2} = -2.5$

[Max: $f(x^*) = 2.5$, with $x^* = (3, 2)$]

let $b_1 = x(o) = (\frac{1}{2}, \frac{1}{2})$

We start with explorative move at $b_1 = (\frac{1}{2}, \frac{1}{2})$

With $f(b_1) = f(\frac{1}{2}, \frac{1}{2}) = -7/8$

E(i) $E(b_1) = f(b_1) = f(\frac{1}{2}, \frac{1}{2}) = -7/8$

        $f(1, \frac{1}{2}) = -1 - \frac{1}{2} + \frac{1}{2} - \frac{1}{2} + \frac{1}{4} = -5/4$ S

        $f(\frac{1}{2}, 1) = -1/2 - 1 + 1/8 - \frac{1}{2} + 1 = -7/8$ F

        $f(1, 1) = 1 + 1 - \frac{1}{2} + 1 - 1 = 3/2$ F

Therefore $b_2 = (1, \frac{1}{2})$, $f(b_2) = f(1, \frac{1}{2}) = -5/4$ is a success. Hence start with pattern move.

P(i) $p_1 = 2b_2 - b_1 = (2, 1) - (\frac{1}{2}, \frac{1}{2}) = (3/2, \frac{1}{2})$

E(b_1) $= f(\varphi_1) = f(3/2, \frac{1}{2})$ $= -3/2 - \frac{1}{2} + 9/8 - \frac{3}{4} + \frac{1}{4}$

$$= -2\,\tfrac{3}{4} + 11/8$$
$$= -22/8 + 11/8 = -11/8 = -1.375$$

Explorative move at $p_1 = (3/2, 1/2)$, $f(p_1) = -1.1/8$

E(i)

$$f(2, 1/2) = -5/4 \quad F$$
$$f(1, 1/2) = -5/4 \quad F$$
$$f(3/2, 1) = -15/8 \qquad S$$
$$f(3/2, 0) = 3/8 \quad F$$
$$f(2, 1) \quad = -2 \quad S$$

$b_3 = (2, 1)$, $f(b3) = f(2, 1) = -2$ is a success We start with pattern move at
$p_2 = 2b3 - b_2 = (4, 2) - (1, 1/2)$
$$= (3, 3/2)$$
$f(p_2) = f(3, 3/2) = -2.25$

Pattern move at $p_2 = (3, 3/2)$ $f(\varphi_2) = -2.25$

P(i) $\quad f(7/2, 3/2) = 69/8 \qquad F$
$\quad f(5/2, 3/2) = -2.375 \qquad F$
$\quad f(3, 2) = -2.5 \quad S$
$\quad f(3,1) = -1.5 \quad F$

$b_4 = (3, 2)$, $f(3,2) = -2.5$ is a success.

Hence start with a pattern move

$P(i) = p_3 = 2b4 - b3 = (6, 4) - (2, 1) = (4, 3)$
$f(p_3) = f(4, 3) = -18$

Explorative move at $p_3 = (4, 3)$

E(i) $\quad f(9/2, 3) = -137/8 = -17\,1/8 \quad F$
$\quad f(4, 7/2) = -69/8 = -8\,5/8 \qquad F$
$\quad f(7/2, 3) = -14\,1/8 \quad F$
$\quad f(7/2, 5/2) = -13\,3/8 \quad F$

All failures at (4, 3).

Therefore the pattern move fails.

Hence the optimal point is already reached at $x^* = b4 = (3, 2)$
with $f(x^*) = f(b_4) = f(3, 2) = -2.5$

Therefore Min. occurs at $x^* = (3, 2)$ with $f(x^*) = -2.5$

Graphical representation

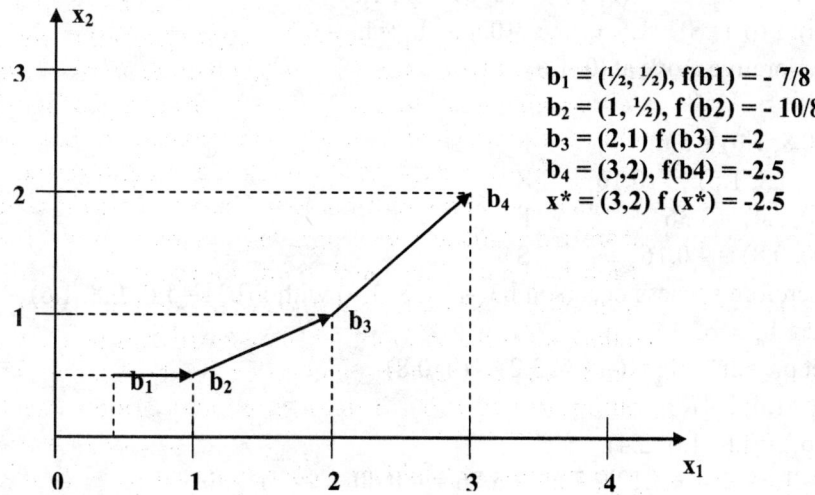

$$b_1 = (½, ½), f(b1) = -7/8$$
$$b_2 = (1, ½), f(b2) = -10/8$$
$$b_3 = (2,1) \, f(b3) = -2$$
$$b_4 = (3,2), f(b4) = -2.5$$
$$x^* = (3,2) \, f(x^*) = -2.5$$

EXAM. 4: Use Hookes and Jeeves search method to solve:

Min $f(x_1, x_2) = x_1 - x_2 + 2x_1^2 + 2x_1x_2 + x_2^2$

Starting with $x(o) = (0, 0)$ and step lengths

$\Delta x_1 = \Delta x_2 = 0.8$ and $\varepsilon = 0.1$

By calculating
$\delta f/\delta x_1 = 1 + 4x_1 + 2x_2 = 0$
$\delta f/\delta x_2 = -1 + 2x_1 + 2x_2 = 0$

at a stationary point.
$4x_1 + 2x_2 = -1 \qquad 1 - 4 + 2x_2 = 0$
$2x_1 + 2x_2 = 1 \qquad 2x_2 = 3$
$2x_1 = -2 \qquad x_2 = 3/2$
therefore $x_1 = -1$
$x^* = (-1, 3/2), f(x^*) = -1 - 3/2 + 2 - 3 + 9/4$
$= ¼ [-4 - 6 - 4 + 9]$
$= ¼ [-14 + 9] = -5/4 = -1.25$
$f(b_1) = f(0, 0) = 0$
Exploratory move:

E(i) $f(0.8, 0) = 0.8 + 1.28 = 2.08$ F
          $f(-0.8, 0) = -0.8 + 1.28 = 0.48$ F
          $f(0, 0.8) = -0.8 + 0.64 = -0.16$ S
          $f(0, -0.8) = 0.8 + 0.64 = 1.44$ F
therefore success occurs at $b_2 = (0, 0.8)$ with $f(b_2) = -0.16$

here $b_2 \neq b_1$

set $p_1 = 2b_2 - b_1 = (0, 1.6) - (0, 0) = (0, 1.6)$

P(i): $f(p_1) = 0 - 1.6 + 2.56 = 0.96$ $h_1 = h_2 = 0.8$

Explorative move at $(0, 1.6)$

E(i)

$f(0.8, 1.6) = 5.6$ F

$f(-0.8, 1.6) = -0.16$ S

$f(0, 2.4) = 3.36$ F

$f(0, 0.8) = -0.16$ S

Therefore success occurs at b3 $=(-0.8, 1.6)$ with $f(b3) = f(-0.8, 1.6) = -0.16$

Here $b_3 \neq b_2$

Set $p_2 = 2b_3 - b_2 = (-1.6, 3.2) - (0, 0.8)$

$= (-1.6, 2.4)$

$f(p_2) = f(-1.6, 2.4)$

$= -1.6 - 2.4 + 5.12 - 7.68 + 5.76 = -0.80$

Explorative moves:

E(i) $f(-0.8, 2.4) = 0$ F

           $f(-2.4, 2.4) = 5.76$ F

           $f(-1.6, 3.2) = 0.32$ F

           $f(-1.6, 1.6) = -0.64$ F

consider the other point b3 $= (0, 0.8)$ with $f(0, 0.8) = 0.16$

Here $b_3 = (0, 0.8) = b_2$

We take the step length as $h_1 = h_2 = 0.4$

Therefore $f(b_3) = f(0, 0.8) = -0.16$

Explorative move at $(0, 0.8)$

E(i) $f(0.4, 0.8) = 2.00$ F

           $f(-0.4, 0.8) = -0.88$ S

           $f(0, 1.2) = 0.24$ F

           $f(0, 0.4) = -0.24$ F

therefore $f(-0.4, 0.8) = -0.88$ is a success

$b_4 = (-0.4, 0.8)$

$f(b_4) = -0.88$

$b_4 \neq b_3$

We start once again the pattern search

P(i)

$P_3 = 2b_4 - b_3 = (-0.8, 0.8)$

$f(p_3) = f(-0.8, 0.8)$

     $= -0.96$

therefore explorative moves are undertaken at

E(i): p3 = (− 0.8, 0.8) with $f(p_3) = -0.96$

$\quad$ f (0.4, 0.8) = − 0.8 $\qquad$ F

$\quad$ f (− 1.2,0.8) = 0.40 $\qquad$ F

$\quad$ f (− 0.8, 1.2) = − 1.20 $\quad$ S

$\quad$ f (− 0.8, 0.4) = − 0.44 $\quad$ F

Success occurs at $b_5 = -1.20$

Here $b_5 \neq b_4$

(− 0.8, 1.2) ≠ (− 0.4, 0.8)

make a pattern move at $b_5 = (-0.8, 1.2)$

with $f(b_5) = f(-0.8, 1.2) = -1.20$

P(i)

Take $p_4 = 2b_5 - b_4$

$\qquad$ = (− 1.6, 2.4) − (− 0.4, 0.8)

$\qquad$ = (− 1.2, 1.6)

therefore $f(p_4) = f(-1.2, 1.6)$

= − 1.2 − 1.6 + 2.88 − 3.84 + 2.56

= 2.2

Explorative moves are undertaken:

E(i)

f (− 0.8, 1.6) = − 1.12 $\qquad$ F

f (− 1.6, 1.6) = − 0.64 $\qquad$ F

f (− 1.2, 2.0) = − 1.12 $\qquad$ F

f (− 1.2, 1.2) = − 0.96 $\qquad$ F

Now $f(p_4) = f(-1.2, 1.6) = 2.2$

The pattern move is now abandoned and commence exploration move from

b5 = (− 0.8, 1.2) with f (b5) = − 1.20

$h_1 = h_2 = 0.4$

E(i)

f (− 0.4, 1.2) = − 0.80 $\qquad$ F

f (− 1.2, 1.2) = − 0.96 $\qquad$ F

f (− 0.8, 1.6) = − 1.12 $\qquad$ F

f (− 0.8, 0.8) = − 0.76 $\qquad$ F

Here all are failures.

Hence take $h_1 = h_2 = 0.2$

$f(b5) = f(-0.8, 1.2) = -1.20$

Explorative move is started

E(i)

$\quad$ f (− 0.6, 1.2) = 0 $\qquad$ F

$\quad$ f (− 1.0, 1.2) = − 0.2 $\qquad$ F

$f(-0.8, 1.4) = -1.2$ S

$f(-0.8, 1.0) = -1.12$ S

therefore $b_6 = (-0.8, 1.4)$ $f(b_6) = -1.2$

$b6 \neq b5$

Start the pattern move

P(i)

Take $p_5 = 2b_6 - b_5 = (-1.6, 2.8) - (-0.8, 1.2)$

$= (-0.8, 1.6)$

$f(p_5) = f(-0.8, 1.6)$

$= -0.8 - 1.6 + 1.28 - 2.56 + 2.56$

$= -1.12$

Start once again the explorative move at $p_5 = (-0.8, 1.6)$ with $f(p_5) = -1.12$

E(i)

$f(-0.6, 1.6)$ $= -0.84$ F

$f(-1, 1.6)$ $= -1.24$ S

$f(-0.8, 1.8)$ $= -0.96$ F

$f(-0.8, 1.4)$ $= -1.2$ S

We take $b_7 = (-1.0, 1.6)$ with $f(b_7) = -1.24$

$b_7 \neq b_6$

[P (i)

Therefore pattern move starts

Take $p_6 = 2b_7 - b_6 = (-2.0, 3.2) - (0.8, 1.4)$

$= (-1.2, 1.8)$

$f(p_6) = -1.2 - 1.8 + 2.88 - 4.32 + 3.24$

$= -1.2$

Explorative move at $p_6 = (-1.2, 1.8)$ with $f(p_6) = -1.2$

E (i)

$f(-1, 1.8) = -1.16$

$f(-1.4, 1.8) = -1.08$

$f(-1.2, 2.0) = -1.12$

$f(1.2, 1.6) = -1.2$

Proceed further till $h_1 = h_2 = 0.1$]

We take the approximate solution as

$X^* = (-1, 1.6)$ with $f(x^*) = -1.24$

EXAM. 5: Hookes' and Geeves' method

Min. $f(x_1, x_2) = x_1^2 - 2x_1x_2 + 2x_2^2 - 2x_1 - 5x_2$

Starting with the point $(2, 2)$ with step length

$h_1 = h_2 = 1$ and till $\varepsilon = h_1 = h_2 = \frac{1}{2}$

$b_1 = (2, 2)$

therefore $f(b_1) = f(2, 2) = -10$

E(i) Exploration starts with $b_1 = (2, 2)$

$f(3, 2) = -11$ S
$f(1, 2) = -7$ F
$f(2, 3) = -9$ F
$f(2, 2) = -10$ F

Therefore $b_2 = (3, 2)$, $f(b_2) = f(3, 2) = -11$

$b_2 \neq b_1$

P(i)

Pattern move is undertaken

$p_1 = 2b_2 - b_1 = (6, 4) - (2, 2) = (4, 2)$

$f(p_1) = f(4, 2) = 16 - 16 + 8 - 8 - 10 = -10$

Exploration search starts with $p_1 = (4, 2)$

$F(p_1) = -10$

E (i)

$f(5, 2) = -7$ F
$f(3, 2) = -11$ S
$f(4, 3) = -13$ S
$f(4, 2) = -10$ S

therefore $b_3 = (4, 2)$, $f(b_3) = f(4, 2) = -13$

Here $b_3 \neq b_2$

P(i) Pattern search starts

$p_2 = 2b_3 - b_2 = (8, 4) - (3, 2) = (5, 2)$

$f(p_2) = f(5, 2) = 25 - 20 + 8 - 10 - 10 = 23 - 40 = -17$

Exploration search starts at $p_2 = (5, 2)$, with $f(p_2) = -17$

E(i)

$f(6, 2) = -2$ F
$f(4, 2) = -10$ F
$f(5, 3) = -12$ F
$f(5, 1) = 2$ F

Pattern move is abandoned and exploration search starts with $b_3 = (4, 2)$ and $f(b_3) =$ -13

E(i)

$f(5, 2) = -7$ F
$f(3, 2) = -11$ F
$f(4, 3) = -13$ S
$f(4, 1) = -5$ F

$b_4 = (4, 3)$ $f(b4) = f(4, 3) = -13$

$b_4 \neq b_3$

Therefore pattern search starts at $p_3 = 2b_4 - b_3$

$= (8, 6) - (4, 2)$

$= (4, 4)$

P(i)

$p_3 = (4, 4), f(4, 4) = 16 - 32 + 32 - 8 - 20$

$= 16 - 28$

$= -12$

Exploration search starts with

E(i) $f(5, 4) = -13$ S

  $f(3, 4) = -9$ F

  $f(4, 5) = -7$ F

  $f(4, 3) = -13$ S

therefore $b_5 = (5, 4), f(5, 4) = -13$

$b_5 = (4, 3)$ with $f(b_5) = -13$

Case i

Take $b_5 = (5, 4)$ $f(5, 4) = -13$

Here $b_5 \neq b_4$

$p_4 = 2b_5 - b_4 = (10, 8) - (4, 3) = (6, 5)$

This pattern search with (5, 4) is discontinued.

Case ii

The other alternative is $b_5 = (4, 3), f(4, 3) = -13$

Here $b_5 = (4, 3) = b_4 = (4, 3)$

Hence we take step length a $h_1 = h_2 = \frac{1}{2}$

Now $f(4, 3) = -13$

E(i) Exploration search starts:

  $f(4.5, 3) = -12\frac{3}{4}$ F

  $f(3.5, 3) = -11\frac{3}{4}$ F

  $f(4, 7/2) = -13$ S

  $f(4, 5/2) = -12$ F

therefore success occurs at $b_6 = (4, 7/2)$

$f(b_6) = f(4, 7/2) = -13$

[Using calculus for the given

$f(x_1, x_2) = x_1^2 - 2x_1x_2 + 2x_2^2 - 2x_1 - 5x_2$ for minimization, we have

$\delta t/\delta x_1 = 2x_1 - 2x_2 - 2 = 0;$

$\delta f/\delta x_2 = -2x_1 + 4x_2 - 5 = 0$

$2x_1 - 2x_2 = 2$

$-2x_1 + 4x_2 = 5$

$2x_2 = 7$

$x_2 = 7/2$

and
$$2x_1 - 7 = 2$$
$$2x_1 = 9$$
$$x_1 = 9/2$$
therefore $x^* = (x_1^* = 9/2, x_2^* = 7/2)$
$f(x^*) = -13$
$b_6 = (4, 7/2), b_5 = (4, 3)$
$b_6 \neq b_5$
We proceed to the pattern search
P(i)
$p_5 = 2b_6 - b_5 = (8, 7) - (4, 3) = (4, 4)$
$f(p_5) = f(4, 4) = 16 - 32 + 32 - 8 - 20 = -12$
Exploration search is done at $p_5 = (4, 4)$
E (i)

$f(9/2, 4) = -12\frac{3}{4}$	S	
$f(7/2, 4) = -10\frac{3}{4}$	F	
$f(4, 9/2) = -10$	F	
$f(4, 7/2) = -13$	S	

$b_7 = (4, 7/2), f(b_7) = -13$
Since $b_7 = b_6 = (4, 7/2)$
Take the step length as $h_1 = h_2 = \frac{1}{4}$
Since the search will have to stop when we reach $h_1 = h_2 = \frac{1}{2}$ as in the hypotheses we stop the searches.
$b_7 = (4, 7/2)$
$x^* = (4, 7/2)$
$f(x^*) = -13$

Graphical representation

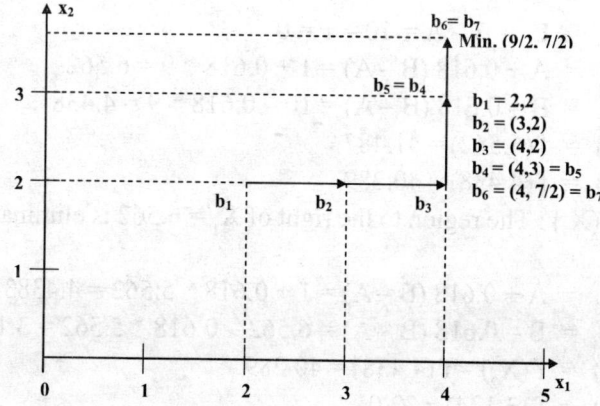

1.10.3 Golden Section Search Method

EXAM. 1: Minimize $f(X) = X^4 - 15X^3 + 72X^2 - 1135X$

In the interval $(1 \leq X \leq 15)$ using the Golden section search and terminate the search when

$$|f(X_n) - f(X_{n-1})| < 0.50.$$

SOLUTION

The first two points are placed symmetrically within the interval $1 \leq X \leq 15$.

The Golden Section search, using the golden section ratio, places these points at

$X_1 = 1 + 0.618 (15 - 1) = 1 + 0.618 * 14 = 9.652$

And $X_2 = 15 - 0.615(15 - 1) = 15 - 0.618 * 14 = 6.348$

$f(X_1) = -9056.29$

$f(X_2) = -6516.82$

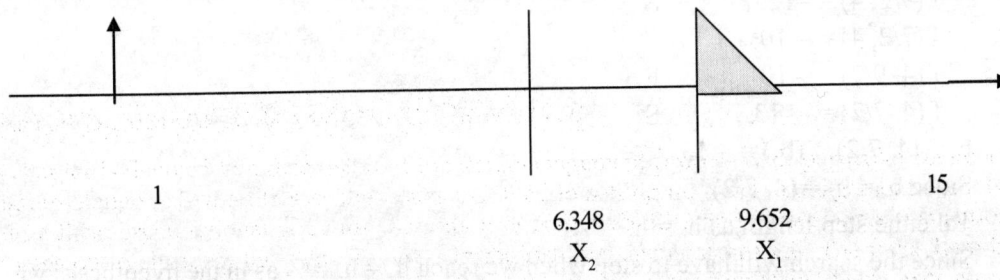

| 1 | | 6.348 X_2 | 9.652 X_1 | 15 |

The region to the right of $X_1 = 9.652$ is eliminated since $f(X_2) > f(X_1)$ and the interval of uncertainty now is given by $6.348 < X < 15$ (Note: this search procedure is the same as Fibonacci search)

Continue the search in the interval $(6.348, 15)$.

EXAM. 2: Use the Golden section search to find the min $f(X) = 4X + 100/X$ in $1 \leq X \leq 10$ with $\varepsilon = \pm 1$

Here
$$L_0 = L_1 = B - A = 10 - 1 = 0$$
$$X_1 = A + 0.618 (B - A) = 1 + 0.618 * 9 = 6.562$$
$$X_2 = B - 0.618 (B - A) = 10 - 0.618 * 9 = 4.438$$
$$f(X_1) = f(6.562) = 41.487$$
$$f(X_2) = f(4.438) = 40.289$$

Here $f(X_2) < f(X_1)$: The region to the right of $X_1 = 6.562$ is eliminated now $A = 1$, $B = 6.562$

$$X_3 = A + 0.618 (B - A) = 1 + 0.618 * 5.562 = 4.4383 = 4.438$$
$$X_4 = B - 0.618 (B - A) = 6.562 - 0.618 * 5.562 = 3.124$$
$$f(X_3) = f(X_2) = f(4.438) = 40.289$$
$$f(X_4) = f(3.124) = 32.01$$

Here $f(X_4) < f(X_3)$ The region new A = 3.124 new B = 6.562

$$X_5 = 3.124 + 0.618 * 3.438 = 5.2481$$
$$X_6 = 6.562 - 0.618 * 3.438 = 4.434$$
$$f(X_5) = 40.04$$
$$f(X_6) = 40.25 > f(x_5)$$

New   A = 4.434 New B = 6.562

$$X_7 = 4.434 + 0.618 (6.562 - 4.434) = 5.2481$$
$$X_8 = 6.562 - 0.618 * 2.128 = 4.2480$$

By Calculus $f(X) = 4X + 100/X, \quad 1 \le X \le 10$

$$f'(X) = 4 - 100/x^2 = 0 \Rightarrow X = 5$$
$$f''(X) = > 0 \quad X = 5$$

makes $f(X)$ is min

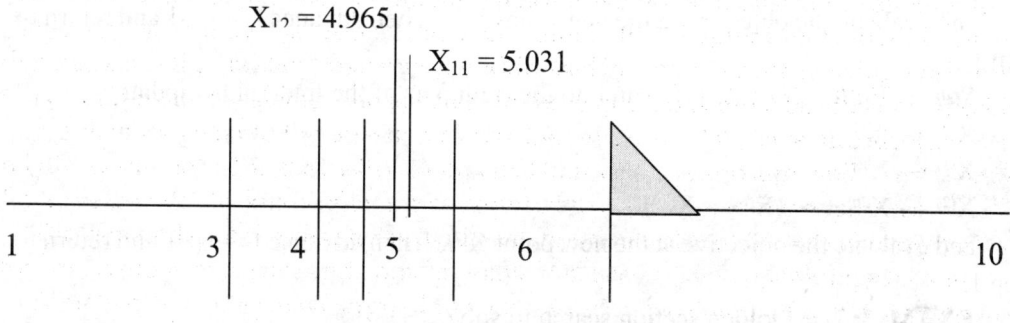

$$X_{12} = 4.965$$
$$X_{11} = 5.031$$

1   3   4   5   6   10

New   A = 4.248 New B = 5.748

$$X_9 = 4.248 + 0.618 (5.748 - 4.248)$$
$$= 4.248 + 0.618 * 1.5 = 5.1750$$
$$X_{10} = 5.748 - 0.618 * 1.5 = 4.821$$

New   A = 4.821, New B = 5.175

$$X_{11} = 4.821 + 0.618 * .354 = 5.031$$
$$X_{12} = 5.175 - 0.618 * .354 = 4.965$$
$$|X_{11} - X_{12}| = |0.066| < 1$$
$$X_{11}{}^* = 5.031 \text{ makes } f(X) = 4X + 100/X \text{ is Min}$$
$$f(X_{11}{}^*) = 5.031 \times 10 + 100/5.031 \approx 40$$

2.3.5 Algorithm for Golden Section search – (from the book on "Optimization in O.R by Ronal L. Rardin – Pearson Edition – Asia Edition Reprint 2002)

Step 0 : Initialization

Choose lower bound $X^{(lo)}$ and upper bound $X^{(hi)}$ on an optimal solution X* along with stopping tolerance $\varepsilon > 0$

Compute

$X^{(1)} \leftarrow X^{(hi)} - \propto (X^{(hi)} - X^{(lo)})$

$X^{(2)} \leftarrow X^{(lo)} + \propto (X^{(hi)} - X^{(lo)})$ for $\propto = 0.618$. evaluate objective function $f(X)$ at all form points and initialize iteration counter $t \leftarrow 0$

Step 1: Stopping: If $(X^{(hi)} - X^{(lo)}) \le \varepsilon$ stop and report as an approximate optimal solution lies

$X^* = \frac{1}{2} (X^{(lo)} + X^{(hi)})$,

The midpoint of the remaining interval, otherwise proceed to step 2, if $f(X^{(1)})$ is superior to $f(X^{(2)})$ (less for a minimize model, greater than for a maximize model) and to step 3 if it is not.

Step2 : Left: Narrow the search to the left part of the interval by updating

$X^{(hi)} \leftarrow X^{(2)}$

$X^{(2)} \leftarrow X^{(1)}$

$X^{(1)} \leftarrow X^{(hi)} - \propto (X^{(hi)} - X^{(lo)})$

and evaluate the objective at the new point $X^{(1)}$. Then advance $t \leftarrow t + 1$ and return to step 1.

Step 3: Right : Narrow the search to the right part of the interval by updating

$X^{(lo)} \leftarrow X^{(1)}$

$X^{(1)} \leftarrow X^{(2)}$

$X^{(2)} \leftarrow X^{(lo)} + \propto (X^{(hi)} - X^{(lo)})$

and evaluate the objective at the new point $X^{(2)}$. Then advance $t \leftarrow t + 1$ and return to step 1.

EXAM. 3: Use Golden section search to solve:

Min $f(X) = 3X^4 + (X - 1)^2, 4 \ge X \ge 0, \varepsilon = 0.10$

SOLUTION

$$\delta f / \delta x = 12X^3 + 2(X - 1) = 2(6X^3 + X - 1)$$
$$\delta^2 f / \delta x^2 = 2(18X^2 + 1)$$

> 0 for $X \varepsilon (0, 4)$

$$f(1/2) = 0.4375$$
$$f(1/3) = 0.52$$
$$f(0.4) = 0.4368$$
$$L_0 = L_1 = B - A = 4 - 0 = 4$$
$$X_1 = A + 0.618 (B - A) \qquad f(X_2) = 49.4809$$
$$= 0 + 0.618(4) = 2.472$$
$$X_2 = B - 0.618 (B - A)$$
$$= 4 - 0.618 (B - A)$$
$$= 4 - 2.472$$
$$= 1.528 \qquad f(X_1) > 49.4809$$

The region to the right of X_1 is discarded

New $\qquad A = 0, B = 2.472$

$$X_3 = A + 0.618\,(2.472) = 0.618 * 2.472 = 1.528$$
$$X_4 = B - 0.618\,(2.472) = 2.472 - 0.618 * 2.472 = 0.944$$

$f(X_3) > 49.480 \qquad f(X_4) = 2.385$

The region to the right of $X_3 = 1.528$ is disearched.

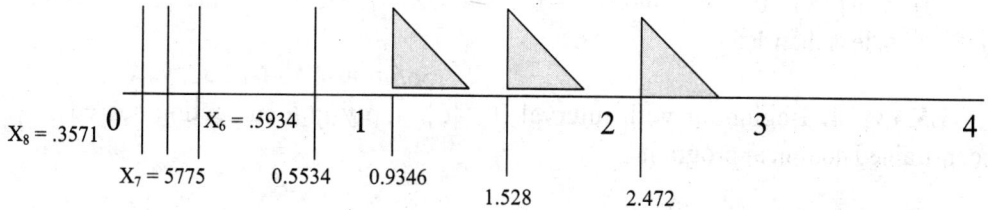

$X_8 = .3571$ 0 $X_6 = .5934$ 1 2 3 4

$X_7 = 5775$ 0.5534 0.9346

1.528 2.472

New $\qquad$ A = 0, New B = 1.528
$$X_5 = A + 0.618\,(B - A) = 0 + 0.618 * 1.528 = 0.9346$$
$$X_6 = B - 0.618\,(B - A) = 1.528 - 0.618 * 1.528 = 0.5934$$
$$f(X_5) = f(.9346) > 0.4262$$
$$f(X_6) = f(0.5934) = 0.2856$$
The region to the right $X_5 = .9346$ is discarded

New $\qquad$ A = 0, New B = 0.9346
$$X_7 = A + 0.618\,(B - A) = 0 + 0.618 \times 0.9346 = 0.5775$$
$$X_8 = B - 0.618\,(B - A) = 0.9346 - 0.618 \times 0.9346 = 0.3571$$
The region to the right of 0.5775 is discarded

The new $\qquad$ A = 0, the new B = 0.5934
$$X_9 = 0 + 0.618 \times 0.5934 = 0.618 \times 0.5934 = 0.3667$$
$$X_{10} = 0.5934 - 0.618 \times 0.5934 = 0.1864$$
Process further x ≈ 0.4 about

Additional Exercises

I. In Golden section search

i. Use Golden section search to find an optimum of the non linear programming problem.

Minimize $f(X) = 10X + 70/X$ within the interval $1 \leq X \leq 10$ to within an error of $\varepsilon = \pm 1$.

Solution

(1) By Calculus $X^* = \sqrt{7} = 2.646$

And $f(X^*) = 52.56$

(ii) By Golden section search

$X^* = 2.646\ f(X^*) = 52.56$

(2) Apply Golden Section search, beginning with the interval [0, 40] to the un constrained non linear problem

Max $f(X) = 2X - (X - 20)4/500$, until the interval containing an optimum is of length at most 10.

(Solution : termination occurs at n = 3 with I_3 = (21.12, 30.56) = 9.44 < ε = 10
$$X^* = 26.95, f(X^*) = 49.25$$

(3) Min $f(X) = 0.75X + 1089.4/(8 - 0.2)\sqrt{400}/X$ in $[8, 32]$ till the interval of uncertainty is less than 1.5

(Solution $X^* = 15.32, f(X^*) = 167.6$)

EXAM. 4: Beginning with interval [0, 40], apply golden section search to the unconstrained nonlinear program

$$\text{Max } f(x) = 2x - \frac{(x - 20)^4}{500}$$

First Evaluation

$x^{(lo)} = 0$, $x^{(hi)} = 40$, $\alpha = 0.618$,

$x^{(1)} \leftarrow x^{(hi)} - \alpha(x^{(hi)} - x^{(lo)}) = 40 - 0.618 \times 40 = 15.28$; $f(15.28) = 29.5 = 29.57$

$= 0 + 0.618 \times 40 = 24.72$; $f(24.72) = 48.45$

Second Evaluation

$x^{(lo)} = 15.28$, $x^{(hi)} = 40$

$x^{(1)} = x^{(hi)} - \alpha(x^{(hi)} - x^{(lo)}) = 40.00 - 0.618 \times 24.72 = 24.72$; $f(24.72) = 48.25$

$x^{(2)} = x^{(lo)} + \alpha(x^{(hi)} - x^{(lo)}) = 15.28 + 0.618 \times 24.72 = 30.56$; $f(30.56) = 36.27$

$f(x^{(hi)}) = f(40) - 240$; $= f(15.28) = 29.57$

Summary of Calculation

t	$x^{(l_0)}$	$x^{(1)}$	$x^{(2)}$	$x^{(hi)}$	$f(x^{(l_0)})$	$f(x^{(1)})$	$f(x^{(2)})$	$f(x^{(hi)})$	$x^{(hi)} - x^{(l_0)}$
0	0.00	15.28	24.72	40.00	-328.00	29.37	48.45	-240.00	40.00
1	15.28	24.72	30.54	40.00	29.57	48.45	36.27	-240.00	24.72
2	15.28	21.12	24.72	30.56	29.57	42.23	48.45	36.27	15.28
3	21.12	24.72	26.55	30.56	42.23	48.45	49.23	36.27	9.44

Termination occurs at t = 3 with $x^{(hi)} - x^{(l_0)} = 9.44 < ∈ = 10$

1.10.4 Fibonacci Search

EXAM. 1: Maximize the function $f(x) = -3x^2 + 21.6x + 10$, with a minimum resolution of 0.50 over six functional evaluations, by Fibonacci search. The optimal value $f(x)$ is assumed to lie in the range $0 \leq x \leq 25$.

The given interval $[0, L_n] = [0, 25]$ $\therefore L_n = 25$

Step 1: $x_1 = r^2 L_n = (0.618)^2(25) = 9.550$ $x^2 = rL_n = (0.618)(25) = 15.450$

Step 2: Evaluate $f(x_1)$ and $f(x_2)$

$f(x_1) = 15.4231; f(x_2) = 9.5769$

The region to the right of $x_1 = 15.42$ is eliminated. Now the interval to be considered is $(0, 15.4231)$. Symmetrically within the present interval of uncertainty, the two new points would be x_2 and x_4. $x_3 = 9.5769; x_4 = 5.8462$. (Note that one of the new functional evaluations corresponds to one of the old functional evaluations. Here $x_3 = x_1 = 9.5769; f(x_3) = f(x_1) = -379.477$)

$f(x_4) = f(5.8462)$

The successive evaluations of the Fibonacci search or shown in the table given below

Functional Evaluations (n)	Interval of Uncertainty	x_{n-1}	$f(x_{n-1})$	x_n	$f(x_n)$
n = 2	$15.4231 \geq x \geq 0$ [15.4231]	9.5769	-67.233	75.4231	-379.477
3	$9.5769 \geq x \geq 0$ [9.5369]	5.8462	24.744	9.5369	-67.233
4	$5.8462 \geq x \geq 0$ [5.8462]	3.731	39.83	5.8462	24.744
5	$5.8462 \geq x \geq 0$ [3.731]	4.115	32.26	3.731	39.83
6	$4.2304 \geq x \geq 0$ [2.115]	3.731	39.83	4.2303	38.688

At the sixth functional evaluation, the interval of uncertainty is established as $I_6 = 2.115$.

The best estimate of the optimal solution is given $x_5 = 3.731$ and $f(x_5^*) = 39.83$

Here $\in = 4.2304 - 3.731 = 0.4994 \approx 0.5$

3.10.5 Newton's Methods

EXAM. 1: Using Newton's method, Min $f(X) = (3x_1 - 1)^3 + 4x_1x_2 + x_2^2$ with initial starting point as $X_0 = (1, 2)$

$$\nabla f(X_0) = (44, 8), \text{ the Hessian is } H(X_0) = \begin{bmatrix} f_{x_1 x_1} & f_{x_1 x_2} \\ f_{x_2 x_1} & f_{x_2 x_2} \end{bmatrix} = 16 \begin{bmatrix} 27 & 1 \\ 1 & 1 \end{bmatrix};$$

$$[H(X_0)]^{-1} = \frac{1}{13} \begin{bmatrix} 1 & -1 \\ -1 & 27 \end{bmatrix}$$

$$X_1 = X_0 - \lambda \times \frac{8}{13} \begin{bmatrix} 1 & -1 \\ -1 & 27 \end{bmatrix} \begin{bmatrix} 44 \\ 8 \end{bmatrix} = \begin{bmatrix} 1 & -\dfrac{288}{13}\lambda \\ 2 & -\dfrac{1372}{13}\lambda \end{bmatrix} = X_I^{(1)}$$

$$f(\lambda): f[X_1^{(1)}] = \left(2 - \frac{864}{13}\lambda\right)^3 + 8\left(1 - \frac{288\lambda}{13}\right)\left(1 - \frac{688\lambda}{13}\right)$$

$$\frac{df}{d\lambda} = 81 \times 864 \times 864 \lambda^2 - 15072 \times 12 \times 13\lambda + 1040 = 0 \text{ for } f \text{ to be a maximum or a}$$

minimum. Etc.

EXAM. 2: Solve by Newton's method

Min $f(X) = 2(x_1 + x_2)^2 + 2(x_1^2 + x_2^2)$ with $X_0 = [5, 2]^T$

$$\nabla f(X_0) = [48, 36]^T; H(X_0) = \begin{bmatrix} 8 & 4 \\ 4 & 8 \end{bmatrix}$$

$$x_1^{(1)} = 5 - \lambda \begin{bmatrix} 8 & 4 \\ 4 & 8 \end{bmatrix}^{-1} \begin{bmatrix} 48 \\ 36 \end{bmatrix} = 5 - 5\lambda$$

$$x_2^{(1)} = 2 - \lambda \begin{bmatrix} 8 & 4 \\ 4 & 8 \end{bmatrix}^{-1} \begin{bmatrix} 48 \\ 36 \end{bmatrix} = 2 - 2\lambda$$

$\therefore f(1): 2(7 - 7\lambda)^2 + 2[(5 - 5\lambda)^2 + (2 - 2\lambda)^2]$. Minimum occurs with $\lambda = 1$

$\therefore X_1 = (x_1^{(1)}, x_2^{(1)}) = (0, 0) \therefore x_1^* = 0, x_2^* = 0, f(x^*) = 0$

13.11 Exercises in Non Linear Optimization

1. Solve by the method of steepest descent the following problems
   (i) Minimize $f(x_1, x_2) = 2x_1^4 + x_2^2 - 4x_1 x_2 + 5x_2$, starting $X^0 = (0,0)$
   (ii) Min $f(x_1, x_2) = 3x_1^2 - 2x_1 x_2 + x_2^2 + x_1$, with $X^0 = (1,1)$
   (iii) $f(x_1, x_2) = x_1 - x_2 + 2x_1^2 + 2x_1 x_2 + x_2^2 +$, with $X^0 = (0,0)$
   (iv) $f(x_1, x_2) = 5x_1^2 - 2x_2^2 + 2x_1 x_2 - 4x_1 x_2 - 4x_1 - 4x_2 + 4$, with $X^0 = (0,0)$
   (v) $f(x_1, x_2) = 2x_1^2 + 2x_1 x_2 + x_2 - 3x_1^2$, with $X^0 = (1/4, -1/4)$

2. Solve the following problems by the steepest ascent method
   - (i) Maximize $f(x_1,x_2) = -x_1 + x_2 - x_1^2 + x_1x_2$, starting $X^0 = (0,0)$
   - (ii) Max $f(x_1,x_2) = -1 + 2x_1 + x_2 - 3x_1^2 - x_1x_2 - 2x_2^2$, with $X^0 = (0,0)$
   - (iii) Max $f(x_1,x_2) = -x_1 - 3x_1^2 - 2x_1x_2 - 2x_2^2$, with $X^0 = (0,0)$
   - (iv) Max $f(x_1,x_2) = 4x_1 + 6x_2 - 2x_1^2 - 2x_1x_2 - 2x_2^2$, with $X^0 = (0,0)$
   - (v) Max $f(x_1,x_2) = -x_1^2 + x_1x_2 - 3x_2^2$, with $X^0 = (1,2)$
   - (vi) Max $f(x_1,x_2) = 2x_1x_2 + 2x_2 - x_1^2 - 2x_2^2$, with $X^0 = (1,2)$

3. Solve the following problems using Newton Raphson method
   - (i) Min $f(x_1,x_2) = -x_1^2 - x_1x_2 - 3x_2^2$, with $X^0 = (1,2)$
   - (ii) Min $f(x_1,x_2) = 3x_1^2 - 2x_1x_2 + x_2^2 + x_1$, with $X^0 = (1,1)$
   - (iii) Min $f(x_1,x_2) = 2x_1^4 - x_2^2 - 4x_1x_2 + 5x_2$, with $X^0 = (1,1)$ using the iterative formula
     - (a) $x_{k+1} = x_k - G_k^{-1}g_k$ $(k = 1,2,3,.....)$ and
     - (b) $x_{k+1} = x_k - \lambda_k {}^*G_k^{-1}g_k$ $(k = 1,2,3,.....)$

     where λ^* is determined by n linear search from x_k in the direction $-G_k^{-1}g_k$.
   - (iv) $f(x) = 4x^4 - x^2 + 5$, for extreme points
   - (v) $f(x_1,x_2) = x_1^3 - x_2^3 - 3x_1x_2$, for extreme points

4. Solve the following problems, using Hooke – Jeeves search. Start your search with unit moves along the coordinates axis and continue until the percent change in the objective function is 2% or less. Plot that the trajectory of search.
   - (i) Min $f(x_1,x_2) = 50 + (2.71 - x_1)^2 + (1 - x_2)^2$. Start at point $X = 0.5, 0$
   - (ii) Min $f(x_1,x_2) = \dfrac{1}{2x_1 + 6(x_2 + 1)^2}$; start from the point $X = (0,0)$
   - (iii) Min $f(x_1,x_2) = x_1 - x_2 + 2x_1^2 + 2x_1x_2 + x_2^2$; start from the point $X_0 = (0,0)$. Take $\Delta x_1 = \Delta x_2 = 0.8, \in = 0.1$
   - (iv) Min $f(x_1,x_2) = x_1^2 + 3x_2^2 + 6x_3^2$ with $X_0 = (2,-11)$; Take $\Delta x_1 = \Delta x_2 = \Delta x_3 = 0.5$
   - (v) Min $f(x_1,x_2) = 50 + (2.7 - x_1)^2 + (1 - x_2)^2$ with $X_0 = (0.5, 0)^T$
   - (vi) Max $f(x_1,x_2) = \dfrac{1}{2x_1 + 6(x_2 + 1)^2}$; with $X_0 = (0, 0)^T$
   - (vii) Min $f(x_1,x_2) = 2x_1^2 + x_2$, with $X_0 = (1,1)^T$ and $\in = 0.1$

5.
   - (i) Find the value of x in the interval $(0,1)$ which maximizes the function $f(x) = x(1.5-x)$ to within ± 0.05 by (a) Golden section method and (b) by Fibonacci method
   - (ii) $f(x) = x^5 - 5x^3 - 20x + 5$ by (a) Fibonacci search in $(0,5)$, (b) Golden section search in $(0,5)$
   - (iii) Min $f(x) = 3x^4 + (x-1)^2$ in $4 \geq x \geq 0$, $\in = 0.10$
   - (iv) Min $f(x) = 4x \sin x$ in $\pi \geq x \geq 0$
   - (v) Min $f(x) = 2(x-3)^2 + \exp(0.5x^2)$ in $100 \geq x \geq 0$

6. Solve the following by Newton's method
   - (i) Min $f(x_1,x_2) = -(x_1^2 + x_2^2 + 2)^{-1}$ with $X_0 = (4,0)^T$.
   - (ii) Min $f(x_1,x_2) = 4x_1^2 + 3x_2^2 - 4x_1x_2 - 8x_1$ with $X_0 = (0,0)^T$.

(iii) $\operatorname{Min} f(x_1, x_2) = 2x_1^2 + x_2^2$ with $X_0 = (1,2)^T$.

(iv) $\operatorname{Min} f(x_1, x_2) = 25(x_1 - 3x_2)^2 + (x_1 - 3)^2$ with $X_0 = (1,2)^T$.

(v) $\operatorname{Min} f(x_1, x_2) = (3x_1 - 1)^2 + 4x_1x_2 + x_2^2$ with $X_0 = (1,2)^T$.

7. Exercises in classical optimization techniques, using the method of Lagrange multipliers.

(i) Find the solution of the problem Minimize $f(x,y) = k.x^{-1}.y^{-2}$, subject to $g(x,y) = x^2 + y^2 - a^2 = 0$

$$\left(\text{Ans: } x^* = \frac{1}{\sqrt{3}}, y^* = \sqrt{\frac{2}{3}}a\right)$$

(ii) Maximize $f(x_1, x_2) = x_1^2 x_2$ subject to $2\pi x_1^2 + 2\pi x_1 x_2 = A_0 = 24\pi$ (Ans: $x_1^* = 2$, $x_2^* = 4, \lambda^* = -1, f^* = 16\pi$)

(iii) Maximize $f(x_1, x_2) = 2x_1 + x_2 + 10$ subject to $g(x_1, x_2) = x_1 + 2x_2^2 - 3 = 0$

(Ans: $x_1^* = 2.97, x_2^* = 0.13, \lambda^* = 2.0, f^* = -6.2972$)

(iv) Maximize $f(x_1, x_2) = 3x_1^2 + x_2^2 + 2x_1x_2 + 6x_1 + 2x_2$ subject to $g(x_1, x_2) = 2x_1 - x_2 - 4 = 0$

$$\left(\text{Ans: } x_1^* = \frac{7}{11}, x_2^* = -\frac{30}{11}, \lambda^* = \frac{24}{11}, f^* = 85.7\right)$$

(iv) Minimize $f(x_1, x_2) = x_1^2 - x_2$ subject to $x_1, x_2 = 6, x_1 \geq 1, x_1^2 + x_2^2 - 26 \leq 0$

(Ans: $x_1^* = 1, x_2^* = 5, \lambda_1 = 0, \mu_1 = 2.2, f^* = -4.0$)

(v) Minimize $f(x_1, x_2, x_3) = x_1^2 + x_2^2 + x_3^2 + 20(x_1 - 50) + 20(x_1 + x_2 - 100)$ subject to $x_1 - 50 \geq 0$, $x_1 + x_2 - 100 \geq 0$ and $x_1 + x_2 + x_3 - 150 \geq 0$

(Ans: The possible solutions are (50, 45, 55); (50, –10,0); (50,50,0); (50,50,50))

(vi) Maximize $f(x_1, x_2) = 8x_1^2 + 4x_2^2 + x_1x_2 - x_1^2 - x_2^2$, subject to $2x_1 + 3x_2 \leq 24$, $-5x_1 + 12x_2 \leq 20, x_2 \leq 5$ by applying Kuhn – tucker condition

(vii) Minimize $f(x_1, x_2) = \frac{1}{2}(x_1^2 + x_2^2 + x_3^2)$, subject to $x_1 - x_2 = 0, x_1 + x_2 + x_3 - 1 = 0$

(viii) Minimize $f(x_1, x_2) = (x_1 - 1)^2 + (x_2 - 5)^2$, subject to $-x_1^2 + x_2 \leq 4, -(x_1 - 2)^2 + x_2 \leq 3$

13.12 ADDITIONAL EXERCISES IN NON LINEAR OPTIMIZATION

1. Prove that any linear function $f(X) = C^T X$ is both concave and convex.

2. Let $f(X) \in C_1$ throught the interior of the convex set S in E^n and suppose that $f(X)$ is concave over S. Then prove that $(X_2 - X_1)^T \nabla f(X_1) \geq f(X_2) - f(X_1)$ for all interior points $X_1 \in S$ and all points $X_2 \in S$.

3. If $f(X)$ is convex function for all $X \geq 0$ and V is the nonempty set $V = \{X: f(X) \leq b, X \geq 0\}$ then prove that V is convex set.

4. If $f \in C'$ is a convex function over a convex set S, then prove that $f(Y) \geq f(X) - \nabla f(X).(Y-X)$ for all $X, Y \in S$.

5. Write down a second order Taylor expansion of $f(X)$ about a point X_0 in some convex region R in E^n in terms of the gradient operator and the Hessian matrix.

6. Determine the gradient vector and the Hessian matrix for the function $f(x_1, x_2, x_3) = x_1^2 + x_2^2 + x_3^2$

7. Prove that if the quadratic form $z = X^TAX$ is convex throughout E^n, then A is positive semi definite.

8. Prove that any positive semi definite quadratic for $f(X) = X^TDX$, where D is symmetric, is a convex function over all of E^n.

9. Write down the Hessian matrix of the function $f(x_1, x_2, x_3) = x_1^3 + x_1x_3 + x_2x_3^2$. For what values of the variables, if any, is the Hessian positive semidefinite, negative semi definite? Identify all local maxima and minima of f.

10. Prove that a convex function is unimodal.

11. Check whether the function $f(x_1, x_2) = 2x_1 + 6x_2 - 2x_1^2 - 3x_2^2 - 4x_1x_2$ is convex or not.

12. Show that the objective function decreases monotonically along the gradient path.

13. Prove that any positive semi definite quadratic $f(X) = X^TDX$, where D is symmetric, is a convex function over all of E^n.

14. Describe Fibonacci search.

15. Show that $\phi^n = F_n\phi + F_{n-1}$ for all integers, when $\{F_n\}$ is a Fibonacci sequence and $\phi = -\left(\dfrac{H\sqrt{5}}{2}\right)$

16. Use the Fibonacci search to approximate the location of the maximum of $f(x) = x(5\pi - x)$ on $[0,20]$ to within $\in = 1$.

17. For the Fibonacci sequence $\{F_n\}$, show that $F_{n-2}^2 - F_{n-1}F_{n-3} = (-1)^n$.

18. Let $\phi(x)$ be a unimodal function defined on a closed interval $[0, L_n]$, where L_n is such that X^* can located with an error of less than unit of making at most n evaluation of $\phi(x)$. Let $F_n = $ l.u.b L_n. Prove that the sequence $\{F_n\}$ is the Fibonacci sequence.

19. Find the number of experiments to be conducted to obtain the value of $\left(\dfrac{L_x}{L_0}\right) = 0.001$ in the

    Fibonacci method.

    [Ans: 15 experiments]

20. Use golden section search to maximize $x\cos x$ over the interval $\left[0, \dfrac{\pi}{2}\right]$ with an error in x^* of

    not more that 0.001.

21. Show that the golden section search and the Fibonacci search have the same asymptotic convergence.

22. Using the golden section search method, find the minimum value of $f(x) = x^2 + 2e^{-x}$, with an error of not more than 0.02 in x^*.

23. Find the minimum of $\phi(x) = (x-2)^2$ with an error in the minimum point $x^* \in R$ of not more than 0.42 by using the golden section search method.

24. Use Newton – Raphson method to minimize the function $f(x_1, x_2) = 100(x_1 - x_2^2) + (1 - x_1)^2$ starting at the point $x^{(0)} = (0,0)^T$.

25. Write down the iterative algorithm for Newton's method and show that the order of convergence is two.

    [Hint: $x_{k+1} = x_k - \lambda^* G^{-1}_k g_k, k = 1, 2, \ldots$]

26. When do you say that the sequence $\{r_k\}$ converges to r^* linearly? Test whether the sequence $r_k = \dfrac{1}{k}$ converges linearly or not

$$[\text{Hint, } \lim_{k \to \infty} \frac{\| x_{k+1} - x^* \|}{\| x - x^* \|} = 0]$$

27. Minimize $f(x_1, x_2) = 4x_1^2 + 3x_2^2 - 5x_1x_2 - 8x_1$ starting from the point $(0,0)$ using Newton's method.

28. Use the steepest ascent method to go two steps towards the maximum of $f(X) = -2x_1^2 - x_2^2 - x_3^2 - 4x_4^2$ starting at the point $(-1, 1, 0, -1)$.

29. Construct a flow – chart for the steepest descent method.

30. What are the relative efficiencies of the unconstrained methods for solving a non linear programming problem?

31. Formulate a non linear programming problem with a quadratic object function and linear constraints.

32. Develop a set of Kuhu – Tucken conditions for the problem:
   Maximize $f(X)$ subject to $g_i(X) \le 0, i = 1, 2, \ldots m, x_j \ge 0, j = 1, 2, \ldots n$

33. Sove the problem using Kuhu – Tucken conditions theory:
   $f(x_1, x_2) = \{25(x_1 - 2)^2 + (x_2 - x)^2\}$ subject to $x_1 + x_2 \ge 2; x_1 - x_2 \ge -3; x_1 + x_2 \le 6; x_1 - 3x_2 \ge 2, x_1, x_2 \ge 0$. Verify the result of this problem graphically.

34. Set up the Kuhu – Tucker conditions in matrix form to solve the quadratic programming problem
   Maximize $f(x_1, x_2) = 6x_1 - 3x_2 - 2x_1^2 - 4x_1x_2 - 3x_2^2$ subject to the conditions $x_1 + x_2 \le 1; 2x_1 + 3x_2 \le 4; x_1, x_2 \ge 0$.

35. Solve the quadratic programming problem using Kuhu – Tucker conditions
   Maximize $f(x_1, x_2) = 3x_1 + 4x_2 - x_1^2 + 2x_1x_2 - 2x_2^2$ subject to $x_1 + 2x_2 \le 7; -x_1 + 2x_2 \le 4; x_1, x_2 \ge 0$.

36. Write down the set of Kuhu – Tucker conditions for the problem:

$$\min_x (x - a)^2 + b \text{ subject to } x \ge c.$$

$[L(x,\lambda) = (x-a)^2 + b + \lambda(x-c)] \quad \dfrac{\partial L}{\partial x} = 2(x-a) + \lambda = 0 \quad \dfrac{\partial L}{\partial \lambda} = x - c \ge 0 \quad \lambda(x-c) = 0, \lambda \le 0, \lambda^*$

$= 0, x^* = c]$

37. Let K be a closed convex set which is bounded from and suppose that $f(X)$ is convex over K in E^n. If the global maximum of $f(X)$ is finite, then prove that it will take on at one or more of the extreme points of K.

[Hint: Case 1: X^* is interior and $X^* = \lambda X_2 + (1-\lambda)X_1$
$f(X^*) \le \lambda f(X_2) + (1-\lambda)f(X_1)$
$\le \lambda f(X_1) + (1-\lambda) f(X^*)$ or $\lambda[f(X^*) - f(X_2)] < 0$
i.e. $f(X^*) < f(X_2)$, a contradiction
Case 2: X^* is a boundary point $T_1 = X \cap H_i$

$T_n = T_{n-1} \cap H_n$

$\dim(T_n) = \text{in } H_n \therefore T_n \supset X^*]$

38. Explain the two – stage programming technique to solve a stochastic linear programming problem.

39. Solve the following stochastic programming problem.

    Maximize $z = 2x_1 + 3x_2$ subject to the constraints

    $3x_1 + 4x_2 \leq 12$, $P(x_1 \leq b_1) \geq 0.5$, $P(2x_2 \leq b_2) \leq 0.75$, and $x_1, x_2 \geq 0$,

    given that b_1 and b_2 are exponentially distributed with mean unity.

40. A manufacturing firm can produce 1,2 or 3 units of a product in month but the demand is uncertain. The demand is a discrete random variable which can take a value 1,2 or 3 with probabilities 0.2, 0.2 and 0.6 respectively. If the unit cost of production is Rs.400, unit revenue is Rs.1000 and unit cost of unfulfilled demand is Rs.10, determine the output that maximizes the expected total profit.

41. Find the point on the plane $ax + by + cz - d = 0$ that is closest to the origin in E^n.

42. Find the equation of the supporting hyper plane at $(2,1)$ to the function $f(x_1, x_2) = x_1^2 + 2x_1 x_2 + 3x_2^2$.

43. Define a quadratic programming problem. Describe any method to solve a quadratic programming problem.

44. Write a note on stochastic programming.

45. Define a stochastic programming problem that can be converted into an equivalent deterministic problem.

46. Explain the structure of primal and dual problems in geometric programming.

47. Determine the minimum of the function $f(x_1, x_2) = 2x_1^3 x_2^{-3} + 4x_1^{-2} x_2 + x_1 . x_2 + 8x_1 x_2^{-1} +$ where x_1 and x_2 are positive quantities.

48. Solve the following geometric programming problem:

    Minimize $z = \dfrac{9}{4} 2x_1^{-1} x_2 x_3^{-1} + x_1^{-1} x_2^{-1}$ subject to the constraint $\dfrac{9}{2} x_1^{1/3} . x^3 + 5x_1 x_2^{-2/3} - 1$

    2 and $x_1, x_2, x_3 > 0$

49. Minimize $f(x_1, x_2, x_3) = x_1^{-1} x_2^{-1} x_3^{-1}$ subject to $x_1^3 + x_2^2 + x_3 \leq 1$, $x_1, x_2, x_3 > 0$

50. Derive the primal – dual relationship and sufficiency conditions in the unconstrained case in a geometric programming problem.

51. Outline the solution of a constrained geometric programming problem.

52. Minimize $f(x_1, x_2) = 4x_1 + 4x_1 x_2^{-2} + 4x_1^{-1} . x_2$, $x_1, x_2 \geq 0$

CHAPTER 2
DYNAMIC PROGRAMMING

2.1 Dynamic Programming

Dynamic programming is a mathematical procedure designed to improve the computational efficiency of certain select mathematical programming problems by decomposing them into smaller sub problems. It solves the problems in stages with each stage involving exactly one optimizing variable.

The different stages are linked by recursive equations which yield a feasible optimal solution to the entire problem. The main unifying theory in dynamic programming is the principles of optimality enunciated by RICHARD BELLAMAN.

2.2 Principle of Optimality

"An optimal policy (or a set of decisions) has the property that, whatever the initial state and the initial decision are, the remaining decisions must constitute an optimal policy with regard to the state resulting from the first decision"

Although the principle of optimality appears to be nothing more than common-sense its application to problem-solving is decidedly non-trivial. The reason for this is that dynamic programming can be applied to a wide variety of problems, and though it is usually obvious how the decision variables should be defined, the same is not always true of the state variables.

2.3 Recursive Computation

A feature common to all dynamic programming solutions is that the given problem is imbedded within a family of similar problems formed by generalizing the data of the given problem.

A STAGE in dynamic programming redefined as the portion of the problem that possesses a set of mutually exclusive alternatives from which the best alternative is to be selected.

A STATE is normally defined to reflect the status of the constraints that bind all the stages together. The successive stages of a problem are separated by using the concept of state. The state at a given stage represents the status of the system and enables us to make feasible decisions for the given stage without having to look back. The state of the process at a stage is usually described by a set of variables, called the state variables.

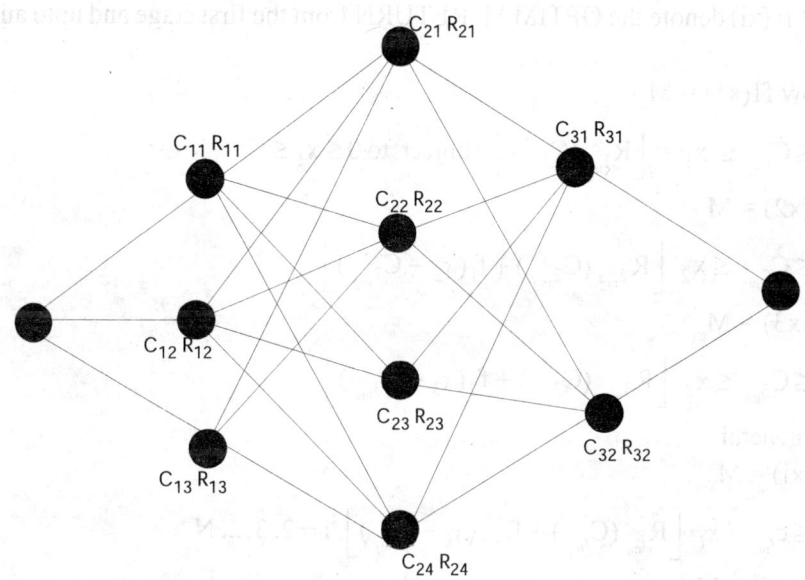

C = Cost
R = Return
Number of possible solutions = $3 \times 4 \times 2 = 24$

2.4 Forward Additive Recursive Relation of Discrete Dynamic Programming for an N-Stage Decision Problem Where The Total Capital Available is given to be C in (Certain) Units

Let there be N stages 1, 2, ...N. Let m_i represent the various alternatives at the ith stage. Let $C_{i_{m_i}}$ represent the cost of investment in the ith stage for the mith alternatives plan. Let $M_{i_{m_i}}$ be the return at the ith stage for the mith alternative plan. Let $M_{i_{m_i}}$ be the amount of investment upto and including the ith stage.

Each stage is independent and mutually exclusive. Each plan in each of the alternatives at the ith stage are independent of one another. It is natural that the return at the ith stage for the mith plan is dependent (or is a function) of the cost of investment

$$\therefore \qquad M_{i_{m_i}} = R_{i_{m_i}}(C_{i_{m_i}}), i = 1, 2, \dots N$$

Let fi(xi|mi) denote the return in the ith stage at the mith alternative plan adopted when the total investment from the first to the ith stage including the ith stage be x_i

For feasibility $0 \leq x_i \leq C$

Therefore, f1(x1|m1) is the return at the first stage.

Let fi (xi) denote the OPTIMAL RETURN from the first stage and upto and including ith stage

Now $f1(x1) = M_{ax}$

$$0 \leq C_{1_{m_1}} \leq x_1 = \left\lfloor R_{1_{m_1}}(C_{1_{m_1}}) \right\rfloor, \text{ subject to } 0 \leq x_1 \leq C$$

$$f_2(x2) = M_{ax}$$

$$0 \leq C_{2_{m_2}} \leq x_2 \left\lfloor R_{2_{m_2}}(C_{2_{m_2}}) + f_1(x_2 - C_{2_{m_2}}) \right\rfloor$$

$$f_3(x3) = M_{ax}$$

$$0 \leq C_{3_{m_3}} \leq x_3 \left\lfloor R_{3_{m_3}}(C_{3_{m_3}}) + f_2(x_3 - C_{3_{m_3}}) \right\rfloor$$

In general

$$f_i(xi) = M_{ax}$$

$$0 \leq c_{i_{m_i}} \leq x_i \left\lfloor R_{i_{m_i}}(C_{i_{m_i}}) + f_{i-1}(x_i - C_{i_{m_i}}) \right\rfloor, i = 2, 3, \dots N$$

$$f_N(xN) = M_{ax}$$

$$0 \leq e_{N_{m_N}} \leq x_N \left\lfloor R_{N_{m_N}}(c_{N_{m_N}}) + f_{N-1}(x_N - c_{N_{m_N}}) \right\rfloor$$

$\therefore$ The forward additive recursive equations are give by

$$f_i(xi) = M_{ax}$$

$$0 \leq c_{i_{m_i}} \leq x_i \left\lfloor R_{i_{m_i}}(C_{i_{m_i}}) + f_{i-1}(x_i - C_{i_{m_i}}) \right\rfloor, i = 2, 3, \dots N,$$

Subject to $0 \leq x_i \leq C$

Similarly the Bakward Inductive Additive Recursive Relation of Discrete Dynamic Programming for an N-Stage Decision Problem when the Total Cost of Investment is C :

$$f_N(x_N) = M_{ax}$$

$$0 \leq x_N \leq C \left\lfloor R_{N_{m_N}}(c_{N_{m_N}}) \right\rfloor$$

and

$$f_i(\text{xi}) = \underset{0 \le x_i \le C}{M_{ax}} \left[R_{i_{m_i}}(C_{i_{m_i}}) + f_{i+1}(x_i - C_{i_{m_i}}) \right], i = N-1, N-2, \dots 3, 2, 1.$$

Subject to $0 \le x_i \le C$

Forward Inductive Multiplicative Recursive Relation of Discrete Dynamic Programming for an N-Stage Decision Problem when the Total Cost of Investment is C

$$f_1(x_1) = \underset{0 \le x_1 \le C}{M_{ax}} \left[R_{1_{m_1}}(C_{1_{m_1}}) \right]$$

$$f_i(x_i) = \underset{0 \le x_1 \le C}{M_{ax}} \left[R_{i_{m_i}}(C_{i_{m_i}}) * f_{i-1}(x_i - C_{i_{m_i}}) \right]$$

$i = 2, 3, \dots N$

Subject is $0 \le x_i \le C$

Backward Approach

$$f_i(x_i) = \underset{0 \le x_i \le C}{M_{ax}} \left[R_{i_{m_i}}(C_{i_{m_i}}) * f_{i+1}(x_i - C_{i_{m_i}}) \right]$$

$i = N-1, N-2, \dots 3, 2, 1$

EXA.: A corporation is entertaining proposals from its 3 plants for possible expansion facilities. The corporation is budgeting Rs. 5 lakhs for allocation to all the there plants. Each plant is requested to submit proposals giving total cost (C) and total revenue (R) for each proposal.

The following table summarizes the costs and revenue in lakhs of rupees. Use D.P. to maximize the total revenue resulting from the allocation of Rs 5 lakhs to the three plants

Proposals	Plant1		Plant2		Plant3	
	C1	R1	C2	R2	C3	R3
1	0	0	0	0	0	0
2	1	5	2	8	1	3
3	2	6	3	9	-	-
4	-	-	4	12	-	-

$$f_i(x_i) = \underset{0 \le x_i \le 5}{M_{ax}} \left[R_{i_{m_i}}(C_{i_{m_i}}) + f_{i-1}(x_i - C_{i_{m_i}}) \right] \quad i = 2, 3$$

FORWARD INDUCTION
FIRST STAGE   **SECOND STAGE**

x_1	$f_1(x_1 \mid m_1) = R_{1_{m_1}}(C_{1_{m_1}})$ i=1 0 0 1 5 2 6 C1 R1 C2 R2 C3 R3	f1(x1)	m 1	x2	$f_2(x_2 \mid m_2) = R_{2_{m_2}}(C_{2_{m_2}}) + f_1(x_2 - C_{2_{m_2}})$ i=2 c2 R2 c2 R2 c3 R3 c4 R4 0 0 2 8 3 9 4 12	f2(x2)
0	- - - - - -	0	1	0	0+0=0 0 0 = - - - - - -	0
1	- - 1 5 - -	5	2	1	0+5=5 0+0 = - - - - - -	5
2	- - 2 5 2 6	6	3	2	0+6=6 2 8+0 = 8 - - - - -	8
3	- - 3 5 3 6	6	3	3	0+6=6 3 8+5 = 13 3 9 + 0 9 - -	13
4	- - 4 5 4 6	6	3	4	0+6=6 4 8+6 = 14 4 9 7 5 14 12 +0	<u>14</u>
5	- - 5 5 5 6	6	3	5	0+6=6 5 8+6 = 14 5 9 +6 15 5 12+5	17

THIRD STAGE

x_3	$f_3(x_3 \mid m_3) = R_{3_{m_3}}(C_{3_{m_3}}) + f_2(x_3 - C_{3_{m_3}})$ i=3 C1 r1 C2 r2 0 0 1 3	f3(x3)	m_3^*
0	0 - - 0 - -	-	-
1	- - - 1 3+0 3	3	2
2	- - - 2 3+0 3	3	2
3	- - - 3 3+8 11	11	2
4	- - - 4 3+13 16	16	2
5	0 + 17 17 5 3+14 17	17	

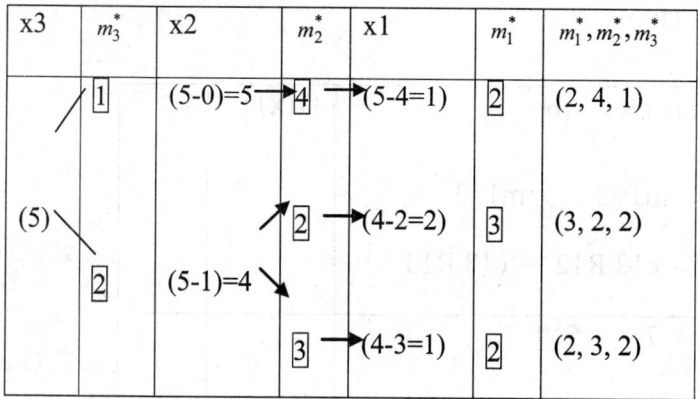

x3	m_3^*	x2	m_2^*	x1	m_1^*	m_1^*, m_2^*, m_3^*
	1	(5-0)=5 ➤ 4 ➤	(5-4=1)		2	(2, 4, 1)
(5)			2 ➤	(4-2=2)	3	(3, 2, 2)
	2	(5-1)=4	3 ➤	(4-3=1)	2	(2, 3, 2)

Ex.: Using Dynamic programming multiplicative forward recursion approach, solve the following reliability problem.

Total cost available for this project is 10 units in 1000s of rupees

mi	i=1		i=2		i=3	
	$R_{1_{m_1}}$	$C_{1_{m_1}}$	$R_{2_{m_2}}$	$C_{2_{m_2}}$	$R_{3_{m_3}}$	$C_{3_{m_3}}$
1	0.5	2	0.7	3	0.6	1
2	0.7	4	0.8	5	0.8	2
3	0.9	5	0.9	6	0.9	3

$$f_1(x_1) = M_{ax} \atop 0 \le x_1 \le C \left[R_{i_{m_i}} (C_{i_{m_i}}) \right]$$

$$f_i(x_i) = M_{ax} \atop 0 \le x_i \le C \left[R_{i_{m_i}} (C_{i_{m_i}}) * f_{i-1}(x_i - C_{i_{m_i}}) \right]$$

$R_{i_{mi}} (c_{i_{mi}}) = $ Reliability in the ith stage

for an investment of Cimi at the mith alternatives

$x_i = $ the amount spent from the stage 1 upto and including the ith stage

FOEWARD INDICTION
FIRST STAGE

x1	$f1(x1)=\text{Max}\,[R_{1_{m_1}}(c_{1_{m_1}})]$			f1(x1)	m_1^*
	m1=1	m1=2	m1=3		
	c11 R11	c12 R12	c13 R13		
..	2 -5	4 .7	5 .9	..	..
0	- -	- -	- -	-	-
1	- -	- -	- -	-	-
2	.2 .5	- -	- -	.5	1
3	.3 .5	- -	- -	.5	1
4	.4 .5	4 .7	- -	.7	2
5	.5 .5	5 .7	5 .9	.9	3
6	.6 .5	6 .7	6 .9	.9	3
7	.7 .5	7 .7	7 .9	.9	3
8	.8 .5	8 .7	8 .9	.9	3
9	.9 .5	9 .7	8 .9	.9	3
10	. 10 .5	10 .7	10 .9	.9	3

SECOND STAGE

x2	$F2(x2\|m2)=\text{Max}\left[R_{2_{m_2}}(c_{2_{m_2}})*f_1(x_2-c_{2_{m_2}})\right]$			f2(x2)	m_2^*
	m2 =1	m2 =1	m2 =3		
	3 .7	5 .8	6 .9		
0	- -	- -	- -	-	-
1	- -	- -	- -	-	-
2	- -	- -	- -	-	-
3	3 .7×0	- -	- -	-	-
4	4 .7×0	- -	- -	-	-
5	5 .7×.5 =.35	5 .8× - = -	- -	.35	1
6	6 .7×.5 =.35	6 .8× - = -	6 .9× -= -	.35	1
7	7 .7×.7 =.49	7 .8×.5 =.40	7 .9× -= -	.49	1
8	8 .7×.9 =.63	8 .8×.5 =.40	8 .9×.5	.63	1
9	9 .7×.9 =.63	9 .8×.7 =.56	9 .9×.5 .9×.5	.63	1
10	10 .7×.9 =.63	10 .8×.9 =.72	10.9×.8=.72	.72	2

Third Stage

$x3$	$f3(x3\|m3)=\text{Max}\left[R_{3_{m_3}}(c_{3_{m_3}}).f_2\left(x_3-c_{3_{m_3}}\right)\right]$			$f3(x3)$	m_3^*
	m3=1 1 .6	m3=2 2 .8	m3=3 3 .9		
0	- -	- -	- -	-	
1	1 .6× -	- -	- -	-	
2	.6× -	2 .8× -	- -	-	
3	.6× -	.8× -	3 .9× -	-	
4	.6× -	.8× -	.9× -	-	
5	.6× -	.8× -	.9× -	-	
6	.6×.35.210	.8× -	.9× -	.210	1
7	.6×.35.210	.8×.35.280	.9× -	.280	2
8	.6×.49.294	.8×.35.280	.9×.35.315	.315	3
9	.6×.63.378	.8×.49.392	.9×.35.315	.392	2
10	.6×.63.378	.8×.63.504	.9×.49.491	.504	2

M_{ax} Reliability $fN(xN) = \text{M}_{ax}$

$$0 \le xN \le c\left[R_{N_{m_N}}(c_{N_{m_N}}) * f_{N-1}(x_N - c_{N_{m_N}})\right]$$

$= 0.504$

Choice of alternative	$m_3^* = 2$	$m_2^* = 1$	$m_1^* = 3$	TOTAL RELIABILITY
Reliability	0.8	0.7	0.9	.8×.7 ×.9 = 0.504
Investment	$C_{3_{m_3}} = 2$	$C_{2_{m_2}} = 3$	$C_{1_{m_1}} = 5$	TOTAL INVESTMENT 10

14.5 Shortest Path Problem

Consider a network consisting of links (arcs) of given lengths, joined together at nodes. The length of a link may represent the distance between its terminal nodes, the time taken to perform a task, the cost of transportation or communication between the nodes, etc. The network is said to be oriented because the admissible paths through them is indicated by the arrows are all in the same general direction from left to right.

Forward Solution

The solution is divided into 4 stages. Let $y(k) = \xi k$ be the state variables where k is the stage variable.

Define the k-stage return function Fk (ξk) as the length of the minimal path from the initial node $(0, \xi 0)$ to the node $(k, \xi k)$. The node $(k, \xi k)$ represents the stage of the network at stage k. Thus ξk is the state variable.

A transformation from one state to the next involves the choice of a link jointing the current node to a node at the next stage.

The decision variable is therefore y the policy function $y(k)$ $(=\xi k)$.

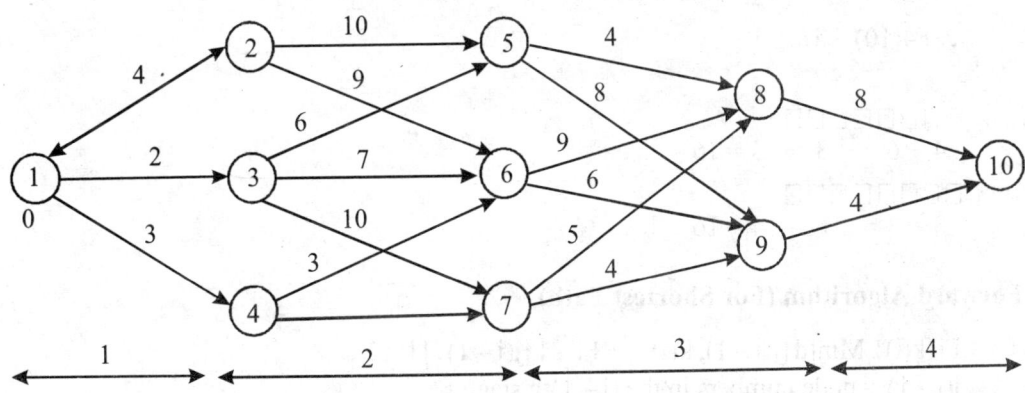

STAGE 1

$F1(2) = d(1, 2) + F0(1) = 4 + 0 = 4$

$F1(3) = d(1, 3) + F0(1) = 2 + 0 = 2$

$F1(4) = d(1, 4) + F0(1) = 3 + 0 = 3$

STAGE 2

$$F2(5) = \text{MIN} \begin{bmatrix} d(2, 5) + F_1(2) \\ d(3, 5) + F_1(3) \end{bmatrix} = \text{MIN} \begin{bmatrix} 10 + 4 = 14 \\ 6 + 2 = 8 \end{bmatrix} = 8$$

$$F2(6) = \text{MIN} \begin{bmatrix} d(2,6) + F_1(2) \\ d(3,6) + F_1(3) \\ d(4,6) + F_1(4) \end{bmatrix} = \text{MIN} \begin{bmatrix} 9+4 = 13 \\ 7+2 = 9 \\ 3+3 = 6 \end{bmatrix} = 6$$

$$F2(7) = \text{MIN} \begin{bmatrix} d(3,7) + F_1(3) \\ d(4,7) + F_1(4) \end{bmatrix} = \text{MIN} \begin{bmatrix} 10+2 = 12 \\ 8+3 = 11 \end{bmatrix} = 11$$

STAGE 3

$$F3(8) = \text{MIN} \begin{bmatrix} d(5,8) + F_2(5) \\ d(6,8) + F_2(6) \\ d(7,8) + F_2(7) \end{bmatrix} = \text{MIN} \begin{bmatrix} 4+8 = 12 \\ 9+6 = 15 \\ 5+11 = 16 \end{bmatrix} = 12$$

$$F3(9) = \text{MIN} \begin{bmatrix} d(5,9) + F_2(5) \\ d(6,9) + F_2(6) \\ d(7,9) + F_2(7) \end{bmatrix} = \text{MIN} \begin{bmatrix} 8+8 = 16 \\ 6+6 = 12 \\ 4+11 = 15 \end{bmatrix} = 12$$

STAGE 4

$$F4(10) = \text{MIN} \begin{bmatrix} d(8,10) + F_3(8) \\ d(9,10) + F_3(9) \end{bmatrix} = \text{MIN} \begin{bmatrix} 8+12 = 20 \\ 4+12 = 16 \end{bmatrix} = 16$$

$\therefore$ F4(10) = 16

□□□□□□□□□
. 4 + 6 + 3 + 3 = 16
□□□□□□□□□
. 3   3   6   4 = 16

Forward Algorithm (For Shortest Path)

Fi{k(i)} Min[d{j(i – 1), k(i)}] + Fi – 1{j(i – 1)}] i = 2, 3...
j(i – 1) = node numbers in the (i – 1)th stage
k(i) = node numbers in the ith stage

Backward Recursive Equation for Shortest Route Problem

Fi{K(i)} = min all feasible routes [d{j(i + 1), k(i)} + Fi + 1{j(i + 1)}]
j(i + 1) = node numbers in the (i + 1)th stage
k(i) = node numbers in the ith stage

The dynamic programming forward recursive computations can also be represented in mathematically as follows:

Let fi(xi) be the shortest distance to node i at stage i and define d(xi – 1, xi) as the distance from node xi − 1 to node xi, then fi is computed from fi − 1 using the following recursive equation.

fi(xi) = min all feasible (xi − 1,xi) routes {d(xi − 1, xi) + fi − 1(xi − 1)}

Associated order of computation is f1 →f2 → ... fN, i = 1, 2, 3 ... N

Similarly the backward recursive equation is given by

fi(xi) = min all feasible (xi, xi + 1) routes {d(xi, xi + 1) + fi + 1(xi + 1)} i = 1, 2, ... N − 1

The associated order of computation is fN → fN − 1→ ... f2 →f1

EXA.: Find the shortest route for the following network (from node 1 to node 7)

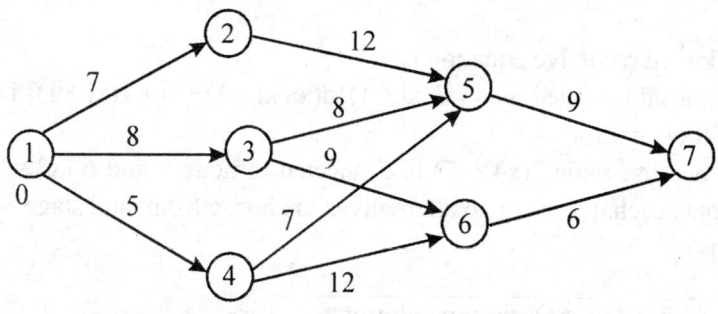

FIGURE

Forward Algorithm

Stage 1: Shortest distance to node 2 = 7 units (from node 1)

It has shortest distance to node 3 = 8 units (from node 1)

3 nodes shortest distance to node 4 = 5 units (from node 1)

Stage 2: It has two end nodes (5 and 6)

$$\begin{pmatrix} \text{Shortest distance} \\ \text{to node 5} \end{pmatrix} = \frac{\min}{i = 2, 3, 4} \left[\begin{pmatrix} \text{Shortest} \\ \text{distance} \\ \text{to node i} \end{pmatrix} + \begin{pmatrix} \text{Distance} \\ \text{from node i} \\ \text{to node 5} \end{pmatrix} \right]$$

$$= \min \begin{cases} 7 + 12 = 19 \\ 8 + 8 = 16 \\ 5 + 7 = 12 \end{cases} = 12 \text{ from node 4}$$

Similarly for node 6 we have

$$\begin{pmatrix} \text{Shortest distance} \\ \text{to node 6} \end{pmatrix} = \frac{\min}{i = 3, 4} \left[\begin{pmatrix} \text{Shortest} \\ \text{distance} \\ \text{to node i} \end{pmatrix} + \begin{pmatrix} \text{Distance} \\ \text{from node i} \\ \text{to node 6} \end{pmatrix} \right]$$

$$= \min \begin{Bmatrix} 8+9=17 \\ 5+13=18 \end{Bmatrix} = 17 \text{ from node 3}$$

Stage 3: The last step is to consider stage 3. The destination node 7 can be reached from either node 5 or 6

$$(\text{shortest distance to node 7}) = \min \begin{Bmatrix} 12+9=21 \\ 17+6=23 \end{Bmatrix} = 21 \text{ from node 5}$$

$\therefore$ Shortest distance to node $7 = 21$ units from node 5
$\therefore$ The shortest route is defined as $1 - 4 - 5 - 7$

The backward recursive equation is

$Fi(xi) = \min$ all feasible routes $(xi, xi + 1)\{d(xi, xi + 1) + fi + 1(xi + 1)\}$ $i = 1, 2, ... N - 1$

Hence $f4(x4) = 0$ for $x4 = 7$

Stage 3 because node $7(x4 = 7)$ is connected to node 5 and 6 $(x3 = 5$ and $6)$ with exactly one route each, there are no alternatives to choose from, and stage 3 results can be summarized as

X3	d(x3, x4) $X4 = 7$	Optimum solution $f3(x3)$	x_4^3
5	9	9	7
6	6	6	7

Stage 2 Route (2, 6) is blocked because it does not exist. Given f3(x3) from state 3, we can compare the feasible alternatives as shown below

x3	d(x2,x3)+f3(x3)		Optimum solution	
	$x4 = 5$	$x3 = 6$	$f2(x2)$	x_3^1
2	12+9=21	-	21	5
3	8+9=17	9+6=15	15	6
4	7+9=16	13+6=19	16	5

The optimum solution of stage 2 reads as follows. If you are in cities 2, 4, the shortest route passes through city 5 and if you are in city 3, the shortest route passes through city 6.

Stage 1: From node 1, we have three alternative routes: (1, 2), (1, 3) and (1, 4) using f2(x2) from stage 2, we can compute the following table

x1	d(x1,x2)+f2(x2)			Optimum solution	
	x1 = 2	x2 = 3	x3 = 4	f1(x1)	x_2^1
1	7+21=28	8+15=23	5+16=21	21	4

The optimum solution at stage 1 shows that city 1 is linked to city 4. Note the optimum solution at stage 2 links city 4 to city 5. Finally the optimum solution at stage 3 connects city 5 to city 7. Thus, the complete route is given on $1 \rightarrow 4 \rightarrow 5 \rightarrow 7$ and the associated distance is 21 units.

Find the maximum path length from node 1 to node 9 in the following network

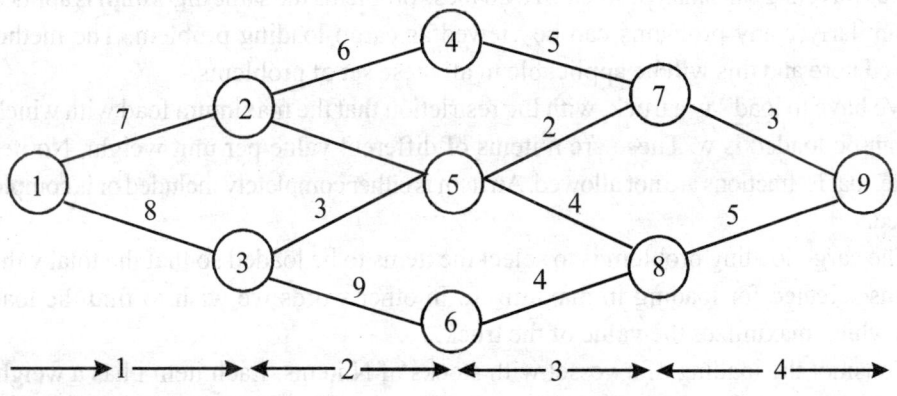

STAGE 1

F1(2) = 7   F1 (3) = 8

STAGE 2

$F2(4) = \text{Max} [d(2, 4) + F1(2)] = \text{Max} [6 + 8] = 13$

$$F2(5) = \text{Max} \begin{bmatrix} d(2, 5) + F_1(2) \\ d(3, 5) + F_1(3) \end{bmatrix} = \text{Max} \begin{bmatrix} 1+7 \\ 3+8 \end{bmatrix} = 11$$

$F2(6) = \text{Max} [d(3, 6) + F1(3)] = \text{Max} [9 + 8] = 17$

STAGE 3

$$F3(7) = \text{Max} \begin{bmatrix} d(4, 7) + F_2(4) \\ d(5, 7) + F_2(5) \end{bmatrix} = \text{Max} \begin{bmatrix} 5+13 \\ 2+11 \end{bmatrix} = 18$$

$$F3(8) = \text{Max} \begin{bmatrix} d(5,8) + F_2(5) \\ d(6,8) + F_2(6) \end{bmatrix} = \text{Max} \begin{bmatrix} 4+11 \\ 4+17 \end{bmatrix} = 21$$

STAGE 4

$$F4(9) = \text{Max} \begin{bmatrix} d(7,9) + F_3(7) \\ d(8,9) + F_3(8) \end{bmatrix} = \text{Max} \begin{bmatrix} 3+18 \\ 5+21 \end{bmatrix} = 26$$

.8 + 9 + 4 + 5

□□□□□□□□□

Maximum path length from node 1 to node 9 is = 26

2.6 CARGO LOADING OR KNAPSACK PROBLEM

KOLESAR applied first, the branch and bound algorithm to cargo loading problem. There are algorithms to solve the traveling salesman problem. Infact many problem can be viewed as traveling salesman problem. To all these problems the same algorithm is applicable.

Similarly many problems can be viewed as cargo loading problem. The method is discussed here and this will be applicable to all these set of problems.

We have to load say a truck, with the restriction that the maximum load with which the truck can be loaded is w. These are n items of different value per unit weight. No item is divisible, that is, fractions are not allowed. An item is either completely included or is completely excluded.

The cargo loading problem is to select the items to be loaded so that the total value of the items selected for loading in maximized. In other words we wish to find the loading pattern which maximizes the value of the truck.

Consider the loading of a vessel with stocks of N items. Each item i has a weight wi and a value vi, i = 1, 2, ... N.

EXA. 1: The maximum cargo weight is W. It is required to determine the most valuable cargo load without exceeding the maximum weight of the vessel.

Specifically, consider the following special case of three items and assume W = 5

i	wi	Ni
1	2	65
2	3	80
3	1	30 ·

Consider the general problem of N items. If ki is the number of units of item i, then the problem becomes

Maximize v1 k1 + v2 k2 + ... + vi ki + ... + vNkN

Subject to w1, k1 + w2 k2 + ... wi ki + ... wNkN = W

ki non negative integer

i = 1, 2, ... N

If ki is not restricted to integer values, then the solution is easily determined by the simplex method. In fact, since there is only one constraint only one variable will be basic and the problem reduces to select the item i for which

$$v_i \left(\frac{W}{w_i} \right) \text{ is maximum.}$$

Since linear programming is not applicable here, the problem will be attempted by dynamic programming; it must be noted that this problem is also typical of the type that can be solved by integer programming techniques.

The dynamic programming model is constructed by first considering its three basic elements.

- Stage it is represented by item j, j = 1, 2, ... N
- State yj at stage j is the total weight assigned to stage j, j + 1, ... N; y1 = W, yj = 0, 1, 2, ... W for j = 2, 3, ... N
- Alternative kj at stage j is the number of units of item j. The value of kj may be as small as zero or as large as [w/wj] where [w/wj] is largest integer included in (w/wj).

Let fj(yj) = optimal value of stage j, j + 1, ... N of item the state yj.

The backward recursive equation is

fN (yN) = max

$$kN = 0, 1, 2... \left[\frac{y_N}{w_N} \right] \{vN \, kN\}$$

yN = 0, 1, 2, ... W

and fj(yj) = max

$$kj = 0, 1, 2... \left[\frac{y_j}{w_j} \right] \{vj \, kj + fj + 1 \, (vj - wjkj)\}$$

yj = 0, 1, 2... W   j = 1,2,..........N-1

Note that the maximum feasible value of kj is given by $\left[\frac{y_j}{w_j} \right]$. This limit will automatically select all infeasible alternatives for a given value of the stage yj.

We start from Stage 3

F3 (y3) = (v3 k3) = {30k3}

And max k3 = $\left[\dfrac{5}{1}\right]$

y3	k3 =0	1	2	3	4	5	fj(yj)	k_3^*
	V3 k3=0	30	60	90	120	150	. .	. .
0	0	-	-	-	-	-	0	0
1	0	30	-	-	-	-	30	1
2	0	30	60	-	-	-	60	2
3	0	30	60	90	-	-	90	3
4	0	30	60	90	120	-	120	4
5	0	30	60	90	120	150	150	5

Stage 2: f2 (y2) = {80 k2 + f3 (y2 – 3k2)}

max k2 = $\left[\dfrac{5}{1}\right]$ = 5

80k2 + f3 (y2 – 3k2)

y2	k2 = 0	k2 = 1	f2(y2)	k_2^*
	v2k2 = 0	80		
0	0+ 0 = 0	-	0	0
1	0+ 30 = 30	-	30	0
2	0+ 60 = 60	-	60	0
3	0+ 90 = 90	80+0 = 80	90	0
4	0+120 =120	80+30 =110	120	0
5	0+150 =150	80+60 =140	150	0

STAGE 1

$F1(y1) = \{65k1 + f2(y1 - 2k1)\}$

$\max = k1 = \left[\dfrac{5}{2}\right] = 2$

$65k1 + f2 (y1 - 2k1)$

y1	k1 = 0 v1k1 =0	k1 =1 65	k2 = 2 130	f1(y1)	k_1^*
0	0+ 0 = 0	-	-	0	0
1	0+ 30 = 30	-	-	30	0
2	0+ 60 = 60	65+ 0 =65	-	65	1
3	0+ 90 = 90	65+30 =95	-	95	1
4	0+120 =120	65+60 =125	130+ 0 =130	130	2
5	0+150 =150	65+90 =155	130+130 =160	160	2

Given $y1 = W = 5$, then associated optimum solution is $(k_1^*, k_2^*, k_3^*) = (2, 0, 1)$. With a total value $= 160$

Note: At stage 1 it is sufficient to construct the value for $y1 = 5$

EXA. 2.

Item number	Weight (kg)	Value (Rs)	Value per unit wt (Rs)
1	40	20	1/2
2	40	80	2
3	44	66	3/2
4	28	70	5/2
5	32	32	1

The maximum allow able weight is w = 105 kg

If fractional part of items are allowed, then the problem is simple because then to find optimum loading, we shall first load the item which has the highest value per unit weight, next the item which has highest value per unit weight and so on. This is best done with the help of the following table.

Index number	Item number	Weight (kg)	Value (Rs)	Value per unit weight (Rs)
1	4	28	70	5/2
2	2	40	80	2
3	3	44	66	3/2
4	5	32	32	1
5	1	40	20	1/2

Assume divisibility the optimum loading is

Index number	Item Number	Weight (kg)	Value (Rs)
1	4	28	70
2	2	40	80
3	3	37	55.50
		105	Rs. 205.50

In the above we take only 37 out of 44 of the third item, so that total does not exceed 105. Thus the optimum value is Rs. 205.50.

But in the problem the items are not divisible that is, we cannot take 37 out of 44 of third item. We should include an item completely or should not include it at all thus if divisibility is not allowed, like the above loading, we will go to index 4 instead of index 3 and shall have

Index number	Item number	Wt	Value
1	4	28	70
2	2	40	80
4	5	32	32
		100	Rs. 182

Comparing 1 and 2, we see that in every case the solution when divisibility is assumed is superior is that when divisibility is not allowed. Thus the value, where divisibility is allowed, is an "Upper Bound (UB)" on feasible solution. Using this fact we shall develop a branch and bound algorithm for the problem under consideration.

A number written inside a node indicates that the item with that index number is completely included in loading, a number above it indicates that the item with the index number is completely not included.

Set of all feasible solution is denoted by S.

First take node S of all feasible solutions. Branch from node S to node (T) and node (1). Node (T) corresponds to loading when item with index number 1 is completely included and node (1) corresponds to which item in the index number 1 is completely included. The following figure gives the relevant details.

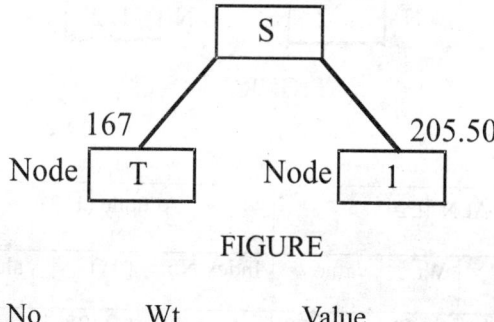

FIGURE

Index No	Wt	Value
2	40	80
3	44	66
4	41	21
	105	167

Remember that under the assumption of divisibility, we select items in descending order of values per unit weight. Since the optimum solution of divisibility problem is an upper bound as the solutions of indivisible problem the UB at node (T) is 167.

Similarly we can determine UB at node (1) (here 1 is completely included) for this we have

Index No	Wt	Value
1	28	70
2	40	80
3	37	55.5
	105	Rs. 205.50

This implies that UB as Node (1) is 205.50. Since the UB on Node (1) is higher we branch off from node (1) into node $(1, \bar{2})$ and node (1, 2). To set get the upper bound UB on node $(1, \bar{2})$, we solve the divisible problem in which index 1 is completely included and index 2 is completely excluded. To set UB on Node (1, 2) we solve the divisible problem in which index 1 and index 2 are completely included.

The figure below gives the real position

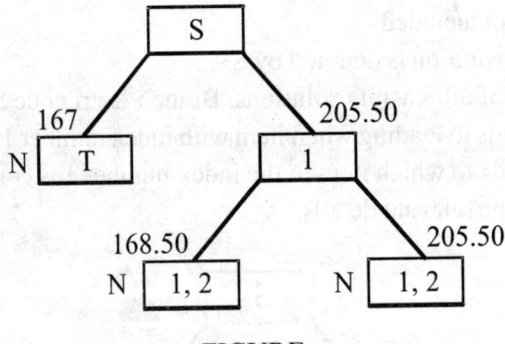

FIGURE

At N $(1,\bar{2})$			At node (1, 2)		
Index No	Wt	value	Index No	Wt	Value
1	28	70	1	28	70
3	44	66	2	40	80
4	32	32	3	37	55
5	1	0.50			
	105	168.50		105	205.50

These imply that UB on $N(1, \bar{2})$ is 168.50 and UB on $N(1, 2)$ is 205.50.

Now the terminal nodes are $N(T)$, N and $N(1, \bar{2})$. We branch off from the terminal node of target UB. Here $N(1, 2)$. So we branch off from $N(1, 2)$ to $N(1, 2, \bar{3})$ and $N(1, 2, 3)$. The figure below indicates the relevant details

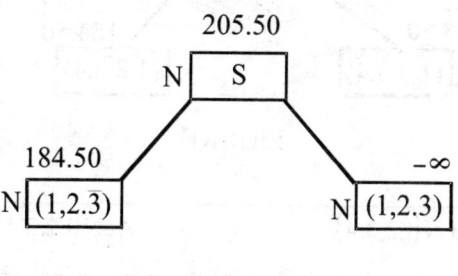

FIGURE

Note that $(1, 2, \bar{3})$ means that items with index numbers 1 and 2 must completely be included and item with index number 3 must be completely be excluded (index 4 and index 5 may be taken in full or in part).

Also $(1, 2, 3)$ means item with index numbers 1, 2 and 3 all must be completely be included (If weight constraint permits index 4 and 5 may be included in full or in part).

At N $(1,2,\bar{3})$			At N$(1, 2, 3)$		
Index No	Wt	value	Index No	Wt	value
1	28	70	1	28	70
2	40	80	2	40	80
4	32	32	3	44	66
5	5	2.50			
	105	184.50		112	216

Here at $N(1, 2, 3)$ the weight constraint is violated, because $112 > 105$. Thus loading $(1, 2, 3)$ is not feasible. This we indicate by UB as $-\infty$. (we branch from the terminal node of largest UB, therefore UB to be $-\infty$ implies no branching.) Now largest UB is with $N(1, 2, \bar{3})$.

So we branch off from that node into $N(1, 2, \bar{3}, \bar{4})$ and $N(1, 2, \bar{3}, 4)$. This is indicated in the following figure

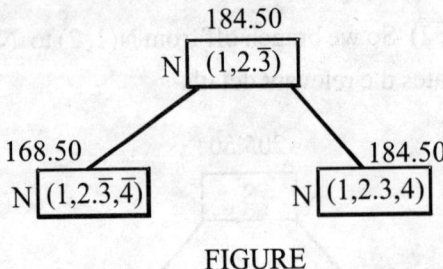

FIGURE

At $N\left(1,2,\bar{3},\bar{4}\right)$			At $N\left(1,2,\bar{3},4\right)$		
Index No	Wt	value	Index No	Wt	Value
1	28	70	1	28	70
2	40	80	2	40	80
5	37	18.50	4	32	32
			5	5	2.5
	105	168.50		105	184.50

This implies that UB on $N(1, 2, \bar{3}, \bar{4})$ is 168.50 and on $N(1, 2, \bar{3}, 4)$ is 184.50.

We branch off from $N(1, 2, \bar{3}, 4)$ into $N(1, 2, \bar{3}, 4, \bar{5})$ and $N(1, 2, \bar{3}, 4, 5)$. The figure below exhibits the details

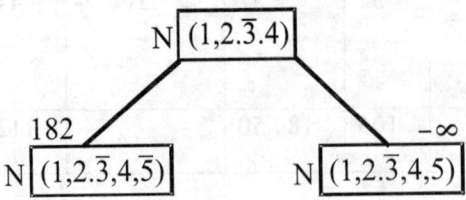

FIGURE

At N$(1,2,\bar{3},4)$			At N$(1,2,\bar{3},4,5)$		
Index No	Wt	value	Index No	Wt	Value
1	28	70	1	28	70
2	40	80	2	40	80
4	32	32	4	32	32
			5	40	20
	100	182		140	202

Hence UB on N$(1, 2, \bar{3}, 4, \bar{5})$ is 182 and UB on N$(1, 2, 3, 4, 5)$ is $-\infty$ because the load constraint is violated.

Now we have considered all the items for loading, so the process leads the final figure is given below.

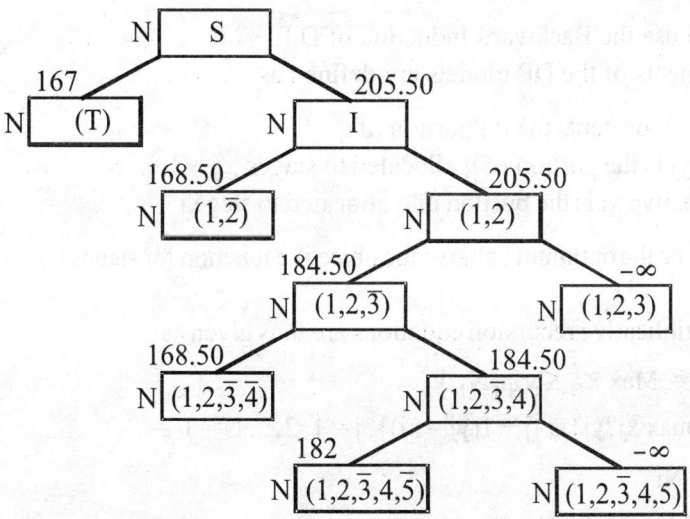

We accept as the optimum solution associated with the terminal index having the target UB. Here N$(1, 2, \bar{3}, 4, \bar{5})$. Thus the optimum loading is

Index No	Item No	Wt	Value
1	4	28	70
2	2	40	80
4	5	32	32
		100	182

2.7 OPTIMAL SUB DIVISION PROBLEM

Divide a quantity q (> 0) into N parts that will maximize the product.
Let xj be the jth portion of q, {i = 1, 2, ... N}
The problem is

Maximize $p = \max \prod_{j=1}^{N} x_j$

Subject to $\sum_{j=1}^{N} x_j = q, x_j \geq 0$ for all j

We shall use the Backward Induction of D.P.
The elements of the DP models are defined as

- Stage j represents the jth portion of q
- State yj is the portion of q allocated to stages j, j + 1, ... N
- Alternative xj is the portion of q allocated to stage j

Let fj(yj) be the optimum value of the objective function for stages j, j + 1, ... N. Given the state yj

The multiplicative recursion equations are thus given as

fN (yN) = Max $x_N \leq y_N \{X_N\}$

fj (yj) = max xj?yj{xj fj + 1(yj – xj)}, j = 1, 2, ... N – 1

STAGE N:

$f_N(y_N) = \text{Max } x_N \leq y_N \{x_N\} = y_N \therefore x_N = y_N = q$

STAGE N–1:

$$\text{fN} - 1 \ (\text{yN} - 1) = \text{Max } x_{N-1} \leq y_{N-1}\{xN - 1 \ f(yN - 1 - xN - 1)\}$$
$$= \text{Max } x_{N-1} \leq y_{N-1}\{xN - 1(yN - 1 - xN - 1)\}$$
$$= \text{Max } x_{N-1} \leq y_{N-1}\{xN - 1 \ yN - 1 - x_{N-1}^2)\}$$

Let $hN - 1 = xN - 1 \ yN - 1 - x_{N-1}^2$

$$\frac{dh_{N-1}}{dx_{N-1}} = y_{N-1} - 2x_{N-1} = 0 \text{ for } hN - 1 \text{ to be a maximum or minimum and}$$

here

$$\frac{d^2 h_{N-1}}{dx_{N-1}^2} = -2 < 0 \text{ for all } N - 1$$

$$\therefore \quad x_{N-1} = \frac{1}{2} y_{N-1}$$

Therefore, $hN - 1$ is maximum where $x_{N-1} = \frac{1}{2} y_{N-1}$

Max

$fN - 1 \ (yN - 1) = x_{N-1} \le y_{N-1} \{xN\text{-}1 \ (yN - 1 - xN - 1)\}$

$$= x_{N-1}^* y_{N-1} - x_{N-1}^{*2}$$

$$= \left\{ \frac{y_{N-1}}{2} \right\}^2 \qquad\qquad \begin{aligned} x_N &= y_N = q \\ x_{N-1} &= \frac{q}{2} \end{aligned}$$

Stage j:

$$fj(yj) = \max x_j \le y_j \{xj \ fj + 1 \ (yj - xj)\}$$

Using induction, the optimum solution at stage j (j = 1, 2, ... N) is

$$f_j(y_j) = \left(\frac{y_j}{N - j + 1} \right)^{N - j + 1}, \quad x_j^* = \frac{y_j}{N - j + 1}$$

$$f_1(y_1) = \left(\frac{y_1}{N} \right)^N, \quad x_1^* = \frac{y_1}{N}, (y_1 = q)$$

$$= \frac{q}{N}$$

is $y1 = q$ and hence $x1 = \dfrac{q}{N}$

$$p = f1(q) = x_1^* . x_2^* ... x_N^* = \frac{q}{N} . \frac{q}{N} ... \text{ the Nth factor}$$

$$= \left(\frac{q}{N} \right)^N$$

Forward Induction

xj is the jth portion of q, j = 1, 2, ... N

yj is the portion of allotment to the stages 1, 2, ... j

$$= x1 + x2 + \dots xj$$

$$\text{Max} \qquad p = \prod_{j=1}^{N} x_j$$

Subject to $\sum xj = q$

FIRST STAGE:

$$f1\ (y1) = \max 0 \le x_1 \le y_1 [x1] = y1$$

$$\therefore\ x_1^* = y_1 = q$$

SECOND STAGE

$$f2\ (y2) = \max 0 \le x_1 \le y_1 \{x2\ f1(y2 - x2)\}$$

$$= x2\ (y2 - x2)$$

$$= x2y2 - x_2^2$$

Let $\qquad F(x2) = x2y2 - x_2^2$

$$\frac{dF}{dx_2} = y_2 - 2x_2 = 0 \rightarrow x_2 = \frac{y_2}{2}$$

$$\frac{d^2F}{dx_2^2} = -2 < 0 \therefore x_2^* = \frac{y_2}{2}$$

$$\therefore \qquad F_2(y_2) = x_2^* f_1\left(y_2 - \frac{y_2}{2}\right)$$

$$= x_2^* f_1\left(\frac{y_2}{2}\right)$$

$$= x_2^* \cdot \frac{y_2}{2}$$

$$= \frac{y_2}{2} \cdot \frac{y_2}{2} = \left(\frac{y_2}{2}\right)^2$$

Third Stage:

$$f3\ (y3) = \max 0 \le x_3 \le y_3 \{x3\ f2(y3 - x2)\}$$

Let $\qquad F(x3) = x_3\left(\frac{y_3 - x_3}{2}\right)^2$

$$= \frac{1}{4} \cdot x_3(y_3 - x_3)^2$$

$$\therefore \quad F'(3) = \frac{1}{4}\left[(y_3 - x_3)^2 + x_3.2(y_3 - x_3) \times -1\right]$$

$$= \frac{1}{4}(y_3 - x_3)\{(y_3 - x_3) + 2x_3\}$$

$$= \frac{1}{4}(y_3 - x_3)(y_3 - 3x_3)$$

Hence $F'' < 0$

$$\therefore \quad x_3^* = \frac{y_3}{3}$$

$$\therefore \quad f_3(y_3) = \frac{y_3}{3}\left[\frac{1}{2}.\frac{2y_3}{3}\right]^2 = \frac{y_3}{3}.\left(\frac{y_3}{3}\right)^2 = \left(\frac{y_3}{3}\right)^3$$

$$\therefore \quad x_3^* = \frac{y_3}{3}$$

Hence by induction

$$x_j^* = \frac{y_j}{j} \text{ and } f_N(y_N) = \left(\frac{y_N}{N}\right)^N = \left(\frac{q}{N}\right)^N$$

$$\left[\therefore x_N^* = \frac{y_N}{N} = \frac{q}{N}\right]$$

The product is maximum when each term in the product is equal

$$x_1^* = x_2^* = ...x_N^* = \frac{q}{N}$$

2.8 Capital Budgeting

A corporation is entertaining proposals from its three plants for possible expansion facilities. The corporation is budgeting Rs. 5 lakhs for allocation to all these plants. Each plant is requested to submit proposals, given the total cost (C) and total revenue (R) for each proposal.

The following table summarizes the cost and revenue in lakhs of rupees. Use dynamic programming to maximize the total revenue resulting from the allocation of Rs. 5 lakhs to the three plants

Proposals	Plant 1 C1 R1	Plant 2 C2 R2	Plant 3 C3 R3
1	0 0	0 0	0 0
2	1 5	2 8	1 3
3	2 6	3 9	- -
4	- -	4 12	- -

The forward additive recursive relation is given by

$$f_i(x_i) = \max_{0 \le x_i \le 5}\left[R_{i_{m_i}}(C_{i_{m_i}}) + f_{i-1}(x_i - C_{i_{m_i}}) \right] \quad i = 2, 3$$

FORWARD INDUCTION : FIRST STAGE i = 1

$f_1(x1|m1) = R_{1_{m_{1i}}}\left(C_{1_{m_{1i}}}\right)$								
	C1	R1	C2	R2	C3	R3	f1(x1)	m_1^*
x1	0	0	1	5	2	6		
0	0	0	0	0	0	0	0	1
1	-	-	1	5	-	-	5	2
2	-	-	2	5	2	6	6	3
3	-	-	3	5	3	6	6	3
4	-	-	4	5	4	5	6	3
5	-	-	5	5	5	6	6	3

SECOND STAGE: i = 2

$x2$	C1 R1	C2 R2	C3 R3	C4 R4	f2(x2)	m_2^*
	$f_2\!\left(x_2 \mid m_2\right) = R_{2_{m_2}}\!\left(C_{2_{m_2}}\right) + f_1\!\left(x_2 - C_{2_{m_2}}\right)$					
	0 0	2 8	3 9	4 12	-	
0	0 0+0	- -	- -	- -	0	1
1	1 0+5	- -	- -	- -	5	1
2	2 0+6	2 8+0	- -	- -	8	2
3	3 0+6	3 8+5	3 9+0	- -	13	2
4	4 0+6	4 8+6	4 9+5	4 12+0	14	2,3
5	5 0+6	5 8+6	5 9+6	5 12+5	17	4

THIRD STAGE:

$x3$	C1 R1	C2 R2	f3(x3)	m_3^*
	$f_3\!\left(x_3 \mid m_3\right) = R_{3_{m_3}}\!\left(C_{3_{m_3}}\right) + f_2\!\left(x_3 - C_{3_{m_3}}\right)$			
	0 0	1 3		
0	0 0 0	0 0	0	-
1	1 -	1 3+0=3	3	2
2	2 -	1 3+0=3	3	2
3	3 -	3 3+8=11	11	2
4	4 -	4 3+13=16	16	2
5	5 -	5 3+14=17	17	1,2

$$x_3 \quad m_3^* \quad x_2 \quad m_2^* \quad x_1 \quad m_1^* \quad (m_1^*, m_2^*, m_3^*) \quad \text{Return}$$

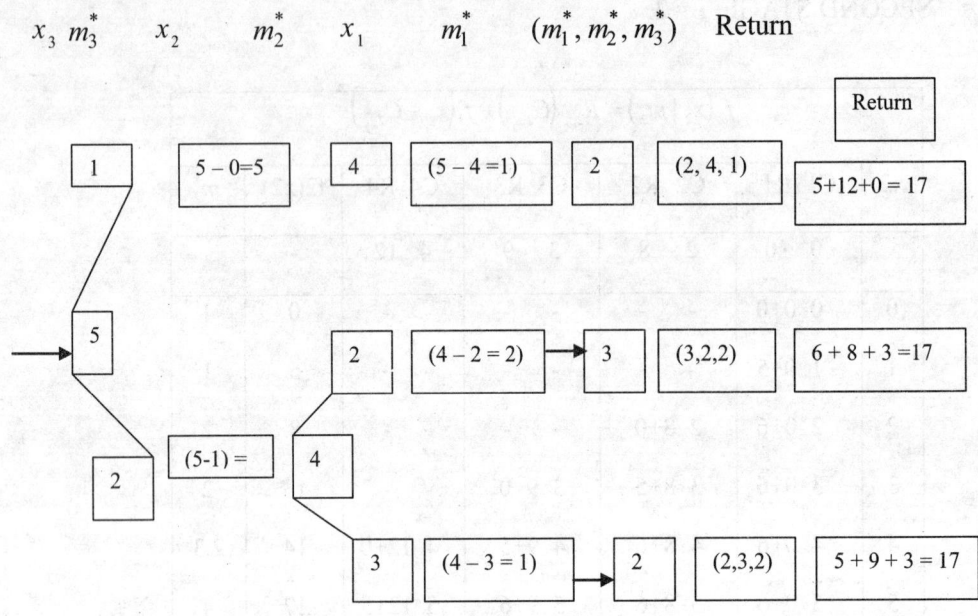

2.9 DIMENSIONALITY IN DYNAMIC PROGRAMMING

In all the Dynamic Programming models the state at any stage is represented by a single variable. For example, in the Knapsack model the problem specifies the weight of the item as the only restriction; more realistically, the volume of the knapsack may also be another variable restriction in such a case, the state at any stage is said to be two dimensional because it consists of two variables: weight and volume.

In optimization problem with n constraints which is solvable by dynamic programming, we define the dimension of the problem to be n. In such a situation we have to define n state variables which correspond to n constraints of the problem. To find the state transformations all alternatives must be examined. It is clear that computational difficulty increases very fast as the number of state variables increases. Consequently the demand made at the computer memory becomes huge and exceeds the capacity of the computer as the computations at each stage increase. This is particularly evident for dynamic programming models in the tabular computations in which the number of rows in each table corresponds to all possible combinations of the state variables. This is the greatest limitation which dynamic programming encounters. This limitations in dynamic programming is known as "dimensionality problem" in the literature as the curse of dimensionality. Hence this is the largest restriction on the any dynamic programming in problems involving even three or four constraints.

2.10 FURTHER APPLICATION OF DISCRETE DYNAMIC PROGRAMMING

1. Work Force Size Models
2. Equipment Replacement Models
3. Investment Models
4. Inventory Models
5. Allocation Models
6. Repairs and Maintenance Models
7. (Including Replacement)
8. Scheduling of Different Products Model
9. Warehouse Problems-models
10. Selection of Elective Courses from Various Departments in a Degree/Postgraduate Degree Programs
11. Solution to Linear Programming Problems by Dynamic Programming
12. Dpp Applications to Oriented or Non Oriented Networks
13. Dpp Applications to the Foarmer's Problem
14. Dpp Models for Scheduling Machines Used Sequentialy to Procedure Manufactured Articles
15. The Travelling Salesman Problem

Exercises

I. Say True Or False in the Following :

1. In dynamic programming, the number of stages equals the number of sub problems. **(T)**
2. The decision alternatives within a stage must be mutually exclusive in the sense that the optimum points for each stage can include exactly one alternation. **(T)**
3. The definition of the stage in D.P models guarantees that feasible decisions can be made independent at each stage. **(T)**
4. In tabular DP computations, the amount of computations at each stage depends on the feasible range of state values. **(T)**
5. In DP it is more difficult to define the stages rather than the states. **(F)**
6. In the network representation of DP models, the node represents the state values at each stage and the arcs represent the flexible alternatives. **(T)**
7. The recursive computations require using the information from each of the previously considered stages in the current stages calculations. **(F)**
8. The principles of optimality guarantees that the future decisions are made independently of the previously made decisions. **(T)**
9. The forward and backward recursive formulations can result in different optimum solution to the same problem. **(F)**
10. Dynamic programming problem can be decomposed either additionally or multiplicationly **(T)**

11. The use of additive and multiplicative decomposition in a single constraint DP problem is decided by whether the different terms comprising the constraint are multiplied or added to one another. **(T)**

12. In DP models the assignment of sub-problems to successive stages is arbitrary, unless the sub problems are specified in a fixed chronological order. **(T)**

13. The problem of dimensionality in DP computations arise because of the increase in the number of stages. **(F)**

14. In any DP model, a reduction in the number of constraints that bind all the stages together can lead to computational savings. **(T)**

15. Dynamic programming provides specific procedures for optimizing the sub-problems at each stage. **(F)**

16. If a linear programme has five variables and three constraints, its equivalent DP model will have three stages and the state at each stage will be designed by a five element vector. **(F)**

17. It is possible to represent and solve a single constraint linear programme as a network model. **(T)**

II. Short Answer Questions

1. State the principle of optimality.
2. What are the three basic elements of a DP Model?
3. Explain each of them.
4. What relationships bind the stages together?
5. What information is needed to make feasible decision at the current stage without considering the decisions made at the previous stages?
6. Explain briefly the "curse of dimensionality" in DP models.
7. Formulate the dynamic programming recursive equation for the problem.

$$\text{Minimize } \sum_{i=1}^{10} y_i^2 \text{ subject to } \prod_{i=1}^{10} y_i = 8,\ y_i > 0$$

8. What is the functional recursive relation assumed in solving the discrete dynamic programming problem?

$$\text{Maximize } \sum_{i=1}^{n} q_i \log q_i \text{ such that } \sum_{i=1}^{n} q_i = c \text{ (constant)}$$

9. Write down the forward recursive equation for solving the following discrete dynamic programming problem.

$$\text{Maximize } p = \prod_{i=1}^{N} z_i \text{ subject to } \sum_{i=1}^{N} z_i = q,\ z_i > 0$$

3. Solve the following model by DP

Maximize $z = y1, y2 \ldots\ldots yn$ subject to $y1 + y2 + \ldots\ldots yn = c$, $yj \geq 0$, $j = 1, 2, n$

(Hint: This problem is similar to problem (II-9) except that the variable yj are continious)

4. Solve the following problem by DP

Minimize $z = y_1^2 + y_2^2 + ... + y_N^2$ subject to $y1, y2 ...... yn = c$

$yi > 0, i = 1, 2, ... n$

III. Long Answers Questions

1. Using multiplicative dynamic programming algorithm, when there are N stages and in each stage there are mi, i = 1, 2, ... N alternative plans, show the reliability problems write the following data

   R = Reliability C = Cost in unite of 1000 rupees

mi	i = 1		i = 2		i = 3	
	R	C	R	C	R	C
1	0.6	2	0.8	3	0.7	1
2	0.7	4	0.8	5	0.8	2
3	0.9	5	0.9	6	0.9	3

   The total capital available in 10 (in units of thousand rupees)

2. An electronic device consists of three components. The three components are in series so that the failure of one component causes the failure of the device. The reliability (probability of no failure) of the device can be improved by installing one or three stand by units in each component. The following table charts the reliability R and the case-c (in thousand of rupees)

Number of		Component 1		Component 2		Component 3	
parallel units		R1	C1	R2	C2	R3	C3
1	..	0.6	1	0.7	3	0.5	2
2	..	0.8	2	0.8	5	0.7	4
3	..	0.9	3	0.9	6	0.9	5

   The total capital available for the construction of the device is Rs. 10,000. How should the device be constructed? (Hint: The objective is to maximize the reliability R1, R2, R3 of the device. This means that the decomposition of the objective function is multiplicative rather than additive).

3. (Capital Budgeting) A manufacturing company has to improve working force of three plants A, B and C with available capital of 6 units in (lakhs of rupees). The possible plans with costs and corresponding return are as follows (in lakhs of rupees)

Plan No	Plant A		Plant B		Plant C	
	C	R	C	R	C	R
1	0	0	0	0	0	0
2	1	4	2	7	1	3
3	3	6	3	10	-	-
4	-	-	4	14	-	-

C = cost, R = return

Find the optimal policy for budgeting the available capital

(A) Use forward in direction

(B) Use backward in direction

4. A business man has four houses in a hill station. He intends to spend 8 lakhs of rupees in converting these houses into hotels. The money required to convert i-th house (i = 1, 2, 3, 4) into c C, B and A type hotel and the corresponding returns are given in the following table. Maximize the total return. Note that house numbers 2 cannot be connected into a "A class hotel". All data are in lakhs of rupees.

	House No.1		House No.2		House No.3		House No.4	
	$x1$	Return	$x2$	Return	$x3$	Return	$x4$	Return
	0	0	0	0	0	0	0	0
C type	2	3	1	2	1	2	3	6
B type	3	5	4	6	2	3	4	7
A type	5	7	-	-	4	6	5	8

5. It is desired to select the shortest high way route between two cities. Draw the network to provide the possible routes between the starting city at node 1 and the destination city at node 7. The routes pass through intermediate cities designated by nodes 2 to 6. Find the shortest route in the network, by assuming the following routes are used.

$d(1, 2) = 5, d(1, 3) = 9, d(1, 4) = 8$

$d(2, 5) = 10, d(2, 6) = 17$

$d(3, 5) = 4, d(3, 6) = 10$

$d(4, 5) = 9, d(4, 6) = 9$

$d(5, 7) = 8, d(6, 7) = 9$

(Distances are given in km)

Find the shortest distance by

(a) Forward algorithm (b) Backward algorithm

Ans $\begin{cases} \text{shortest distance} = 17 \text{ km} \\ \text{Route } 1-3-5-7 \end{cases}$

6. For the network drawn below, it is designed to determine the shortest route between cities 1 to 7. Define the stages and the states using (a) the forward recursion and (b) backward

recursion and then solve the problem for both the cases (a) and (b).
(Distances are given in kilometers)

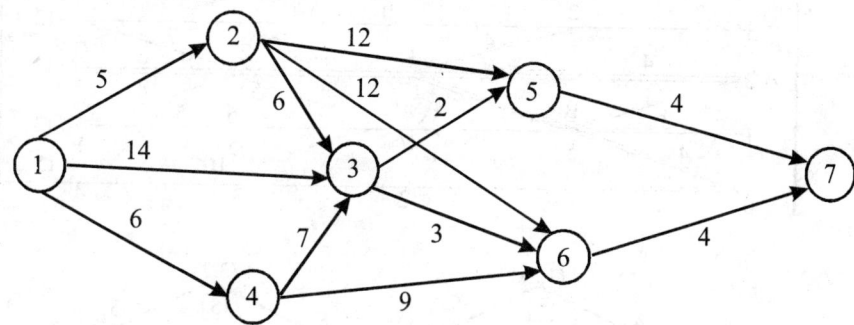

Route is
Ans $1 - 2 - 3 - 5 - 7$
Shortest distance is 17 km.

7. A 4 ton vessel is loaded with one of the more of three items. The following table gives the unit weight w in tons and unit revenue in lakhs of rupees. Ri, for items i. How should the vessel be loaded to maximize the total return?

Item i	Wi	Ri
1	2	31
2	3	47
3	1	14

Because the unit weight wi and the maximum weight W wi integer the stage xi must assume integer values only

(a) use forward recursive $m_1^* = 2, m_2^* = 0, m_3^* = 0$
(b) use backward recursive
(c) what would happen if the vessel capacity is 3 tons in the place of 4 tons? Find the solution here by (a) and (b)
(d) what would happen in the vessel capacity in 5 tons?
(e) what would happen if the vessel capacity in 6 tons?

8. Solve the cargo loading problem for each of the following sets of data
   (i) $w1 = 4, r1 = 70, w2 = 1, r2 = 20, w3 = 3, r3 = 40, W = 6$
   (ii) $w1 = 1, r1 = 30, w2 = 2, r2 = 60, w3 = 3, r3 = 80, W = 4$

   [solution for (i) value $= 120$, $m_1^*, m_2^*, m_3^* = (0, 0, 3)$ or $(0, 2, 2)$ or $(0, 4, 1)$ or $(0, 6, 0)$]

9. Find the path of minimum length from x = 0 to x = 4 in the network given below
   (a)

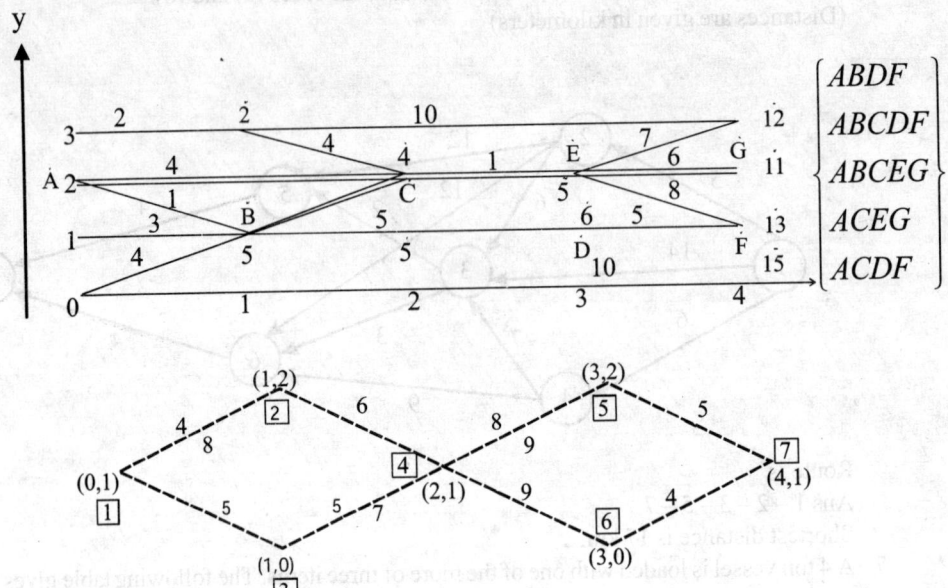

Ans: Max. path length = 23

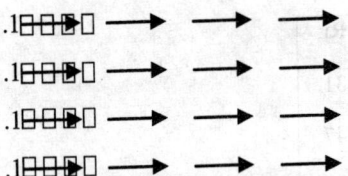

Consider the loading of a vessel with stocks of N items. Each unit of item i has weight wi and a value vi, i = 1, 2, ... N. The maximum cargo weight is W. Formulate this problem at a dynamic programming model to determine the most valuable cargo without exceeding the maximum cargo weight of the vessel. Find the optimal solution with following state

i	Wi	Vi
1	1	20
2	3	90
3	2	70

CHAPTER 3

SEPARABLE PROGRAMMING

3.1 INTRODUCTION

A function $f(x_1, x_2, \ldots x_n)$ is separable if it can be expressed as the sum of n single variable functions $f_1(x_1), f_2(x_2) \ldots \ldots f_n(x_n)$ that is

$$f(x_1, x_2, \ldots x_n) = f_1(x_1) + f_2(x_2) + \ldots + f_n(x_n)$$

For example, the linear function $h(x_1, x_2, \ldots x_n) = a_1x_1 + a_2x_2 + \ldots + a_nx_n$ is separable (the parameter a_i, $i = 1, 2, \ldots n$ are constants). On the other hand, the function

$$h(x_1, x_2, x_3) = x_1^2 + x_1 \sin (x_2+x_3) + x_2e^{x3} \text{ is not separable.}$$

Some non linear functions are not directly de-separable but can be made so by appropriate substitution. Consider, for example, the case of maximizing

$z = x_1, x_2$. Letting $y = x_1, x_2$, then

$\log y = \log x_1$, $\log x_2$ and the problem becomes

Maximize $z = y$

subject to $\log y = \log x_1$, $\log x_2$

which is separable. Here the substitution assumes that x_1 and x_2 are positive variables; otherwise, the logarithmic function is undefined.

In the case where x_1 and x_2 assume zero values (i.e. $x_1, x_2 \geq 0$) may be handled in the following manner. Let δ_1 and δ_2 be positive constants and define

$w_1 = x_1 + \delta_1$

$w_2 = x_1 + \delta_2$

The new variables are strictly positive.

Now

$$x_1x_2 = w_1w_2 - \delta_2w_1 - \delta_1w_2 + \delta_1\delta_2$$

Letting $y = w_1 w_2$, the problem is repressed as

Maximize $z = y - \delta_2 w_1 - \delta_1 w_2 + \delta_1 \delta_2$

subject to $\log y = \log w_1, \log w_2$

$y \geq 0, w_1 \geq \delta_1, w_2 \geq \delta_2$,

The new problem is separable.

3.2 SEPARABLE PROGRAMMING

Separable programming deals with non linear problems in which the objective function and the constraints are separable. An approximate solution can be obtained for any separable problem by linear approximation and the simplex method of linear programming.

The single – variable function $f(x)$ can be approximated by a piecewise linear function using mixed integer programming. Suppose that $f(x)$ is to be approximated over an interval $[a, b]$. Determined a_k, $k = 1, 2, \ldots K$, as the kth breaking point on the x – axis such that $a_1 < a_2 < \ldots < a_k$. The points a_1 and a_k coincide with the end points a and b of the interval under investigation. Thus, $f(x)$ is approximated as follows.

$$f(x) = \sum_{k=1}^{K} f(a_k) y_k$$

$$x = \sum_{k=1}^{K} a_k t_k$$

where t_k is a non negative weight associated with the kth breaking point such that

$$\sum_{k=1}^{K} t_k = 1$$

Mixed integer programming ensure the validity of the approximation. Specifically, the piecewise linear approximation is valid if

- At most two t_k are positive.
- If t_k is a positive, then only an adjacent t_k as t_{k+1} can assume a positive value.

To show how the conditions are satisfied, consider the separable problem.

$$\text{Maximize (or minimize) } z = \sum_{i=1}^{n} f_i(x_i)$$

subject to $\sum_{i=1}^{n} g_i^j(x_i) \leq b_j, j = 1, 2, \ldots m$.

This problem can be approximated by a mixed integer problem as follows: (**Restricted**

Basis Method).

Let the number of breaking points for the ith variable x_i equal K_i and let a_i^k be its kth breaking value. Let t_i^k be the weight associated with kth breaking point of variable i. Then the equivalent mixed problem is

$$\text{Maximize (or minimize) } z = \sum_{i=1}^{n} \sum_{k=1}^{K_i} f_i(a_i^k)\, a_i^k$$

$$\text{subject to } \sum_{i=1}^{n} \sum_{k=1}^{K_i} g_i^j(a_i^k)\, t_i^k \leq b_j, \ j = 1, 2, \dots m$$

$$0 \leq t_i^1 \leq y_i^1, \ i = 1, 2, \dots n$$

$$0 \leq t_i^k \leq y_i^{k-1} + y_i^k, \ k = 2, 3, \dots K_i - 1$$

$$0 \leq t_i^{Ki} \leq y_i^{Ki-1}, \ i = 1, 2, \dots n$$

$$\sum_{k=1}^{K_i - 1} y_i^k = 1$$

$$\sum_{k=1}^{K_i} t_i^k = 1$$

$$y_i^k = (0, 1), \ k = 1, 2, \dots K_i, \ i = 1, 2, \dots n.$$

The variables for the approximating problem are t_i^k and y_i^k.

This formulation shows how any separable problem can be solved, at least in principle, by mixed integer programming. The difficulty is that the number of constraints increases rather rapidly with the number of breaking point. In particular, the computational feasibility of procedure is questionable because there are no reliable computer codes for solving large mixed integer programming problems.

Another method for solving the approximate model is the regular simplex method using restricted basis. In this case the additional constraints involving y_i^k are not needed. The restricted basis specifies that no more than two positive t_i^k can appear in the basis. Moreover, two t_i^k can be positive only if they are adjacent. Thus, strict optimality condition of the simplex method is used to select the entering variable t_i^k only if it satisfies the foregoing conditions. Otherwise, the variable t_i^k having the next best optimality indicator $(z_i^k - c_i^k)$ is considered for entering the solution. The process is repeated until the optimality condition is satisfied or until it is impossible to introducer t_i^k without violating the restricted basis condition, which ever occurs first. The last tableau gives the approximate optimal solution to the problem.

The mixed integer programming method yields a global optimum to the approximate problem, but the restricted basis method can only guarantee local optimum. Additionally, in the two methods, the approximate solution may not be feasible for the original problem. In fact, the approximate model may give rise to additional extreme points that are not part of the solution space of the original problem.

Example

Use separable programming to solve the following problem. Verify approximate answer graphically.

$$\text{Maximize} \quad z = 3x_1^2 + 2x_2^2$$
$$\text{subject to} \quad = x_1^2 + x_2^2 \leq 9$$
$$x_1 + x_2 \leq 3$$
$$x_1, x_2 \geq 0$$

3.2.1 *Separable Convex Programming*

A special case of separable programming occurs when $g_i^j(x_i)$ is convex for all i and j, thus ensuring a convex solution space. Additionally, if $f_i(x_i)$ is convex for (minimization) or concave/for (maximization) for all i, then the problem has a global optimum. Under such conditions, a simplified approximation can be used.

Consider a minimization problem and let $f_i(x_i)$ be shown as in the adjoining figure.

The breaking points of the function $f_i(x_i)$ are $x_i = a_{ki}$, $k = 0, 1, 2, ... K_i$. Let x_{k_i} define the increment of the variable x_i in the range $(a_{k-1,i}, a_{ki})$, $k = 1, 2, ... k_i$ and let ρ_{ki} be the corresponding slops of the line in the same range.

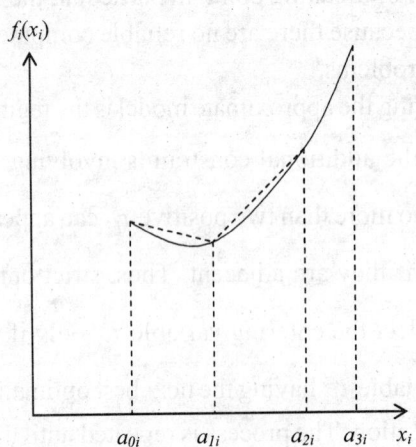

Then

$$f_i(x_i) = \sum_k^{k_i} \rho_{ki} x_{ki} + f_i(a_{0i})$$

$$x_i = \sum_k^{K_i} x_{ki}$$

$0 \le x_{ki} \le a_{ki} - a_{k-1, i}, k = 1, 2, \ldots K_i.$

The fact that $f(x_i)$ is convex ensures that $\rho_{1i} < \rho_{2i} < \ldots < \rho_{Ki}$. Thus, in a minimization problem, for $p < q$, the variable x_{pi} is more attractive than x_{qi}, which means that x_{pi} will always enter solution before x_{qi}.

The convex constraint functions $g_i^j(x_i)$ are approximated in essentially the same way. Let ρ_{ki}^j be the slope of the kth line segment corresponding to $g_i^j(x_i)$. It follows that the ith function is approximated as

$$g_i^j(x_i) = \sum \rho_{ki}^j x_{ki} + g_i^j(a_{0i})$$

The complete problem is thus approximated as

Minimize $z = \sum_{i=1}^{n} \left(\sum_{k=1}^{Ki} \rho_{ki} x_{ki} + f_i(a_{0i}) \right)$

subject to

$$\sum_{i=1}^{n} \left(\sum_{k=1}^{Ki} \rho_{ki}^j x_{ki} + g_i^j(a_{0i}) \right) \le b_j, j = 1, 2, \ldots m.$$

$0 \le x_{k-i} \le a_{ki} - a_{k-1, i}, k = 1, 2, \ldots K_{i-1}, i = 1, 2, \ldots n$
where

$$\rho_{ki} = \frac{f_i(a_{ki}) - f(a_{k-1, i})}{a_{ki} - a_{k-1, i}},$$

$$\rho_{ki}^j = \frac{g_i^j(a_{ki}) - g_i^j(a_{k-1, i})}{a_{ki} - a_{k-1, i}}$$

The **maximization** problem is treated essentially the same way. In this case, $\rho_{1i} > \rho_{2i} > \ldots > \rho_{Ki}$, which means, that for $p < q$, the variable x_{pi} will always enter the solution before x_{qi}.

The new problem can be solved by the simplex method with upper bounded variables. The restricted basis concept is not needed because the convexity (concavity) of the functions guarantees correct selection of basic variables.

3.2.2 Examples

EXAM. 1: Consider the example

Maximize $\quad z = 3x_1^2 + 2x_2^2$

subject to $x_1^2 + x_2^2 \leq 9$

$x_1 + x_2 \leq 3$, $x_1, x_2 \geq 0$

The separable functions of this problem are

$f_1(x_1) = 3x_1^2 \qquad f_2(x_2) = 2x_2^2$

$g_1^1(x_1) = x_1^2 \qquad g_2^1(x_2) = x_2^2$

$g_1^2(x_1) = x_1 \qquad g_2^2(x_2) = x_2$

These functions satisfy the concavity condition required for the maximization problem.
Range for the variables

$0 \leq x_1 \leq 3$ and $0 \leq x_2 \leq 3$

Thus x_1 and x_2 are partitioned in these ranges. Let $k_1 = 3$ and $k_2 = 3$ with $a_{01} = a_{02} = 0$.
The slopes corresponding to the separable functions are as follows

For $i = 1$	k	a_{k1}	r_{k1}	ρ^1_{k1}	ρ^2_{k1}	x_{k1}
	0	0	-	-	-	-
	1	1	3	1	1	x_{11}
	2	2	9	5	1	x_{21}
	3	3	15	5	1	x_{31}
For $i = 2$	k	a_{k2}	r_{k2}	ρ^1_{k2}	ρ^2_{k2}	x_{k2}
	0	0	-	-	-	-
	1	1	2	1	1	x_{12}
	2	2	6	3	1	x_{22}
	3	3	10	5	1	x_{32}

The complete problem thus becomes

Maximize $z \approx 3x_{11} + 9x_{21} + 15x_{31} + 2x_{12} + 6x_{22} + 10x_{32}$

subject to $x_{11} + 3x_{21} + 5x_{31} + x_{12} + 3x_{22} + 5x_{32} \leq 9$

$x_{11} + x_{21} + x_{31} + x_{12} + x_{22} + x_{32} \leq 3$

$0 \leq x_{k1} \leq 1$, $k = 1, 2, 3$

$0 \leq x_{k2} \leq 1$, $k = 1, 2, 3$

$s_1 = 9 - x_{11} - 3x_{21} - 5x_{31} - x_{12} - 3x_{22} - x_{39}$

$\therefore s_2 = 3 - x_{11} - x_{21} - x_{31} - x_{12} - x_{22} - x_{32}$

	Const	x_{11}	x_{21}	x_{31}	x_{12}	x_{22}	x_{32}	
z	0	3	9	15	2	6	10	
s_1	9	-1	-3	-5	-1	-3	-5	Replace s_1 by x_{31}
s_2	3	-1	-1	-1	-1	-1	-1	

	Const	x_{11}	x_{21}	s_1	x_{12}	x_{22}	x_{32}	
z	27	0	0	-3	-1	-3	-5	all ≤ 0
x_{31}	$\dfrac{9}{5}$	$-\dfrac{1}{5}$	$-\dfrac{3}{5}$	$-\dfrac{1}{5}$	$-\dfrac{1}{5}$	$-\dfrac{3}{5}$	-1	
s_2	$\dfrac{6}{5}$	$-\dfrac{4}{5}$	$-\dfrac{2}{5}$	$\dfrac{1}{5}$	$-\dfrac{4}{5}$	$\dfrac{2}{5}$	0	Replace s_2 by x_{21}

	Const	x_{11}	s_2	s_1	x_{12}	x_{22}	x_{32}
z	27	0	0	0	-1	-3	-5
x_{31}	0	1	$\dfrac{3}{2}$	$-\dfrac{1}{2}$	1	0	-1
x_{21}	3	-2	$-\dfrac{5}{2}$	$\dfrac{1}{2}$	-2	-1	0

$\therefore x_1 = x_{11} + x_{21} + x_{31} = 0 + 3 + 0$ and $x_{k2} = 0$, for $k = 1, 2, 3, z = 27$

Alternate solution

Replace s_2 by x_{11}

	Const	s_2	x_{21}	s_1	x_{12}	x_{22}	x_{32}
z	27	0	0	-3	-1	-3	-5
x_{31}	$\dfrac{3}{2}$	$\dfrac{1}{4}$	$-\dfrac{1}{2}$	$-\dfrac{1}{4}$	0	$-\dfrac{1}{4}$	-1
x_{21}	$\dfrac{3}{2}$	$-\dfrac{5}{4}$	$-\dfrac{1}{2}$	$\dfrac{1}{4}$	-1	$-\dfrac{1}{2}$	

$\therefore x_{11} = \dfrac{3}{2}$, $x_{31} = \dfrac{3}{2}$ and $x_1 = \dfrac{3}{2} + \dfrac{3}{2} = 3$; $x_2 = 0$ and $z = 27$

EXAM. 2: Solve the separable programming problem

$$\text{Max } f(X) = 4x_1 + 6x_2 - x_1^3 - 2x_2^2$$

subject to $= x_1 + 3x_2 \le 8$

$5x_1 + 2x_2 \le 14$, $x_1, x_2 \ge 0$

Here the objective function and the two constraints are concave functions. Hence the problem has a global optimization.

The separable functions for this problem are

$$f_1(x_1) = 4x_1 - x_1^3 \qquad\qquad f_2(x_2) = 6x_2 - 2x_2^2$$

$$g_1^1(x_1) = x_1 \qquad\qquad g_2^1(x_2) = 3x_2$$

$$g_1^2(x_1) = 5x_1 \qquad\qquad g_2^2(x_2) = 2x_2$$

Range for the variable are

$0 \le x_1 \le 3$ $0 \le x_2 \le 3$

Take $\qquad k_1 = 3 \qquad\qquad k_2 = 3$

and $\qquad a_{01} = 0 \qquad\qquad a_{02} = 0$

For $i = 1$	k	a_{k1}	r_{k1}	ρ_{k1}^1	ρ_{k1}^2	x_{k1}
	0	0	-	-	-	-
	1	1	3	1	5	x_{11}
	2	2	-3	1	5	x_{21}
	3	3	-15	1	5	x_{31}

For $i = 2$	k	a_{k2}	r_{k2}	ρ_{k2}^1	ρ_{k2}^2	x_{k2}
	0	0	-	-	-	-
	1	1	4	3	2	x_{12}
	2	2	0	3	2	x_{22}
	3	3	-4	3	2	x_{32}

The complete problem becomes

Maximize $f(X) = 3x_{11} - 3x_{21} - 15x_{31} + 4x_{12} + 0x_{22} - 4x_{32}$

subject to

$x_{11} + x_{21} + x_{31} + 3x_{12} + 3x_{22} + 3x_{32} \le 8$

$5x_{11} + 5x_{21} + 5x_{31} + 2x_{12} + 2x_{22} + 2x_{32} \le 14$

$0 \le x_{k1} \le 1 \qquad 0 \le x_{k2} \le 1$

$k = 1, 2, 3 \qquad k = 1, 2, 3$

Introducing the slack variables s_1 and s_2 the constraints become

$s_1 = 8 - x_{11} - x_{21} - x_{31} - 3x_{12} - 3x_{22} - 3x_{32}$

$s_2 = 14 - 5x_{11} - 5x_{21} - 5x_{31} - 2x_{12} - 2x_{22} - 2x_{32}$

	Const	x_{11}	x_{21}	x_{31}	x_{12}	x_{22}	x_{32}	
z	0	3	-3	-15	4	0	-4	
s_1	8	-1	-1	-1	-3	-3	-3	Replace s_1 by x_{31}
s_2	14	-5	-5	-5	-2	-2	-2	

	Const	x_{11}	x_{21}	x_{31}	s_1	x_{22}	x_{32}	
z	$\dfrac{32}{3}$	$\dfrac{5}{3}$	$-\dfrac{13}{3}$	$-\dfrac{49}{3}$	$-\dfrac{4}{3}$	-4	-8	
x_{12}	$\dfrac{8}{3}$	$-\dfrac{1}{3}$	$-\dfrac{1}{3}$	$-\dfrac{1}{3}$	$-\dfrac{1}{3}$	-1	-1	
s_2	$\dfrac{26}{3}$	$-\dfrac{13}{3}$	$-\dfrac{13}{3}$	$-\dfrac{13}{3}$	$\dfrac{2}{3}$	0	0	Replace s_2 by x_{11}

	Const	s_2	x_{21}	x_{31}	s_1	x_{22}	x_{32}	
z	14	$-\dfrac{5}{13}$	-6	-18	$-\dfrac{4}{3}$	-4	-8	Here all $z_j - c_j \leq 0$
x_{12}	2	$\dfrac{1}{13}$						
x_{21}	2	$-\dfrac{3}{13}$	-1	-1	$\dfrac{2}{13}$	0	0	

$\therefore x_{12} = 2, \; x_{11} = 2 \quad x_1 = x_{11} + x_{21} + x_{31} = 2 + 0 + 0 = 2$

$\qquad\qquad\qquad x_2 = x_{12} + x_{22} + x_{32} = 2 + 0 + 0 = 2$

$\therefore z^* = f(X^*) = 4 \times 2 + 6 \times 2 - 8 - 8 = 4$

EXAM. 3: Show how the following problem can be made separable.

Maximize $z = x_1 x_2 + x_3 + x_1 x_3$

subject to $x_1 x_2 + x_2 + x_1 x_3 \leq 10, \; x_1, x_2, x_3 \geq 0$

Let $u = x_1 x_2, \quad v = x_1 x_3$

The given problem becomes

Maximize $z = u + x_3 + v$

subject to $u + v + x_2 \leq 10$

$u, v, x_2 \geq 0$

$\therefore$ The problem is written as

Max $z = u + v + x_3$

subject to $u + v + x_2 \leq 10$

$u, v, x_2, x_3 \geq 0$

$f_1(u) = u \quad f_2(v) = v \qquad f_3(x_3) = x_3$

$g_1^1(u) = u \quad g_2^1(v) = v \qquad g_2^1(x_2) = x_2$

$\log u = \log x_1 + \log x_2 \qquad \log v = \log x_1 + \log x_2$

Hence the problem is made separable.

EXAM. 4: Show how the following problem can be made separable

Minimize $z = e^{2x_1 + x_2^2} + (x_3 - 2)^2$

$$= e^{2x_1} e^{x_2^2} + (x_3 - 2)^2$$

subject to $x_1 + x_2 + x_3 \leq 6, \quad x_1, x_2, x_3 \geq 0$

Let $\qquad u = e^{2x_1}, \ v = e^{x_2^2},$

$\therefore \qquad\qquad 2x_1 = \log u \qquad\qquad\qquad\qquad\qquad \log v = x_2^2$

$x_1 = \dfrac{1}{2} \log u \ \ x_2 = (\log v)^{\frac{1}{2}}$

subject to $\dfrac{1}{2} \log u + (\log v)^{\frac{1}{2}} + x_3 \leq 6$

$u > 0, v > 0, x_3 \geq 0$

$\therefore$ The given problem becomes.

let $y = uv \log y = \log u + \log v$

$\therefore$ Min $z = y + (x_3 - 2)^2$

subject to $\dfrac{1}{2} \log u + \sqrt{\log v} + x_3 \leq 6$

$\log y = \log u + \log v$

$u > 0, v > 0, x_3 \geq 0$

Hence the problem given is separable

EXAM. 5: Show how the following problem can be made separable.

Max $z = e^{x_1 x_2} + x_2^2 x_3 + x_4$

subject to $x_1 + x_2 x_3 + x_3 \leq 10, \quad x_1, x_2, x_3 \geq 0, x_4$ unrestricted in sign

Let $u = x_1 x_2, \ v = x_2 x_3$

$\therefore \ \log u = \log x_1 + \log x_2$ and $\log v = \log x_2 + \log x_3$

Min $z = e^u + v x_2 + x_4$

subject to $x_1 + v + x_3 \leq 10,$ let $w = v . x_2, \qquad \log w = \log v . \log x_2$

$\therefore$ The problem becomes

Minimize $z = e^u + w + x_4$

subject to $x_1 + v + x_3 \leq 10$

$\log u = \log v + \log x_2$

$\qquad = \log x_2 + \log x_3 + \log x_2 = 2 \log x_2 + \log x_3$

x_2, x_3, w are ≥ 0 and x_4 is unrestricted in sign

∴ The given problem is separable.

EXAM. 6: Maximize $z = x_1 + x_2^4$

Subject to $3x_1 + 2x_2^2 \leq 9$

$\qquad x_1 x_2 \geq 0$

The exact optimum solution to this problem, obtained by inspection, is $x_1 = 0, x_2 = 2.12$ and $z = 20.2$. To show how the approximating method is used, consider the separable functions

$f_1(x_1) = x_1, f_2(x_2) = x_2^4$

$g_1^1(x_1) = 3x_1 \qquad g_1^2(x_2) = 2x_2^2$

The functions $f_1(x_1)$ and $g_1^1(x_2)$ are left in their present form because they are already linear. In this case x_1 is treated as one of the variables. Considering $f_2(x_2)$ and $g_1^2(x_2)$, we assume that there are four breaking points ($k_2 = 4$). Because the value x_2 cannot exceed (2 + 1), it follows that

k	a_2^k	$f_2(a_2^k)$	$g_1^2(a_2^k)$
1	0	0	0
2	1	1	2
2	1	16	8
4	3	81	18

This yields:

$f_2(x_2) = t_2^1 f_2(a_2^1) + t_2^2 f(a_2^2) + t_2^3 f(a_2^3) + t_2^4 f(a_2^4) = 0t_2^1 + 1t_2^2 + 16t_2^3 + 81t_2^4$

Similarly,

$g_1^2(x_2) = 2t_2^2 + 8t_2^3 + 18t_2^4$

The approximating thus becomes

Maximize $z = x_1 + t_2^2 + 16t_2^3 + 81t_2^4$

subject to $3x_1 + 2t_2^2 + 8t_2^3 + 18t_2^4 \leq 9$

$t_2^1 + t_2^2 + t_2^3 + t_2^4 = 1$

$t_2^k \geq 0, \; k = 1, 2, 3, 4$

and $\qquad x_1 \geq 0$

Hence the given problem becomes separable.

The solution must satisfy the restricted basis conditions.

The initial simplex tableau with rearranged columns to give a starting solution is given by

Basic	x_1	t_2^2	t_2^3	t_2^4	s_1	t_2^1	Solution
z	1	+1	+16	81	0	0	0
s_1	-3	-2	-8	-18	1	0	9
t_2^1	0	-1	-1	-1	0	1	1

The variable s_1 (≥ 0). (This problem happened to have an obvious starting solution. In general one may have to use the artificial variables techniques.)

From the z row coefficients, t_2^4 is the entering variable. Because t_2^1 is currently basic at positive level, the restricted basic condition dictates that it must leave before t_2^4 can enter the solution. By the feasibility condition s_1 must be the leaving variable. This means that t_2^4 cannot enter the solution; the next best entering variable, t_2^3 requires t_2^1 to leave the basic solution, condition that happens to be guaranteed by the feasibility conditions. The new table becomes

Basic	x_1	t_2^2	t_2^3	t_2^4	s_1	t_2^3	Solution
z	1	-15	0	65	0	-16	-16
s_1	3	6	0	-10	-1	8	-1 Replace s_1 by t_2^4
t_2^3		0	-1	-1	-1	0	-1 -1

Basic	x_1	t_2^2	t_2^3	t_2^4	s_1	t_2^1	Solution
z	$-\dfrac{17}{2}$	24	0	0	$-\dfrac{13}{2}$	36	$22\dfrac{1}{2}$
t_2^4	$+\dfrac{3}{10}$	$-\dfrac{6}{10}$	0	1	$-\dfrac{1}{10}$	$\dfrac{8}{10}$	$-\dfrac{1}{10}$
t_2^3	$\dfrac{3}{10}$	$-\dfrac{16}{10}$	-1	0	$\dfrac{1}{10}$	$\dfrac{18}{10}$	$-\dfrac{9}{10}$

The above tableau shows that t_2^1 and t_2^2 are candidates for the entering variable. Because

t_2^1 is not adjacent to basic t_2^3 or t_2^4 it cannot enter. Similarly t_2^2 cannot enter because t_2^4 cannot leave.

The process ends at this point, and the solution given is the best feasible solution for the approximate problem.

The optimum solution to the original problem is

$$x_1 = 0$$

$$x_2 \approx 2t_2^3 + 3t_2^4 = 2\left(\frac{9}{10}\right) + 3\left(\frac{1}{10}\right) = 2.1$$

$$z = 0 + (2.1)^4 = 19.45$$

The approximate optimum value of $x_2 (= 2.1)$ approximately equals the true optimum value $(= 2.12)$

The previous example is also worked out as follows:

Maximize $z = x_1 + X_2 + 16X_3 + 81X_4$

subject to $3x_1 + 2X_2 + 8X_3 + 18X_4 \le 9$

$X_1 + X_2 + X_3 + X_4 = 1$

$X_i \ge 0,\ i = 1, 2, 3, 4;\ x_1 \ge 0$

$3x_1 + 2X_2 + 8X_3 + 18X_4 + s_1 = 9$

$X_1 + X_2 + X_2 + X_4 = 1$

$X_i \ge 0,\ i = 1, 2, 3, 4;\ x_1 \ge 0,\ s_1 \ge 0$

$\therefore\ s_1 = 9 - 3x_1 - 2X_2 - 8X_3 - 18X_4$

$X_1 = 1 - 0.X_1 - X_2 - X_3 - X_4$

	Const	x_1	X_2	X_3	X_4
z	0	1	1	16	81
s_1	9	-3	-2	-8	-18
X_1	1	0	-1	-1	-1

$\longrightarrow$

	Const	x_1	X_2	X_1	X_4
z	16	1	-15	-16	65
s_1	1	-3	6	8	-10
X_3	1	0	-1	-1	-1

Replace s_1 by X_4

Replace X_1 by X_3

	Const	x_1	X_2	X_1	s_1
z	22.5	$-\dfrac{37}{2}$	24	36	$-\dfrac{13}{2}$
X_4	$\dfrac{1}{10}$	$-\dfrac{3}{10}$	$\dfrac{6}{10}$	$\dfrac{8}{10}$	$-\dfrac{1}{10}$
X_3	$\dfrac{9}{10}$	$\dfrac{3}{10}$	$-\dfrac{16}{10}$	$-\dfrac{18}{10}$	$\dfrac{1}{10}$

The last tableau shows $X_1 = t_2^1$ and $X_2 = t_2^2$ are candidates for the entering variable. Because $X_1 = t_2^1$ is not adjacent to basic $X_3 = t_2^3$ or $X_4 = t_2^4$, it cannot enter. Similarly $X_2 = t_2^2$ cannot enter because $X_4 = t_2^4$ cannot leave. The process ends at this point and the solution given is the best feasible solution for the approximation problem. The optimum solution is

$$x_1 = 0, \ x_2 = 2t_2^4 + 3t_2^4 = 2X_3 + 3X_4$$

$$x_1 = 0, \ x_2 \approx 2\frac{9}{10} + 3\frac{1}{10} = \frac{18}{10} + \frac{3}{10} = \frac{28}{10} \approx \frac{21}{10} \approx 2.1$$

$$z = 0 + (2.1)^4 = 19.45$$

The approximate optimum value of $x_2 \ (= 2.1)$ approximately equals the true optimum value $(= 2.12)$

EXAM. 7: Show how the problem given below can made separable

Maximize $z = x_1 x_2 + x_3 + x_1 x_3$

subject to $x_1 x_2 + x_2 + x_1 x_3 \le 10$, $x_1, x_2, x_3 \ge 0$

SOLUTION

Let $w_j = x_j + 1$, $j = 1, 2, 3$, $v_1 = w_1 w_2$, $v_2 = w_1 w_3$

Then, we have

Max $z = v_1 + v_2 - 2w_1 - w_2 + 1$

Subject to $v_1 + v_2 - 2w_1 - w_2 \le 9$, $\log v_1 = \log w_1 \log w_2 = 0$.

$\log v_2 - \log w_1 - \log w_2 = 0$, all variables are non negative.

EXAM. 8: Solve by separable convex programming method

Minimize $z = (x_1 - 4)^2 + (x_2 - 3)^2$ subject to $x_1^2 + 4x_2^2 \le 16$, $x_1, x_2 \ge 0$

$f_1(x_1) = (x_1 - 4)^2 \quad f_2(x_2) = (x_2 - 3)^2$

$g_1^1(x_1) = x_1^2 \qquad g_2^1(x_2) = 4x_2^2$

Both the objective function and the constraint are convex.

For $i = 1$

k	a_{k1}	ρ_{k1}	ρ_{k1}^1	x_{k1}
0	0	-	-	-
1	1	-7	1	x_{11}
2	2	-5	3	x_{21}
3	3	-3	5	x_{31}
4	4	-1	7	x_{41}

For $i = 2$

k	a_{k2}	ρ_{k2}	ρ_{k2}^1	x_{k2}
0	0	-	-	-
1	1	-5	4	x_{12}
2	2	-3	12	x_{22}

The problem becomes

Min $z = -7x_{11} - 5x_{21} - 3x_{31} - x_{41} - 5x_{12} - 3x_{22}$

subject to $x_{11} + 3x_{21} + 5x_{31} + 7x_{41} + 4x_{12} + 12x_{22} \leq 16$

$\therefore s_1 = 16 - x_{11} - 3x_{21} - 5x_{31} - 7x_{41} - 4x_{12} - 12x_{22}$

	Const	x_{11}	x_{21}	x_{31}	x_{41}	x_{12}	x_{22}
z	0	-7	-5	-3	-1	-5	-3
s_1	16	-1	-3	-5	-7	-4	-12

by x_{11} Replace s_1

	Const	s_1	x_{21}	x_{31}	x_{41}	x_{12}	x_{22}
z	-112	7	16	32	48	23	81
x_{11}	16	-1	-3	-5	-7	-4	-12

$\therefore x_1 = x_{11} + x_{21} + x_{31} + x_{41} = 16$

$\therefore x_2 = 0$

$\therefore$ Min $z = -112$

EXAM. 9: Solve by separable programming technique by approximating $f(x) = z$ by a piecewise linear function using mixed integer programming technique.

Maximize $z = 4x_1 - x_1^2 + 2x_2$

subject to $2x_1^2 + 3x_2^2 \leq 24$, $x_1, x_2 \geq 0$

SOLUTION

$f_1(x_1) = 4x_1 - x_1^2$ $f_2(x_2) = 2x_2$ (linear already)

$g_1^1(x_1) = 2x_1^2$ $g_1^2 = 3x_2^2$

The optimum solution of this problem by inspection is $x_1^* = 0$, $x_2^* = 2\sqrt{2} = 2.83$ and the maximum value is $z^* = 0 + 2(2.83) \approx 5.66$. Assume that there are five breaking points for x_1 and four breaking points for x_2 i.e. $k_1 = 5$, $k_2 = 4$.

k_1	a_1^{k1}	$f_1(a_1^{k1})$	$g_1(k_1)$
1	0	0	0
2	1	3	2
3	2	4	8
4	3	3	18
5	4	0	32

$\rightarrow$

k_2	a_2^{k2}	$g_1^2(a_2^{k2})$
1	0	0
2	1	3
3	2	12
4	3	27

Now

$$f_1(x_1) + f_2(x_2) = 0t_1^1 + 3t_1^2 + 4t_1^3 + 3t_1^4 + 0.t_1^5 + 2x_2 = 3t_1^2 + 4t_1^3 + 3t_1^4 + 2x_2$$

$$\therefore \text{ Max } z = 3t_1^2 + 4t_1^3 + 3t_1^4 + 2x_2$$

subject to $0t_1^1 + 2t_1^2 + 8t_1^3 + 18t_1^4 + 32t_1^5 + 3t_1^2 + 12t_2^3 + 27t_2^4 \le 24$

with $t_1^1 + t_1^2 + t_1^3 + t_1^4 + t_1^5 = 1$ and $t_2^1 + t_2^2 + t_2^3 + t_2^4 + 0 = 1$

15.2.3 Exercises

I. Short answer questions in separable programming

1. Define a separable programming problem.
2. Define separable convex programming problem.
3. What are the two methods for solving approximately a nonlinear separable programming problem?

II. Long answer questions in separable programming

1. Solve as a separable convex programming problem
   Minimize $z = x_1^4 + 2x_2 + x_3^2$
   Subject to $x_1^2 + x_2 + x_3^2 \le 4, |x_1 + x_2| \le 0, x_1, x_3 \ge 0$ and x_2 unrestricted in sign.
2. Solve the following as a separable convex programming problem
   Minimize $z = (x_1 - 2)^2 + 4(x_2 - 6)^2$
   Subject to $6x_1 + 3(x_2 + 1) \le 12, x_1, x_2 \ge 0$
3. Use separable convex programming to solve the following (verify approximate answer graphically)
   (a) Max $x_0 = (x_1 - 1)^2 + (x_2 - 2)^2$ (b)   Max $x_0 = 3x_1^2 + 2x_2^2$
       s.t $x_1 + x_2 \le 6$ s.t $x_1^2 + x_2^2 £ 9$
       $3x_1 + 2x_2 \le 7$ $x_1 + x_2 \le 3$
       $x_1, x_2 \ge 0$ $x_1, x_2 \ge 0$
   Determine the upper bound an x_1 and x_2 graphically.
4. Min $f(X) = 2x_1 - 3x_2$ subject to $4x_1^2 + 9x_2^2 \le 36, x_1, x_2 \ge 0$

   $$\left[\text{Ans: } t_1^1 = 1, t_2^3 = 1, (x_1 = 0, x_2 = 0) \right]$$

   Hint: Take in the above the breaking points for x_1 to be 0, 1, 2, 3 and for x_2 to be 0, 1, 2.
5. Solve the following problem as a separable convex programming
   Maximize $f(X) = 3x_1 + 2x_2$ subject to $4x_1^2 + x_2^2 \le 16, x_1, x_2 \ge 0$

   $$\left[\text{Ans}: x_1^* = 1, x_2^* = \frac{24}{7}, \text{max } z = z^* = \frac{69}{7} \right]$$

6. Use separable convex programming to solve the following problem
   Max $f(X) = 16 - 2(x_1 - 3)^2 - (x_2 - 7)^2$, subject to $x_1^2 + x_2^2 \le 16, x_1, x_2 \ge 0$
   [Hint: Take $f_1(x_1) = 8 - 2(x_1 - 3)^2$ and $f_2(x_2) = 8 - (x_2 - 7)^2$]

7. Solve the following separable programming problem

   (i) Max $z = 2\sqrt{x_1} - x_2^2$ subject to $x_1 - 2x_2 \le 6, x_1 - x_2 \le 0, x_1, x_2 \ge 0$

      Take $x_1 = 0(1)6, x_2 = 0(1)3$

   (ii) Min $z = (x_1 - 3)^2 + (x_2 - 3)^2$, subject to $x_1 + x_2 \le 4, x_1, x_2 \ge 0$

   (iii) Max $z = x_1^2 - x_2^2$ subject to $2x_1 - x_2 \le 0, -x_1 + 2x_2 \le 3, x_1, x_2 \ge 0$

8. Use the convex simplex method to solve the separable programming problem

   Max $z = 18x_1 - x_1^2 - x_1 x_2 - x_2^2$ subject to $x_1 + x_2 \le 12, -x_1 + x_2 \le 6, x_1, x_2 \ge 0$

9. Solve the convex separable programming problems given below

   (i) Max $z = (x_1 - 4)^2 + (x_2 - 4)^2$, subject to $x_1 + x_2 \le 6, -x_1 + 3x_2 \le 6, x_1, x_2 \ge 0$

   (ii) Min $z = (x_1 - 4)^2 + (x_2 - 3)^2$, subject to $x_2^2 + 4x_2^2 \le 16, x_1, x_2 \ge 0$

   (iii) Max $z = x_1^2 - x_2^2$ subject to $2x_1 - x_2 \le 0, -x_1 + 2x_2 \le 3, x_1, x_2 \ge 0$

10. Solve the following problem by separable programming

   Maximize $z = 4x_1 + 2x_2 - x_1^2$ subject to $2x_1 + 3x_2 \le 12, 2x_1 + x_2 \le 8, x_1, x_2 \ge 0$

   Actual solution $x_1^* = 1.333$ and for $x_2^* = 3.111, z^* = 4.777$

   Choose for $x_1 = 0(1)6$ and for $x_2 = 0(1)4$

11. Solve the separable convex programming problem

   Max $f(X) = 4x_1 + 6x_2 - x_1^3 - 2x_2^2$ subject to $x_1 + 3x_2 \le 8, 5x_1 + 2x_2 \le 14, x_1, x_2 \ge 0$

12. Using the restricted basis method, formulate to the approximate LPP of the following separable programming problem

   Maximize $z = (x_1 - 2)^2 - (x_2 - 2)^2$, subject to $x_1 + 2x_2 \le 4, x_1, x_2 \ge 0$ by taking breaking points for x_1 to be 0, 1, 2, 3, 4 and for x_2 to be 0, 1, 2.

13. Solve the separable programming technique

   Maximize $z = 4x_1 - x_1^2 + 2x_2^2$ subject to $2x_1^2 + 3x_2^2 \le 24, x_1, x_2 \ge 0$

14. Solve by separable convex programming problem

   Minimize $z = (x_1 - 4)^2 - (x_2 - 3)^2$, subject to $x_1^2 + 4x_2^2 \le 16, x_1, x_2 \ge 0$

   [Ans: $x_{11} = 16, z_0 = -112$]

CHAPTER 4

QUADRATIC PROGRAMMING

4.1 THE KARUSH–KHUN–TUCKER (KKT) CONDITIONS

The development of KKT necessary conditions for identifying stationary points of a non linear constrained problem subject to inequality constraints is developed here. This is based on the Lagrangean method. These conditions are also sufficient under certain rules that will be stated later.

Consider the problem

Maximize $z = f(X)$ subject to $g(X) \leq 0$

The inequality constraints may converted into equations by using non-negative slack variables. Let $s_i^2 (\geq 0)$ be the slack quantity added to ith constraint $g_i(X) \leq 0$

and define

$$S = (S_1, S_2, \dots S_m)^T, \ S^2 = (S_1^2, S_2^2, \dots S_m^2)^T,$$

where m is the total number of inequality constraints. The Lagrangean function is thus given by

$$L(X, S, \lambda) = f(X) - \lambda[g(X) + S^2]$$

Given the constraints $g(X) \leq 0$

a necessary condition for optimality is that λ be non negative (non positive) for maximization (minimization) problems. This result is justified as follows. The vector λ measures the rate of variation of f with respect to g, that is

$$\lambda = \frac{\partial f}{\partial g}$$

In the maximization case, as the right hand side of the constraints $g(X) \le 0$ changes from 0 to ∂g (> 0), the solution space becomes less constrained and hence f cannot decrease. This means that $\lambda \ge 0$. Similarly for minimization, as the right hand side of the constraints increases, f cannot increase, which implies that $\lambda \le 0$. If the constraints are equalities, that is $g(X) = 0$, then λ becomes unrestricted in sign.

The restrictions on λ are part of the KKT necessary conditions. The remaining conditions will now be derived.

Taking partial derivative of L with respect to X, S and λ, we obtain

$$\frac{\partial L}{\partial X} = \nabla f(X) - ?\nabla g(X) = 0$$

$$\frac{\partial L}{\partial S_i} = -2\lambda_i S_i = 0, \ i = 1, 2, \dots m$$

$$\frac{\partial L}{\partial ?} = -(g(X) + S^2) = 0$$

The second set of equations reveals the following results

- If $\lambda_i \ne 0$, then $s_i^2 = 0$, which means that the corresponding resources is scarce and, hence, it is consumed completely (equality constraint)

- If $s_i^2 > 0$, then $\lambda_i = 0$. This means resource i is not scarce, and, consequently, it has

  no effect on the values of f $\left(\text{i.e. } \lambda_i = \dfrac{\partial f}{\partial g_i} = 0\right)$

From the second and third sets of equations, we obtain
$\lambda g_i(X) = 0, \ i = 1, 2, \dots m$
This new condition essentially repeats the foregoing argument, because if $\lambda_i > 0$, $g_i(X) = 0$ or $S_i^2 = 0$ and if $g_i(X) < 0$, $S_i^2 > 0$ and $\lambda_i = 0$.

The KKT necessary conditions for the maximization problem can now be summarized as follows
$\lambda \ge 0$

$\nabla f(X) - ?\nabla g(X) = 0$

$\lambda_i g_i(X) = 0, \ i = 1, 2, \dots m$

$g(X) \le 0$

These conditions apply to the minimization case as well, except that λ must be non positive. In both maximization and minimization, the Lagrange multipliers corresponding to equality constraints must be unrestricted in sign.

Sufficiency of the KKT Conditions

The Kuhn–Tucker necessary conditions are also sufficient if the objective function and the solution space satisfy the following

• For maximization

Objective function must be concave and the solution space is a convex set.

• For minimization

Objective function must be convex and the solution space is a convex set.

We define the generalized non linear problems as

Maximize or Minimize $z = f(X)$

subject to $g_i(X) \leq 0$, $i = 1, 2, \dots r$

$g_i(X) \geq 0$, $i = r + 1, \dots p$

$g_i(X) = 0$, $i = p + 1, \dots m$

$$L(X, S, \lambda) = f(X) - \sum_{i=1}^{r} \lambda_i[g_i(X) + S_i^2] - \sum_{i=r+1}^{p} \lambda_i[g_i(X) - S_i^2] - \sum_{i=p+1}^{m} \lambda_i g_i(X)$$

Where l_i is the Lagrangean multiplier associated with the constraint i. The conditions for establishing the sufficiency of KKT conditions are:

Source of optimization	$f(X)$	Required Conditions	
		$g(X)$	λ_i
Maximization	Concave	Convex	$\geq 0 \quad 1 \leq i \leq r$
		Concave	$\leq 0 \quad r+1 \leq i \leq p$
		Linear	unrestricted $p+1 \leq i \leq m$
Minimization	Convex	Convex	$\leq 0 \quad 1 \leq i \leq r$
		Concave	$\geq 0 \quad r+1 \leq i \leq p$
		Linear	unrestricted $p+1 \leq i \leq m$

The conditions specified in the table are valid because the given conditions yield a concave Lagrangian function $L(X, S, \lambda)$ in case of maximization and convex $L(X, S, \lambda)$ in case of minimization. This result is verified by noticing that if $g_i(x)$ is convex, then $\lambda_i g_i(x)$ is convex if $\lambda_i^3 0$ and concave $\lambda_i \leq 0$. Similar interpretations can be established for all the remaining conditions. It has to be noted that a linear function is both convex and concave. Also if a function f is concave, then $(-f)$ is convex and vice versa.

Solution to the Quadratic programming using Kuhn–Tucker conditions in matrix form:

Maximize $z = CX + X^TDX$

subject to $AX \le b, X \ge 0$, or $-X \le 0$,

where $X = (x_1, x_2, \ldots x_j, \ldots x_n)^T$

$C = (c_1, c_2, \ldots c_j, \ldots c_n)$

$b = (b_1, b_2, \ldots b_j, \ldots b_m)^T$

$$A(m \times n) \text{ matrix} = \begin{bmatrix} a_{11} & a_{12} & \cdots & a_{1j} & \cdots & a_{1n} \\ a_{21} & a_{22} & \cdots & a_{2j} & \cdots & a_{2n} \\ \cdots & & & & & \\ a_{i1} & a_{i2} & \cdots & a_{ij} & \cdots & a_{in} \\ \cdots & & & & & \\ a_{m1} & a_{m2} & \cdots & a_{mj} & \cdots & a_{mn} \end{bmatrix} = (a_{ij}),\ i = 1, 2, \ldots m, j = 1, 2, \ldots n$$

$$D(n \times n) \text{ matrix} = \begin{bmatrix} d_{11} & d_{12} & \cdots & d_{1j} & \cdots & d_{1n} \\ d_{21} & d_{22} & \cdots & d_{2j} & \cdots & d_{2n} \\ \cdots & & & & & \\ d_{i1} & d_{i2} & \cdots & d_{ij} & \cdots & d_{in} \\ \cdots & & & & & \\ d_{n1} & d_{n2} & \cdots & d_{nj} & \cdots & d_{nn} \end{bmatrix} = (d_{ij}),\ \text{and} \ge 0$$

and $d_{ij} = d_{ji}, i = 1, 2, \ldots n,\ i = 1, 2, \ldots n$.

Since D is a symmetric matrix of order $(n \times n)$

z is strictly **concave** for maximization and z is strictly **convex** for minimization.

The problem is given in the matrix notation as follows

Maximize $z = CX + X^TDX$

subject to $G(X) = \begin{bmatrix} A \\ -I \end{bmatrix} X - \begin{bmatrix} b \\ 0 \end{bmatrix} \le 0$

Let $\lambda = (\lambda_1, \lambda_2, \ldots \lambda_j, \ldots \lambda_m)^T$

$\mathbf{U} = (\mu_1, \mu_2, \ldots \mu_j, \ldots \mu_m)^T$

The Kuhn–Tucker conditions yields

$\lambda \ge 0, \mathbf{U} \ge 0,$

$\nabla Z - (\lambda^T, U^T)\nabla G(X) = 0$

$\lambda_i \left(b_i - \sum_{j=1}^{n} a_{ij}x_j \right) = 0,\ i = 1, 2, \ldots i, \ldots m$ or $\lambda_i S_i = 0, i = 1, 2, \ldots m,$

where S_i is the slack variable corresponding to the ith constraint,

$\mu_j x_j = 0, j = 1, 2, \ldots n$

$AX \leq b$

$-X \leq 0$

$$\nabla G(X) = \begin{bmatrix} A \\ -I \end{bmatrix}$$

$\nabla Z = C + 2X^T D$

Let $S = b - AX \geq 0$ be the slack variables of the m constraints.

The above K.T condition reduce to

$-2X^T D + \lambda^T A - U^T = 0$

or its transpose

$-2DX^T + A^T \lambda - U = C^T$

$AX + S = b$

$\lambda_j S_j = \mu_j x_j = 0$ for all i and j

$\lambda, U, X, S \geq 0, D^T = D$ (due to symmetry)

∴ The K.T. conditions can be stated neatly in matrix notation as

$$\left[\begin{array}{c|c|c|c} -2D & A^T & -I & 0 \\ \hline A & 0 & 0 & I \end{array} \right] \begin{bmatrix} X \\ \lambda \\ U \\ S \end{bmatrix} = \begin{bmatrix} C^T \\ b \end{bmatrix}$$

$\lambda_j S_j = m_j x_j = 0$ for all i and j

$\lambda, U, X, S \geq 0$

Use the artificial variable technique for the Phase I of the general simplex method which yields the optimal solution to the quadratic programming problem.

4.2 EXAMPLES

EXAM. 1: Solve the QPP: Max $Z = 8x_1 + 10x_2 - x_1^2 - x_2^2$ subject to the condition $3x_1 + 2x_2 \leq 6$

SOLUTION

$$A = (3, 2), A^T = \begin{pmatrix} 3 \\ 2 \end{pmatrix}, -2D = \begin{pmatrix} 2 & 0 \\ 0 & 2 \end{pmatrix}$$

Hence the KJ conditions are given by

$$\left[\begin{array}{c|c|c|c} -2D & A^T & -I & 0 \\ \hline A & 0 & 0 & I \end{array} \right] \begin{bmatrix} X \\ \lambda \\ U \\ S \end{bmatrix} = \begin{bmatrix} 8 \\ 10 \\ 6 \end{bmatrix}$$

$$\text{or } \begin{bmatrix} 2 & 0 & 3 & -1 & 0 & 0 \\ 0 & 2 & 2 & 0 & -1 & 0 \\ 3 & 2 & 0 & 0 & 0 & 1 \end{bmatrix} \begin{bmatrix} x_1 \\ x_2 \\ \lambda_1 \\ \mu_1 \\ \mu_2 \\ S_1 \end{bmatrix} = \begin{bmatrix} 8 \\ 10 \\ 6 \end{bmatrix}$$

or $2x_1 + 0x_2 + 3\lambda_1 - \mu_1 + 0.\mu_2 + 0.S_1 + R_1 = 8$
$0x_1 + 2x_2 + 2\lambda_1 + 0.\mu_1 - \mu_2 + 0.S_1 + R_2 = 10$
$3x_1 + 2x_2 + 0.\lambda_1 + 0.\mu_1 + 0.\mu_2 + S_1 = 6$
or $R_1 = 8 - 2x_1 + 0.x_2 - 3\lambda_1 + \mu_1 + 0.\mu_2$
$R_2 = 10 - 0.x_1 - 2x_2 - 2\lambda_1 + 0.\mu_1 + 0.\mu_2$
$S_1 = 6 - 3x_1 - 2x_2 + 0.\lambda_1 + 0.\mu_1 + 0.\mu_2$
The Phase I of the Simplex method is
Min $W = R_1 + R_2$

Basis	Const	x_1	x_2	λ_1	μ_1	μ_2
W	18	-2	-2	-5	1	1
R_1	8	-2	0	-3	1	0
R_2	10	0	-2	-2	0	1
S_1	6	-3	-2	0	0	0

$\rightarrow$

	Const	x_1	x_2	R_1	μ_1	μ_2
W	$\frac{14}{3}$	$\frac{4}{3}$	-2	$\frac{5}{3}$	$-\frac{2}{3}$	1
λ_1	$\frac{8}{3}$	$-\frac{2}{3}$	0	$-\frac{1}{3}$	$\frac{1}{3}$	0
R_2	$\frac{14}{3}$	$\frac{4}{3}$	-2	$\frac{2}{3}$	$-\frac{2}{3}$	1
S_1	6	-3	-2	0	0	0

Basis	Const	x_1	R_2	R_1	μ_1	μ_2
W	0	0	1	1	0	0
λ_1	$\frac{8}{3}$	$-\frac{2}{3}$	0	$-\frac{1}{3}$	$\frac{1}{3}$	0
x_2	$\frac{7}{3}$	$\frac{2}{3}$	$-\frac{1}{2}$	$\frac{1}{3}$	$-\frac{1}{3}$	$\frac{1}{2}$
S_1	$\frac{4}{3}$	$-\frac{13}{3}$	1	$-\frac{2}{3}$	$\frac{2}{3}$	-1

Phase I is over $\qquad$ $W = 0$

$\therefore$ The optimal solution is $x_1{}^* = 0$, $x_2{}^* = \dfrac{7}{3}$, $z^* = 8 \times 0 + 10 \times \dfrac{7}{3} - 0 - \dfrac{49}{9} = \dfrac{161}{9}$

EXAM. 2: Solve the QPP: Max $z = 2x_1{}^2 + 2x_2{}^2 - 4x_1 - 4x_2$ subject to $2x_1 + 3x_2 \le 6$, x_1, $x_2 \ge 0$

SOLUTION

The problem is put in the matrix form and we get

$$\text{Min } z = \begin{bmatrix} -4 & -4 \end{bmatrix} \begin{bmatrix} x_1 \\ x_2 \end{bmatrix} + \begin{bmatrix} x_1 & x_2 \end{bmatrix} \begin{bmatrix} 2 & 0 \\ 0 & 2 \end{bmatrix} \begin{bmatrix} x_1 \\ x_2 \end{bmatrix} \text{ subject to } \begin{bmatrix} 2 & 3 \end{bmatrix} \begin{bmatrix} x_1 \\ x_2 \end{bmatrix} \le 6, \ x_1, x_2 \ge 0$$

Hence the KT conditions are given by

$$\left[\begin{array}{c|c|c|c} -2D & A^T & -I & 0 \\ \hline A & 0 & 0 & I \end{array} \right] \begin{bmatrix} X \\ \lambda \\ U \\ S \end{bmatrix} = \begin{bmatrix} C^T \\ b \end{bmatrix}, \ \lambda_i S_i = \mu_j x_j = 0 \text{ for all } i, \ j \text{ and } X, \lambda, \ U, \ S \ge 0$$

$$\text{or} \quad \left[\begin{array}{cc|c|cc|c} -4 & 0 & 2 & -1 & 0 & 0 \\ 0 & -4 & 3 & 0 & -1 & 0 \\ \hline 2 & 3 & 0 & 0 & 0 & 1 \end{array} \right] \begin{bmatrix} x_1 \\ x_2 \\ \lambda_1 \\ \mu_1 \\ \mu_2 \\ S_1 \end{bmatrix} = \begin{bmatrix} -4 \\ -4 \\ 6 \end{bmatrix}$$

or $-4x_1 + 0.x_2 + 2\lambda_1 - \mu_1 + 0.\mu_2 + 0.S_1 = -4$

$0x_1 - 4x_2 + 3\lambda_1 + 0.\mu_1 - \mu_2 + 0.S_1 + 0 = -4$

$2x_1 + 3x_2 + 0.\lambda_1 + 0.\mu_1 + 0.\mu_2 + 0.S_1 + R_1 = 4$

or $4x_1 + 0.x_2 - 2\lambda_1 + \mu_1 + 0.\mu_2 + 0.S_1 + R_1 = 4$

$0x_1 + 4x_2 - 3\lambda_1 + 0.\mu_1 + \mu_2 + 0.S_1 + R_2 = 6$

$2x_1 + 3x_2 - 2\lambda_1 + \mu_1 + 0.\mu_2 + S_1 = 6$

or $R_1 = 4 - 4x_1 + 0.x_2 + 2\lambda_1 - \mu_1 + 0.\mu_2$

$R_2 = 4 - 0.x_1 - 4x_2 + 3\lambda_1 + 0.\mu_1 - \mu_2$

$S_1 = 6 - 2x_1 - 3x_2 + 0.\lambda_1 + 0.\mu_1 + 0.\mu_2$

The Phase I of the Simplex tableau is

I

Basis	Const	x_1	x_2	λ_1	μ_1	μ_2
W	8	-4	-4	5	-1	-1
R_1	4	-4	0	2	-1	0
R_2	4	0	-4	3	0	-1
S_1	6	-2	-3	0	0	0

II

	Const	R_1	x_2	λ_1	μ_1	μ_2
W	4	1	-4	3	0	-1
x_1	1	$-\dfrac{1}{4}$	0	$\dfrac{1}{2}$	$-\dfrac{1}{4}$	0
R_2	4	0	-4	3	0	-1
S_1	4	$\dfrac{1}{2}$	-3	-1	$\dfrac{1}{2}$	0

$\longrightarrow$

Basis	Const	R_1	R_2	λ_1	μ_1	μ_2
W	0	-3	1	1	0	0
x_1	1	$-\dfrac{1}{4}$	0	$\dfrac{1}{2}$	$-\dfrac{1}{4}$	0
x_2	1	0	$-\dfrac{1}{4}$	$\dfrac{3}{4}$	0	$-\dfrac{1}{4}$
S_1	1	$\dfrac{1}{2}$	$\dfrac{3}{4}$	$\dfrac{5}{4}$	$\dfrac{1}{2}$	$\dfrac{3}{4}$

Here $W = 0$

$\therefore$ The solution is $W = 0$

$x_1^* - 1, x_2^* - 1, z^* = 2x_1^{*2} + 2x_2^{*2} - 4x_1^* - 4x_2^* = 2 + 2 - 4 - 4 = -4$

EXAM. 3: Solve the following QPP: Min $z = 2x_1^2 - x_1 - 3x_2$ subject to $x_1 + x_2 \leq 1$, $x_1 + 4x_2 \leq 2$, $x_1, x_2 \geq 0$

SOLUTION

The problem is put in the matrix form and we get

$$\text{Min } z = \begin{bmatrix} -1, & -3 \end{bmatrix} \begin{bmatrix} x_1 \\ x_2 \end{bmatrix} + \begin{bmatrix} x_1, & x_2 \end{bmatrix} \begin{bmatrix} 2 & 0 \\ 0 & 0 \end{bmatrix} \begin{bmatrix} x_1 \\ x_2 \end{bmatrix} \text{ subject to}$$

$$\begin{bmatrix} 1, & 1 \end{bmatrix} \begin{bmatrix} x_1 \\ x_2 \end{bmatrix} \leq 1, \begin{bmatrix} 1, & 4 \end{bmatrix} \begin{bmatrix} x_1 \\ x_2 \end{bmatrix} \leq 2, \ x_1, x_2 \geq 0$$

KT conditions are

$$\left[\begin{array}{c|c|c|c} -2D & A^T & -I & 0 \\ \hline A & 0 & 0 & I \end{array} \right] \begin{bmatrix} X \\ \lambda \\ U \\ S \end{bmatrix} = \begin{bmatrix} C^T \\ b \end{bmatrix}$$

or

$$\begin{bmatrix} -4 & 0 & 1 & 1 & -1 & 0 & 0 & 0 \\ 0 & 0 & 1 & 4 & 0 & -1 & 0 & 0 \\ 1 & 1 & 0 & 0 & 0 & 0 & 1 & 0 \\ 1 & 4 & 0 & 0 & 0 & 0 & 0 & 1 \end{bmatrix} \begin{bmatrix} x_1 \\ x_2 \\ \lambda_1 \\ \mu_1 \\ \mu_2 \\ S_1 \\ S_2 \end{bmatrix} = \begin{bmatrix} -1 \\ -3 \\ 1 \\ 2 \end{bmatrix}$$

or

$-4x_1 + 0.x_2 + \lambda_1 + \lambda_2 - \mu_1 + 0.\mu_2 + 0.S_1 + 0.S_2 = -1$

$0.x_1 + 0.x_2 + \lambda_1 + 4\lambda_2 + 0\mu_1 - \mu_2 + 0.S_1 + 0.S_2 = -3$

$x_1 + x_2 + 0.\lambda_1 + 0.\lambda_2 + 0.\mu_1 - 0.\mu_2 + 1.S_1 + 0.S_2 = 1$

$x_1 + 4x_2 + 0.\lambda_1 + 0.\lambda_2 + 0.\mu_1 + 0.\mu_2 + 0.S_1 + 1.S_2 = 2$

or

$-4x_1 + 0.x_2 - \lambda_1 - \lambda_2 + \mu_1 + 0.\mu_2 + 0.S_1 + 0.S_2 + R_1 = 1$

$0.x_1 + 0.x_2 - \lambda_1 - 4\lambda_2 + 0.\mu_1 + \mu_2 + 0.S_1 + 0.S_2 + R_2 = 3$

$x_1 + x_2 + 0.\lambda_1 + 0.\lambda_2 + 0.\mu_1 + 0.\mu_2 + 1.S_1 + 0.S_2 = 1$

$x_1 + 4x_2 + 0.\lambda_1 + 0.\lambda_2 + 0.\mu_1 + 0.\mu_2 + 0.S_1 + S_2 = 2$

or

$R_1 = 1 - 4x_1 + 0.x_2 + \lambda_1 + \lambda_2 - \mu_1 - 0.\mu_2$

$R_2 = 3 - 0.x_1 + 0.x_2 - \lambda_1 - 4\lambda_2 + 0.\mu_1 - \mu_2$

$S_1 = 1 - x_1 - x_2 + 0.\lambda_1 + 0.\lambda_2 + 0.\mu_1 + 0.\mu_2$

$S_2 = 2 - x_1 - 4x_2 + 0.\lambda_1 + 0.\lambda_2 + 0.\mu_1 + 0.\mu_2$

Min $W = R_1 + R_2$

I

Basis	Const	x_1	x_2	λ_1	λ_2	μ_1	μ_2
W	4	-4	0	2	5	-1	-1
R_1	1	-4	0	1	1	-1	0
R_2	3	0	0	1	4	0	-1
S_1	1	-1	-1	0	0	0	0
S_2	2	-1	-4	0	0	0	0

II

Basis	Const	R_1	x_2	λ_1	λ_2	μ_1	μ_2
W	3	1	0	1	4	0	-1
x_1	$\frac{1}{4}$	$-\frac{1}{4}$	0	$\frac{1}{4}$	$\frac{1}{4}$	$\frac{1}{4}$	0
R_2	3	0	0	1	4	0	-1
S_1	$\frac{3}{4}$	$\frac{1}{4}$	-1	$-\frac{1}{4}$	$-\frac{1}{4}$	$\frac{1}{4}$	0
S_2	$\frac{7}{4}$	$\frac{1}{4}$	-4	$-\frac{1}{4}$	$-\frac{1}{4}$	$\frac{1}{4}$	0

Basis	Const	R_1	x_2	λ_1	λ_2	μ_1	R_2
W	0	1	0	0	0	0	1
x_1	$\dfrac{1}{4}$	$-\dfrac{1}{4}$	0	$\dfrac{1}{4}$	$\dfrac{1}{4}$	$\dfrac{1}{4}$	0
μ_2	3	0	0	1	4	0	-1
S_1	$\dfrac{3}{4}$	$\dfrac{1}{4}$	-1	$-\dfrac{1}{4}$	$-\dfrac{1}{4}$	$-\dfrac{1}{4}$	0
S_2	$\dfrac{7}{4}$	$\dfrac{1}{4}$	-4	$-\dfrac{1}{4}$	$-\dfrac{1}{4}$	$-\dfrac{1}{4}$	0

$W = 0$

$\therefore$ The phase I is now over and verification of the constraints the solution is

$$x_1^* = \frac{1}{4}, \ x_2^* = 0, \ \mu_2 = 3,$$

$$S_1^* = \frac{3}{4}, \ S_2^* = \frac{7}{4}, \ \mu_1 = 0$$

$$x_1^* = \frac{1}{4}, \ x_2^* = 0, \ z^* = -\frac{1}{8}$$

EXAM. 4: Solve the QPP: Min $z = 4x_1 + 6x_2 - 2x_1^2 - 2x_1 x_2 - 2x_2^2$ subject to $x_1 + 2x_2 \le 2, x_1, x_2 \ge 0$

$$\text{Min } z = \begin{bmatrix} 4 & 6 \end{bmatrix} \begin{bmatrix} x_1 \\ x_2 \end{bmatrix} + \begin{bmatrix} x_1 & x_2 \end{bmatrix} \begin{bmatrix} -2 & -1 \\ -1 & -2 \end{bmatrix} \begin{bmatrix} x_1 \\ x_2 \end{bmatrix} \text{ subject to } \begin{bmatrix} 1 & 2 \end{bmatrix} \begin{bmatrix} x_1 \\ x_2 \end{bmatrix} \le 2, \ x_1, x_2 \ge 0$$

KT conditions are

$$\begin{bmatrix} -2D & A^T & -I & 0 \\ \hline A & 0 & 0 & I \end{bmatrix} \begin{bmatrix} X \\ \lambda \\ U \\ S \end{bmatrix} = \begin{bmatrix} C^T \\ b \end{bmatrix}$$

or

$$\begin{bmatrix} 4 & 2 & 1 & -1 & 0 & 0 \\ 2 & 4 & 2 & 0 & -1 & 0 \\ 1 & 2 & 0 & 0 & 0 & 1 \end{bmatrix} \begin{bmatrix} x_1 \\ x_2 \\ \lambda_1 \\ \mu_1 \\ \mu_2 \\ S_1 \end{bmatrix} = \begin{bmatrix} 4 \\ 6 \\ 2 \end{bmatrix}$$

$$-4x_1 + 2x_2 + \lambda_1 - \mu_1 + R_1 = 4$$
$$2x_1 + 4x_2 + 2\lambda_1 - \mu_2 + R_2 = 6$$
$$x_1 + 2x_2 + S_1 = 2$$
$$R_1 = 4 - 4x_1 - 2x_2 - \lambda_1 + \mu_1 \quad \lambda_1 S_1 = 0, \ \mu_1 x_1 = 0, \ \mu_2 x_2 = 0$$
$$R_2 = 6 - 2x_1 - 4x_2 - 2\lambda_1 + \mu_2$$
$$S_1 = 2 - x_1 - 2x_2$$

Phase 1 of LPP is

Min $W = R_1 + R_2$ subject to the above condition

I

Basis	Const	x_1	x_2	λ_1	μ_1	μ_2
W	10	-6	-6	-3	1	1
R_1	4	-4	-2	-1	1	0
R_2	6	-2	-4	-2	0	1
S_1	2	-1	-2	0	0	0

II

Basis	Const	R_1	x_2	λ_1	μ_1	μ_2
W	3	$\frac{3}{2}$	-3	$-\frac{3}{2}$	$-\frac{1}{2}$	1
x_1	1	$-\frac{1}{4}$	$-\frac{1}{2}$	$-\frac{1}{4}$	$\frac{1}{4}$	0
R_2	4	$\frac{1}{2}$	-3	$-\frac{3}{2}$	$-\frac{1}{2}$	1
S_1	1	$\frac{1}{4}$	$-\frac{3}{2}$	$\frac{1}{4}$	$-\frac{1}{4}$	0

→

Basis	Const	R_1	S_2	λ_1	μ_1	μ_2
W	2	1	$\frac{1}{2}$	-2	0	1
x_1	$\frac{2}{3}$	$-\frac{1}{3}$	$\frac{1}{3}$	$-\frac{1}{3}$	$\frac{1}{3}$	0
R_2	2	0	2	-2	0	1
x_2	$\frac{2}{3}$	$\frac{1}{6}$	$-\frac{2}{3}$	$\frac{1}{6}$	$-\frac{1}{6}$	0

Basis	Const	R_1	x_2	λ_1	μ_1	μ_2
W	0			1		
x_1	$\frac{1}{3}$			$\frac{1}{6}$		
λ_1	1	0	1	$-\frac{1}{2}$	0	$\frac{1}{2}$
x_2	$\frac{5}{6}$			$-\frac{1}{12}$		

→

Here $W = 0$ ∴ The phase I is over. Solution is

$$x_1^* = \frac{1}{4}, \ x_2^* = \frac{5}{6}$$

$$\text{Min } z = z^* = \frac{4}{3} + 5 - \frac{2}{9} - \frac{5}{9} - \frac{25}{18} = \frac{25}{6}$$

EXAM. 5: Solve the QPP: Min $z = x_1^2 - x_1 - 3x_2$ subject to $x_1 + x_2 \le 1$, $x_1 + 4x_2 \le 2$, $x_1, x_2 \ge 0$

$$\text{Min } z = [-1 \ \ -3]\begin{bmatrix} x_1 \\ x_2 \end{bmatrix} + [x_1 \ \ x_2]\begin{bmatrix} 1 & 0 \\ 0 & 0 \end{bmatrix}\begin{bmatrix} x_1 \\ x_2 \end{bmatrix} \text{ subject to } \begin{bmatrix} 1 & 1 \\ 1 & 4 \end{bmatrix}\begin{bmatrix} x_1 \\ x_2 \end{bmatrix} \le \begin{bmatrix} 1 \\ 2 \end{bmatrix}, \ x_1, x_2 \ge 0$$

KT conditions are

$$\left[\begin{array}{c|c|c|c} -2D & A^T & -I & 0 \\ \hline A & 0 & 0 & I \end{array}\right]\begin{bmatrix} X \\ \lambda \\ U \\ S \end{bmatrix} = \begin{bmatrix} C^T \\ b \end{bmatrix}$$

or

$$\left[\begin{array}{cc|cc|cc|cc} -2 & 0 & 1 & 1 & -1 & 0 & 0 & 0 \\ 0 & 0 & 1 & 4 & 0 & -1 & 0 & 0 \\ \hline 1 & 1 & 0 & 0 & 0 & 0 & 1 & 0 \\ 1 & 4 & 0 & 0 & 0 & 0 & 0 & 1 \end{array}\right]\begin{bmatrix} x_1 \\ x_2 \\ \lambda_1 \\ \mu_1 \\ \mu_2 \\ S_1 \\ S_2 \end{bmatrix} = \begin{bmatrix} -1 \\ -3 \\ 1 \\ 2 \end{bmatrix}$$

or $-2x_1 + \lambda_1 + \lambda_2 - \mu_1 = -1$ or $2x_1 - \lambda_1 - \lambda_2 + \mu_1 + R_1 = 1$
$\lambda_1 + 4\lambda_2 - \mu_2 = -3$ $-\lambda_1 - 4\lambda_2 + \mu_2 + R_2 = 3$
$x_1 + x_2 + S_1 = 1$ $x_1 + x_2 + S_1 = 1$
$x_1 + 4x_2 + S_2 = 2$ $x_1 + 4x_2 + S_2 = 2$

$R_1 = 1 - 2x_1 + \lambda_1 + \lambda_2 - \mu_1$
$R_2 = 3 + \lambda_1 + 4\lambda_2 - \mu_2$
$S_1 = 1 - x_1 - x_2$
$S_2 = 2 - x_1 - 4x_2$

I

Basis	Const	x_1	x_2	λ_1	λ_2	μ_1	μ_2
W	4	-2	0	2	5	-1	-1
R_1	1	-2	0	1	1	-1	0
R_2	3	0	0	1	4	0	-1
S_1	1	-1	-1	0	0	0	0
S_2	2	-1	-4	0	0	0	0

II

Basis	Const	R_1	x_2	λ_1	λ_2	μ_1	μ_2
W	3	1	0	1	4	0	-1
x_1	$\dfrac{1}{2}$	$-\dfrac{1}{2}$	0	$\dfrac{1}{2}$	$\dfrac{1}{2}$	$-\dfrac{1}{2}$	0
R_2	3	0	0	1	4	0	-1
S_1	$\dfrac{1}{2}$	$\dfrac{1}{2}$	-1	$-\dfrac{1}{2}$	$-\dfrac{1}{2}$	$\dfrac{1}{2}$	0
S_2	$\dfrac{3}{2}$	$\dfrac{1}{2}$	-4	$-\dfrac{1}{2}$	$-\dfrac{1}{2}$	$\dfrac{1}{2}$	0

Basis	Const	R_1	x_2	λ_1	λ_2	μ_1	R_2
W	0	1	0	0	0	0	1
x_1	$\dfrac{1}{2}$						0
μ_2	3	0	0	1	4	0	1
S_1	$\dfrac{1}{2}$						0
S_2	$\dfrac{3}{2}$						0

Here $W = 0$ ∴ The phase I is over. Solution is

$$x_1^* = \frac{1}{2}, \; x_2^* = 0 \quad \text{Min } z = z^* = \frac{1}{4} + \frac{1}{2} = -\frac{1}{4}$$

EXAM. 6: Solve the QPP: Min $z = x_1^2 - 2x_1x_2 + 2x_2^2 - 2x_1 - 5x_2$ subject to $2x_1 + 3x_2 \leq 20$, $3x_1 - 5x_2 \leq 4$, $x_1, x_2 \geq 0$

$$\text{Min } z = \begin{bmatrix} -2 & -5 \end{bmatrix} \begin{bmatrix} x_1 \\ x_2 \end{bmatrix} + \begin{bmatrix} x_1 & x_2 \end{bmatrix} \begin{bmatrix} 1 & -1 \\ -1 & 2 \end{bmatrix} \begin{bmatrix} x_1 \\ x_2 \end{bmatrix} \text{ subject to } \begin{bmatrix} 2 & 3 \\ 3 & -5 \end{bmatrix} \begin{bmatrix} x_1 \\ x_2 \end{bmatrix} \leq \begin{bmatrix} 20 \\ 4 \end{bmatrix}$$

KT conditions are

$$\left[\begin{array}{c|c|c|c} -2D & A^T & -I & 0 \\ \hline A & 0 & 0 & I \end{array} \right] \begin{bmatrix} X \\ \lambda \\ U \\ S \end{bmatrix} = \begin{bmatrix} C^T \\ b \end{bmatrix}$$

or
$$\begin{bmatrix} -2 & 2 & 2 & 3 & -1 & 0 & 0 & 0 \\ 2 & -4 & 3 & -5 & 0 & -1 & 0 & 0 \\ 2 & 3 & 0 & 0 & 0 & 0 & 1 & 0 \\ 3 & -5 & 0 & 0 & 0 & 0 & 0 & 1 \end{bmatrix} \begin{bmatrix} x_1 \\ x_2 \\ \lambda_1 \\ \mu_1 \\ \mu_2 \\ S_1 \\ S_2 \end{bmatrix} = \begin{bmatrix} -2 \\ -5 \\ 20 \\ 4 \end{bmatrix}$$

$-2x_1 + 2x_2 + 2\lambda_1 + 3\lambda_2 - \mu_1 = -2$
$2x_1 - 4x_2 + 3\lambda_1 - 5\lambda_2 - \mu_2 = -5$
$2x_1 + 3x_2 + S_1 = 20$
$3x_1 - 5x_2 + S_2 = 4$

or $2x_1 - 2x_2 - 2\lambda_1 - 3\lambda_2 + \mu_1 + R_1 = 2$
$-2x_1 + 4x_2 - 3\lambda_1 + 5\lambda_2 + \mu_2 + R_2 = 5$
$2x_1 + 3x_2 + S_1 = 20$
$3x_1 - 5x_2 + S_2 = 4$
$\therefore R_1 = 2 - 2x_1 + 2x_2 + 2\lambda_1 + 3\lambda_2 - \mu_1$
$R_2 = 5 + 2x_1 - 4x_2 + 3\lambda_1 - 5\lambda_2 - \mu_2$
$S_1 = 20 - 2x_1 - 3x_2$
$S_2 = 4 - 3x_1 + 5x_2$

I

Basis	Const	x_1	x_2	λ_1	λ_2	μ_1	μ_2
W	7	0	$-\dfrac{2}{2}$	5	-2	-1	-1
R_1	2	$-\dfrac{2}{2}$	2	2	3	-1	0
R_2	5	2	$-\dfrac{4}{4}$	3	-5	0	-1
S_1	20	$-\dfrac{2}{2}$	$-\dfrac{3}{3}$	0	0	0	0
S_2	4	$-\dfrac{3}{3}$	5	0	0	0	0

II

Basis	Const	x_1	R_2	λ_1	λ_2	μ_1	μ_2
W	$\dfrac{9}{2}$	-1	$\dfrac{1}{2}$	$\dfrac{7}{2}$	$\dfrac{1}{2}$	-1	$-\dfrac{1}{2}$
R_1	$\dfrac{9}{2}$	-1	$-\dfrac{1}{2}$	$\dfrac{7}{2}$	-2	-1	$-\dfrac{1}{2}$
x_2	$\dfrac{5}{4}$	$\dfrac{1}{2}$	$-\dfrac{1}{4}$	$\dfrac{3}{4}$	$-\dfrac{5}{4}$	0	$-\dfrac{1}{4}$
S_1	$\dfrac{65}{4}$	$-\dfrac{7}{2}$	$\dfrac{3}{4}$	$-\dfrac{9}{4}$	$\dfrac{15}{4}$	0	$\dfrac{3}{4}$
S_2	$\dfrac{41}{4}$	$-\dfrac{1}{2}$	$-\dfrac{5}{4}$	$\dfrac{15}{4}$	$-\dfrac{25}{4}$	0	$-\dfrac{5}{4}$

Basis	Const	R_1	R_2	λ_1	λ_2	μ_1	μ_2
W	0	1	1	0	$\frac{5}{2}$	0	0
x_1	$\frac{9}{2}$	-1	$-\frac{1}{2}$	$\frac{7}{2}$	-2	-1	$-\frac{1}{2}$
x_2	$\frac{7}{2}$	$-\frac{1}{2}$					
S_1	$\frac{1}{2}$	$\frac{7}{2}$					
S_2	8	$\frac{1}{2}$					

Here $W = 0$ ∴ The phase I is over. Solution is

$$x_1^* = \frac{9}{2}, \; x_2^* = \frac{7}{2} \quad \text{Min } z = z^* = -\frac{59}{4}$$

EXAM. 7: Solve the QPP: Max $z = 8x_1 - 10x_2 - x_1^2 - x_2^2$, subject to $3x_1 + 2x_2 \leq 6$, x_1, $x_2 \geq 0$

$$\text{Max } z = \begin{bmatrix} 8 & 10 \end{bmatrix} \begin{bmatrix} x_1 \\ x_2 \end{bmatrix} + \begin{bmatrix} x_1 & x_2 \end{bmatrix} \begin{bmatrix} -1 & 0 \\ 0 & -1 \end{bmatrix} \begin{bmatrix} x_1 \\ x_2 \end{bmatrix} \text{ subject to } \begin{bmatrix} 3 & 2 \end{bmatrix} \begin{bmatrix} x_1 \\ x_2 \end{bmatrix} \leq 6, \; x_1, x_2 \geq 0$$

KT conditions are

$$\left[\begin{array}{c|c|c|c} -2D & A^T & -I & 0 \\ \hline A & 0 & 0 & I \end{array} \right] \begin{bmatrix} X \\ \lambda \\ U \\ S \end{bmatrix} = \begin{bmatrix} C^T \\ b \end{bmatrix}$$

or
$$\left[\begin{array}{cc|c|cc|c} 2 & 0 & 3 & -1 & 0 & 0 \\ 0 & 2 & 2 & 0 & -1 & 0 \\ \hline 3 & 2 & 0 & 0 & 0 & 1 \end{array} \right] \begin{bmatrix} x_1 \\ x_2 \\ \lambda_1 \\ \mu_1 \\ \mu_2 \\ S_1 \end{bmatrix} = \begin{bmatrix} 8 \\ 10 \\ 6 \end{bmatrix}$$

$$2x_1 + 3\lambda_1 - \mu_1 + R_1 = 8$$
$$2x_2 + 2\lambda_1 - \mu_2 + R_2 = 10$$

$3x_1 + 2x_2 + S_1 = 6$

Min $W = R_1 + R_2$ subject to

$\therefore R_1 = 8 - 2x_1 - 3\lambda_1 + \mu_1$

$R_2 = 10 - 2x_2 - 2\lambda_1 + \mu_2$

$S_1 = 6 - 3x_1 - 2x_2 \quad \lambda_1 S_1 = 0, \; \mu_1 x_1 = 0, \; \mu_2 x_2 = 0$

I

Basis	Const	x_1	x_2	λ_1	μ_1	μ_2
W	18	-2	-2	-5	1	1
R_1	8	-2	0	-3	1	0
R_2	10	0	-2	-2	0	1
S_1	6	-3	-2	0	0	0

$\longrightarrow$

II

Basis	Const	R_1	x_2	λ_1	μ_1	μ_2
W	10	1	-2	-2	0	1
x_1	4	$-\dfrac{1}{2}$	0	$-\dfrac{3}{2}$	$\dfrac{1}{2}$	0
R_2	10	0	-2	-2	0	1
S_1	-6	$\dfrac{3}{2}$	-2	$\dfrac{9}{2}$	$-\dfrac{3}{2}$	0

Basis	Const	R_1	R_2	λ_1	μ_1	μ_2
W	0	1	1	0	0	0
x_1	4	$-\dfrac{1}{2}$	0	$-\dfrac{3}{2}$	$\dfrac{1}{2}$	0
x_2	5	0	$-\dfrac{1}{2}$	-1	0	$\dfrac{1}{2}$
S_1	-16	$\dfrac{3}{2}$	1	$\dfrac{13}{2}$	$-\dfrac{3}{2}$	-1

$\therefore$ The phase I is over. Solution is

$x_1^* = 4, \; x_2^* = 5 \quad$ Min $z = z^* = 31$

EXAM. 8: Solve the QPP: Max $z = -2x_1 - x_2 - x_2^2$, subject to $x_1 + x_2 \le 1, \; x_1 + 2x_2 \ge 1, \; x_1, x_2 \ge 0$

In matrix notation, the problem is written as

$$\text{Max } z = \begin{bmatrix} -2 & -1 \end{bmatrix} \begin{bmatrix} x_1 \\ x_2 \end{bmatrix} + \begin{bmatrix} x_1 & x_2 \end{bmatrix} \begin{bmatrix} 0 & 0 \\ 0 & -1 \end{bmatrix} \begin{bmatrix} x_1 \\ x_2 \end{bmatrix} \text{ subject to } \begin{bmatrix} 1 & 1 \\ -1 & -2 \end{bmatrix} \begin{bmatrix} x_1 \\ x_2 \end{bmatrix} \le \begin{bmatrix} 1 \\ -1 \end{bmatrix},$$

$x_1, x_2 \ge 0$

KT conditions are

$$\begin{bmatrix} -2D & A^T & -I & 0 \\ \hline A & 0 & 0 & I \end{bmatrix} \begin{bmatrix} X \\ \lambda \\ U \\ S \end{bmatrix} = \begin{bmatrix} C^T \\ b \end{bmatrix}$$

or

$$\begin{bmatrix} 0 & 0 & 1 & -1 & -1 & 0 & 0 & 0 \\ 0 & 2 & 1 & -2 & 0 & -1 & 0 & 0 \\ \hline 1 & 1 & 0 & 0 & 0 & 0 & 1 & 0 \\ -1 & -2 & 0 & 0 & 0 & 0 & 0 & 1 \end{bmatrix} \begin{bmatrix} x_1 \\ x_2 \\ \lambda_1 \\ \mu_1 \\ \mu_2 \\ S_1 \\ S_2 \end{bmatrix} = \begin{bmatrix} -2 \\ -1 \\ 1 \\ -1 \end{bmatrix} \quad \lambda_1 S_1 = \lambda_2 S_2 = 0 \; \mu_1 x_1 = \mu_2 x_2 = 0$$

or

$$\lambda_1 - \lambda_2 - \mu_1 = -2$$
$$2x_2 + \lambda_1 - 2\lambda_2 - \mu_2 = -1$$
$$x_1 + x_2 + S_1 = 1$$
$$-x_1 - 2x_2 + S_2 = -1$$

or

$$-\lambda_1 + \lambda_2 + \mu_1 + R_1 = 2$$
$$-2x_2 - \lambda_1 + 2\lambda_2 + \mu_2 + R_2 = 1$$
$$x_1 + x_2 + S_1 = 1$$
$$x_1 + 2x_2 - S_2 + R_3 = 1$$

I

Basis	Const	x_1	x_2	λ_1	λ_2	μ_1	μ_2	S_2
W	4	-1	0	2	-3	-1	-1	1
R_1	2	0	0	1	-1	-1	0	0
R_2	1	0	2	1	-2	0	-1	0
R_3	1	-1	-2	0	0	0	0	1
S_1	1	-1	-1	0	0	0	0	0

λ_2 enter with the basis

II

Basis	Const	x_1	x_2	λ_1	R_2	μ_1	μ_2	S_2
W	$\frac{5}{2}$	-1	-3	$\frac{1}{2}$	$\frac{3}{2}$	-1	$\frac{1}{2}$	1
R_1	$\frac{3}{2}$	0	-1	$\frac{1}{2}$	$\frac{1}{2}$	-1	$\frac{1}{2}$	0
λ_2	$\frac{1}{2}$	0	1	$\frac{1}{2}$	$-\frac{1}{2}$	0	$\frac{1}{2}$	0
R_3	1	-1	-2	0	0	0	0	1
S_1	1	-1	-1	0	0	0	0	0

x_2 enter with the basis

Basis	Const	x_1	R_3	λ_1	R_2	μ_1	μ_2	S_2
W	1	$\frac{1}{2}$	$\frac{3}{2}$	$\frac{1}{2}$	$\frac{3}{2}$	-1	$\frac{1}{2}$	$-\frac{1}{2}$
R_1	1	$\frac{1}{2}$	$\frac{1}{2}$	$\frac{1}{2}$	$\frac{1}{2}$	-1	$\frac{1}{2}$	$-\frac{1}{2}$
λ_2	1	$-\frac{1}{2}$	$-\frac{1}{2}$	0	$-\frac{1}{2}$	0	$\frac{1}{2}$	$\frac{1}{2}$
x_2	$\frac{1}{2}$	$-\frac{1}{2}$	$-\frac{1}{2}$	0	0	0	0	$\frac{1}{2}$
S_1	$\frac{1}{2}$	$-\frac{1}{2}$	$\frac{1}{2}$	0	0	0	0	$-\frac{1}{2}$

$\rightarrow$

Basis	Const	x_1	R_3	λ_1	R_2	μ_1	μ_2	R_2
W	0	0	1	0	1	0	0	1
S_2	2	1	1	1	1	-2	1	-2
λ_2	2							-1
x_2	$\frac{3}{2}$							-1
S_1	$-\frac{1}{2}$							1

Now S_2 enters into the basis

∴ Solution is

$$x_1{}^* = 0, \ x_2{}^* = \frac{3}{2} \qquad \text{Min } z = z^* = -\frac{15}{4}$$

Exam. 9: Solve the QPP: Minimize $z = x_1{}^2 - x_1 - x_2$ subject to $x_1 + x_2 \le 1$, $x_1, x_2 \ge 0$

$$\text{Min } z = \begin{bmatrix} -1 & -1 \end{bmatrix} \begin{bmatrix} x_1 \\ x_2 \end{bmatrix} + \begin{bmatrix} x_1 & x_2 \end{bmatrix} \begin{bmatrix} 1 & 0 \\ 0 & 0 \end{bmatrix} \begin{bmatrix} x_1 \\ x_2 \end{bmatrix} \text{ subject to } \begin{bmatrix} 1 & 1 \end{bmatrix} \begin{bmatrix} x_1 \\ x_2 \end{bmatrix} \le 1$$

KT conditions are

$$\left[\begin{array}{c|c|c|c} -2D & A^T & -I & 0 \\ \hline A & 0 & 0 & I \end{array} \right] \begin{bmatrix} X \\ \lambda \\ U \\ S \end{bmatrix} = \begin{bmatrix} C^T \\ b \end{bmatrix}$$

or

$$\left[\begin{array}{cc|c|cc|c} -2 & 0 & 1 & -1 & 0 & 0 \\ 0 & 0 & 1 & 0 & -1 & 0 \\ \hline 1 & 1 & 0 & 0 & 0 & 1 \end{array} \right] \begin{bmatrix} x_1 \\ x_2 \\ \lambda_1 \\ \mu_1 \\ \mu_2 \\ S_1 \end{bmatrix} = \begin{bmatrix} -1 \\ -1 \\ 1 \end{bmatrix}$$

$$-2x_1 + \lambda_1 - \mu_1 = -1 \qquad 2x_1 - \lambda_1 + \mu_1 + R_1 = 1$$
$$\lambda_1 - \mu_2 = -1 \qquad\qquad -\lambda_1 + \mu_2 + R_2 = 1$$
$$x_1 + x_2 + S_1 = 1 \qquad\quad x_1 + x_2 + S_1 = 1$$

$$\therefore \left. \begin{array}{l} R_1 = 1 - 2x_1 + \lambda_1 - \mu_1 \\ R_2 = 1 - \lambda_1 - \mu_2 \\ S_1 = 1 - x_1 - x_2 \end{array} \right\} \qquad\qquad \ldots(1)$$

Min $W = R_1 + R_2$ subject to (1) $\lambda_1 S_1 = m_1 x_1 = m_2 x_2 = 0$, X, λ, U, $S \geq 0$

I

Basis	Const	x_1	x_2	λ_1	μ_1	μ_2
W	2	-2	0	0	-1	-1
R_1	1	-2	0	1	-1	0
R_2	1	0	0	-1	0	-1
S_1	1	-1	-1	0	0	0

II

Basis	Const	R_1	x_2	λ_1	μ_1	μ_2
W	1	1	0	-1	0	1
x_1	$\dfrac{1}{2}$	$-\dfrac{1}{2}$	0	$\dfrac{1}{2}$	$-\dfrac{1}{2}$	0
R_2	1	0	0	-1	0	-1
S_1	$\dfrac{1}{2}$	$\dfrac{1}{2}$	-1	$-\dfrac{1}{2}$	$\dfrac{1}{2}$	0

λ_1 can't enter, since S_1 is in the basis. μ_2 enters

Basis	Const	R_1	x_2	λ_1	μ_1	R_2
W	0	1	0	0	0	1
x_1	$\dfrac{1}{2}$	$\dfrac{1}{2}$	0	$\dfrac{1}{2}$	$-\dfrac{1}{2}$	0
μ_2	1	0	0	-1	0	-1
S_1	$\dfrac{1}{2}$	$\dfrac{1}{2}$	-1	$-\dfrac{1}{2}$	$\dfrac{1}{2}$	0

Here $W = 0$ The iteration is over. Solution is

$$x_1^* = \frac{1}{2}, \ x_2^* = 0 \ \ \text{Min } z = z^* = -\frac{1}{4}$$

EXAM. 10: Solve the QPP: Max $z = 4x_1 + 8x_2 - x_1^2 - x_2^2$, subject to $x_1 + x_2 \le 2$, $x_1, x_2 \ge 0$

$$\text{Max } z = \begin{bmatrix} 4 & 8 \end{bmatrix}\begin{bmatrix} x_1 \\ x_2 \end{bmatrix} + \begin{bmatrix} x_1 & x_2 \end{bmatrix}\begin{bmatrix} -1 & 0 \\ 0 & -1 \end{bmatrix}\begin{bmatrix} x_1 \\ x_2 \end{bmatrix} \text{ subject to } \begin{bmatrix} 1 & 1 \end{bmatrix}\begin{bmatrix} x_1 \\ x_2 \end{bmatrix} \le 2, \ x_1, x_2 \ge 0$$

KT conditions are

$$\left[\begin{array}{c|c|c|c} -2D & A^T & -I & 0 \\ \hline A & 0 & 0 & I \end{array}\right]\begin{bmatrix} X \\ \lambda \\ U \\ S \end{bmatrix} = \begin{bmatrix} C^T \\ b \end{bmatrix}$$

or $\left[\begin{array}{cc|c|cc|c} 2 & 0 & 1 & -1 & 0 & 0 \\ 0 & 2 & 1 & 0 & -1 & 0 \\ \hline 1 & 1 & 0 & 0 & 0 & 1 \end{array}\right]\begin{bmatrix} x_1 \\ x_2 \\ \lambda_1 \\ \mu_1 \\ \mu_2 \\ S_1 \end{bmatrix} = \begin{bmatrix} 4 \\ 8 \\ 2 \end{bmatrix}$

or $2x_1 + \lambda_1 - \mu_1 + R_1 = 4$ or $R_1 = 4 - 2x_1 - \lambda_1 + \mu_1$
   $2x_2 + \lambda_1 - \mu_2 + R_2 = 8$ $R_2 = 8 - 2x_2 - \lambda_1 + \mu_2$
   $x_1 + x_2 + S_1 = 1$ $S_1 = 2 - x_1 - x_2$
Min $W = R_1 + R_2$ subject to $\lambda_1 S_1 = \mu_1 x_1 = \mu_2 x_2 = 0$

		I								II				
Basis	Const	x_1	x_2	λ_1	μ_1	μ_2		Basis	Const	R_1	x_2	λ_1	μ_1	μ_2
W	12	-2	-2	-2	1	1		W	8	1	-2	-1	0	1
R_1	4	-2	0	-1	1	0		x_1	2	$-\frac{1}{2}$	0	$\frac{1}{2}$	$-\frac{1}{2}$	0
R_2	8	0	-2	-1	0	1	$\rightarrow$	R_2	8	0	-2	-1	0	1
S_1	2	-1	-1	0	0	0		S_1	0	$\frac{1}{2}$	-1	$\frac{1}{2}$	$-\frac{1}{2}$	0

Basis	Const	R_1	S_1	λ_1	μ_1	μ_2
W	8	0	2	-2	1	1
x_1	2	$-\dfrac{1}{2}$	0	$\dfrac{1}{2}$	$-\dfrac{1}{2}$	0
R_2	8	-1	2	-2	1	1
x_2	0	$\dfrac{1}{2}$	-1	$\dfrac{1}{2}$	$-\dfrac{1}{2}$	0

$\rightarrow$

Basis	Const	R_1	S_1	R_2	μ_1	μ_2
W	0			1		
x_1	0			-1		
λ_1	4	$-\dfrac{1}{2}$	1	$-\dfrac{1}{2}$	$\dfrac{1}{2}$	$\dfrac{1}{2}$
x_2	2			$-\dfrac{1}{4}$		

$W = 0$

$\therefore$ Solution is

$x_1^* = 0, x_2^* = 2 \qquad$ Max $z = z^* = 12$

EXAM. 11: Solve the QPP: Min $z = x_1^2 - x_1 - 3x_2$ subject to $x_1 + x_2 \le 1$, $x_1 + 4x_2 \le 2$, $x_1, x_2 \ge 0$

$$\text{Min } z = \begin{bmatrix} -1 & -3 \end{bmatrix}\begin{bmatrix} x_1 \\ x_2 \end{bmatrix} + \begin{bmatrix} x_1 & x_2 \end{bmatrix}\begin{bmatrix} 1 & 0 \\ 0 & 0 \end{bmatrix}\begin{bmatrix} x_1 \\ x_2 \end{bmatrix} \text{ subject to } \begin{bmatrix} 1 & 1 \\ 1 & 4 \end{bmatrix}\begin{bmatrix} x_1 \\ x_2 \end{bmatrix} \le \begin{bmatrix} 1 \\ 2 \end{bmatrix}$$

KT conditions are

$$\left[\begin{array}{c|c|c|c} -2D & A^T & -I & 0 \\ \hline A & 0 & 0 & I \end{array}\right]\begin{bmatrix} X \\ \lambda \\ U \\ S \end{bmatrix} = \begin{bmatrix} C^T \\ b \end{bmatrix}$$

or

$$\left[\begin{array}{cc|cc|cc|cc} -2 & 0 & 1 & 1 & -1 & 0 & 0 & 0 \\ 0 & 0 & 1 & 4 & 0 & -1 & 0 & 0 \\ \hline 1 & 1 & 0 & 0 & 0 & 0 & 1 & 0 \\ 1 & 4 & 0 & 0 & 0 & 0 & 0 & 1 \end{array}\right]\begin{bmatrix} x_1 \\ x_2 \\ \lambda_1 \\ \mu_1 \\ \mu_2 \\ S_1 \\ S_2 \end{bmatrix} = \begin{bmatrix} -1 \\ -3 \\ 1 \\ 2 \end{bmatrix}$$

$-2x_1 + 0x_2 + \lambda_1 + \lambda_2 - \mu_1 = -1$

or $2x_1 - 0x_2 - \lambda_1 - \lambda_2 + \mu_1 + R_1 = 1$

$0x_1 + 0x_2 + \lambda_1 + 4\lambda_2 + 0\mu_1 - \mu_2 = -3$

or $-\lambda_1 - 4\lambda_2 + \mu_2 + R_2 = 3$

$x_1 + x_2 + S_1 = 1$

$x_1 + 4x_2 + S_2 = 2$

$\therefore R_1 = 1 - 2x_1 + \lambda_1 + \lambda_2 - \mu_1$

$R_2 = 3 + \lambda_1 + 4\lambda_2 - \mu_2$

$S_1 = 1 - x_1 - x_2$

$S_2 = 2 - x_1 - 4x_2$

Min $W = R_1 + R_2$

I

Basis	Const	x_1	x_2	λ_1	λ_2	μ_1	μ_2
W	4	-2	0	2	5	-1	-1
R_1	1	-2	0	1	1	-1	0
R_2	3	0	0	1	4	0	-1
S_1	1	-1	-1	0	0	0	0
S_2	2	-1	-4	0	0	0	0

II

Basis	Const	R_1	x_2	λ_1	λ_2	μ_1	μ_2
W	3	1	0	1	4	0	-1
x_1	$\dfrac{1}{2}$	$-\dfrac{1}{2}$	0	$\dfrac{1}{2}$	$\dfrac{1}{2}$	$-\dfrac{1}{2}$	0
R_2	3	0	0	1	4	0	-1
S_1	$\dfrac{1}{2}$	$\dfrac{1}{2}$	-1	$-\dfrac{1}{2}$	$-\dfrac{1}{2}$	$\dfrac{1}{2}$	0
S_2	$\dfrac{3}{2}$	$\dfrac{1}{2}$	-4	$-\dfrac{1}{2}$	$-\dfrac{1}{2}$	$-\dfrac{1}{2}$	0

Basis	Const	R_1	x_2	λ_1	λ_2	μ_1	R_2
W	0						1
x_1	$\dfrac{1}{2}$						0
x_2	3	0	0	1	4	0	-1
S_1	$\dfrac{1}{2}$						
S_2	$\dfrac{3}{2}$						

Here $W = 0$ $\therefore$ The phase I is over. Solution is

$$x_1^* = \frac{1}{2}, \ x_2^* = 0 \quad \text{Min } z = z^* = -\frac{1}{4}$$

Exercises

1. Max $z = f(x_1, x_2) = 6x_1 + 3x_2 + x_1x_2 - 0.5x_1^2 - x_2^2 + 10$ subject to $x_1 \leq 15, x_2 \leq 10, x_1, x_2 \geq 24, x_1, x_2 \geq 0$.
   Ans: $z^* = 68.5, x_1^* = 15, x_2^* = 9$

2. Min $z = f(x_1 + x_2) = 100 - 1.2x_1 - 1.5x_2 + 0.03x_1^2 + 0.05x_2^2$ subject to $x_1 + x_2 \geq 15, x_1 - x_2 \geq 6, x_1, x_2 \geq 0$.
   Ans: $x_1^* = 20.5, x_2^* = 14.5, \lambda_1^* = 0.01, \lambda_2^* = 0.04, z^* = 76.77$

3. Max $z = f(x_1 + x_2) = 100x_1 - 0.3x_1^2 + 80x_2 - 0.2x_2^2$ subject to $2x_1 + 4x_2 \leq 1040, x_1, x_2 \geq 0$.
   Ans: $x_1^* = 160, x_2^* = 100, z^* = 16240$

4. Solve the following quadratic programming problem Minimize $z = -x_1^2 - 6x_1 + 2x_2 + 9$ subject $x_1 \geq 3, x_2 \geq 3$.

5. Solve the following quadratic programming problem Max $z = -x_1^2 + 2x_1x_2 - 2x_2^2 + 2x_1 + 5x_2$ subject to $2x_1 + 3x_2 \leq 20, 3x_1 - 5x_2 \leq 4, x_1, x_2 \geq 0$.

6. Consider the problem Max $z = 6x_1 + 3x_2 - 4x_1x_2 - 2x_1^2 - 3x_2^2$ subject to $x_1 + x_2 \leq 1, 2x_1 + 3x_2 \leq 4, x_1, x_2 \geq 0$.
   Show that z is strictly concave and then solve the problem using quadratic programming algorithm.

7. Solve the following QPP Min $f(x_1 + x_2) = 1.2x_1 + 2x_2 + 0.001x_1^2 + 0.005x_2^2 + 100$, subject to $2x_1 + 3x_2 \leq 2500, x_1 + 2x_2 \leq 1500, x_1, x_2 \geq 0$.

   Ans: $x_1^* = \dfrac{200}{3}, \ x_2^* = 0, z^* = 184.44$

8. Solve the following QPP Max $z = 100x_1 - 0.3x_1^2 + 80x_2 - 0.2x_2^2$, subject to $2x_1 + 4x_2 \leq 1040, x_1, x_2 \geq 0$.
   Ans: $x_1^* = 20.5, x_2^* = 14.5, z^* = 76.77, \lambda_1^* = 0.01, \lambda_2^* = 0.04$

9. Solve the following QPP Max $z = -x_1^2 + 3x_1x_2 - 3x_2^2 + 2x_1 + 5x_2$, subject to $x_1 + 4x_2 \leq 7, 3x_1 + x_2 \leq 4, x_1, x_2 \geq 0$.

   Ans: $x_1^* = \dfrac{71}{74}, \ x_2^* = \dfrac{83}{74}, \ z^* = \dfrac{897}{148}$

10. Solve the following QPP Max $z = x_1^2 - 6x_1 + 2x_2 + 9$, subject to $x_1 \geq 3, x_2 \geq 3$.
    Ans: $x_1^* = 3, x_2^* = 3, z^* = 6$

11. Solve the following QPP Max $z = 10x_1^2 + 4x_2 - x_1^2 + 4x_1x_2 - 5x_2^2$, subject to $x_1 + x_2 \leq 6, 4x_1 + x_2 \leq 18, x_1, x_2 \geq 0$.

    Ans: $x_1^* = 0, x_2^* = \dfrac{3}{22}, \ z^* = \dfrac{219}{484}$

12. Solve the following QPP Max $z = 6x_1 + 3x_2 - 4x_1x_2 - 2x_1^2 - 3x_2^2$, subject to $x_1 + x_2 \leq 1, 2x_1 + 3x_2 \leq 4, x_1, x_2 \geq 0$.

    Ans: $x_1^* = 1, x_2^* = 0, z^* = 4$

13. Solve the following QPP Min $f(x_1 + x_2) = 100 - 1.2x_1 - 1.5x_2 + 0.03x_1^2 + 0.05x_2^2$ subject to $x_1 + x_2 \geq 35, x_1 - x_2 \geq 6, x_1, x_2 \geq 0.$

$$\textbf{\textit{Ans:}} \ x_1^* = 20.5, x_2^* = 14.5, \lambda_1^* = 0.01, \lambda_2^* = 0.04 \text{ and } f^* = 76.77.$$

14. Solve the following QPP Max $z = 100x_1 - 0.3x_1^2 + 80x_2 - 0.2x_2^2$, subject to $2x_1 + 4x_2 \leq 1040, x_1, x_2 \geq 0.$

$$\textbf{\textit{Ans:}} \ x_1^* = 160, x_2^* = 180, z^* = 16240$$

15. Solve the following QPP Max $z = -2x_1 - x_2 - x_2^2$ subject to $x_1 + x_2 \leq 1, x_1 + 2x_2 \geq 1, x_1, x_2 \geq 0.$

$$\textbf{\textit{Ans:}} \ x_1^* = 0, x_2^* = \frac{1}{2}.$$

16. Solve the following QPP Min $z = -10x_1 - 25x_2 + 10x_1^2 + x_2^2 + 4x_1x_2$, subject to $x_1 + 2x_2 \leq 10, x_1 + x_2 \leq 9, x_1, x_2 \geq 0.$

$$\textbf{\textit{Ans:}} \ x_1^* = 0, x_2^* = 5, z^* = -100$$

17. Solve the following QPP Max $z = 2x_1 - x_2 - 3x_1^2 - x_2^2$, subject to $-x_1 + x_2 \leq 10, x_1 + 2x_2 \geq 4, x_1, x_2 \geq 0.$

$$\textbf{\textit{Ans:}} \ x_1^* = \frac{9}{13}, x_2^* = \frac{43}{26}$$

18. Use QPP technique to solve the following problem Min $z = -x_1 - 2x_2 + \dfrac{x_1^2}{2} + \dfrac{x_2^2}{2}$, subject to $2x_1 + 3x_2 \leq 6, x_1 + 4x_2 \leq 5, x_1, x_2 \geq 0.$

$$\textbf{\textit{Ans:}} \ x_1^* = \frac{13}{17}, x_2^* = \frac{18}{17}$$

19. Maximize $z = x_1 + x_2 - \dfrac{1}{2}x_1^2 + x_1x_2 - x_2^2$, subject to $x_1 + x_2 \leq 3, 2x_1 + 3x_2 \leq 6, x_1, x_2 \geq 0.$

$$\textbf{\textit{Ans:}} \ x_1^* = \frac{9}{5}, x_2^* = \frac{6}{5}$$

20. Solve the following QPP Minimize $z = x_1^2 - 2x_1x_2 + 2x_2^2 - 2x_1 - 5x_2$, subject to $2x_1 + 3x_2 \leq 20, 3x_1 - 5x_2 \leq 4, x_1, x_2 \geq 0.$

$$\textbf{\textit{Ans:}} \ x_1^* = \frac{9}{2}, x_2^* = \frac{7}{2}$$

21. Solve the following QPP Minimize $z = 2x_1 + 3x_2 - 2x_2^2$, subject to $x_1 + 4x_2 \leq 4, x_1 + x_2 \leq 2, x_1, x_2 \geq 0.$

22. Use QPP technique to solve the following Max $z = x_1^2 + x_2^2$, subject to $3x_1 + 2x_2 \leq 6, x_1, x_2 \geq 0.$

$$\textbf{\textit{Ans:}} \ x_1^* = \frac{9}{2}, x_2^* = \frac{7}{2}$$

23. Solve the following QPP Minimize $z = 4x_1^2 + 3x_2^2$, subject to $x_1 + 3x_2 \geq 5, 2x_2 + 0.5x_1 \geq 2, x_1, x_2 \geq 0.$

24. Solve the following QPP Maximize $z = 2x_1 + 3x_2 - x_1^2$, subject to $x_1 - x_2 \geq 0, -x_1 + 2x_2 \leq 2, x_1, x_2 \geq 0.$

25. Solve the following QPP Maximize $z = -x_1^2 + x_1x_2 - 2x_2^2 + x_1 + x_2$, subject to $x_1 - x_2 \geq 3, x_1 + x_2 = 4, x_1, x_2 \geq 0.$

$$\textbf{\textit{Ans:}} \ x_1^* = \frac{7}{2}, x_2^* = \frac{1}{2}, \lambda_1^* = -4 < 0, \lambda_2^* = -\frac{3}{2} < 0, z^* = -7$$

26. Solve the following QPP Maximize $z = -x_1^2 + x_1 x_2 - 2x_2^2 + x_1 + x_2$, subject to $2x_1 + x_2 \le 1, x_1, x_2 \ge 0$.

Ans: $x_1^* = \dfrac{4}{11}$, $x_2^* = \dfrac{3}{11}$, $z^* = \dfrac{5}{11}$

27. Solve the following QPP Maximize $z = -x_1^2 + 2x_1 x_2 - 2x_2^2 + 2x_1 + 5x_2$, subject to $2x_1 + 3x_2 \le 20$, $3x_1 - 5x_2 \le 4, x_1, x_2 \ge 0$.

Ans: $x_1^* = \dfrac{9}{2}$, $x_2^* = \dfrac{7}{2}$, $z^* = \dfrac{53}{4}$

28. Solve the following QPP
(i) Max $z = -x_1^2 + 3x_1 x_2 - 3x_2^2 + 2x_1 + 5x_2$, subject to $x_1 + 4x_2 \le 7, 3x_1 + x_2 = 4, x_1, x_2 \ge 0$.

Ans: $x_1^* = \dfrac{71}{74}$, $x_2^* = \dfrac{83}{74}$, $z^* = \dfrac{897}{148}$

(ii) Max $z = x_1^2 - 2x_1 - x_2$, subject to $x_1 + 5x_2 \le 12, x_1, x_2 \ge 0$
Ans: $x_1^* = 12, x_2^* = 0, z^* = 120$
(iii) Max $z = -3x_1^2 + 2x_1 x_2 - 2x_2^2 + 2x_1 + 3x_2$, subject to $x_1 + x_2 \ge 1, 3x_1 + 4x_2 \ge 12, x_1, x_2 \ge 0$.

Ans: $x_1^* = \dfrac{7}{10}$, $x_2^* = \dfrac{11}{10}$, $z^* = \dfrac{47}{20}$

29. Solve the following QPP
(i) Minimize $f(x_1 + x_2) = x_1^2 - x_1 x_2 + 3x_2^2 - 4x_2 + 4$, subject to $x_1 + x_2 \le 1, x_1, x_2 \ge 0$.
(ii) Minimize $f(x_1 + x_2) = 2x_2^2 + 3x_1^2 - 3x_1 x_2 - 25(x_1 + x_2)$ subject to $2x_1 + x_2 \le 5, x_1, x_2 \ge 0$.
Ans: $x_1^* = 12, x_2^* = 0, z^* = 120$
30. Solve the following QPP Min $z = 10x_1 - 25x_2 + 10x_1^2 + 4x_1 x_2 + x_2^2$, subject to $x_1 + 2x_2 \le 10, x_1 + x_2 \le 9, x_1, x_2 \ge 0$

Ans: $x_1^* = 0, x_2^* = 5, z^* = -100$

31. Solve the following QPP Min $z = \dfrac{1}{2} x_1^2 - 2x_2^2$, subject to $3x_1 + 4x_2 \ge 13, x_1, x_2 \ge 0$.

Ans: $x_1^* = 3, x_2^* = 1, z^* = \dfrac{13}{2}$

32. Solve the following QPP Min $z = -x_1 - 2x_2 + \dfrac{1}{2} x_1^2 + \dfrac{1}{2} x_1^2$ subject to $x_1 + 4x_2 \le 5, x_1, x_2 \ge 0$.

Ans: $x_1^* = \dfrac{13}{17}$, $x_2^* = \dfrac{18}{17}$

33. Solve the following QPP Min $z = x_1^2 - x_1 x_2 + 2x_2^2 - x_1 - x_2$ subject to $2x_1 + x_2 \le 1, x_1, x_2 \ge 0$.

Ans: $x_1^* = \dfrac{4}{11}$, $x_2^* = \dfrac{3}{11}$, $z^* = -\dfrac{5}{11}$

CHAPTER 5

GEOMETRIC PROGRAMMING

5.1 INTRODUCTION

This technique was developed by R. DUFFIN and C. ZENER in 1964. It finds the solution by considering an associated dual problem. This dual problem is much simpler computationally.

Geometric programming deals with the problems in which the objective and constraint functions are of the following type.

$$Z = f(X) = \sum_{j=1}^{N} U_j$$

where $\quad U_j = c_j \prod_{i=1}^{n} x_i^{a_{ij}}, \; j = 1, 2, \ldots N$

It is assumed that all a_{ij} are un restricted. The function f(X) takes the form of a polynomial except that the exponents a_{ij} may be negative. For this reason, and because all $c_i > 0$, DUFFIN and ZENER give the function f(X) the name POSYNOMIAL.

We deal with unconstrained problems. Consider the minimization of the function f(X) as defined in the posynomial form given. This problem is referred to as the PRIMAL. The variables xi are assumed to be STRICTLY POSITIVE so that the region xi = 0 represents an infeasible space ($x_i \neq 0$ plays an important part). Let

$$Z = f(X) = \sum_{j=1}^{N} U_j = \sum_{j=1}^{N} c_j \prod_{i=1}^{n} x_i^{a_{ij}}$$

Taking logarithms on both sides, we see

$$\frac{\partial Z}{\partial x_*} = \sum_{j=1}^{N} \frac{\partial U_i}{\partial x} \qquad \qquad ...(1)$$

$$= \sum_{j=1}^{N} \frac{\partial U_i}{\partial x}$$

$$= \sum_{j=1}^{N} c_j a_{ij} \left(x_k\right)^{a_{kj}-1} \pi_{i \neq k} \left(x_i\right)^{a_{ij}} = 0$$

for k = 1, 2, ... n

$$= \frac{1}{x_k} \sum_{j=1}^{N} a_{ij} c_j (x_k)^{akj} \pi_{i \neq k} (x_i)^{a_{kj}} = 0$$

$$= \frac{1}{x_k} \sum_{j=1}^{N} c_j \pi_{i=k}^{n} (x_i)^{a_{kj}} = 0$$

$$= \frac{1}{x_k} \sum_{j=1}^{N} c_j \pi_{i=k}^{n} (x_i)^{a_{kj}} = 0$$

$$= \frac{1}{x_k} \sum_{j=1}^{N} a_{kj} U_j = 0$$

$$= \sum_{j=1}^{N} a_{kj} U_j = 0 \quad k = 1, 2, ... n \qquad ...(2)$$

Now $\qquad U_j = c_j \pi_{i=1}^{n} x_i^{a_{ij}}$

$$\therefore \qquad \log U_j = \log c_j + \sum_{i=1}^{n} a_{ij} . \log x_i$$

Differentiate both sides with respect to xk

$$\frac{1}{U_i} . \frac{\partial U_j}{\partial x_k} = 0 + \frac{a_{kj}}{x_k}, \; j = 1, 2, ... N$$

∴ (1) becomes

$$\frac{\partial Z}{\partial x_k} = \sum_{j=1}^{N} \frac{\partial U_j}{\partial x_k} = \sum_{j=1}^{N} U_j \cdot \frac{1}{x_k} \cdot a_{kj}$$

$$= \frac{1}{x_k} \sum_{j=1}^{N} a_{kj} U_j, \quad x = 1, 2, \ldots n$$

$$\therefore \qquad \frac{\partial Z}{\partial x_k} = 0 \text{ for } k = 1, 2, \ldots n \qquad \qquad \ldots(3)$$

$Z^* = $ Minimum of Z

$Z^* > 0$ since Z is a posynomial and each $x_k^* > 0$. Define

$$y_j = \frac{U_j^*}{Z^*} > 0 \text{ and } \sum_{j=1}^{N} y_j = 1$$

The value of y_j represents the relative contribution of the jth term U_j to the optimal value of the objective function Z^*.

The NECESSARY conditions can now be written as

$$\left.\begin{cases} \displaystyle\sum_{j=1}^{N} a_{kj} y_j = 0, \ k = 1, 2, \ldots n \\ \displaystyle\sum_{j=1}^{N} y_j = 1, \ y_j > 0 \text{ for all } j \end{cases}\right\} \qquad \ldots(4)$$

These are (n + 1) constraints as (n + 1) equations these are known as ORTHOGONALITY and NORMALITY CONDITIONS and will yield an UNIQUE SOLUTION for y_j, n + 1 = N and all the (n + 1) equations are independent.

The problems becomes more complex when N > (n + 1) because the value of y_j are no longer UNIQUE.

$\sum y_j$ can be determined UNIQUELY for minimizing Z.

We shall now determine Z^* and x_i^*, $i = 1, 2, \ldots n$

Consider $\quad Z^* = \displaystyle\sum_{i=1}^{n} y_i^*$

$$\left(Z^*\right)$$

$$= \left(\frac{U_1^*}{y_1^*}\right)^{y_1^*} \left(\frac{U_2^*}{y_2^*}\right)^{y_2^*} \cdots \left(\frac{U_N^*}{y_N^*}\right)^{y_N^*}$$

$$= \left\{ \prod_{j=1}^{N} \left(\frac{c_j}{y_j} \right)^{y_i^*} \right\} \left\{ \prod_{j=1}^{N} \left(\prod_{i=1}^{N} x_i^{*a_{ij}} \right)^{y_i^*} \right\}$$

$$= \left\{ \prod_{j=1}^{N} \left(\frac{c_j}{y_j^*} \right)^{y_i^*} \right\} \left\{ \prod_{i=1}^{N} \left(x_i^* \right)^{\sum_{i=1}^{N} a_{ij} y_i^*} \right\} = \prod \left(\frac{c_j}{y_j^*} \right)^{y_i^*} \quad \text{...(4)}$$

The above is justified, since, $= 1$

$\sum a_{ij} y_j^* = 0$ (exponent of x_i^*) in (4).

The value of Z^* is thus terms determined as soon as all y_j^* and Z^*, $U_j^* = y_j^* Z^*$ can be determined.

Since $\qquad U_j^* = c_j \prod_{j=1}^{N} (x_i^*)^{a_{ij}}, \; j = 1, 2, \ldots N$ \qquad ...(5)

are simultaneous solution of these equations, they should yield x_i^*.

Solution is thus transform into the solution of a set of linear equations in y_j. It can be shown that these form as the necessary conditions (for minimization) and are also sufficient conditions. (Please see BEIGHTLER et al 1979, page 33).

The variables y_j actually define the dual variables associated with the Z-primal problem. To see this relationship, consider the primal problem in the form:

$$Z = \sum_{j=1}^{N} y_j \left(\frac{U_j}{y_j} \right)$$

Defines the function

$$W = \prod_{j=1}^{N} \left(\frac{U_j}{y_j} \right)^{y_j}$$

$$= \prod_{j=1}^{N} \left(\frac{c_j}{y_j} \right)^{y_j}$$

Since $\sum_{j=1}^{N} y_j = 1$ and $y_j > 0$, by **Cauchy Inequality**, we have [Cauchy's inequality

states that for $Z_i > 0$ $\sum_{j=1}^{N} w_j z_j \geq \prod_{j=1}^{N} (Z_j)^{W_j}$, where $w_j > 0$, and $\sum_{j=1}^{N} W_j = 1$. This is called

Arithmetic Geometric Mean Inequality]

$\therefore W \leq Z$

The function W with its variables $y_1, y_2 \ldots y_N$ defines the dual problem to the primal. Since W represents the Lower Bound on Z and since Z is associated with the minimization problems, it follows, by maximizing W, that

$$W^* = \max_{y_j} W = \min_{x_i} = Z^*$$

∴ **Maximum value of W (=W*) over the values of y_j. is equal to the minimum value of Z = (Z*) over the values of x_i.**

5.2 EXAMPLE

EXAM. 1: For the case when $N = n + 1$, consider the problem

Min $Z = 7x_1x_2^{-1} + 3x_2x_3^{-2} + 5x_1^{-3}x_2x_3 + x_1x_2x_3$

This function is written by

$Z = 7x_1^{1}\,x_2^{-1}\,x_3^{0} + 3x_1^{0}x_2^{1}x_3^{-2} + 5x_1^{-3}x_2^{1}x_3^{1} + x_1^{1}x_2^{1}x_3^{1}$

so that $(c_1, c_2, c_3, c_4) = (7, 3, 5, 1)$ and

$$\begin{bmatrix} a_{11} & a_{12} & a_{13} & a_{14} \\ a_{21} & a_{22} & a_{23} & a_{24} \\ a_{31} & a_{32} & a_{33} & a_{34} \end{bmatrix} = \begin{bmatrix} 1 & 0 & -3 & 1 \\ -1 & 1 & 1 & 1 \\ 0 & -2 & 1 & 1 \end{bmatrix}$$

The orthogonality and normality conditions are given by

$$\begin{bmatrix} 1 & 0 & -3 & 1 \\ -1 & 1 & 1 & 1 \\ 0 & -2 & 1 & 1 \\ 1 & 1 & 1 & 1 \end{bmatrix}\begin{bmatrix} y_1 \\ y_2 \\ y_3 \\ y_4 \end{bmatrix} = \begin{bmatrix} 0 \\ 0 \\ 0 \\ 1 \end{bmatrix}$$

This yields the unique solution

$$y_1^* = \frac{12}{24}, y_2^* = \frac{4}{24}, y_3^* = \frac{5}{24}, y_4^* = \frac{3}{24} \text{ as follows}$$

$$\begin{bmatrix} 1 & 0 & -3 & 1 \\ -1 & 1 & 1 & 1 \\ 0 & -2 & 1 & 1 \\ 1 & 1 & 1 & 1 \end{bmatrix} \begin{bmatrix} 0 \\ 0 \\ 0 \\ 1 \end{bmatrix} \begin{matrix} \\ r_2 + r_1 \\ r_4 - r_1 \\ \end{matrix} = \begin{bmatrix} 1 & 0 & -3 & 1 \\ 0 & 1 & -2 & 2 \\ 0 & -2 & 1 & 1 \\ 0 & 1 & 4 & 0 \end{bmatrix} \begin{bmatrix} 0 \\ 0 \\ 0 \\ 1 \end{bmatrix} \begin{matrix} \\ r_3 + 2r_2 \\ r_4 - r_2 \\ \end{matrix}$$

$$\begin{bmatrix} 1 & 0 & -3 & 1 \\ 0 & 1 & -2 & 2 \\ 0 & 0 & -3 & 5 \\ 0 & 0 & 6 & -2 \end{bmatrix} \begin{bmatrix} 0 \\ 0 \\ 0 \\ 1 \end{bmatrix} \begin{matrix} \\ \\ r_4 + 2r_3 \\ \end{matrix} \begin{bmatrix} 1 & 0 & -3 & 1 \\ 0 & 1 & -2 & 2 \\ 0 & 0 & -3 & 5 \\ 0 & 0 & 0 & 8 \end{bmatrix} \begin{bmatrix} 0 \\ 0 \\ 0 \\ 1 \end{bmatrix}$$

$$y_4^* = \frac{1}{8} = \frac{3}{24}, -3y_3^* + 5y_4^* = 0, \; 3y_3^* = 5y_4^* = \frac{5}{24}y_3^*$$

$$y_2^* - \frac{10}{24} + \frac{6}{24} = 0 \Rightarrow y_2^* = \frac{4}{24}, \; y_2^* = 1 - \frac{4}{24} - \frac{5}{24} - \frac{3}{24} = \frac{12}{24}$$

$$\therefore \qquad Z^* = \left(\frac{e_1}{y_1^*}\right)^{y_1}\left(\frac{e_2}{y_2^*}\right)^{y_2}\left(\frac{e_3}{y_3^*}\right)^{y_3}\left(\frac{e_4}{y_4^*}\right)^{y_4}$$

$$= \left(\frac{7}{12/24}\right)^{12/24}\left(\frac{3}{4/24}\right)^{4/24}\left(\frac{5}{5/24}\right)^{5/24}\left(\frac{1}{3/24}\right)^{3/24}$$

$$= \left(\frac{7\times24}{12}\right)^{1/2}\left(\frac{3\times24}{4}\right)^{1/6}\left(\frac{5\times24}{5}\right)^{5/24}\left(\frac{24}{3}\right)^{1/8}$$

$$= (14)^{1/2}(18)^{1/6}(24)^{5/24}(8)^{1/8}$$

$$Z^* = 15.22$$

From the equation $U_j^* = y_j^*.z^*$ if follows that

$$7x_1x_2^{-1} = U_1 = \frac{1}{2}\times15.22 = 7.61 \qquad\qquad \text{I}$$

$$3x_2x_3^{-2} = U_2 = \frac{1}{6}\times15.22 = 2.54 \qquad\qquad \text{II}$$

$$5x_1^{-3}x_2x_3 = U_3 = \frac{5}{24}\times15.22 = 3.17 \qquad\qquad \text{III}$$

$$x_1x_2x_3 = U_4 = \frac{1}{8}\times15.22 = 1.90 \qquad\qquad \text{IV}$$

$$\frac{x_1x_2x_3}{5x_1^{-3}.x_2x_3} = \frac{u_4}{u_3} = \frac{1.90}{3.17} \therefore x_1^4 = \frac{9.5}{3.17} \qquad\qquad \text{V}$$

$$4\log_{10}x_1^* = \log_{10}9.5 - \log_{10}3.17 \Rightarrow x_1^* = 1.315 \qquad\qquad \text{VI}$$

$$7x_1x_2^{-1} = 7.61 \Rightarrow \frac{7\times1.315}{x_2} = 7.61, \; x_2 = \frac{7\times1.315}{7.61} \qquad\qquad \text{VII}$$

$$\frac{3(1.21)}{x_3^2} = 2.54 \; x_3^* = \frac{3\times1.21}{2.54} = 1.24 \, or \, x_3^* = 1.20 \qquad\qquad \text{VIII}$$

$$\therefore \qquad x_1^* = 1.315, x_2^* = 1.21, x_3^* = 1.2 \qquad\qquad\text{IX}$$

IX is the optimal solution of the PRIMAL.

EXAM. 2: Min $Z = 5x_1 + 20x_2 + 10x_1^{-1}x_2^{-1}$, $x_1, x_2 > 0$

$$= 5x_1^1 x_2^0 + 20x_1^0 x_2^1 + 10x_1^{-1}x_2^{-1}$$

$(c_1, c_2, c_3, c_4) = (5, 20, 10)$

$$\begin{pmatrix} a_{11} & a_{12} & a_{13} \\ a_{21} & a_{22} & a_{23} \end{pmatrix} = \begin{pmatrix} 1 & 0 & -1 \\ 0 & 1 & -1 \end{pmatrix} \begin{pmatrix} 1 & 0 & -1 \\ 0 & 1 & -1 \\ 1 & 1 & 1 \end{pmatrix} \begin{pmatrix} y_1 \\ y_2 \\ y_3 \end{pmatrix} = \begin{pmatrix} 0 \\ 0 \\ 1 \end{pmatrix}$$

$$\begin{bmatrix} 1 & 0 & -1 \\ 0 & 1 & -1 \\ 1 & 1 & 1 \end{bmatrix}\begin{bmatrix} 0 \\ 0 \\ 1 \end{bmatrix} \underset{r_3-r_1}{} \begin{bmatrix} 1 & 0 & -1 \\ 0 & 1 & -1 \\ 1 & 1 & 1 \end{bmatrix}\begin{bmatrix} 0 \\ 0 \\ 1 \end{bmatrix} \underset{r_3-r_2}{} \begin{bmatrix} 0 & 0 & -1 \\ 0 & 1 & -1 \\ 0 & 0 & 3 \end{bmatrix}\begin{bmatrix} 0 \\ 0 \\ 1 \end{bmatrix}$$

$$\Rightarrow 3y_3^* = 1, y_3^* = \frac{1}{3}, y_2^* - y_3^* = 0, y_2^* = y_3^* = \frac{1}{3}, y_1^* - y_3^* = 0$$

or $\qquad y_1^* = y_3^* = \frac{1}{3} \quad \therefore y_1^* = y_3^* = y_3^* = \frac{1}{3}$ ...(1)

$$\therefore \qquad Z^* = \left(\frac{5}{\frac{1}{3}}\right)^{\frac{1}{3}}\left(\frac{20}{\frac{1}{3}}\right)^{\frac{1}{3}}\left(\frac{10}{\frac{1}{3}}\right)^{\frac{1}{3}}$$

$$= [(15)(60)(30)]^{\frac{1}{3}} = (15^3 \times 4 \times 2)^{\frac{1}{3}}$$

$$Z^* = 15.(2^3)^{\frac{1}{3}} = 15 \times 2 = 30 \qquad\qquad\text{...(2)}$$

From the equation $u_j^* = y_j^* . z^*$ we have

$$5x_1 = U_1 = \frac{1}{3} \times 30 = 10 \text{ or } x_1^* = 2$$

$$20x_2 = U_2 = \frac{1}{3} \times 30 = 10 \text{ or } x_2^* = \frac{1}{2}$$

$$x_1^* = 2 \text{ and } x_2^* = \frac{1}{2} \qquad\qquad\text{...(3)}$$

EXAM. 3: Min $Z = 4x_1^2 x_2^{-3} + 5x_1^{-3}x_2 + 7x_1 x_2$, $x_1 x_2 > 0$

$(c_1, c_2, c_3) = (4, 5, 7)$

$$\begin{bmatrix} a_{11} & a_{12} & a_{13} \\ a_{21} & a_{22} & a_{23} \\ y_1 & y_2 & y_3 \end{bmatrix}\begin{bmatrix} 0 \\ 0 \\ 1 \end{bmatrix}\begin{bmatrix} y_1 \\ y_2 \\ y_3 \end{bmatrix} \approx \begin{bmatrix} 2 & -3 & 1 \\ -3 & 1 & 1 \\ 1 & 1 & 1 \end{bmatrix}\begin{bmatrix} 0 \\ 0 \\ 1 \end{bmatrix}\begin{matrix} r_2 + \dfrac{3}{2}r_1 \\ \\ r_3 - \dfrac{1}{2}r \end{matrix}$$

$$\approx \begin{bmatrix} 2 & -3 & 1 \\ 0 & -3.5 & 2.5 \\ 0 & 2.5 & .5 \end{bmatrix}\begin{bmatrix} 0 \\ 0 \\ 1 \end{bmatrix} \approx \begin{bmatrix} 2 & -3 & 1 \\ 0 & 7 & 5 \\ 0 & 5 & 1 \end{bmatrix}\begin{bmatrix} 0 \\ 0 \\ 2 \end{bmatrix} r_3 + \dfrac{5}{7}r_2 \approx \begin{bmatrix} 2 & -3 & 1 \\ 0 & -7 & 5 \\ 0 & 0 & \dfrac{32}{7} \end{bmatrix}\begin{bmatrix} 0 \\ 0 \\ 2 \end{bmatrix}$$

$$\therefore \qquad \frac{32}{7}y_3^* = 2 \ \therefore y_3^* = \frac{7}{16}$$

$$-7y_2^* + 5y_3^* = 0,\ 7y_2^* = 5y_3^* = \frac{5 \times 7}{16}y_2^* = \frac{5}{16}$$

$$2y_1^* - \frac{5}{16} + \frac{7}{16} - 0,\ 2y_1^* = \frac{1}{2},\ y_1^* = \frac{1}{4}$$

$$y_3^* = \frac{7}{16},\ y_2^* = \frac{5}{16},\ y_1^* = \frac{1}{4}$$

$$y_1^* = \frac{4}{16},\ y_2^* = \frac{5}{16},\ y_3^* = \frac{7}{16}$$

$$\therefore \qquad Z^* = \left(\frac{4}{4/16}\right)^{4/16}\left(\frac{5}{5/16}\right)^{5/16}\left(\frac{7}{7/16}\right)^{7/16}$$

$$= (16)^{\frac{1}{4}}(16)^{\frac{5}{16}}(16)^{\frac{7}{16}} = (16)^{\frac{1}{4}+\frac{5}{16}+\frac{7}{16}}$$

$$(16)^{\frac{4+5+7}{16}} = (16)^1 = 16$$

From $u_j^* = y_j^*.z^*$ we have

$$4x_1^2 x_2^{-3} = u_1 = y_1^* z^* = \frac{1}{4} \times 16 = 4$$

$$5x_1^{-3} x_2^2 = u_2 = y_2^* z^* = 5$$

$$7x_1 x_2 = u_3 = y_3^* z^* = \frac{7}{16} \times 16 = 7$$

$$\therefore \quad x_1^2 x_2^{-3} = 1 \frac{x_1^2}{x_2^3} = 1 = \frac{x_2}{x_1^3} = x_1 x_2$$

$$x_1^{-3} x_2 = 1 \; x_2 = x_1^3 = x_2 = \frac{1}{x_1} \therefore x_1^4 = 1 \qquad \text{...(2)}$$

$$x_1 x_2 = 1$$

or $\qquad x_1^* = 1 \; x_2^* = 1 \qquad \text{...(3)}$

and

$$x_1^* = x_2^* = 1 \; Z^* = 4 + 5 + 7 = 16 \qquad \bullet \qquad \text{...(4)}$$

EXAM. 4: Min $Z = 5x_1^2 x_2^{-1} x_3 + x_1^{-1} x3^{-2} + 10x_2^3 + 2x_1^{-3}.x_2.x_3^{-1} \; x_1, x_2, x_3 > 0$

Min $Z = 5x_1 2x_2^{-1} x_3^0 + 1x_1^{-1} x_2^0 x_3^{-2} + 10x_1^0 x_2^3 x_3^0 + 2x_1^{-3}.x_2^1.x_3^{-1}$

$(c_1, c_2, c_3, c_4) = (5, 1, 10, 2)$

$$\begin{bmatrix} a_{11} & a_{12} & a_{13} & a_{14} \\ a_{21} & a_{22} & a_{23} & a_{24} \\ a_{31} & a_{32} & a_{33} & a_{34} \end{bmatrix} = \begin{bmatrix} 2 & -1 & 0 & -3 \\ -1 & 0 & 3 & 1 \\ 1 & -2 & 0 & -1 \end{bmatrix}$$

$$\begin{bmatrix} 2 & -1 & 0 & -3 \\ -1 & 0 & 3 & 1 \\ 1 & -2 & 0 & -1 \\ 1 & 1 & 1 & 1 \end{bmatrix} \begin{bmatrix} y_1 \\ y_2 \\ y_3 \\ y_4 \end{bmatrix} = \begin{bmatrix} 0 \\ 0 \\ 0 \\ 1 \end{bmatrix}$$

$$\approx \begin{bmatrix} 2 & -1 & 0 & -3 \\ -1 & 0 & 3 & 1 \\ 1 & -2 & 0 & -1 \\ 1 & 1 & 1 & 1 \end{bmatrix} \begin{bmatrix} 0 \\ 0 \\ 0 \\ 1 \end{bmatrix} \begin{matrix} \\ r_2 + \frac{1}{2}r_1 \\ r_3 - \frac{1}{2}r_1 \\ r_4 - \frac{1}{2}r_1 \end{matrix} \approx \begin{bmatrix} 2 & -1 & 0 & -3 \\ 0 & -1/2 & 3 & -1/2 \\ 0 & -3/2 & 0 & 1/2 \\ 0 & 3/2 & 1 & 5/2 \end{bmatrix} \begin{bmatrix} 0 \\ 0 \\ 0 \\ 1 \end{bmatrix}$$

$$\approx \begin{bmatrix} 2 & -1 & 0 & -3 \\ 0 & -1 & 6 & -1 \\ 0 & -3 & 0 & 1 \\ 0 & 3 & 2 & 5 \end{bmatrix} \begin{bmatrix} 0 \\ 0 \\ 0 \\ 2 \end{bmatrix} \approx \begin{bmatrix} 2 & -1 & 0 & -3 \\ 0 & -1 & 6 & -1 \\ 0 & 0 & -18 & 4 \\ 0 & 0 & 2 & 6 \end{bmatrix} \begin{bmatrix} 0 \\ 0 \\ 0 \\ 1 \end{bmatrix} \begin{matrix} \\ \\ \\ r_4 + \frac{1}{9}.r_3 \end{matrix}$$

$$= \begin{bmatrix} 2 & -1 & 0 & -3 \\ 0 & -1 & 6 & -1 \\ 0 & -3 & 0 & 1 \\ 0 & 3 & 2 & 5 \end{bmatrix} \begin{bmatrix} 0 \\ 0 \\ 0 \\ 2 \end{bmatrix} \begin{matrix} \\ \\ r_3 - 3r_2 \\ r_4 + r_3 \end{matrix} \approx \begin{bmatrix} 2 & -1 & 0 & -3 \\ 0 & -1 & 6 & -1 \\ 0 & 0 & -18 & 4 \\ 0 & 0 & 2 & 6 \end{bmatrix} \begin{bmatrix} 0 \\ 0 \\ 0 \\ 1 \end{bmatrix} \begin{matrix} \\ \\ \\ r_4 + \frac{1}{9}.r_3 \end{matrix}$$

$$= \begin{bmatrix} 2 & -1 & 0 & -3 \\ 0 & -1 & 6 & -1 \\ 0 & -3 & 0 & 1 \\ 0 & 3 & 2 & 5 \end{bmatrix} \begin{bmatrix} 0 \\ 0 \\ 0 \\ 2 \end{bmatrix} \begin{matrix} \\ \\ r_3 - 3r_2 \\ r_4 + r_3 \end{matrix} \approx \begin{bmatrix} 2 & -1 & 0 & -3 \\ 0 & -1 & 6 & -1 \\ 0 & 0 & -18 & 4 \\ 0 & 0 & 2 & 6 \end{bmatrix} \begin{bmatrix} 0 \\ 0 \\ 0 \\ 1 \end{bmatrix} r_4 + \frac{1}{9} \cdot r_3$$

$$= \begin{bmatrix} 2 & -1 & 0 & -3 \\ 0 & -1 & 6 & -1 \\ 0 & 0 & -18 & 4 \\ 0 & 0 & 2 & \dfrac{58}{9} \end{bmatrix} \begin{bmatrix} 0 \\ 0 \\ 0 \\ 2 \end{bmatrix} \begin{bmatrix} y_1 \\ y_2 \\ y_3 \\ y_4 \end{bmatrix}$$

$$y_4^* = \frac{18}{58} = \frac{9}{29}, 18y_3^* = 4 \times \frac{9}{29} = \frac{36}{29} y_3^* = \frac{2}{29}$$

$$-y_2^* + 6y_3^* - y_4^* = 0$$

$$y_2^* = 6y_3^* - y_4^* = \frac{12 - 9}{29} = \frac{3}{29}$$

$$2y_1^* = y_2^* + 3y_4^* = \frac{3}{29} + \frac{27}{29}$$

$$\therefore \quad y_1^* = \frac{15}{29}$$

$$y_1^* = \frac{15}{29}, y_2^* = \frac{3}{29}, y_3^* = \frac{2}{29}, y_4^* = \frac{9}{29}$$

$$\sum_{j=1}^{4} y_j^* = 1 \text{ is verified.}$$

$$Z^* = \left(\frac{4}{15/29} \right)^{15/29} \left(\frac{1}{3/29} \right)^{3/29} \left(\frac{10}{2/29} \right)^{2/29} \left(\frac{2}{9/29} \right)^{9/29}$$

$$= \left(\frac{116}{15} \right)^{15/29} \left(\frac{29}{3} \right)^{3/29} \left(\frac{290}{2} \right)^{2/29} \left(\frac{58}{9} \right)^{9/29}$$

$$= \left(\frac{116}{15} \right)^{15/29} \left(\frac{29}{3} \right)^{3/29} (145)^{2/29} \left(\frac{58}{9} \right)^{9/29}$$

etc.

EXAM. 5: What is to be transported from one side of a river to the other side in an open rectangular box of length x1, width x2, and height x3 all in meters. The four sides cost Rs. 20 per square meter and the bottom cost is Rs. 80 per square meter. The transportation cost per round trip is Rs. 2. The box will have no salvage value. Find the minimum total of transporting 160 cubic meters of wheat.

$$\text{Min } Z = 80x_1x_2 + 20\left[x_1x_3 + 2x_2x_3\right] + \frac{2 \times 160}{x_1x_2x_3}$$

$$= 80x_1^1x_2^1 + 40x_1^1x_3^1 + 40x_2^1x_3^1 + 320x_1^{-1}x_2^{-1}x_3^{-1}$$

$$(c_1, c_2, c_3, c_4) = (80, 40, 40, 320)$$

$$\begin{bmatrix} 1 & 1 & 0 & -1 \\ 1 & 0 & 1 & -1 \\ 0 & 1 & 1 & -1 \\ 1 & 1 & 1 & -1 \end{bmatrix} \begin{bmatrix} 0 \\ 0 \\ 0 \\ 1 \end{bmatrix} \begin{bmatrix} y_1 \\ y_2 \\ y_3 \\ y_4 \end{bmatrix}$$

SOLUTION

$$y_1^* = \frac{1}{5}, y_2^* = \frac{1}{5}, y_3^* = \frac{1}{5}, y_4^* = \frac{2}{5}, z^* = 400$$

$$x_1^* = x_2^* = 1, x_3^* = 2 \text{ Min } Z^* = 400$$

EXAM. 6: A cylindrical water tank of capacity 900π cubic meters is to be constructed. Each square meter of bottom, top and curved side costs respectively Rs. 17, Rs. 10, and Rs. 30 per square meter. Find the dimensions of the tank so that the construction cost is minimized.

Bottom area = πx_1^2

Surface area = $2\pi x_1 x_2$

Top area = πx_1^2

Volume of the tank = $\pi x_1^2 x_2 = 400\pi$

Min $Z = 17\pi x_1^2 + 10\pi x_1^2 + 60\pi x_1 x_2$

$= 27\pi x_1^2 + 60\pi x_1 x_2$

subject to $p x_1^2 x_2 \leq 900\pi$

or $x_1^2 x_2 \leq 900$

The dual of the above problem is obtained as

$$4(\lambda) = \left(\frac{27\pi x_1^2}{\lambda_1}\right)^{\lambda_1} \left(\frac{60\pi x_1 x_2}{\lambda_2}\right)^{\lambda_2} \left(\frac{x_1^2}{900\lambda_3}\right)^{\lambda_3}$$

This is called the predual. The dual problem is

$$\text{Max } \phi(\lambda) = \left(\frac{27\pi}{\lambda_1}\right)^{\lambda_1} \left(\frac{60\pi}{\lambda_2}\right)^{\lambda_2} \left(\frac{1}{900\lambda_3}\right)^{\frac{1}{\lambda_3}},$$

$$2\lambda_1 + \lambda_2 + 2\lambda_3 = 0 \quad ...(1)$$

$$\lambda_1 + \lambda_2 + \lambda_3 = 0 \quad \lambda_3 = -\lambda_2 \quad ...(2)$$

$$\lambda_1 + \lambda_2 = 1 ...(3)$$

$$\lambda_1, \lambda_2, \lambda_3 = 0$$

$$\therefore \text{ becomes } 2\lambda_1 + \lambda_2 + 2\lambda_3 = 0$$

$$\text{or } 2\lambda_1 + \lambda_2 - 2\lambda_2 = 0$$

$$2\lambda_1 - \lambda_2 = 0$$

$$\lambda_1 + \lambda_2 = 1 \text{ add}$$

$$3\lambda_1 = 1$$

$$\text{or} \qquad \lambda_1 = \frac{1}{3}$$

$$\lambda_2 = \frac{2}{3}, \lambda_3 = -\lambda_2 = -\frac{2}{3}$$

$$\lambda_1 = \frac{1}{3}, \lambda_2 = \frac{2}{3}, \lambda_3 = -\frac{2}{3}$$

$$\phi(\lambda) = \left(\frac{27\pi}{\lambda_1}\right)^{\lambda_1} \left(\frac{60\pi}{\lambda_2}\right)^{\lambda_2} \left(\frac{1}{900\lambda_3}\right)^{\frac{1}{\lambda_3}} (\lambda_3)^{\lambda_3}$$

$$= (81\pi)^{\frac{1}{3}} (90\pi)^{\frac{2}{3}} \left(-\frac{1}{600}\right)^{\frac{-2}{3}} \left(\frac{-2}{3}\right)^{\frac{-2}{3}}$$

$$= (81\pi)^{\frac{1}{3}} (90\pi)^{\frac{2}{3}} \left(\frac{1}{600} \times \frac{2}{3}\right)^{\frac{-2}{3}}$$

$$= 3^{4 \times \frac{1}{3}} 3^{2 \times \frac{2}{3}} .10^{\frac{2}{3}} \pi \times (900)^{\frac{2}{3}}$$

$$= \pi.3^{\frac{4}{3}} .3^{\frac{4}{3}} .10^{\frac{2}{3}} .3^{2 \times \frac{2}{3}} .10^{2 \times \frac{2}{3}}$$

$$= \pi 3^{\frac{10}{3}} .10^{\frac{6}{3}}$$

$$= \pi 3^{\frac{10}{3}} .10^2$$

$$= \phi(\lambda^*)$$

$$\therefore \qquad f(\lambda^*) = 100\pi 3^{\frac{10}{3}} = 100\pi.3^3.3^{\frac{1}{3}}$$

$$= 2700\pi\sqrt[3]{3}$$

EXAM. 7: Min $Z = 5x_1x_2^{-1} + 2x_1^{-1}x_2 + 5x^1 + x2^{-1}$

The orthogonality and normality conditions are given by

$$
\begin{bmatrix}
1 & -1 & 1 & 0 \\
-1 & 1 & 0 & -1 \\
1 & 1 & 1 & 1
\end{bmatrix}
\begin{bmatrix}
y_1 \\ y_2 \\ y_3 \\ y_4
\end{bmatrix}
=
\begin{bmatrix}
0 \\ 0 \\ 1
\end{bmatrix}
$$

Because $N > n + 1$, then equations do not yield y_j directly. Solving y_1, y_2 and y_3 in terms of y_4 we see

$$
\begin{bmatrix}
1 & -1 & 1 \\
-1 & 1 & 0 \\
1 & 1 & 1
\end{bmatrix}
\begin{bmatrix}
y_1 \\ y_2 \\ y_3
\end{bmatrix}
=
\begin{bmatrix}
0 \\ y_4 \\ 1 - y_4
\end{bmatrix}
$$

$y_1 - y_2 + y_3 = 0$...(1)

$-y_1 + y_2 = y_4$...(2)

$y_1 + y_2 + y_3 = 1 - y_4$ ...(3)

$(1) + (2) \Rightarrow y_3 = y_4$

$(3) \Rightarrow y_1 + y_2 = 1 - 2 y_4$

$-y_1 + y_2 = y_4$

$2y_2 = 1 - y_4$

$$y_2 = \frac{1}{2}(1 - y_4)$$

$\therefore \qquad y_1 = y_2 - y_4 = \frac{1}{2}(1 - y_4) - 4y$

$$= \frac{1}{2}[1 - y_4 - 2y_4]$$

$$= \frac{1}{2}[1 - 3y_4]$$

$\therefore \qquad y_1 = \frac{1}{2}[1 - 3y_4], y_2 = \frac{1}{2}[1 - y_4], y_3 = y_4$

$\therefore$ The associated dual problem is

$$\text{Max } W = \left[\frac{5}{.5(1 - 3y_4)}\right]^{.5(1 - 3y_4)} \left[\frac{2}{.5(1 - y_4)}\right]^{.5(1 - y_4)} \left(\frac{5}{y_4}\right)^{y_4} \left(\frac{1}{y_4}\right)$$

Maximization of W is equivalated to maximization of log W.

$\therefore \log W = .5(1-3y_4)\left[\log 10 - \log(1-3y_4)\right] + .5(1-y_4)\left[\log 4 - \log(1-y_4)\right]$

$+ y_4\left[\log 5 - \log 4 - \log 4\right]$

The value of y_4 maximize. W must be unique (because the optimal problem has a unique minimum). Hence

$$\frac{\partial}{\partial y_4}(\log W) = \left(-\frac{3}{2}\log 10 - \frac{1}{2}\log 4 + \log 5\right) + \frac{3}{2}\log(1-3y_4) + \frac{1}{2}\log(1-y_4) - 2\log y_4 = 0$$

This gives after simplification

$$-\log\left[\frac{2\times 10^{3/2}}{5}\right] + \log\left[\frac{(1-3y_4)^{3/2}(1-y_4)^{1/2}}{y_4^2}\right] = 0$$

$$\frac{\sqrt{(1-3y_4)^3(1-y_4)}}{y_4^2} = 12.6$$

This gives $y_4^* = 0.16$

$\therefore \; y_3^* = 0.16, \; y_2^* = 0.42, \; y_1^* = 0.26$

$\therefore$ The value of Z^* is optimal given

$$Z^* = W^* = \left(\frac{5}{.26}\right)^{.26}\left(\frac{3}{.42}\right)^{.42}\left(\frac{5}{.16}\right)^{.16}$$

$= 9.661$

Hence $u_3 = 0.16 \, (9.661) = 1.546 = 5x^1$

$u_4 = 0.16 \, (9.661) = 0.1546 = x_2^{-1}$

The above given

$x_1^* = 0.309$ and $x_2^* = 0.647$.

CHAPTER 6

STOCHASTIC PROGRAMMING

6.1 Introduction

Stochastic programming deals with situations when some or all parameters of the problems are described by random variables. Such cases are seen typical of real life problems, where it is difficult to determine the values of the parameters exactly. The sensitivity analysis of the linear programming can be used to study the effects of changes in problem's parameters of optimal solution. This, however represents only a partial answer to the problem, especially when the parameters are actually random variables. The OBJECTIVE OF STOCHASTIC PROGRAMMING is to consider these random effects explicitly in the solution of the model.

The basic idea of all stochastic programming models. is to convert the probabilistic nature of the problem into an equivalent determinist in situation. General models have been developed to handle special case of the general problem.

We introduce the techniques of Chance—Constrained Programming Problems

6.2 Technique of Chance—Constrained Programming

A chance constrained model is defined generally as

Maximize $Z = \sum\limits_{j=1}^{n} c_j x_j$ subject to

$$P\left\{ \sum_{j=1}^{n} a_{ij} x_j \leq b_i \right\} \geq 1 - \alpha_i \quad i = 1, 2, \ldots n, \; xj \geq 0 \text{ for all } j$$

The name chance constrained follows from each constraint.

$$\sum_{j=1}^{n} a_{ij} x_j \le b_i$$

being realized with a minimum probability of $1 - \propto_i,\ 0 < \propto_i < 1$

In the general case, it is assumed that cj, aij and bi are all random variables. The fact that cj is a random variable can always be treated by replacing it by its expected value.

In what flows we consider Three Cases:

The first two correspond to the separate considerations of aij and bi as random variables. The third combines the random effects of aij and bi. In all cases, it is assumed that the parameters are normally distributed with known means and variances.

Case 1

In this case, each aij is normally distributed with mean E{aij} and variance var{aij}. Also the Covariance of aij and ai¹j¹ is given by cov{aij, ai¹j¹}.

Consider the ith constraint.

$$P\left\{ \sum_{j=1}^{n} a_{ij} x_j \le b_i \right\} \ge 1 - \propto_i$$

Define $h_i = \sum_{j=1}^{n} a_{ij} x_j$

Then hi is normally distributed with $E\{h_i\} = \sum_{j=1}^{n} E\{a_{ij}\} x_j$ and $\text{var}\{h_i\} = X^T D_i X$,

where $X = (x_1, x_2, \dots x_j, \dots x_n)^T$

Di = the ith covariance matrix.

$$\begin{bmatrix} \text{var}\{a_{ij}\} \dots \text{cov}\{a_{i1}, a_{ir}\} \\ \text{cov}\{a_{ir}, a_{i1}\} \dots \text{var}\{a_{ir}\} \end{bmatrix}$$

Now $P\{h_i \le b_i\} = P\left[\left\{ \dfrac{h_i - E\{h_i\}}{\sqrt{\text{var}\{h_i\}}} \right\} \le \left\{ \dfrac{h_i - E\{h_i\}}{\sqrt{\text{var}\{h_i\}}} \right\} \right] \ge 1 - \propto_i$

where $\dfrac{\{h_i - E\{h_i\}\}}{\sqrt{\text{var}\{h_i\}}}$ is standard normal with mean Zero and Variance One.

This means that $P\{h_i \leq b_i\} = \phi\left(\dfrac{h_i - E\{h_i\}}{\sqrt{\mathrm{var}\{h_i\}}}\right)$, where ϕ represents the CDF of the

standard normal distribution. Let k_{∞_i} be the standard normal value such that

$$\phi\{k_{\infty_i}\} = 1 - \infty_i$$

Then the statement $P\{hi \leq bi\} \geq 1 - \infty_i$ is realized if and only is

$$\frac{b_i - E\{h_i\}}{\sqrt{\mathrm{var}\{h_i\}}} \geq k_{\infty_i}$$

This yields the following non linear deterministic constraint:

$$\sum_{j=1}^{n} E\{a_{ij}\}x_j + k_{\infty_i}\sqrt{X^T D_i X} \leq b_i$$

which is equivalent to the original stochastic constraint.
For the special case when the normal distributions are independent.
Cov $\{ai, j, ai^1, j^1\} = 0$
and the last constraint reduces to

$$\sum_{j=1}^{n} E\{a_{ij}\}x_j + k_{2_i}\sqrt{\sum_{j=1}^{n} \mathrm{var}\{a_{ij}\}x_j^2} \leq b_i$$

This constraint can now be put in the separate programming form, using the substitution.

$$y_1 = \sqrt{\sum_{j=1}^{n} \mathrm{var}\{a_{ij}\}x_j^2}, \text{ for all i.}$$

Thus the original constraint is equivalent to

$$\sum_{j=1}^{n} E(a_{ij}).x_i + k_{\infty_i}y_i \leq b_i \text{ and } \sum_{j=1}^{n} \mathrm{var}\{a_{ij}\}x_j^2 - y_i^2 = 0 \text{ where yi} \geq 0$$

Case 2

In this case only bi is normal with mean E{bi} and variance var{bi}. The analysis is very much similar to Case 1.
Consider the stochastic constraint.

$$P\left\{ b_i \geq \sum_{j=1}^{n} a_{ij} x_j \right\} \geq \propto_i$$

As in Case 1

$$P\left\{ \frac{b_i - E\{b_i\}}{\sqrt{\text{var}\{b_i\}}} \geq \frac{\sum_{j=1}^{n} a_{ij} x_j - E\{b_i\}}{\sqrt{\text{var}\{b_i\}}} \right\} \geq \propto_i$$

This can hold only if

$$\frac{\sum a_{ij} x_j - E\{b_i\}}{\sqrt{\text{var}\{b_i\}}} \leq k_{\propto_i}$$

Thus the stochastic constraint is equivalent to the deterministic linear constraint

$$\sum_{j=1}^{n} a_{ij} x_j = E\{b_i\} + k_{\propto_i} \sqrt{\text{var}\{b_i\}}$$

Thus in Case 2, the chance constrained model, can be converted into are equivalent linear programming problem.

Case 3

In this case all aij and bi are normal random variables. Consider the constraint.

$$\sum_{j=1}^{n} a_{ij} x_j \leq b_i$$

This may be written as

$$\sum_{j=1}^{n} a_{ij} x_j - b_i \leq 0$$

Because all aij and bi are normal, it follows from the theory of statistics that

$\displaystyle\sum_{j=1}^{n} a_{ij} x_j - b_i$ is also normal. This shows that the chance constraint reduces to the

situation given in Case 1 and is treated in a similar manner.

6.3 EXAMPLES

EXAM. 1: Consider the chance constrained problem.

Maximize $Z = 5x_1 + 6x_2 + 3x_3$

Subject to $P\{a_{11}x_1 + a_{12}x_2 + a_{13}x_3 \leq 8\} \geq 0.95$

$P\{5x_1 + x_2 + 6x_3 \leq b_2\} \geq 0.10$

$x_1, x_2, x_3 \geq 0$

Assume that all x_{ij}s are independent and normally distributed random variables with the following means and variances.

$E\{a_{11}\} = 1, E\{a_{12}\} = 3, E\{a_{13}\} = 9$

$\text{var}\{a_{11}\} = 25, \text{var}\{a_{12}\} = 16, \text{var}\{a_{13}\} = 4$

The parameter b_2 is normally distributed with mean 7 and variance 9.

From the standard normal tables

$k_{\infty_i} = k_{0.05} \approx 1.645, k_{\infty 2} = k_{0.10} \approx 1.285$

For the first constraint the equivalent deterministic constraint is given by

$$x_1 + 3x_2 + 9x_3 + 1.645\sqrt{25x_1^2 + 16x_2^2 + 4x_3^2} \leq 8$$

and the second constraint

$5x_1 + x_2 + 6x_3 \leq 7 + 1.285(3) = 10.855$

If we let $y_2 = 25x_1^2 + 16x_2^2 + 4x_3^2$

The complete problem then becomes

Max $Z = 5x_1 + 6x_2 + 3x_3$

Subject to $x_1 + 3x_2 + 9x_3 + 1.645y \leq 8$ ...(1)

$25x_1^2 + 16x_2^2 + 4x_3^2 - y_2 = 0$ ...(2)

$5x_1 + x_2 + 6x_3 \leq 10.855$ ...(3)

$x_1, x_2, x_3, y \geq 0$

which can be solved by separable programming.

The separable functions of then problem are

$f_1(x_1) = 5x_1; f_2(x_2) = 6x_2; f_3(x_3) = 3x_3$

$g_1^{~1}(x_1) = x_1; g_1^{~2}(x_2) = 3x_2; g_1^{~3}(x_3) = 9; g_1^{~4}(y) = 1.645y$

$g_1^{~2}(x_1) = 25x_1^2; g_2^{~2}(x_2) = 16x_2^2; g_2^{~3}(x_3) = 4x_3^2, g_2^{~4}(y) = -y^2$

$g3^1(x_1) = 5x_1; g_3^{~2}(x_2) = x_2; g_3^{~3}(x_3) = 6x_3$

$(3) \times 3 - (1) \Rightarrow 15x_1 + 3x_2 + 18x_3 - x_1 - 3x_2 - 9x_3 - 1.645y \leq 32.565 - 10 = 24.565$

$14x_1 + 9x_3 - 1.645y \leq 24/565$ etc.

Case 4

C is the only random variable; let c_j ($j = 1, 2, ... n$) be the normal variable given by $N(\mu_j, \sigma_j^2)$, where μ_j and σ_j^2 are the expected value and variance of the random variable c_j, $j = 1, 2, ... n$. The $\displaystyle\sum_{j=1}^{n} c_j x_j$ is also a random variable distributed normally with expectation.

$$E\{f(X)\} = \sum_{j=1}^{n} \mu_j x_j$$ As A and b are deterministic, there is no need of chance constrained technique.

$\therefore$ The deterministic equivalent to the given stochastic programming is the LPP given by

$$\text{Max } E\{f(X)\} = \sum_{j=1}^{n} \mu_j x_j$$

subject to A X $\leq$ b, X $\geq$ 0.

Case 5

When c, b and A we random variables, since c occurs only in the objective function, the deterministic objective function is

$$Z = E\{f(X)\} = \sum_{j=1}^{n} \mu_j x_j \text{ where } c_j \text{ is } N(\mu_j, \sigma_j^2), \text{ for } j = 1, 2, ... n.$$

Consider ith stochastic constraint

$$P\left\{\sum_{j=1}^{n} a_{ij} x_j - b_i \leq 0\right\} \geq 1 - \propto_i, i = 1, 2 ... n \quad x_{ij} \geq 0, j = 1, 2, ... n$$

Let each aij be $N\{E\{a_{ij}\} = \mu_{ij}, \sigma_{ij}^2 = \text{var } \{a_{ij}\}\}$ for all i, j
And let bi be $N\{E\{b_i\} = \mu b_i$ and $\text{var}\{b_i\} = \sigma^2 b_i\}$

Let $h_i = \displaystyle\sum_{j=1}^{n} a_{ij} x_j - b_i = \sum_{j=1}^{n+1} a_{ij} x_j, i = 1, 2 ... n$ where a_i, n + 1 = b_i and $x_{n+1} = -1$.

Then hi is a normal variable into $E\{h_i\} = \displaystyle\sum_{j=1}^{n} \mu_j x_j - \mu_{bi}$

and $\text{var}\{h_i\} = X^T V_i X$, where $X = (x_1, x_2, ... x_n, x_{n+1})^T$

and $v_i = \begin{bmatrix} \text{var}\{a_{i1}\}\text{cov}\{a_{i1},a_{i2}\}...\text{cov}\{a_{i1},a_{i,n}\} \\ \cdot \\ \cdot \\ \text{cov}\{a_{i,n+1},a_{i1}\}, \text{cov}\{a_{i,n+1},a_{i2}\}...\text{var} \end{bmatrix} \{a_{i,n}, a_{i,n}\}$

with this the ith stochastic constraint becomes

$P\{h_i \leq 0\} = 1 - \infty_i$

or $\phi\left\{E\{h_i\}\Big/\sqrt{\text{var}\{h_i\}}\right\} \leq \infty_i = \phi(k_{\infty_i})$

Thus the deterministic non linear problem is equivalent the given Stochastic Linear Programming Problem and is given by

$$\text{Max } E\{f(X)\} = \sum_{j=1}^{n} \mu_j x_j$$

such that $E\{h_i\} - k_{\infty i}\sqrt{\text{var}\{h_i\}} \leq 0, i = 1, 2...m$ and all $x_j \geq 0$

EXAM. 2: Convert the following stochastic problem into an equivalent deterministic model.

Max $Z = x_1 + 2x_2 + 5x_3$

Such that $P\{a_1x_1 + 3x_2 + a_3x_3 \leq 10\} \geq 0.9$

$P\{7x_1 + 5x_2 + x_3 \leq b_2\} \geq 0.1$

$x_1, x_2, x_3 \geq 0$

Assume that a_1 and a_3 are independent and normally distributed random variables with means $E\{a_1\} = 2$; and $E\{a_3\} = 16$ and var$\{a_1\} = 1$; var $\{a_3\} = 16$. Assume further that b_2 is normally distributed with mean 15 and variance 25.

$k_{\infty_1} = k0.05 = 1.645, \ k_{\infty_2} = k0.10 = 1.285$

For the two constraints the equivalent deterministic constraints are given by

$2x_1 + 3x_2 + 5x_3 + 1.645\sqrt{9x_1^2 + 0x_2^2 + 16x_3^3} \leq 10$

$7x_1 + 5x_2 + x_3 = 15 + 1.285 \times 5 = 15 + 6.425 = 21.425$

If we let $y^2 = 9x_1^2 + 0x_2^2 + 16x_3^2$.

The complete problem because

$x_1 + 2x_2 + 5x_3$

Ma $X = x_1 + 2x_2 + 5x_3$ subject to

$2x_1 + 3x_2 + 5x_3 + 1.645y \leq 10$

$9x_1^2 + 0x_2^2 + 16x_3^2 - y^2 = 0$

$7x_1 + 5x_2 + x_3 \leq 21.425, x_1, x_2, x_3, y \geq 0$

The above can be solved by Separable Programming.

EXAM. 3: A company manufactures three products. Each product has to pass through three different operations. The time each component takes in each operation is shown in the following table. The table also gives the maximum time for which an operation can work. The profit per item sold is also shown there. Assume that all items manufactured are sold. Formulate thus problem as an LPP to maximize the profit.

Operations	Product			Operations capacity (minutes/day)
	I	II	II	
1	1	0	1	480
2	0	3	2	500
3	2	4	0	680
PROFIIT PER UNIT IN RS	4	3	6	...

Let x_1, x_2, x_3 units be produced in products I, II and III respectively. The LPP becomes

Max $Z = 4x_1 + 3x_2 + 6x_3$,

Subject to $x_1 + x_3 \leq 480$

$3x_2 + 2x_3 \leq 500$

$2x_1 + 4x_2 \leq 360$

$x_1, x_2, x_3 \geq 0$ and integer.

The company decides to manufacturing a Fourth product on the same three operations. The new product takes 4, 6, 1 minutes per unit on the three operations respectively. The profit/unit is Rs. 8. The company decides to use the entire capacity of operations 2. Incorporate these things and formulate as a new LPP.

Max $Z = 4x_1 + 3x_2 + 6x_3 + 8x_4$

s.t. $x_1 + x_3 + 4x_4 \leq 480$

$2x_2 + 2x_3 + 6x_4 \leq 500$

$2x + 4x_2 + x_4 \leq 360$

$x_1, x_2, x_3, x_4 \geq 0$ and integers.

In this example, let the operations capacity on different operations be normal variates $hj, j = 1, 2, 3$, ones with expected values 1000, 2000 and 1500 and standard deviations 200, 400 and 300 respectively. It is required that all the constraints must be satisfied with a probability of at least 0.9. Formulate the equivalent deterministic LPP.

From the normal probability table, $i = 1 - 0.9 = 0.1$. From the table of standard normal distribution $N(0, 1)$, we have $k \propto 1 = -1.28$.

Hence the equivalent deterministic LPP is

$4x_1 + 3x_2 + 6x_3$

Max $Z = 4x_1 + 3x_2 + 6x_3$ subject to

$$\sum_{j=1}^{n} a_{ij}x_j \leq E\{h_i\} + k_{\infty i}\sqrt{\text{var}\{h_i\}}_{ij} \quad \text{i.e. subject to}$$

$x_1 + x_3 \leq 1000 + (-1.28)200 = 744$

$3x_1 + 2x_3 \leq 2000 + (-1.28)400 = 1488$

$2x_1 + 4x_2 \leq 1500 + (-1.28)300 = 1116$

$x_j \geq = 0, j = 1, 2, 3.$

EXAM. 4: Convert the following stochastic problem into an equivalent deterministic LPP model.

Max $Z = 2x_1 - x_2 + x_3$

Subject to $P\{a_{11}x_1 + a_{12}x_2 + a_{13}x_3 \leq 5\} \geq 0.9$

$P\{2x_1 + 3x_2 + 4x_3 \leq b_2\} \geq 0.8$

all $x_j \geq 0$

where a_{11}, a_{12}, a_{13} and b_2 are independent normal variables with expected values and variances 1, 5, 8, 10 and 36, 25, 16, 9 respectively. For the first constraint, since

$\propto_1 = 0.1, K_{\infty_1} = K_{0.1} = 1.285$

$\therefore$ the first constraint (probabilistic) is replaced by the following deterministic constraint.

$x_1 + 5x_2 + 8x_3 + (1.285)y_1 \leq 5$

$36x_1^2 + 25x_2^2 + 16x_3^2 - y_1^2 \leq 0$

For the second constraint $k_{\infty_2} = k_{0.2} = -0.84$ and hence the second constraint becomes

$2x_1 + 3x_2 + 4x_3 \leq 10 + (-0.84)\sqrt{9} = 7.40$

Hence the deterministic LPP is

Max $Z = 2x_1 - x_2 + x_3$

Subject to $x_1 + 5x_2 + 8x_3 + 1.28y_1 \leq 5$

$$36x_1^2 + 25x_2^2 + 16x_3^2 - y_1^2 \leq 0$$
$$2x_1 + 3x_2 + 4x_3 \leq 7.48$$
$$x_1, x_2, x_3, y_1 \geq 0$$

This problem can be solved by the method of separable programming.

EXAM. 5: Find the solution of the following problem by maximizing the expected value of f(X)

Max $f(X) = c_1 x_1 + a x_2$ s.t.

$2x_1 + 3x_2 \leq 50;\ 2x_1 + 5x_2 \leq 80;\ 3x_1 + 3x_2 \leq 70,\ x_1, x_2 \geq 0$

when c_1 is a discrete random variable which takes the values 6, 7, 8, 9, 10 with probabilities 0.1, 0.2, 0.4, 0.2, and 0.1 respectively.

$$E\{c_1\} = \frac{1}{1.0}[6 \times .1 + 7 \times .2 + 8 \times .4 + 9 \times .2 + 10 \times .1]$$

$$= 8 \therefore \overline{c} = 8$$

$$\text{var}\{c_1\} = E\left[\{c - \overline{c}\}^2\right] = 1.095$$

c1	p1	c1p1	c1²p1
6	.1	.6	3.6
7	.2	1.4	9.8
8	.4	3.2	25.6
9	.2	1.8	16.2
10	.1	1.0	10.0
	1.00	8.00	65.2

$$\overline{c} = \frac{8.0}{1.00} = 8$$

$$\sigma_{c1} = \sqrt{(65.2) - 8^2}$$

$$= \sqrt{1.2} = 1.095$$

$$\therefore \text{ Max } E\{f(X)\} = E\{c_1\}x_1 + 9x_2$$
$$= 8x_1 + 9x_2$$

s.t. $2x_1 + 3x_2 \leq 50 \quad 2x_1 + 3x_2 + s_1 = 50$

$2x_1 + 5x_2 \leq 80 \qquad 2x_1 + 5x_2 + s_2 = 80$

$3x_1 + 3x_2 \leq 70 \qquad 3x_1 + 3x_2 + s_3 = 70$

$x_1, x_2 \geq 0$

	const	x1	x2
Z	0	8	9
s1	50	-2	-3
s2	80	-2	-5
s3	70	-3	-3

	const	x1	s1
Z	150	2	-3
x2	50/3	-2/3	-1/3
s2	-10/3	4/3	5/3
s3	20	-1	1

	const	s2	s1
Z	155	-3/2	-1/2
x2	15	1/2	1/2
x1	-5/2	3/4	5/4
	35/2	3/4	

$$x_1^* = \frac{5}{2}, x_2^* = \frac{35}{2}, Z^* = 155$$

$$E\{f(X)\} = 8x_1^* + 9x_1^* = 8 \times \frac{5}{2} + 9 \times \frac{35}{2} = 29 + 135 = 155$$

EXAM. 6: Convert the following stochastic programming problem into equivalent deterministic LP model.

Max $f(X) = -x_1 + 3x_2 - 2x_3$

s.t. $P\{a_1x_1 + a_2x_2 + 2x_3 \le b_1\} \ge 0.95$

$P\{- 2x_1 + 4x_2 + 0x_3 \le b_2\} \ge 0.10$

$- 4x_1 + 3x_2 + 8x_3 \le 10$, all $x_1, x_2, x_3 \ge 0$

where a_1, a_2, b_1, b_2 are independent value random variable with distribute.

$N(3, 16), N(- 1, 25), N(7, 1)$ and $N(12, 1)$ respectively.

$E\{a_1\} = 3, E\{a_2\} = - 1, E\{b_1\} = 7, E\{b_2\} = 12$

$k_{\infty_1} = k_{0.05} = 1.645 \; k_{\infty_2} = k_{0.10} = 1.285$

The first constraint is equivalent to the deterministic. Constraint and is give by

$$3x_1 - x_2 + 2x_3 \le 7 - 1.645\sqrt{16x_1^2 + 25x_2^2 + 0x_3^2 + 1(-1)^2}$$

second value $\le 7 + 1.645\sqrt{16x_1^2 + 25x_2^2 + 1}$

$- 2x_1 + 4x_2 \le 12 + 1.285(1) = 13.285$

$y^2 = 16x_1^2 + 25x_2^2 + 1$

$\therefore$ Max $f(X) = - x_1 + 3x_2 - 2x_3$

s.t. $3x_1 - x_2 + 2x_3 + 1.645y \le 7$

$16x_1^2 + 25x_2^2 + 1 - y_2 \le 0$

$- 2x_3 + 4x_2 \le 13.285$

$- 4x_1 + 3x_2 + 8x_3 \le 10$

$x_1, x_2, x_3 \; y \ge 0$

This can be solved by separable programming method.

EXAM. 7: (a) Explain the structure of a stochastic programming problem that can be converted into a deterministic LPP.

(b) solve the following stochastic programming problem.

Max $Z = 5x_1 + 6x_2$ s.t. $P\{x_1 + 2x_2 \le b_1\} \ge 0.975$

$P\{2x_1 + x_2 \le b_2\} \ge 0.5, x_1, x_2 \ge 0$

b_1 and b_2 are stochastic variables which are normally distributed with means 1 and 4 and variance 4 and 2 respectively.

If t is a standard normal variable in

$P\{t / \ge 1.96\} = 0.05$

Max $Z = 5x_1 + 6x_2$

s.t. $x_1 + 2x_2 \le E(b_1) + K_{\infty_1}\sqrt{\text{var}(b_1)}$

$\le 10 + 1.96 \times 2 = 13.92$

$2x_1 + x_2 = E(b_2) + K_{\infty_2}\sqrt{\text{var}(b_2)}$

$= 4 + 0\sqrt{2} = 4$

∴ The LPP is Max $Z = 5x_1 + 6x_2$

s.t. $x_1 + 2x_2 \le 13.92 \cdot 2x_1 + x_2 = 4$, $x_1, x_2 \ge 0$

EXAM. 8 : Find the optimal solution to the problem

Min $X = 2x - 5x_2$ s.t.

$P\{2x_1 + x_2 \ge b_1\} \ge 0.729$

$P\{2x_1 + 4x_2 \ge b_2\} \ge 0.9772$ $\qquad 0 \le x, \le 1, x_2 \le 0$

Where b_1 is having the CDF given by $F(x) \le 0 \le x \le 1 + b_2$ is N(5, 1)

$$E(b_1) = \frac{2}{3}, \text{bar}(b_1) = \frac{1}{\sqrt{18}}$$

$$\sum a_{ij} x_j \le t\{x_i\} + k_{\infty i} \sqrt{\text{var}(b_i)}$$

Ans. $x_1^* = 0$, $x_2^* = 0.84$, $Z^* = -4.20$

6.4. LINEAR COMBINATION METHOD OF SEARCH

This method deals with a constrained problem in which all constraints are linear. Specifically, the problem is given as

Max $Z = f(X)$

subject to $AX \le b$, $X \ge 0$, when A is a matrix of order mX_n and X is a vector with x_1, x_2, ... x_n as components and b is a vector of m components b_1, b_2 ... b_m.

The procedure is based on the general idea of the steepest ascent (gradient) method. However, the direction specified by the gradient vector may not yield a feasible solution for the constrained problem. Also the gradient vector will not necessarily be null at the optimum (Constrained) point. The steepest ascent method must be modified to handle the constrained case.

Let X_k be the feasible trial point at the kth iteration. The objective function f(X) can be expanded in the neighbour of X^k using Taylor's series. This gives as

$$f(X) \approx f(X^k) + \nabla f(X^k)(X - X^k) + ... \approx f(X^k) - \nabla f(X^k).X^k + \nabla f(X^k).X$$

The procedure calls for determining a feasible point $X = X^k$ such that f(X) is a maximized subject to the (linear) constraints of the problem. Since

$\{f(X^k) - \nabla f(X^k).X^k\}$ is a constant.

the problem of determining X* becomes

Max $Wk(X) = \nabla f(X^k).X$

subject to $A X \le b$, $X \ge 0$

This is a linear programming problem in X that can be used to determine X*.

Since Wk is constructed from the gradient of f(X) at X^k an improved solution point can be secured if and only if Wk(X*) > Wk(X^k). From Taylor's expansion, this does not guarantee that f(X*) > f(X^k) unless X* is in the neighbour hood of X^k.

However, given Wk(X*) > Wk(X^k) there must exist a point X^{k+1} or the line segment (Xk, X*) such that

f(X^{k+1}) > f(X^k)

The objective is to determine X^{k+1}.

Define X^{k+1} = (1 – ?)X^k + rX* = X^k + r(X* – X^k) 0 < r ≤ 1. This means that X^{k+1} is a linear combination of X^k and X*. Since X^k and X* are two feasible points in a Convex Solution Space, X^{k+1} is also feasible. By comparison with the steepest ascent method, the parameter r may be regarded as a step size.

The point X^{k+1} is determined such that f(X) is maximized. Since X^{k+1} is a function of r only, the determination of X^{k+1} is secured by maximizing.

h(r) = f[X^k + r(X* – X^k)] in terms of r.

The procedure just described is repeated until at the kth iteration the condition.

Wk(X*) ≤ Wk(X^k) is satisfied. At this point, no further improvements are possible. The process is thus terminated with X^k as the best solution point.

The linear programming problems generated at the successive iterations differ only in the coefficients of the objective function. The sensitivity analysis procedures may be used to carry out calculations efficiently.

6.5 EXAMPLES

EXAM. 1: Consider the quadratic programming problem

Max $f(X) = 4x + 6x_2 - 2x_1^2 - 2x_1x_2 - 2x_2^2$

Such that $x_1 + 2x_2 \leq 2$, $x_1, x_2 \geq 0$

Let the initial trial point be $X^k = \left(\dfrac{1}{2}, \dfrac{1}{2}\right)$ which is feasible. Now

$\nabla f(X) = (4 - 4x_1 + 2x_2, 6 - 2x_1 - 4x_2)$

First Iteration

The associated linear program is to max $W_1 = x_1 + 3x_2$ such that the same constraints as in the original problem are used.

This gives the optimal solution as X* = (0, 1)

$(w_1)(0, 0) = 0$, $(w_1)(0, 1) = 3$, $(w_1)(2, 0) = 2$

The value of w_1 at X^0 and X* are $\dfrac{1}{2} + \dfrac{3}{2} = 2$ and 3 respectively.

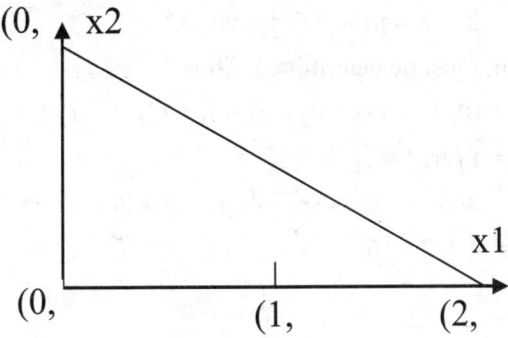

Hence a new trial point must be determined. Thus

$$X^1 = \left(\frac{1}{2}, \frac{1}{2}\right) + r\left[(0,1) - \left(\frac{1}{2}, \frac{1}{2}\right)\right]$$

$$= \left(\frac{1}{2}, \frac{1}{2}\right) + r\left(-\frac{1}{2}, \frac{1}{2}\right)$$

$$= \left(\frac{1-r}{2}, \frac{1+r}{2}\right)$$

Now Max $h(r) = f\left\{\frac{1-r}{2}, \frac{1+r}{2}\right\}$

$$= 2(1-r) + 3(1+r) - \frac{1}{2}(1+r)^2 - \frac{1}{2}(1-r)(1+r)$$

$$= \frac{7}{2} + r - \frac{1}{2}r^2 - \frac{1}{2}(1+r)^2$$

$$\frac{dh(r)}{dr} = 1 - r = 0 \Rightarrow r = 1 \quad r^{(1)} = 1$$

with $f(X^{(1)}) = \left(\frac{1-r}{2}, \frac{1+r}{2}\right) = (0,1)$

and $\nabla f(X^{(1)}) = 6 - 2 = 4$

Second Iteration: commences

$$\nabla f(X^{(1)}) = (2,2)$$

$w2 = 2x_1 + 2x_2$ such that $x_1 + 2x_2 \le 2$

$(w_1)(0, 0) = 0$, $(w_1)(0, 1) = 2$, $(w_1)(2, 0) = 4$

(w_2) at $X^{(1)}$ is $= 0 + 2 = 2$, and $w_2* = 4$; max $X*$

∴ A new trial point must be determined. Thus

$$X^{(2)} = (0, 1) + r[(2, 0) - (0, 1)] = (2r, 1 - r)$$

∴
$$h(r) = f[2r, 1 - r]$$
$$= 8r + 6 - 6r - 8r^2 - 4r(1 - r) - 2(1 - r)^2$$
$$= 4 + 2r - 6r^2$$

∴
$$\frac{dh(r)}{dr} = 2 - 12r = 0 \Rightarrow r = \frac{1}{6}$$

$$h''(r) = -12 < 0 \therefore r = \frac{1}{6} \text{ maximum } h(r)$$

∴
$$X^{(2)} = \left(\frac{1}{3}, \frac{5}{6}\right) \text{with } f(X^{(2)}) \approx 4.16$$

Third Iteration

$$\nabla f(X^{(2)}) = \left(4 - \frac{4}{3} - \frac{5}{3}, 6 - \frac{2}{3} - \frac{10}{3}\right)$$

$$= \left(\frac{12 - 9}{3}, \frac{18 - 12}{3}\right) = (1, 2)$$

The corresponding objective function is $w_3 = x_1 + 2x_2$

Here, the optimal solution of thus problem yield two alternative solutions.

$X* = (0, 1)$ and $X* = (2, 0)$

The values of w_3 for both the value of $X*$ equals, its value at $X^{(2)}$. Consequently, no further improvements are possible.

The Approximate Optimum Solution Is

∴ $X^{(2)} = \left(\frac{1}{3}, \frac{5}{6}\right)$ with $f(X^{(2)}) = 4.16$ This happens (here) the Exact Optimum.

EXAM. 2: Solve the following quadratic programming problem by gradient method (linear combination method)

Max $Z = f(X) = 10x_1 + 4x_2 - x_1^2 + 4x_1x_2 - 5x_2^2$

Subject to $x_1 + x_2 \le 6$ ...(1)

$4x_1 + x_2 \le 18$ ...(2)

$x_1, x_2 \ge 0$

$\nabla f(X^{(1)})^* = (10, 4)$

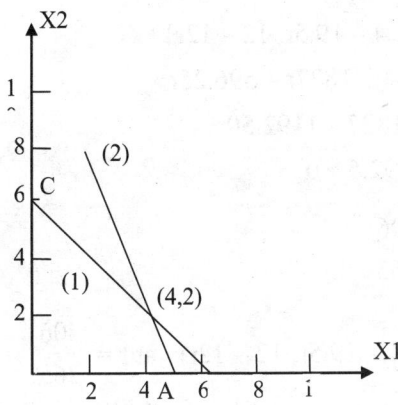

The associated LPP is max $w_1 = 10x_1 + 4x_2$

Subject to $x_1 + x_2 \leq 6$, $4x_1 + x_2 \leq 18$, $x_1, x_2 \leq 0$

$(w_1)(0,0) = 0$, $(w_1)(4.5,0) = 45$, $(w_1)(4,2) = 48$, $(w_1)(0,6) = 24$

$X^* = (4, 2)$

Value of w_1 at $X_0 = (0, 0)$ is 0

Value of w_1 at $X^* = 8$

Hence a new trial point must be determined

$\therefore \qquad X^1 = (0, 0) + r[(4, 2) - (0, 0)]$

$\qquad\qquad = (4r, 2r)$

$\therefore \qquad h(r) = f(4r, 2r)$

$\qquad\qquad = 40r + 8r - 16r2 + 32r^2 - 20r^2$

$\qquad\qquad = 48r - 4r^2$

$\therefore \qquad \dfrac{dh(r)}{dr} = 48 - 8r = 0 \Rightarrow r = 6 = (10, 4)$

$h''(r) = -8 < 0 \therefore r = 6$ makes h(r) a max.

$\therefore \qquad X^{(1)} = (24, 12)$

Second Iteration

$\qquad \nabla f(X^{(1)}) = \{10 - 48 + 48, 4 + 96 - 120\} = (10, -20)$

$\qquad\qquad w_2 = 10x_1 - 20x_2$

such that $x_1 + x_2 \leq 6$

$4y_1 + x_2 \leq 18$

$\therefore$ A New Trial Point must be determined

(w2)(0,0) =0
(w2)(4.5,0) = 45
(w2)(4,2) = 40 − 40 = 0
(w2) (0,6) = − 120

Hence $\quad X^{(2)} = (24, 12) + r\,[(4.5, 0) - (24, 12)]$

$\quad\quad\quad\quad = (24 - 19.5r, 12 - 12r)$

$\therefore \quad\quad h(r) = f\{24 - 19.5r, 12 - 12r\}$

$\quad\quad\quad\quad = 144 - 1827r - 596.25r^2$

$\quad h'(r) = -1827 - 1192.50r$

$\quad h''(r) = 1192.5 > 0$

$$r = \frac{406}{265}$$

$$X^{(2)} = (24 - 19.5r, 12 - 12r) \,/\, \text{at } r = \frac{406}{265}$$

EXAM. 3: Solve the following problem by the linear combination method

Min $f(X) = x_1^{\,3} + x_2^{\,3} - 3x_1 x_2$

Subject to $3x_1 + x_2 = 3,\ 5x_1 - 3x_2 = 5,\ x_1\, x_2 \geq 0$

Let the initial point be $X^0 = \left(\dfrac{1}{2}, \dfrac{1}{2}\right)$

$$\nabla f(X) = \left(3x_1^2 - 3x_2,\ 3x_2^2 - 3x_1\right)$$

$$\nabla f(X^{(1)}) = \left(\frac{-3}{4}, \frac{-3}{4}\right)$$

The associated linear program is to minimize

$$w_1 = \frac{-3}{4} x_1 = \frac{-3}{4} x_2$$

s.t. $3x_1 + x_2 \leq 3$

$5x_1 - 3x_2 \leq 3$

$x_1, x_2 \geq 0$

$$(w_1)_{0,0} = 0,\ (w_1)_{(1,0)} = \frac{-3}{4},\ (w_1)_{(0,3)} = \frac{-9}{4}.X^* = (0, 3)$$

$\therefore$ The values of w1 at X^0 and $X^{(1)}$ are found to be 0 and $-\dfrac{9}{4}$.

Hence a new trial point must be determined

$$\therefore \quad X^{(1)} = \left[\frac{1}{2}, \frac{1}{2}\right] + r\left[(0, 3) - \left(\frac{1}{2}, \frac{1}{2}\right)\right]$$

$$= \left[\frac{1}{2}, \frac{1}{2}\right] + r\left[-\frac{1}{2}, \frac{5}{2}\right]$$

$$= \left[\frac{1-r}{2}, \frac{5\Pi+1}{2}\right]$$

$$\therefore \qquad h(r) = f\left[\frac{1-r}{2}, \frac{1+5\Pi}{2}\right]$$

$$= \frac{1}{8}\left[-4 - 12r + 108r^2 + 124r^3\right]$$

$$\frac{dh(r)}{dr} = \frac{1}{8}\left[-12 + 216r + 248r^2\right] = 0$$

$$\frac{d^2h}{dr^2} = \frac{1}{8}\left[216r + 744r\right]$$

$$372r^2 + 216r - 12 = 0$$

$$r = \frac{3.1}{62} = \frac{1}{20}$$

$$\therefore \qquad X^{(1)} = \left[\frac{1-r}{2}, \frac{5+r}{2}\right] = \left(\frac{19}{40}, \frac{5}{8}\right)$$

$$\therefore \qquad f(X^{(1)}) = 0.1072 + 0.24 + 1.78 = 2.1272$$

Second Iteration

$$\nabla f(X^{(1)}) = (-1.187, -13.078)$$

$$\text{Min } w_2 = -1.187x_1 - 13.078x_2$$

$$= 0.564 - 8.174 = -7.610$$

$$(w_2)(0, 0) = 0, (w_2)(1, 0) = -1.187 (w_2)(0, 3) = -39.234$$

$\therefore$ The new trial point

$$X^{(1)} = \left(\frac{19}{40}, \frac{5}{8}\right) + r\left[(0, 3) - \left(\frac{19}{40}, \frac{5}{8}\right)\right]$$

$$= \left\{\frac{19}{40}(1-r), \frac{1}{6}(5+19r)\right\}$$

$$\therefore \qquad h(r) = f\left[\frac{19}{40}(1-r), \frac{1}{5}(5+19r)\right]$$

Tedious

$\therefore$ Take the trial point at $X^0 = (0, 0)$ we get the Soln. quickly.

$X^0 = (0, 0)$

$$\nabla f(X) = \left(3x_1^2 - 3x_2,\ 3x_2^2 - 3x_1\right)$$

$$\nabla f(X^{(0)}) = (0, 0)$$

$\therefore$ The associated LPP is

Min $w1 = 0x_1 + 0x_2 = 0$

s.t. $3x_1 + 3x_2 \le 3$, $5x - 3x_2 \le 5$, $x_1, x_2 \ge 0$

$(w_1)(0, 0) = 0$, $(w_1)(1, 0) = 0$, $(w_2)(0, 3) = 0$

$\therefore w_1{}^* =$ is Min at $X^0 = (0, 0)$

$\therefore X^* = X^{(0)} = (0, 0)$ is the OPTIMAL SOL

$\therefore$ Min $f(x) = f(x^*) = f(0, 0) = 0$.

6.6 SUMT Algorithm

It is a more general gradient method. It is assumed that $f(X)$ is concave and each constraint function $gi(X)$ is convex. The solution space is assumed to have an interior. This rules out both implict and explicit use of the Equality Constraints.

The SUMT (Sequential Unconstrained Maximization Technique) algorithm is based on transforming the constrained problem into an unconstrained problem. The procedure is more or less similar to the use of the Lagrange Multipliers Method

Sumt Method Contd

The transformed problem can then be solved by using the Steepest Ascent Method. Consider the new function

$$p(X, t) = f(X) + t\left(\sum_{i=1}^{n} g_i(x) - \sum_{j=1}^{n} \frac{1}{x_j}\right) \quad \text{where t is a non negative parameter. The}$$

second summation sign is based on the non negativity constraints which must be part in the form $- x_j \le 0$ to conform with the original constraints $g_i(X) \le 0$. Since $g_i(X)$ is convex,

$\dfrac{1}{g_i(x)}$ is Concave. This means that $p(X, t)$ is concave in X. Consequently, $p(X, t)$ processes a Unique Maximum. It is now shown that optimization of the original constrained problem is equivalent to the optimization of $p(X, t)$.

This algorithm is initiated by arbitrarily selecting an initial Non Negative Value for t. The initial point Xo is selected as the first true solution. This point must be an Interior Point, that is it must not lie on the boundaries of the solution space. Given the value of t, the steepest ascent method is used to determine the corresponding optimal solution [maximum of $p(X, t)$].

The new solution point will always be an interior point because if the solution point is close to be boundaries, at last are of the functions will acquire a very large negative value (Boundary to $-M$) (or called a Penalty Funciton). This provides the gradient search procedure from crossing the BOUNDARY of the feasible region where one more $g_i(X) = 0$. Since the objective function is to maximize $p(X, t)$ such solution points are automatically excluded. The main result is that successive solution points will always be interior points. Consequently the problem can always be treated as an Unconstrainded case.

Once the optimum solution corresponding to a given value of t is reached, a new value of t is generated and the optimization process using the steepest ascent method is repeated. Thus if t′ is the current value of t, the next value of t, namely, t″ must be selected such that $0 < t″ < t′$.

The SUMT procedure is terminated if for 2 successive values of t, the corresponding optimum values of X obtained by maximizing the function $p(X, t)$ we approximately the same. At this point further trials will produce little improvement.

6.7. EXAMPLES

EXAM. 1: To illustrate the SUMT algorithm, consider the two variable problem

$$\text{Min } f(X) = \frac{(x_1 + 1)^3}{3} + x_2$$

Subject to $x_1 \geq 0$ 1, $x_2 \geq 0$
Thus $g_1(X) = (x_1 - 1)$ and $g_2(X) = x_2$

$$\therefore p(X, t) = \frac{(x_1 + 1)^3}{3} + x_2 + t\left[\frac{1}{x_1 - 1} + \frac{1}{x_2}\right]$$

This problem is sufficiently simple enough, so that minimizing the solution of $p(x, t)$ can be derived analytically.

$$\frac{\partial p}{\partial x_1} = (x_1 + 1)^2 - \frac{t}{(x_1 - 1)^2} = 0$$

or $(x_1 - 1)^2 (x_1 + 1)^2 = t$

is $(x_1^2 - 1)^2 = t \ldots (1)$

$$\frac{\partial p}{\partial x_2} = 1 - \frac{t}{x_2^2} = 0 \text{ or } x_2^2 = t \qquad \ldots(2)$$

$$\therefore \qquad \overline{x_1} = \left(\sqrt{t} + 1\right)^2 \text{ and } \overline{x_2} = \sqrt{t} \qquad \ldots(3)$$

Then if the gradient search procedure were still used, it would obtain essentially this solution at each iteration.

A typical sequence of values of t for the iterations would be $t = 1, 10^{-2}, 10^{-4}, 10^{-6}, \ldots$ Using this sequence gives the required results summarized in the table below.

Iteration	t	$\overline{x_1}$	$\overline{x_2}$	$f(\overline{X}) - t\sum I_i(\overline{X})$ lower bound	Upper bound
1	1	1.4142	1	2.27	5.69
2	10^{-2}	1.0488	-1	2.66	2.97
3	10^{-4}	1.0050	0.01	2.67	2.70
4	10^{-6}	1.0005	0.001	2.67	2.67
		?	?	?	?
		1	0	2.67	2.67

Where the last two columns give the lower and upper bound on f(x).

X* tends to the boundary values of the two variables x_1 and x_2. The optimum solution must be X* = (1, 0) (This can also be verified by the Kuhn–Tucker conditions)

l.b. = 5.69 – 3.414 = 2.27 upper bound = 65.69

EXAM. 2: Use SUMT algorithm to solve the problem

Min $f(X) = x_1^2 - 6x_1 + 9 + 2x$

Subject to $x_1 > 3, x_2 \geq 3$,

Now $\quad g_1(X) = x_1 - 3, g_2(X) = x_2 - 3$

$$p(X, t) = x_1^2 - 6x_1 + 9 + 2x_2 + t\left[\frac{1}{x_1 - 3} + \frac{1}{x_2 - 3}\right]$$

$$\frac{\partial p}{\partial x_1} = 2x_1 - 6\frac{-t}{(x_1 - 3)^2} = 0 \Rightarrow t = 2(x_1 - 3)^3$$

$$\frac{\partial p}{\partial x_2} = 2\frac{-t}{(x_1-3)} = 0 \Rightarrow t = 2(x_2-3)^2$$

$$\therefore \quad x_1 - 3 = \sqrt[3]{t/2} \quad x_2 - 3 = \sqrt{t/2}$$
$$x_1 = 3 + \sqrt[3]{t/2} \quad x_2 = 3 + \sqrt{t/2}$$

$$\left(\overline{x}_1, \overline{x}_2\right) = \left[3 + \left(\frac{t}{2}\right)^{1/3}, 3 + \left(\frac{t}{2}\right)^{1/2}\right]$$

The above minimizes p(X, t) so that as $t \to \infty$, $\left(\overline{x}_1, \overline{x}_2\right) \to (3,3)$ and it is optimal.

Iteration	t	$\overline{x}_1$	$\overline{x}_2$	$f\left(\overline{X}\right) - t\sum_{i=1}^{n}\frac{1}{I_i\left(\overline{X}\right)}$	$f\left(\overline{X}\right)$
1	1	3.7938	3.7071	..	..
2	10^{-2}	3.1710	3.0707	..	..
3	10^{-4}	3.03681	3.007071	..	..
4	10^{-6}	?	?	..	..
		3	3		

(1) $\quad x_1 = 3 + \left(\frac{t}{2}\right)^{\frac{1}{3}} \quad x_2 = 3 + \left(\frac{t}{2}\right)^{\frac{1}{2}}$ $\qquad$ (1) $x_2 = 3.07074$

$\qquad\qquad = 3 + 2^{-1/3} = 3 + \left(\frac{1}{200}\right)^{\frac{1}{3}}$ $\qquad$ (2) $x_2 = 3.007071$

$\qquad\qquad = 3 + 0.7938 = 3 + 200^{-1/3}$ $\qquad$ $\therefore \overline{x}_1 * = 3, \overline{x}_2 * = 3$

(1) $\quad x_2 = 3 + .7071$

(1) $\quad x_3 = 3.007071 = 3.007071$

EXAM. 3: Use SUMT to solve the problem

$$\text{Min } f(X) = x_1^2 + x_2^2 + x_1 + x_2 - x_1 x_2$$

s.t. $\quad x_i \geq 0, i = 1, 2$

$$p(X, t) = x_1^2 - x_2^2 + x_1 + x_2 - x_1 x_2 + t\left[\frac{1}{x_1} + \frac{1}{x_2}\right]$$

$$\frac{\partial p}{\partial x_1} = 2x_1 + 1 - x_2 - \frac{t}{x_1^2} = 0 \Rightarrow t = x_1^2 \ (1 + 2x_1 - x_2)$$

$$\frac{\partial p}{\partial x_1} = 2x_2 + 1 - x_1 - \frac{t}{x_2^2} = 0 \Rightarrow t = x_2^2 \ (1 + 2x_2 - x_1)$$

$x_1^2 + 2x_1^3 - x_1^2 x_2 = t$

$x_2^2 + 2x_2^3 - x_1 x_2^2 = t$

$x_1^2 - x_2^2 + 2(x_1^3 - x_2^3) - x_1 x_2 (x_1 - x_2) = 0$

$(x_1 - x_2)[x_1 + x_2 + 2(x_1^2 + x_1 x_2 + x_2^2)] - x_1 x_2 = 0$

is $x_1 = x_2$ or $x_1 + x_2 + 2x_1^2 + 2x_2^2 + x_1 x_2 = 0$

$$\nabla p(X, t) = \left(1 + 2x_1 - x_2 - \frac{t}{x_1^2}, \ 1 + 2x_2 - x_1 - \frac{t}{x_2^2}\right)$$

$$X^0 = (1, 1), t = 1$$

$$\therefore \quad \nabla p(X^0, t) = (1 + 2 - 1 - 1, \ 1 + 2 - 1 - 1)$$

$$= (1, 1)$$

$$\therefore \quad X^{(1)} = X^{(0)} + t^{(1)} \nabla f(X^{(0)})$$

$$= (1, 1) + 1(1, 1) = (2, 2)$$

Second Iteration

$$x^{(1)} = (2, 2), t = 10^{-2} = \frac{1}{100}$$

$$\nabla p(X^{(t)}, t) = \left(1 + 2x_1 - x_2 - \frac{t}{x_1^2}, \ 1 + 2x_2 - x_1 - \frac{t}{x_2^2}\right)$$

$$= \left(1 + 4 - 2 - \frac{1}{400}, \ 1 + 4 - 2 - \frac{1}{400}\right)$$

$$= \left(3 - \frac{1}{400}, \ 3 - \frac{1}{400}\right)$$

Third Iteration

$$X^{(2)} = x^{(1)} + t^{(2)} \nabla p(X^{(1)}, t)$$

$$= (1,4) + \frac{1}{100}\left(3 - \frac{1}{400}, 3 - \frac{1}{400}\right)$$

$$= \left(2 + \frac{3}{100} - \frac{1}{40000}, 2 + \frac{3}{100} - \frac{1}{40000}\right)$$

$$= 2 + 0.03 - .000025, 2.03 - .000025$$

$$\mathbf{X}^{(2)} = (2.03, 2.03)$$

Fourth Iteration

$$X^{(3)} = X(4) + 10^{-4} \ \nabla p(X^{(2)}, t)$$

$$(2.03, 2.03) + 10^{-4}\left[1 + 4.06 - 2.03 - \frac{1}{10^4} \cdot \frac{1}{(2.03)^2}, 1 + 4.06 - 2.03 - \frac{1}{10^4} \cdot \frac{1}{(2.03)^2}\right]S$$

$$= (2.03 + .000303, 2.03 + .000303)$$

$$= (2.0303, 2.0303)$$

$$\therefore \qquad X^{(3)} = (2.0303, 2.0303)$$

CHAPTER 7

GENETIC ALGORITHMS

7.1 INTRODUCTION

Genetic algorithms are search algorithms based on the mechanics of natural selection and natural genetics. They combine string structures with a structured yet randomized information exchange to form a search algorithm with some of the innovative flair of human search. While randomized, genetic algorithms are no simple random walk, they efficiently exploit historical information to speculate on new search points with expected improved performance.

Genetic algorithms have been developed by John Holland at the University of Michigan. The goals of their research are (i) to abstract and rigorously explain the adaptive processes of natural systems and (ii) to design artificial systems software that retains the important mechanisms of natural systems.

The research studies on genetic algorithms are: robustness and the balance between efficiency and efficacy necessary for survival in many different environments. [robust: vigorous or healthy; efficiency: able to perform well; efficacy: quality of producing the desired result]

If artificial systems can be made more robust costly re design can be reduced or eliminated. If higher levels of adaptation can be achieved, existing systems can perform their functions longer and better. Designers of artificial systems—both software and hardware, whether engineering systems, computer systems or business systems—can only marvel at the robustness, the efficiency and the flexibility of biological systems. Features for self repair, self guidance and reproduction are the rule in biological systems. Such features do not exist even in most sophisticated artificial systems. Nature possesses such secrets of adaptation and survival. These can be learned from the careful study of biological example. For problems requiring efficient and effective search, the genetic algorithms are finding more widespread applications, in business, and engineering circles. The reasons are

The G.As are computationally simple and powerful in their search for improvement.

The G.As are not limited by assumptions like continuity, existence of derivatives, unimodality etc.

7.1.1 *Understanding Genetics, Evolution and Biodiversity*

1. **Concept and Process of Photosynthesis: Photosynthesis** is the source of energy and organic materials for other organisms besides plants.

The ultimate source of all metabolic energy is the sun and photosynthesis is responsible for the maintenance of life on Earth.

Photosynthesis in green plants is **the process** in which energy from the sun is transformed into chemical bond energy in organic molecules. Oxygen is produced in the transformation process.

Classification of describing the distribution of organisms in their habitats is done based on (i) the generic name, which states the genus and is common to a group of closely related organisms and (ii) a specific name, starting the species, which is unique to a particular organism and often descriptive of one of its characteristics.

2. **Currently Margulis** and **Schnarts** classified the living organisms into the five kingdoms:

(i) **Prokaryotae**; (ii) **Protoctista**; (iii) **Frengi**; (iv) **Plantae** and (v) **Animalia**

3. **Taxonomy** is the study of groups of organisms, called **taxa** into hierarchies, which attempt to take into account their supposed evolutionary descent.

4. A **species** is defined as a group of individuals with a large number of features in common. This definition is restricted to sexually reproducting organisms and cannot apply where reproduction behaviour has not been observed like fossils or parthenogenetic forms etc.

5. A number of physical factors affect the distribution of organisms in their habitats. The physical factors can be divided into:

- **Climatic :** temperature, light, wind and water availability
- **Soil :** edaphic (or soil condition) factors
- **Topographic :** altitude, aspect (whether north facing or south facing) and inclination.
- **Others :** such as wave action.

6. **The PH value** is a measure of the **hydrogen ion concentration** of an aqueous solution and indicates the level of **acidity** or **alkalinity**.

7. A **population** is defined as a group of individuals of one species found in the same habitat.

8. **Variation:** When organisms reproduce their offspring are of the same species and

the members of one family are all similar to one another and to their parents in their specific characteristics. It is recognizable that individuals within one species vary in small ways. These small differences constitute variation, which may be the result of genetic changes taking place during the formation of the gametes, or of the influence of the environment, or a combination of both. It is possible to recognize two forms of variation:

- discontinuous
- continuous

These two forms of variation describe differences in appearance of the characters and do not describe the genetic differences.

9. **Genetics:** The science that deals with the mechanisms responsible for similarities and differences in a species is called Genetics.

The word "genetics" is derived from the Greek word "genesis" meaning "to grow" or "to become".

10. **Heredity:** Heredity refers to the transmission of characters, resemblances as well as differences from one generation to the next. It explains how off springs in a family resemble their parents.

11. **Gene:** A gene is a physical and functional unit of heredity. It carries information from one generation to the next. Gene is also defined as a nucleotide sequence that is responsible for the production of specific protein.

12. A **gene** represents the specific length of DNA coding for a polypeptide occupying a specific site on a chromosome. The specific site is called a **locus**.

13. **Locus** The genes are present at a specific position on the chromosome called locus.

14. **Alleles** A single gene may have more than one functional state or form. The functional states are referred to as alleles. The alleles may be dominant or recessive.

15. An **allele** is an alternative form of a gene occupying the same locus on a chromosome as other alternative alleles of the same gene.

16. **Mutations** are sudden heritable changes in the structure of the genetic material.

Mutations constitute the principal raw material with which nature work to bring about evolution.

17. **Mutation** Discontinuous variations refer to large, conspicuous differences of the offsprings from the parents. These variations are also called Mutations or sports or saltations.

Individuals showing these types of variations are called mutants.

18. **Mutation** During the process of replication DNA is normally copies exactly so that the genetic material remains the same from generation to generation. Occasionally changes can occur so that an organism may inherit altered genetic material. Such inherited changes are known as **mutations**.

Mutation is another source of genetic variability in a population and it can occur spontaneously.

19. **Chromosomes** Chromosomes are the physical carriers of genes, which are made up of DNA and associated proteins.

Chromosomes occur in all living organisms.

20. **Genome** Genome may be defined as the totality of the DNA sequences of an organism or organelle such as mitochondria and chloroplasts. Each species has a characteristic number of chromosomes in the nuclei of its gametes and somatic cells. The gametic chromosome number constituter a basic set of chromosomes called genome. In all organisms it is made up of DNA but in viruses, it is made up of either DNA or RNA.

21. **Linkage** The tendency of genes or characters to be inherited together because of their location on the same chromosome is called linkage.

22. **Coupling** Two genes which inherit together are called linked genes. This aspect is called coupling.

23. **Evolution** The process by which changes have occurred over geological time is termed **evolution**.

24. **Palaeontology** It is a branch of science concerned with the study of biological events in the past.

25. **The theory of evolution by natural selection** consists of three key points.

- The individual characteristics of an organism, such as its height, colour or speeed of movement, are vitally important for its ability to survive and to breed.
- Individual within a given species of organism vary in many characteristics. Individuals with certain advantageous characteristics, such as ability to avoid a predator, are likely to live longer and to produce more off spring.
- Only a small proportion of the offspring will survive. The numbers of individuals less well adapted to the environment will correspondingly decrease, that is, they will be selected against. This is the basis of evolutionary change.

26. **Fitness** The term fitness means the ability of an organism to survive and produce off spring which themselves can survive and produce offspring.

27. **Robustness (maintaining or healthyvigour)** It indicates maintaining the strength or vigorousness or healthy in any environment.

28. **Archaeology** is the study of recent events in human history.

29. **Fossils** The preserved remains of living organisms are usually described of fossils when they are more than about 10000 years old.

30. **Replication** Genes undergo duplication by a phenomenon called replication.

31. **Recombinations** New combinations of linked genes are called recombinations.

32. **Crossing over** The process which produces recombination of genes by interchanging the corresponding segments between non-sister chromatids of homologous chromosomes is called crossing over.

33. **Crossingover** is the exchange of DNA between homologous chromosomes.

34. Significance of crossing over

- Crossing over leads to the production of new combination of genes and provides basis for obtaining new varieties of plants.
- It plays an important role in the process of evolution.
- The crossing over frequency helps in the construction of genetic maps of the chromosomes.
- It gives the evidence for linear arrangement of linked genes in a chromosome.

35. **Crossingover** Recombination of linked genes is accomplished through a process which homogeneous chromosomes exchange parts. This process is called crossing over.

Certain parts necessary to the explanation of crossing over

- The genes in chromosomes are in linear order, some what like beads on a string.
- When a gene (X) and its allele (x) are present in different members of a pair of homologous chromosomes, the gene and its allele occupy corresponding places in the homologues.
- In order to produce recombinations between two different alletic pairs situated in the same chromosome pair, crossing over must occur between the locations of the genes involved.
- Crossing over characteristically occurs in the first division of meiosis. It is this meiotic crossing over that concern mostly. However, somatic crossing over is also known to occur.

36. **Dominance** always refer to the modification of the expression of one member of a pair of alleles by the other, never to an interaction between different genes.

37. **Dominatly inherited traits**, like all traits, may be passed on to off spring.

38. **DNA is the genetic material**. It encodes instructions for proteins in discrete units known as genes. These genes are responsible for all our traits.

39. **Genes** are made of DNA and are arranged into long strands known as chromosomes.

40. **Ecology** is the study of the interactions between organisms and their environment.

41. A community consists of all the populations of various species that live and interact in a particular environment.

42. **An ecosystem** consists of a community and its physical environment.

43. **Biosphere** is the entire portion of Earth inhabited by life.

44. **Ecological Succession** is the process that involves the introduction of organisms into a disturbed environment and their gradual replacement by other species, which in turn are replaced by still other species.

45. **Genetic** Engineering is the deliberate alteration of the genetic makeup of organism by the manipulation of its DNA molecule to effect a change in heredity traits.

7.2 Comparative Study of Robustness in Search and Classical Optimization Techniques

Conventional types of the search methods are

- Calculus based
- Enumerative
- and random

(i) *Calculus based methods* are divided into two classes: indirect and direct.

Indirect methods seek local extrema by solving the usually non linear set of equations resulting from setting the indirect of the objective function equal to zero. This is the multi dimensional of the elementary calculus notion of extrimal points.

Given a smooth, un constrained function, finding a possible peak starts by restructuring search to those points with slopes of zero in all directions.

Direct search methods seek local optima by hopping on the function and moving in a direction related to the local gradient. This corresponds to the notion of hill elimbing: to find the local best, climb the function in the steepest permissible direction.

Both methods lack robustness because they are local in scope.

The optima they seek are best in a neighbour hood of the current point.

Complete domain of interest may have sub domains having different peaks. Once a lower peak is reached, further improvement can be made through random restart. In addition, calculus based methods depend on the existence of derivatives (well-defined slope values). There is also a short coming even by the numerical approximation of the derivatives. The notion of a derivative and the smoothness cannot be guaranteed in many practical parameter spaces. The world of mathematics takes for granted a clean world of quadratic objective functions, ideal constraints and ever present derivatives. But the real world of search is with discontinuities and vast multi modal, noisy search spaces unsuitable fraught to calculus. Hence calculus based search methods are to be rejected for want of robustness in unintended.

(ii) *Enumerative method*: Though enumerative schemes are fairly straight forward in a finite search space or in a descritized infinite search space, the search algorithm starts looking at objective function values at every point in the space, one at a time. The search algorithm is attractive due its simplicity, the enumerative is a taxing human kind of search, this method is discounted in the robustness due to the lack of efficiency. The only exception is dynamic programming as the problem can be broken up into sub problems of moderate size and complexity. But it suffers from the "curse of dimensionality".

(iii) *Random search:* Random search algorithms have achieved increasing popularity as researchers have recognized the short comings of calculus based and enumerative schemes. Yet, random walks and random schemes that search and save the best must also be discounted because of the efficiency requirement. But the randomized techniques are to be separated from random schemes because the genetic algorithm is a search procedure that uses random choice as a tool to guide a highly exploitative search through a coding parameter space.

Another currently popular include technique, simulated annealing , uses random processes to help guide its form of search for minimal energy states. The recent book by Davis explores the connection between simulated annealing and genetic algorithms. It has to be borne in mind that randomized search does not necessarily imply directionless search. [anneal: cool very slowly after heating in order to toughen temper]

Many traditional schemes work well in a narrow problem domain. Enumerative schemes and random walks work equally inefficiently across a broad spectrum. A robust method works well across a broad spectrum of problems.

7.3 Goals of Optimization

Optimization theory encompasses the quantitative study of optima and methods for finding them. Thus optimization seeks to improve the performance toward some optimal point or points. This definition has two parts.

We seek improvement to approach some

Optimal point

There is a clear distinction between the process of improvement and destination or optimum itself.

Optimization procedures focus upon convergence to the best. But in business and in other walks of life, we desire in better relative to others.

Hence the most important goal of optimization is improvement

7.4 Genetic Algorithms vs Traditional Methods

Genetic algorithms surpass the traditional methods in the quest for robustness. GAs are different from more normal optimization and search procedures in four ways.

GAs work with a coding of the parameter set, not the parameters themselves.

GAs search from a population of points, not a single point.

GAs use pay off (objective function) information, not derivatives or other auxiliary knowledge.

GAs use probability transition rules, not deter ministic rules.

The genetic algorithms require that natural parameter set of the optimization problem to be coded as a finite length string over some finite alphabet.

Consider the optimization problem $f(x) = x^2$ on the integer interval $[0, 31]$

The highest value of the objective function is $f(x) = 31^2 = 961$

Consider the device with 5 input switches For every setting of the five switches, there is an output signal f. Mathematically $f = f(s)$. The code generated by considering a string of five 1's and 0's where each of the five switches is represented by a 1 if the switches is on and a 0 if the switches is off.

The genetic algorithm starts with a population of strings and then after generates succesive populations of strings.

A random start using successive coin flips (head = 1, tail = 0) might generate the strings like 01101, 11000, 01000, 10011, ... etc. when the initial population size is n = 4.

The genetic algorithms only require two things: a coding and a pay off measure. To perform an effective search for better and better structures, the GAs only require pay off values (objective function values) associated with the individual strings.

Unlike many methods, GAs use probabilistic transition rules to guide their search. GAs use random choice as a tool to guide a search toward regions of the search space with likely improvement.

The direct use of a coding, search from a population blindness to auxiliary information, and randomized operators contribute to a genetic algorithm's robustness and resulting advantage over other commonly used techniques.

7.5 A Simple Genetic Algorithm

The mechanics of a simple genetic algorithm are simpler involving copying strings and swapping partial strings.

Simplicity of operation and power of effect are two of the main attractions of the genetic algorithm approach.

The black box (with 5 switches) switching problem, the initial population has five strings: 01101, 11000, 01000, 10011

The above population was chosen at random through 20 successive flips of an unbiased coin. We generate successive populations that improve over time.

A simple genetic algorithm that fields good results in many practical problems in composed of three operators.

Reproduction

Cross over

Mutation selection

Reproduction is a process in which individual strings are copied according to the objective function values, f [biologists call it the fitness function]. This function, f, is taken as a measure of profit utility or goodness that is to be maximized. Copying strings according to their fitness values means that strings with a higher value, have a higher probability of contributing one or more off spring in the next generation. This operator is an artificial version of natural selection, a Darwinian survival of the fittest among string creatures. Fitness is determined by a creator's ability to survive predators, pestilence and other obstacles to adult-hood and subsequent reproduction. In the artificial setting, the objective function is the final arbiter of the string creature's life or death.

Simple reproduction allocates off spring strings using a roulette wheel with slots sized according to fitness. The sample wheel is sized for the problem as the table below represents:

No	String	Fitness	% of total
1	01101	169	14.4
2	11000	576	49.2
3	01000	64	5.5
4	10011	361	30.9
	1170	10.00	

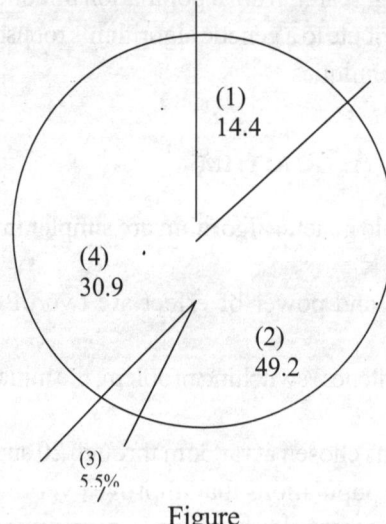

Figure

7.6 EVOLUTIONARY OPERATORS

When certain techniques which have minimal resembleness to natural genetics these techniques are called EVOLUTIONARY OPERATION. To start with the such evolutionary operation is less an algorithm than more a management technique to permit less technically minded industrial workers to execute a regular plan of experimentation about the consent operating scenario. The aim was to use the experiments to improve some desirable process metric.

Consider a process dependent on three variables: carbon treatment, air blowing and rate of temperature rise. Taking a hyper cube and beginning with the current operating point to represent as the starting vertex of the hyper cube. A systematic schedule of visiting the vertices of the cube was undertaken and if significant improvement was found by visiting any of the neighbouring points, a decision was made by the evolutionary operations committee

Sr. No.	Initial Population Randomly Generated	x-Value unsigned	f(x) x²	Percentage selection= ■	Expected count= ■	Actual Count (from Roulette Wheel)	Mating pool before Reprod-uction	Mate (Random selection)	Crossover site (Random selection)	New population	x-val	f(x) = x²
1.	01101	13	169	0.14	0.58	1	01101	2	4	01100	12	144
2.	11000	24	576	0.49	1.97	2	11000	1	4	11001	25	625
3.	01000	8	64	0.06	0.22	0	11000	4	2	11011	27	729
4.	10011	19	361	0.31	1.23	1	10011	3	2	10000	16	256
Sum			1170	1.00	4.00	4.0						1754
Average			293	0.25	1.00	1.0						439
Max			576	0.49	1.97	2.00						729

to change the operating point and a new hyper cube was created and a new experimentation process was started. Subsequent automated simpler schemes have proved to provide useful for local search called EVOUTIONARY.

BOX provided an useful way to interpret that living things advance by means of two mechanisms:

- Genetic variability due to various agencies such as mutation and
- Natural selection

Chemical processes and new discovery of route for manufacture correspond to the concept of a "mutation".

Adjustment of the process variables to their best levels, once the route is agreed involves a process of natural selection in which compromising combinations of the levels of process variables are neglected in favour of processing ones.

Even though biological populations are rarely so orderly or regular, the lack of a recombination operator prevents the innovative information exchange between pairs of structures.

Although the method of evolutionary operation may be an useful tool and an ancestor of other local search techniques, it is not considered as a genetic algorithm in the modern sense.

7.7 OTHER EVOLUTIONARY OPTIMIZATION TECHNIQUES

Bremerman, Friedman and their research students suggested a binary string coding. The scheme combined individual by individual generation, mutation and save the better selection.

Bledsoe extended the above scheme by generating successive population of strings using selection and mutation. He also used a recombination operator.

The techniques under the name

"Evolution- Strategic" were independently developed evolving an airfoil shape using a physical apparatus that permitted local perturbation of airfoil geometry. Computer simulations of similar processes were performed following the early experiments.

However the use of real parameters limits the schema processing inherent in these methods. Nonetheless the Evolution-Strategic has gained in engineering and scientific circles.

7.8 EVOLUTIONARY PROGRAMMING-DEVELOPED BY FOGEL, OWENS AND WALSH

Evolutionary operations and the techniques of evolutionary optimization are followed by the Evolutionary Programming Techniques by the above three persons.

In their work a sequential symbol prediction tasks were performed by searching through a space of small finite state machines.

Consider a state diagram of a three state machine as shown here

The Greek letters are output symbols and capital letters are states and the symbols 0 and 1 are input symbols.

If the machine is in state A and the machine receives an input symbol of 1, a ? is output and the machine remains in state A.

On the other hand, if the machine is in state A and receives a 0 as input, a ? is written and the machine moves to state B. A complete state transition description for this machine is contained in the following table.

This type of machine was trained to predict repeating cycles of output symbols using the techniques evolved by the 3 authors and it consists of 2 operators.

- Selection
- Mutation

STATE TRANSITION TABLE FOR FINITE STATE MACHINE

Present state	Input symbols	Next state	Output symbol
C	0	B	β
B	1	C	∝
C	1	A	√
A	1	A	β
A	0	B	β
A	1	C	∝

Here the selection operator chose the better of two machines, the parent or the mutated off spring machine.

The mutation operator is only the structure - mystifying mechanism used in this set up. The mutation is the single mystification of the finite state machine state diagram in the following sense:

An off spring of these machine is then produced through mutation, that is, through a single modification of the parent machine in accordance with some mutation noise distribution. The mode of mutation is determined by the interval within which a number selected from a random number table lies. The intervals are chosen in accordance with a probability distribution over the permitted modes of mutation.

Additional numbers are then selected in order to determine the specific details of the mutation. Thus, off-spring is made w differ from its parent either by an output symbol a state transition, the number of states or the initial state.

Here the evolutionary programming of the 3 authors with its random alternation of a finite state machine diagram and save the best selection was insufficiently powerful to search either other than small problem spaces quickly.

Here we turn away from these and similar studies that ignore fundamental importance of structure recombination and now consider a powerful study of genetic algorithms that combine theoretical analysis with easily controlled computational experiments.

7.9 De Jong and Function Optimization

De Jong constructed a test environment of 5 problems in function minimization. The following characteristics are included in feature of optimization

- Continuous /discontinuous
- Convex/non convex
- Uni modal/multi modal
- Quadratic/non quadratic
- Low dimensionality/ high dimensionality
- Deterministic/stochastic

His five function test bed are:

No. function	Limits
1. $\quad f_1(x_i) = \sum_1^4 x_i^2$	$-5.12 \le x_i \le 5.12$
2. $\quad f_2(x_i) = 100(x_1^2 - x_2^2) + (1 - x)^2$	$-2.048 \le x_i \le 2.048$
3. $\quad f_3(x_i) = \sum_1^5 \text{integer}(x_i)$	$-5.12 \le x \le 5.12$
4. $\quad f_4(x_i) = \sum_1^{50} ix_i^4 + \text{gauss}(0,1)$	$-1.28 \le x_i \le 1.28$
5. $\quad f_i(x_i) = 0.002 + \sum_{i=1}^{25} \dfrac{1}{j + \sum (x_i - a_{ij})^6}$	$(-65.536 \le x_i \le 65.526)$

He devised two measures one to gauge convergence and the other to gauge on going performance respectively.

In an off line (convergence) application many function evaluations can be simulated and the best alternative so far saved and used after the achievement of some stopping criteria.

An on line (on going) application does not afford this luxury and function evaluations are achieved through real experimentation on line, as a result, a premium is placed on getting to acceptable performance quickly.

He defined the on line performance $x_e(s)$ of strategy on environment e as follows:

$$x_e(s) = \frac{1}{T}\sum_{1}^{T} f_e(t)$$

where $f_e(t)$ is the objective function value for environment e on trial t.

In words, the on-line performance is an average of all function evaluations upto and including the current trial.

He also defined the performance measure $x_e^*(s)$, the off-line performance of strategy s on environment e as follows:

$$x_e^*(s) = \frac{1}{T}\sum_{1}^{T} f_e(t) \text{ when it is equal to the best } \{f_e(1), f_e(2)....f_e(t)\}$$

In words, the off line performance is a running average of the best performance values to a particular time. He used uniform trial weighting with a test bed of 5 trial functions and two criteria of goodness defined. De Jong setout to investigate variations of what we have come to call the simple genetic algorithm.

He started with a version which he called R1 (reproductive plan) which consisted of the three operators:

1. Roule the wheel selection
2. simple cross-over (with random mating)
3. simple mutation

All operators were applied is successive populations of binary strings coded as mapped, concatenated and unsigned binary integers.

De Jong was aware that the plan R1 was not just a single plan but rather a family of plans depending upon four parameters.

n = population size

p_c = cross are probability

p_m = mutation probability

G = generation gap

Generation gap G was introduced by De Jong to permit overlapping populations. It was defined between 0 and 1 as follows:

G = 1 for non overlapping population

0 < G < 1 for overlapping population

In the overlapping populations, n.G. individuals are selected for further genetic action. Resulting off spring are placed in the existing population by choosing n.G. population slots uniformly at random (without replacement)

TRIAL FUNCTION F1 is a smooth, quadratic function of 3 variables.

Population size experiments are function F_1

Larger populations lead to better ultimate off-line performance (convergence) because of the larger pool of diverse schemates are available in a larger population. It leads us to expect the poorer initial online performance.

Smaller populations have the ability to change more rapidly and then exhibit better initial online performance.

To combat premature allele loss mutation rate increases are often suggested as a way to maintain sufficient diversity for continued improvement.

De Jong also experimented with cross over probabilities and pgeneration gap values.

The studies of generation gap suggested that the non overlapping population model was best in most optimization studies, where off-line performance tends to be the over riding concern. Online performances not severely degraded by using smaller generation gap values.

This fact is useful in machine learning where learning while performing well is important.

7.10 GENETIC OPERATORS

Reproduction, Corssover, Mutation

The three operators or the simple tripartite algorithm can each be implemented in a straight forward code segments. There is a common thread running through the three operators; each depends on the random choice. We assume the existence of three random choice routines:

RANDOM—returns a real psedo-random number between zero and one (a uniform random variable on the real interval [0, 1])

FLIP—returns a Boolean true value with specified probability (a Bernoulli random variable)

RND—returns an integer value between specified lower and upper limits (a uniform random variable over a subset of adjacent integers)

In the simple genetic algorithm, *reproduction* is implemented in the function select as a linear search through a roulette wheel with slots weighted in proportion to string fitness values.

Select returns the population index value corresponding to the selected individual. To do this, the partial sum of the fitness values is accumulated in the real variable part-sum; the real variable *rand* contains the location where the wheel has landed after a random spin according to the composition.

Rand: random * sum fitness. Repeat—until construct searches through the weighted roulette wheel until the partial sum in greater than or equal to the slopping point rand. The function returns with the current population index value j assigned to select.

[A binary search will certainly speed up things.]

The code segment *select* given us a straight forward way of choosing off spring for the next generation. The next step is *crossover*.

In SGA, the *crossover* operator is implemented in a providence that we call crossover.

The routine *crossover* takes two parent strings called parent1 and parent2 and generates two off springs called child1 and child2.

The probabilities of crossover and mutation, p-cross and p mutation are passed to crossover, along with the string length *lcbrom*, a crossover count accumulator *n cross* and a mutation count accumulator *n mutation*.

Within crossover, the operations mirror our description (already done); at the top of the routine, we determine whether we are going to perform crossover on the current pair of parent chromosomes; specifically, we toss a biased coin that comes up heads (true) with probability p cross. The coin toss is simulated in the Boolean function flip where flip in turn calls on the pseudo random number routine random.

If a cross is called for, a crossing site is selected between 1 and the last cross site. The crossing site is selected in the function rand, which returns a pseudo random integer between specified lower and upper limits (between 1 and lchrom -1)

If no crossover is to be performed, the cross site is selected as lchrom (the full length string length l) so a bit-by-bit mutation will take place despite the absence of a cross. Finally, the partial exchange of crossover is carried out in the two *for-do* constructs at the end of the code. The first for – do handly the partial transfer of bits between parent1 and child1 and between parent2 and child2. The second *for-do* construct handles the transfer and partial exchange of material between parent1 and child1 and between parent2 and child2. In all the cases, a bit-by-bit mutation is carried out by the Boolean (or allelean) function mutation.

Mutation at a point is carried out by *mutation*. This function uses the function flip (the biased coin toss) to determine whether or not to change a true to a false (a_1 to a_0) or vice-versa. The function flip will only come up heads (true) p mutation percent of the time as a result of the call to the pseudo random number generator *random* with *flip* itself. The function also keeps tabs on the number of mutations by incrementing the variable n-mutation. As with reproduction, there are ways to improve our simple mutation operator.

It is possible to avoid much random number generation if we decide when the next mutation should occur rather than calling flip each time.

The three main pieces of our genetic algorithm puzzle have proven to be none too puzzling (the three* may be easily coded and easily understood)

(*reproduction, crossover, mutation)

A Time to Reproduce, a Time to Cross

The above material are taken from the Book Genetic Algorithms—In Search, Optimization And Machine Learning—By David E. Goldberg—University of Alabama Pearson Education, ASIA, Sixth Indian reprint, 2003; Rs. 260, ISBN-81-7808-130-X.

7.11 SIMULATED ANNEALING SEARCH

Another method of improving search is termed Simulated Annealing because its analogy to the annealing process of slowly cooling metals to improve strength.

Simulated annealing algorithms control cycling by accepting non improving moves according to probabilities tested with computer generated random numbers.

Improving and accepted non improving moves are pursued; reficted ones are not.

Algorithm described below provides details. The move selection process at each iteration begins with random choice of a provisional feasible move totally ignoring its objective function impact. Next, the net objective function improvement ? obj (non-positive for non improving moves) is computed for the chosen moves. The move is always accepted if it improves (Δ obj > 0), otherwise.

Probability of acceptance = $e^{\Delta obj/q}$. That is, all improving moves and some non improving ones are accepted. The probability of accepting a non improving move delines as net objective improvement Δ obj. becomes more negative. The parameter q above, is a temperature controlling the randomness of the search. If q is large, the exponent in the above iteration approaches zero, implying that the probability of accepting non improving moves approximates $e^0 = 1$. If q is small, the probability of accepting very bad moves decreases dramatically. Simulated annealing searches usually begin with q relatively large and decrease it every few iterations.

Here the search makes non improving moves, an incumbent solution must $\hat{X}$ be maintained to keep track of the best feasible solution found so far. When computation stops, is $\hat{X}$ output as are approximately optimal solution.

Algorithm: Simulated Annealing Search

Step 0: Initialization: Choose any starting feasible solution $X^{(0)}$, an iteration limit tmax and a relatively large initial temperature q > 0. Then set incumbent solution $\hat{X} \leftarrow X^{(0)}$ and solution index t $\leftarrow$ 0.

Step 1: Stopping: If no move ΔX in move set M leads to a feasible neighbour of current solution $X^{(t)}$, or if t = tmax, then stop. Incumbent solution is an approximate optimum.

Step 2: Provisional Move: Randomly choose a feasible move $\Delta X \epsilon M$ as a provisional $\Delta X^{(t+1)}$, and compute the (possibly negative) net objective function improvement Δobj for moving from $X^{(t)}$ to $(X^{(t)} + \Delta X^{(t+1)})$ (increase for a maximize, decrease for a minimize.)

Step 3: Acceptance: If $\Delta X^{(t+1)}$ improves, or with probability $e^{\Delta obj/q}$, if $\Delta obj \le 0$, accept

$\Delta X^{(t+1)}$ and update $X^{(t+1)} \leftarrow X^{(t)} + \Delta X^{(t+1)}$.

Otherwise, return to step II

Step 4: Incumbent Solution: If the objective function value of $X^{(t+1)}$ is superior to that of incumbent solution $\hat{X}$ replace $\hat{X} \leftarrow X^{(t+1)}$.

Step 5: Temperature Reduction: If a sufficient number of iterations have passed since the last temperature change, reduce temperature q.

Step 6: Increment: Increment $t \leftarrow t + 1$ and return to step 2.

7.12 Genetic Algorithms

A popular method of avoiding local optima in improving search is known as Genetic Algorithms because it attempts to parallel the process of biological evolution to find better and better solutions.

Genetic Algorithms evolve good heuristic optima by operations combining members of an improving population of individual solutions.

The best single solution encountered so far will always be part of the population, but each generation will also include a spectrum of other solutions. Ideally, all will be feasible, and some may be nearly as good in the objective function as the best. Other may have quite poor solution values.

New solutions are created by combining pairs of individuals in the population. Local optima are less frequent because this combining process does not center entirely on the best current solution.

Cross over operations in Genetic algorithms: The standard genetic algorithm method of combining solutions of the population is known as crossover.

Crossover combines a pair of "parent" solutions to produce a pair of "children" by breaking both parent vectors at the same point and reassembling the first part of one parent solution with the second part of the other and vice-versa.

We can illustrate with two binary solution vectors $X^{(1)}$ and $X^{(2)}$.

$X^{(1)} = (1, 0, 1, 1, 0 \mid 0, 1, 0, 0)$

$X^{(2)} = (0, 1, 1, 0, 1 \mid 1, 0, 0, 1)$

Cross over after component $j = 5$ leads to children.

$X^{(3)} = (1, 0, 1, 1, 0 \mid 1, 0, 0, 1)$

$X^{(4)} = (0, 1, 1, 0, 1 \mid 0, 1, 0, 0)$

One child $X^{(3)}$ combines the initial part of $X^{(1)}$ with the final part of $X^{(2)}$. Child $X^{(4)}$ does the opposite. Both become members of the new population, and the search continues.

There is no guarantee that cross overs rather arbitrary manipulation of parent solutions will yield improvement. Still, if does lead to fundamentally new solutions that preserve significant parts of their parents. Experience shows that this is often enough to produce very good results.

7.13 Managing Genetic Algorithms with Elites, Immigrants and Crossovers

The principal differences in the various implementations concern how to select pairs of current solutions to produce new ones via crossover, how to decide which new and/or old solutions will survive in the next population and how to maintain diversity in the population as the search advances from generation to generation. The only requirement is that better solution have greater chance to breed.

We consider now only a single elitist method of population management. Each new generation will be composed of a combination of elite, immigrant and crossover solutions.

The Elitest strategy for implementation of genetic algorithms forms each new generation as a mixture of elite (best) solutions held over from the previous generation, immigrant solutions added arbitrarily to increase diversity, and children of crossover operations on non-overlapping pairs of solutions in the previous population.

Maintenance of the elite solutions from the proceeding of generation assures that the best solutions known so far will remain in the population and have more opportunities to produce off spring. Addition of new immigrant solutions will help to maintain diversity as solutions are combined. The bulk of the new solutions will be the product of crossovers, with elites in the preceding population allowed to serve as parents.

The procedure is given in the algorithm furnished below.

Genetic Algorithm Search

Step 0: Initialization: Choose a population size p, initial starting feasible solutions $X^{(1)}$, $X^{(2)}$, ... $X^{(p)}$ a generation limit tmax, and population sub divisions pe for elites, pi for immigrants and pc for cross over. Also, set generation index $t \leftarrow 0$.

Step 1: Stopping: If t = tmax, stop and report the best solution of the current population as an approximate optimum.

Step 2: Elite: Initialize the population of generation t + 1 with copies of he pe best solutions in the current generation.

Step 3: Immigrants: Arbitrarily choose pi new immigrant feasible solutions and include them in the t + 1 population.

Step 4: Crossovers: Choose pe/2 non overlapping pairs of solutions from the generation t population and execute crossover on each pair at an independently chosen random cut point to complete the generation t + 1 population.

Step 5: Increment: Increment $t \leftarrow t + 1$ and return to step 1.

7.14 Solution Encoding for Genetic Algorithm Search

[The move set M of discrete improving search must be compact enough to be checked at each iteration for improving feasible neighbours.

The solution produced by a discrete improving search depends on the move set (or neighbour-hood) employed, with large move sets generally resulting in superior local optima.

Multi start or keeping the best of several local optima obtained by searches from different starting solutions is one way to improve the heusistic solutions produced by improving search.

Non improving moves will lead to infinite cycling of improving search unless some provision is added to prevent repeating solutions.

Tabu search deals with cycling by temporarily forbidding moves that would return to a solution recently visited.]

Just as design of our ordinary improving search requires careful construction of a move set, implementations of genetic algorithms require judicious choice of a scheme for encoding solutions in a vector.

To see the difficulty, we have to consider the NCB drilling example.

If solutions are encoded merely by displaying the drilling sequence, two that might come together as crossover parents would be.

$(3, 1, 2, 4, 7, 8, 5, 9, 10, 6)$ and $(7, 2, 3, 1, 6, 5, 8, 4, 9, 10)$

Crossing over the solutions after component $j = 6$ would yield children

$(3, 1, 2, 4, 7, 8, 8, 4, 9, 10)$ and $(7, 2, 3, 1, 6, 5, 9, 10, 6)$

Neither is a feasible drilling sequence because some holes are visited more than once and some never. A poor choice of solution encoding has made it almost impossible for crossover to produce useful new solutions.

Effective Genetic Algorithm Search Requires A Choice for Encoding Problem Solutions that Often, if not Always, Preserves Solution Feasibility After Crossover

We can obtain a better encoding in the NCB example by a technique known as random keys. Sequences are encoded in directly under random keys as a vector of random numbers such as

$(0.32, 0.56, 0.91, 0.44, 0.21, 0.68, 0.51, 0.07, 0.12, 0.39)$

The drilling sequence implied is the one obtained by all aligning hole 1, with the lowest random component, hole 2 at the next lowest, and so on. That is, the given random vector, encodes the drilling sequence,

$(4, 8, 10, 6, 3, 9, 1, 2, 5)$

But notice that cross over as two vectors of random numbers produces two others. Thus every crossover operation of these random key encodings will yield two new feasible solutions. Also, it is very easy to generate arbitrary new solutions for the initial population and immigration simply by generating random vector of random numbers.

7.15 GENETIC ALGORITHM SEARCH OF NCB EXAMPLE

The figure given below shows objective function values in a 30 generation Algorithm [(Genetic Algorithm Search) – (Already Given)] search of the NCB example. The population has size 20, with the initial population generated randomly. Following the principle given in the elitist strategy principle, each new generation contained the 6 best (elite) solutions, of the preceding generation, 4 randomly generated immigrant solution and 10 solutions obtained from crossover of 5 pairs of parents in the preceding generation. All solutions were encoded by random keys.

Bars in the figure extend from the lowest to the highest route length of solutions in each population.

The lowest ones converge systematically to 81.8, which we know is the optimal value of this example. Still, the population always contains a variety of solutions, some with rather poor objective values. This diversity makes it possible for the search to range into distinctly new parts of he feasible space as it seeks improved solutions.

Example 1

Consider solving approximately the ILP
Max $12x_1 + 7x_2 + 9x_3 + 8x_4$
s.t. $3x_1 + x_2 + x_3 + x_4 \leq 3$
$x_1 + x_4 \leq 1$, $x_1, x_2, x_3, x_4 = 0$ or 1
Improve the search problem given above starting from $x^{(0)} = (0, 0, 0, 1)$. Compute an approximate optimum by simulated annealing algorithm, using a temperature of $q = 20$, limiting the search to tmax = 4 moves, and resolving probabilistic decisions with (uniform [0, 1]) random numbers 0.65, 0.10, 0.40, 0.53, 0.33, 0.98, 0.88, 0.37.

[Ans. $\overset{n}{X} = (0, 1, 1, 0)$]

Example 2

Use the ILP problem given in below.
Min $50x_1 + 30x_2 + 20x_3 + 15x_4$
s.t. $x_1 + x_2 \geq 1$, $x_1 + x_4 \geq 1$, $x_1, x_2, x_3, x_4 = 0$ or 1
use random numbers 0.66, 0.87, 0.77, 0.43, 0.18, 0.13, 0.21, 0.48, 0.71, 0.83, 0.29.
start at $x^{(0)} = (1, 0, 0, 0)$

Example 3

Return to the improving search problem of the exercise (1) above

(a) show that the solutions $x^{(1)} = (0, 0, 1, 0)$ and $x^{(2)} = (0, 0, 0, 1)$ are eligible to belong to a genetic algorithm population for the problem.

(b) Construct all crossover results (all cut points) for the $x^{(1)}$ and $x^{(2)}$ of part a

(c) determine whether all your resulting solutions in part (b) are feasible and, if not, explain what difficulty this preseuts for effective application of genetic algorithm search.

Example 4

Do the example in Example 3 on the model of Example 2 using $x^{(1)} = (0, 1, 1, 1)$ and $x^{(2)} = (1, 0, 1, 1)$.

[Ans. to Example 3]

(1) check both feasible

(2) after 1 or 2

$x^{(3)} = x^{(1)}$, $x^{(4)} = x^{(2)}$, after

$x^{(3)} = (0, 0, 1, 1)$, $x^{(4)} = (0, 0, 0, 0)$,

(c) all feasible except $x^{(3)}$ cutting after 3, infeasible must either be excluded from the population or included with a large negative objective value.

7.16 IMPROVING SEARCH HEURISTICS FOR DISCRETE OPTIMIZATION INLP'S

Suitable adaptations of improving search methods can often yield effective heuristics algorithms. That is, we can still find good feasible solutions even though we will not be able to guarantee their optimality or even be sure about how close they come to be optimal.

Rudimentary Improving Search Algorithm

Algorithm—Discrete Improvigng Search shows a rudimentary adaptation of improving search to discrete modéls. the process begins with an initial feasible solution $x^{(0)}$. Each iteration t considers neighbours of current solution $x^{(t)}$ and tries to advance to one that is feasible and superior in objective value. If no feasible neighbour is improving, the process stops in the local optimum and heuristics optimum $x^{(t)}$.

Discrete Neighbour Hoods and Move Sets

What is new about discrete form of improving search is that we must explicitly define the neighbour hood of a current solution. Unlike the continuous case, where there are infinitely many points near a current solution, discrete search must advance to a binary or integer point. Explicitly move sets (denoted $-M$) control of what solutions are considered neightbours of current $x^{(t)}$.

Improving searches over discrete variables define neighbour hoods by specifying a move set M of moves allowed. The current solution and all reachable from it in a single move $\Delta x \in M$ comprise its neighbour hood.

Algorithm: Discrete Improving Search

Step 0: Initialization: Choose only staring feasible solution $x^{(0)}$ and set solution index $t \leftarrow 0$

Step 1: Local Optimum: If no move Δx in move set M is both improving and feasible of current solution $x^{(t)}$, stop. Point $x^{(t)}$ is a local optimum.

Step 2: Move Choose some improving feasible move as $\Delta x \in M$ as $\Delta x^{(t+1)}$.

Step 3: Update; $X^{(t+1)} \leftarrow x^{(t)} + \Delta x^{(t+1)}$.

Step 4: Increment: Increment $t \leftarrow t + 1$, and return to step 1.

Example 1

Defining Move Sets

Consider the discrete optimization model

Max $20x_1 - 4x_2 + 14x_3$

s.t. $2x_1 + x_2 + 4x_3 \leq 5$

$x_1, x_2, x_3 = 0$ or 1

and assume that an improving search begins at $x^{(t)} = (1, 1, 0)$

(a) List all neighbours of $x^{(t)}$ under move set

$$M = \left\{ \begin{pmatrix} 1 \\ 0 \\ 0 \end{pmatrix}, \begin{pmatrix} -1 \\ 0 \\ 0 \end{pmatrix}, \begin{pmatrix} 0 \\ 1 \\ 0 \end{pmatrix}, \begin{pmatrix} 0 \\ -1 \\ 0 \end{pmatrix}, \begin{pmatrix} 0 \\ 0 \\ 1 \end{pmatrix}, \begin{pmatrix} 0 \\ 0 \\ -1 \end{pmatrix} \right\}$$

(b) determine which members of the neighbours hood are both improving and feasible

Analysis

(a) Following searches definition, the neighbours of x(0) under the specified move set are

$(1, 1, 0) + (1, 0, 0) = (2, 1, 0)$

$(1, 1, 0) + (-1, 0, 0) = (0, 1, 0)$

$(1, 1, 0) + (0, 1, 0) = (1, 2, 0)$

$(1, 1, 0) + (0, -1, 0) = (1, 0, 0)$

$(1, 1, 0) + (0, 0, 1) = (1, 1, 1)$

$(1, 1, 0) + (0, 0, -1) = (1, 1, -1)$

(b) at the neighbours in part (a), only x = (0,1,0) which has objective value – 4, and x = (1, 0, 0), which has objective value 20, of the model. Current point $x^{(t)}$ = (1, 1, 0) has objective value 20, are feasible for all constraints of the model. Current point $x^{(t)}$ = (1, 1, 0) has objective value 16. Thus x = (1, 0, 0) is the only neighbour that is both improving and feasible i.e. these are to which improving search would advance.

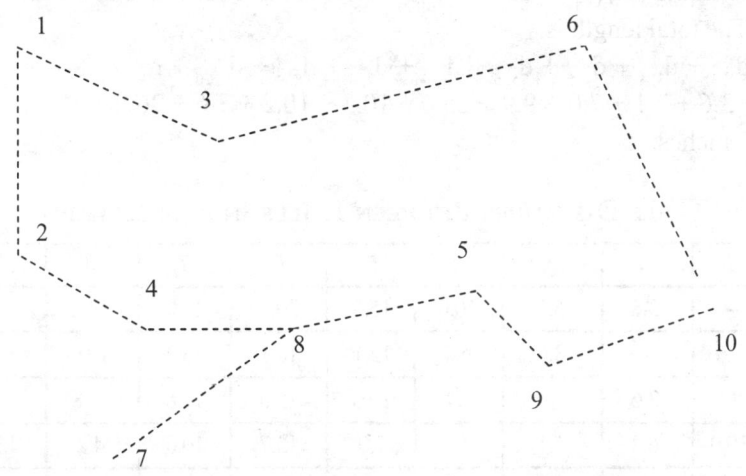

NCB Example Revisited

The above example is taken to illustrate improving search in discrete optimization. We seek a shortest—distance routing the rough 10 points in a circuit brand that must be drilled. The table given below details hole—to hole travel distances.

For improving search, it will be most convenient to employ the quadratic assignment formulation.

$$\text{Min} \sum_{k=1}^{10} \sum_{i=1}^{10} \sum_{j=1}^{10} d_{i,j} \cdot y_{k,j} \cdot y_{k+1,j} \quad \text{(total distance)}$$

$$\sum_{i=1}^{10} \cdot y_{k,i} = 1 \text{ for all } k = 1, 2 \dots 10 \text{ (some hole each k)}$$

$$\sum_{k=1}^{10} \cdot y_{k,i} = 1 \text{ for all } i = 1,2,\dots10 \text{ (each i assigned)}$$

$y_{k,i}$ = 0 or 1 for all k = 1,2,...10, i = 1, 2, ...10

where $y_{k^2 i}$ = 1 if kth hold drilled is ij0, otherwise, and is understood we can $y_{i,j}$ in the objective function summations. We will absiss notation in the usual ways to think of solutions as vectors y even though components have two subscripts. The NCB optimal solution y*, which is depicted in the figure drawn has length 81.8" and non-zero components.

$y^*1, 1 = y^*2, 3 = y^*3, 6 = y^*4, 10 = y^*5, 9 = y^*6, 5 = y^*7, 8 = y^*8, 7 = y^*9, 4 = y^*10, 3 = 1$

we begin one improving searches with initial feasible solution.

$y^{(0)} = (1, 0,...,0; 0, 1, 0...0; ...; 0,...0, 1)$

Corresponding to $y_{1, 1} = y_{2, 2} = ...y_{10, 10} = 1$ That is, hole 1 is dricted first, then hole 2; and son on. The total length is

$d_{1, 2} + d_{2, 3} + d_{3, 4} + d_{4, 5} + d_{5, 6} + d_{6, 7} + d_{7, 8} + d_{8, 9} + d_{9, 10} + d_{10, 1}$
$= 3.6 + 3.6 + 7.1 + 7.0 + 9.9 + 25.0 + 10.3 + 10.2 + 3.6 + 26.4$
$= 106.7$ inches.

TABLE DISTANCES BETWEEN HOLES IN NCB EXAMPLE

j/i	1	2	3	4	5	6	7	8	9	10
1	-	3.6	5.1	10.0	15.3	20.0	16.0	14.2	23.0	26.4
2	3.6	-	3.6	6.4	12.1	18.1	13.2	10.6	19.7	23.0
3	5.1	3.6	-	7.1	10.6	15.0	15.8	10.8	18.4	21.9
4	10.0	6.4	7.1	-	7.0	15.7	10.0	4.2	13.9	17.0
5	15.3	12.1	10.6	7.0	-	9.9	15.3	5.0	7.3	11.3
6	20.0	18.1	15.0	15.7	9.9	-	25.0	14.9	12.0	15.0
7	16.0	13.2	15.8	10.0	15.3	25.0	-	10.3	19.2	21.0
8	14.2	10.6	10.8	4.2	5.0	14.9	10.3	-	10.2	13
9	23.0	19.7	18.4	13.9	7.3	12.0	19.2	10.2	-	3.6
10	26.4	23.0	21.9	17.0	11.3	15.0	21.0	13	3.6	-

Choosing A Move Set

The critical element of a discrete improving search heuristics is its move set. If it were possible, we would make every solution a neighbour of every other. Then the search would yield global optima because a stop implies that no solution at all is feasible and superior in objective value to the current.

In a practical search, however, we must accept much less.

The Move Set M of a Discrete Improving Search Must Be Compact Enough to be Checked at Each Iteration for Improving Feasible Neighbours.

In the other hand, we would not want too limited a move set.

The Solution Produced by a Discrete Improving Search Depends on the Move Set (Or Neighbour) Employed, with karger move sets Generally Resulting in Superior Local Optima.

If move set M of a discrete improving search is too restrictive very few solutions will be considered at each iteration, and poor – quality local optima will result.

We will adopt one of the simplest move sets for our NCB example. Specifically, our M will consist of pairwise interchanges swapping one position k with another l. Corresponding move vectors Δy have two (-1) components at deleted assignments; two $(+1)$ components at revised ones, and all other components 0. For example, if the hole in route position $k = 3$ is now number 7, and the whole in position $l = 5$ is now number 1, the corresponding interchange move direction has

$$\Delta y_{3,7} = -1, \Delta y_{5,1} = -1, \Delta y_{3,1} = +1, \Delta y_{5,7} = +1$$

Changing hole 1 to position 3 and whole 7 to position 5:

In all, this pairwise interchange M contains a Δy for $(10 \times 9)/2 = 45$ choices of k and l, each with $10 \times 9 = 90$ possible current hole assignment pairs a total of $45 \times 90 = 4050$ moves. However, at any particular solution y, only 45 moves interchanging its specific assignments lead to a feasible neighbour. In all searches we adopt the one such move most improving the objective function.

Example 2

In the example 1 at initial point $X^{(0)} = (1, 1, 0)$
(a) show that $X^{(0)}$ is not locally optimal under the move set of example 1.
(b) show that $X^{(0)}$ is weakly optimal over smaller move set
$M = (1, 0, 0), (0, 1, 0), (0, 0, 1)$

Solution

(a) In example 1(b) established that $X = (1, 0, 0)$ is a feasible neighbour of $X^{(0)}$ with superior objective value. The $X^{(0)}$ is not best in its neighbour hood, and so not locally optimal. Algorithm—Discrete Improving Search would advance to
$X^{(1)} = (1, 0, 0)$ and repeat
(b) over this more restricted move set, neighbours are
$(1, 1, 0) + (1, 0, 0) = (2, 1, 0)$
$(1, 1, 0) + (0, 1, 0) = (1, 2, 0)$
$(1, 1, 0) + (0, 0, 1) = (1, 1, 1)$
None is feasible, so $X^{(0)}$ is locally optimal.

CHAPTER 8

MULTIVARIATE ANALYSIS

8.1 INTRODUCTION

Multivariate analysis is concerned with gathering measurable data by experimentation or observation on simultaneous occurrence of many variables and their relationships with statistical methods designed to elicit information on data sets. The objectives of scientific investigations on multivariable methods are :

- Data reduction and simplification
- Classifying the measured characteristics into well defined groups.
- Relationships among the variables of their dependence or independence.
- Prediction of the values of one or more variables on the basis of observations on the other variables.
- Formulation of statistical hypothesis in terms of the parameters of multivariate populations and testing their validity.

8.1.1 Mean Vectors and Convariance Matrices

Let $X = \{x_i\}$ be a random matrix (p x 1) with p rows and one column. It is also regarded as a column vector. Usually it is written in the form of a row vector $[x_1, x_2, \ldots\ldots, x_p]^T$ where T represents the transpose of the column vector for simplicity. Each element of X is regarded as a random variable with its own probability distribution with its mean and variance defined as

$\mu_i = E[X_i]$ and $\sigma i^2 = E[(x_i - \mu_i)^2]$, i = 1, 2 ... p
respectively and μ_i and σi^2 are given by

$$\mu_i = \begin{cases} \int\limits_{-\infty}^{\infty} x_i f_i(xi)\, dxi & \text{if } X_i \text{ is a continuous random variable with proabability density} \end{cases}$$

function $f_i(x_i)$

$$= \begin{cases} \sum\limits_{\text{all } x_i} x_i\, p_i(x_i) & \text{if } X_i \text{ is a discrete random variable with probability function } p_i(x_i) \end{cases} \quad ...(1)$$

$$\sigma_i^2 = \begin{cases} \int\limits_{-\infty}^{\infty} (x_i - \mu_i)^2\, f_i(x_i)\, dx_i & \text{for } X_i \text{ continous} \\[2em] \sum\limits_{\text{all } x_i} (x_i - \mu_i)^2\, p_i(x_i) & \text{for } X_i \text{ discrete} \end{cases} \quad ...(2)$$

σ_i^2 is invariably written as σ_{ii} for multinomial distributions.

The behavior pattern of any pair of random variables X_i and X_k is described by their joint probability function and a measure of the linear association between them is specified by the covariance σ_{ik}, where

$$\sigma_{ik} = E\big[(x_i - \mu_i)(x_k - \mu_k)\big]$$

$$= \int\limits_{-\infty}^{\infty} \int\limits_{-\infty}^{\infty} (x_i - \mu_i)(x_k - \mu_k)\, f_{ik}(x_i, x_k)\, dx_i\, dx_k$$

If X_i, X_k are continuous random variables with joint density function $f_{ik}(X_i, X_k)$,

$$\sum\limits_{\text{all } x_i}\sum\limits_{\text{all } x_k} (x_i - \mu_i)(x_k - \mu_k)\, p_{ik}(x_i, x_k),$$

if X_i, X_k, are discrete random variables with joint probability function $p_{ik}(x_i, x_k)$ ...(3)

and μ_i, μ_k, i, k = 1, 2, ... p are the corresponding means. When i = k the covariance becomes the variance σ_{ii}

The collective behavior of the p random variables $X_1, X_2, ... X_p$ or equivalently the random vector $X^T = [X_1, X_2, ..., X_p]$ is described by a joint probability density function $f[X_1, X_2, ... , X_p] = f(x)$.

If the joint probability $P[X_i \le x_i]\, P[X_k \le x_k]$ can be written as the corresponding marginal probabilities so that

$$P[X_i \le X_i \text{ and } X_k \le X_k] = P[X_i \le X_i]\, P[X_k \le X_k] \qquad ...(4)$$

for all pains of values x_i, x_k, then X_i and X_k are said to be **statistically independent** when X_i and X_k are continuous random variables with joint density $f_{ik}(x_i, x_k)$ and marginal densities $f_i(x_i)$ and $f_k(x_k)$, the independent condition becomes

$$f_{ik}(x_i, x_k) = f_i(x_i)f_k(x_k) \text{ for all pairs } (x_i, x_k) \quad \text{...(5)}$$

More generally, the p continuous random variables $X_1, X_2, ..., Xp$ are **mutually** statistically independent if their joint density functions as

$$f_{1,2,...,p}(x_1, x_2, ... x_p) = f_1(x_1)f_2(x_2) ... f_p(x_p) \quad \text{...(6)}$$

for all p tuples $x_1, x_2, ...x_p$)

Statistical independence has an important implication for covariance. The factorization mentioned above implies $\text{Cov}(X_1 . X_k) = 0$.

Thus $\text{Cov}(X_i . X_k) = 0$ if X_i and X_k are independent ...(7)

[Note: The converse of (7) is not true in general. There are situations where $\text{Cov}(X_i, X_k) = 0$ and X_i and X_k are not independent]

8.2 MULTIVARIATE NORMAL DISTRIBUTION

8.2.1 Introduction

Real data are not exactly multivariate normal; the normal density is useful as an approximation to the true population distribution. The advantages of the multivariate normal distribution are the following :

- It is mathematically tractable and nice results can be obtained when compared to the other data – generating distributions.
- Normal distribution has an honest intention to model any genuine population.
- The sampling distributions of many multivariate statistics are approximately normal regardless of the form of the parent population, because of the central limit theorem.
- Many real – would problems come under the frame work of normal theory.

8.2.2 The Multivariate Normal Density

The multivariate normal density is a generalization of the univariate normal density to the dimensions $p \geq 2$. The univariate normal distribution with mean m and variance σ^2 has the probability density function

$$f(x) = \frac{1}{\sqrt{2\pi\sigma^2}} \exp\left[-\left(\frac{x-\mu}{\sigma}\right)^2 / 2\right], -\infty < x < \infty$$

$$= \frac{1}{\sqrt{2\pi\sigma^2}} \exp\left[\frac{1}{2}(x-\mu)(\sigma^2)^{-1}(x-\mu)\right] \quad \text{...(1)}$$

While plotting this function, yields the familiar bell – shaped curve and the curve within ± 1 standard deviations and ± 2 standard deviations of the mean cover areas represented by means of probabilities specified by

$$P\,(\mu - \sigma \le X \le \mu + \sigma) = 0.68$$

and $P(\mu - 2\sigma \le X \le \mu + 2\sigma) = 0.95$ \qquad\qquad ...(2)

where X denotes a random variable. The notation to denote the normal density function with mean μ and variance σ^2 is by N (μ, σ^2).

This notation can be generalized for a p.1 vector X of observations on several variables as

$$(X - \mu)^T \Sigma (X - \mu) \qquad\qquad ...(3)$$

The px1 vector μ represents the expected value of the random vector X, and the pxp matrix Σ is its variance – covariance matrix. It is assumed that the symmetric matrix Σ is positive definite, so the expression in (3) is the squared generalized distance from X to μ.

The multivariate normal density is obtained by replacing the univariate distance $(x - \mu)(\sigma^2)^{-1}(x - \mu)$ by the multivariate generalized distance of (3)is the density function of (1). When this replacement is made, the univariate normalized constant $(2\pi)^{-1}(\sigma^2)^{-1/2}$ must be changed to a more general constant that makes the volume under the surface of the multivariate density function unity for any p. This is necessary because in the multivariate case, probabilities are represented by volumes under the surface over regions defined by intervals of the x_i values. It can be shown that this constant is $(2\pi)^{-\frac{p}{2}} |\Sigma|^{-1/2}$ and consequently a p-dimensional normal density for the random vector $X = [X_1, X_2, X_3, ..., X_p]^T$ has the form

$$f(X) = \frac{1}{(2\pi)^{p/2} |\Sigma|^{1/2}} \exp 1/2 \left\{ (X - \mu)^T \Sigma^{-1} (X - \mu)_{/2} \right\} \qquad\qquad ...(4)$$

where $-\infty < x_i < \infty$, i = 1, 2, ..., p. The p-dimensional normal density is denoted by $N_p(\mu, \Sigma)$, which is analogous to the univariate case.

For the bivariate normal density in terms of the individual parameters

$$\mu_1 = E(X_1), \mu_2 = E(X_2), \ s_{11} = Var(X_1), s_{22} = Var(X_2)$$

and $\rho_{12} \dfrac{\sigma_{12}}{\sqrt{\sigma_{11}} \sqrt{\sigma_{22}}} Cov\,(X_1, X_2)$; also

the inverse of the covariance matrix

$$\Sigma = \begin{bmatrix} \sigma_{11} & \sigma_{12} \\ \sigma_{12} & \sigma_{22} \end{bmatrix} \text{ is given by}$$

$$\Sigma^{-1} = \frac{1}{\sigma_{11}\sigma_{22} - \sigma^2_{12}} \begin{bmatrix} \sigma_{22} & -\sigma_{12} \\ -\sigma_{12} & \sigma_{11} \end{bmatrix}$$

Introducing the correlation coefficient ρ_{12} by writing $\sigma_{12} = \rho_{12}\sqrt{\sigma_{11}}\sqrt{\sigma_{22}}$, we obtain

$\sigma_{11}\sigma_{22} - \sigma^2_{12} = \sigma_{11}\sigma_{22}(1 - \rho^2_{12})$ and the squared distance becomes

$$(X - \mu)^T \Sigma^{-1}(X - \mu)$$

$$= [x_1 - \mu_1, x_2 - \mu_2] \frac{1}{\sigma_{11}\sigma_{22}(1-\rho_{12}^2)} \begin{bmatrix} \sigma_{22} & -\rho_{12}\sqrt{\sigma_{11}}\sqrt{\sigma_{22}} \\ -\rho_{12}\sqrt{\sigma_{11}}\sqrt{\sigma_{22}} & \sigma_{11} \end{bmatrix} \begin{bmatrix} x_1 & -\mu_1 \\ x_2 & -\mu_2 \end{bmatrix}$$

$$= \frac{1}{1-\rho_{12}^2} \left[\left(\frac{x_1 - \mu_1}{\sqrt{\sigma_{11}}} \right)^2 + \left(\frac{x_2 - \mu_2}{\sqrt{\sigma_{22}}} \right)^2 - 2\rho_{12} \left(\frac{x_1 - \mu_1}{\sqrt{\sigma_{11}}} \right) \left(\frac{x_2 - \mu_2}{\sqrt{\sigma_{22}}} \right) \right] \qquad \text{...(5)}$$

Consequently the expression for the bivariate ($p = 2$) normal density becomes

$$f(x_1, x_2) = \frac{1}{2\pi\sqrt{\sigma_{11}\sigma_{22}(1-\rho_{12}^2)}}$$

$$x \exp \left\{ -\frac{1}{2(1-\rho_{12}^2)} \left[\left(\frac{x_1 - \mu_1}{\sqrt{\sigma_{11}}} \right)^2 + \left(\frac{x_2 - \mu_2}{\sqrt{\sigma_{22}}} \right)^2 - 2\rho_{12} \left(\frac{x_1 - \mu_1}{\sqrt{\sigma_{11}}} \right) \left(\frac{x_2 - \mu_2}{\sqrt{\sigma_{22}}} \right) \right] \right\} \qquad \text{...(6)}$$

If the random variables X_1 and X_2 are uncorrelated so that $\rho_{12} = 0$ the joint density function

$$f(x_1, x_2) = \frac{1}{2\pi} \frac{1}{\sqrt{\sigma_{11},\sigma_{22}}} \exp \left\{ \left[\left(\frac{x_1 - \mu}{\sqrt{\sigma_{11}}} \right)^2 + \left(\frac{x_2 - \mu_2}{\sqrt{\sigma_{22}}} \right)^2 \right] x \frac{-1}{2} \right\}$$

$$= \frac{1}{\sqrt{2\pi}} \frac{1}{\sigma_{11}} \exp \left\{ \left(-\frac{1}{2} \right) \left(\frac{x_1 - \mu_1}{\sqrt{\sigma_{11}}} \right)^2 + \left(-\frac{1}{2} \right) \left(\frac{x_2 - \mu_2}{\sqrt{\sigma_{22}}} \right)^2 \right\}$$

Here $\sigma_{11} = \sigma_1^2$ and $\sigma_{22} = \sigma_2^2$, the above joint density function $f(x_1, x_2)$ becomes

$$f(x_1, x_2) = \frac{1}{\sqrt{2\pi}.\sigma_1} \exp \left\{ -\frac{1}{2} \left(\frac{x_1 - \mu_1}{\sigma_1} \right)^2 \right\} x \frac{1}{\sqrt{2\pi}.\rightleftharpoons \sigma_2} \exp \left\{ -\frac{1}{2} \left(\frac{x_2 - \mu_2}{\sigma_2} \right)^2 \right\}$$

$$= f(x_1) f(x_2)$$

and hence X_1 and X_2 are independent.

[Note : The multivariate normal density is constant on the surface where the squared distance $(X - \mu)^T \Sigma^{-1} (X - \mu)$ is constant. These paths are called contours.

• Constant probability density contour
• = {all X such that $(X - \mu)^T \Sigma^{-1} (X - \mu) = c^2$} ...(7)
• = surface of the ellipsoid centered at μ
• The axes of each ellipsoid of constant density are in the direction of the eigen – vectors of Σ^{-1} and their lengths are proportional to the reciprocals of the square

roots of the eigen values of Σ^{-1}.

- These ellipsoids are also determined by the eigen values and eigen vectors of Σ.

- If Σ is positive definite, then Σ^{-1} exists and $\Sigma e = \lambda e$ implies $\Sigma^{-1}e = \left(\dfrac{1}{\lambda}\right) e$, so (λ, e)

is an eigen value – eigen vector pair for Σ corresponding to the pair $\left(\dfrac{1}{\lambda}, e\right)$ for Σ^{-1}; so Σ^{-1} is positive definite].

8.2.3 Examples

EXAM. 1: Obtain the axes of constant probability density contours for a bivariate normal distribution when $\sigma_{11} = \sigma_{22}$.

SOLUTION:

From $(X - \mu)^T \Sigma^{-1} (X - \mu) = c^2$, these axes are given by the eigen values and eigen vectors of Σ. Here $|\Sigma - \lambda I| = 0$ becomes

$$O = \begin{vmatrix} \sigma_{11} - \lambda & \sigma_{12} \\ \sigma_{12} & \sigma_{11} - \lambda \end{vmatrix} = (\sigma_{11} - \lambda)^2 - \sigma_{12}^2 = (\lambda - \sigma_{11} - \sigma_{12})(\lambda - \sigma_{11} + \sigma_{12})$$

Consequently, the eign values are $\lambda_1 = \sigma_{11} + \sigma_{12}$ and $\lambda_2 = \sigma_{11} - \sigma_{12}$. The eigen vector e_1 is determined from

$$\begin{bmatrix} \sigma_{11} & \sigma_{12} \\ \sigma_{12} & \sigma_{11} \end{bmatrix}\begin{bmatrix} e_1 \\ e_2 \end{bmatrix} = \begin{bmatrix} \sigma_{11}e_1 & + & \sigma_{12}e_2 \\ \sigma_{12}e_1 & + & \sigma_{11}e_2 \end{bmatrix} = (\sigma_{11} + \sigma_{12})\begin{bmatrix} e_1 \\ e_2 \end{bmatrix}$$

This implies $e_1 = e_2$ and normalizing then yields $\lambda_1 = \sigma_{11} + \sigma_{12}, e_1 = \left(\dfrac{1}{\sqrt{2}}, \dfrac{1}{\sqrt{2}}\right)^T$ and λ_2

$= \sigma_{11} - \sigma_{12}$ yields $e_2 = \left(\dfrac{1}{\sqrt{2}}, \dfrac{1}{\sqrt{2}}\right)^T$

When the covariance σ_{12} (or correlation ρ_{12}) is positive, $\lambda = \sigma_{11} + \sigma_{12}$ is the largest eigenvalue and its associated eigenvector $e_1 = \left(\dfrac{1}{\sqrt{2}}, \dfrac{1}{\sqrt{2}}\right)^T$ lies along the 45° line through the point $\mu = (\mu_1, \mu_2)^T$. The eigenvectors have each length unity and the major axis will be associated with the largest eigenvalue.

When the covariance (or correlation) is negative, $\lambda_2 = \sigma_{11} - \sigma_{12}$ will be the largest – eigenvalue and the major axis will lie along a line at right – angle to the 45° line through μ

[Note: The choice of $c^2 = \chi_p^2(\alpha)$, (a), where $\chi^2(\alpha)$ is the upper (100α) percentile of a chi-square distribution with p degrees of freedom, leads to the contours that contain $(1 - \alpha)100\%$ of the probability.

The solid ellipsoid of X values satisfying $(X - \mu)^T \Sigma^{-1} (X - \mu) \le \chi_p^2(\alpha)$ has probability $(1 - \alpha)]$

8.2.4 Key Properties of the Multivariate Normal Distribution

Let X be a random vector having a multivariate normal distribution. It has the following properties.

- Linear Combinations of the components of X are normally distributed.
- All subsets of the components of X have a multivariate normal distribution
- Zero covariance implies that the corresponding components are independently distributed.
- The conditional distributions of the components are multivariate normal.

8.2.5 Examples

EXAM. 1: For X distributed as $N_3(\mu, \Sigma)$ find the distribution of

$$\begin{bmatrix} x_1 - x_3 \\ -x_2 + x_3 \end{bmatrix} = \begin{bmatrix} 1 & 0 & -1 \\ 0 & -1 & 1 \end{bmatrix} \begin{bmatrix} x_1 \\ x_2 \\ x_3 \end{bmatrix} = AX$$

The distribution of AX is multivariate normal with mean

$$A\mu = \begin{bmatrix} 1 & 0 & -1 \\ 0 & -1 & 1 \end{bmatrix} \begin{bmatrix} \mu_1 \\ \mu_2 \\ \mu_3 \end{bmatrix} = \begin{bmatrix} \mu_1 & -\mu_3 \\ -\mu_2 & +\mu_3 \end{bmatrix}$$

and covariance matrix

$$A\Sigma A^T = \begin{bmatrix} 1 & 0 & -1 \\ 0 & -1 & 1 \end{bmatrix} \begin{bmatrix} \sigma_{11} & \sigma_{12} & \sigma_{13} \\ \sigma_{12} & \sigma_{22} & \sigma_{23} \\ \sigma_{13} & \sigma_{23} & \sigma_{33} \end{bmatrix} \begin{bmatrix} 1 & 0 \\ 0 & -1 \\ -1 & 1 \end{bmatrix}$$

$$= \begin{bmatrix} \sigma_{11} - \sigma_{13} - \sigma_{13} + \sigma_{33} & \sigma_{12} + \sigma_{13} + \sigma_{23} - \sigma_{33} \\ -\sigma_{12} + \sigma_{13} + \sigma_{23} - \sigma_{33} & \sigma_{22} - 2\sigma_{23} + \sigma_{33} \end{bmatrix}$$

Alternately the mean vector Am and covariance matrix $A\Sigma A^T$ may be verified by direct calculation of the means and covariance of the two random variables

$Y_1 = X_1 - X_3$ and $Y_2 = -X_2 + X_3$

EXAM. 2: If X is distributed as $N_5(\mu, \Sigma)$, find the distribution of $\begin{bmatrix} X_2 \\ X_3 \end{bmatrix}$,

SOLUTION:

We set $X_1 = \begin{bmatrix} X_2 \\ X_3 \end{bmatrix}$, $\mu_1 = \begin{bmatrix} \mu_2 \\ \mu_3 \end{bmatrix}$, $\Sigma_{11} = \begin{bmatrix} \sigma_{11} & \sigma_{23} \\ \sigma_{23} & \sigma_{33} \end{bmatrix}$ X, μ and Σ can be rearranged and

partitioned as

$$X = \begin{bmatrix} X_2 \\ X_3 \\ X_1 \\ X_4 \\ X_5 \end{bmatrix}, \mu = \begin{bmatrix} \mu_2 \\ \mu_3 \\ \mu_1 \\ \mu_4 \\ \mu_5 \end{bmatrix}, \Sigma \begin{bmatrix} \sigma_{22} & \sigma_{23} & \sigma_{12} & \sigma_{24} & \sigma_{25} \\ \sigma_{23} & \sigma_{33} & \sigma_{13} & \sigma_{34} & \sigma_{35} \\ \sigma_{12} & \sigma_{13} & \sigma_{11} & \sigma_{14} & \sigma_{15} \\ \sigma_{24} & \sigma_{34} & \sigma_{14} & \sigma_{44} & \sigma_{45} \\ \sigma_{25} & \sigma_{35} & \sigma_{15} & \sigma_{45} & \sigma_{55} \end{bmatrix}$$

$$\text{or } X = \begin{bmatrix} X_1 \\ 2x_1 \\ X_2 \\ 3x_1 \end{bmatrix}, \mu = \begin{bmatrix} \mu_1 \\ 2x_1 \\ \mu_2 \\ 3x_1 \end{bmatrix}, \Sigma = \begin{bmatrix} \Sigma_{11} & \Sigma_{12} \\ 2x2 & 2x3 \\ \Sigma_{21} & \Sigma_{22} \\ (3x2) & (3x3) \end{bmatrix}$$

For $X_1 = \begin{bmatrix} X_2 \\ X_3 \end{bmatrix}$, we have the distribution

$$N_2[\mu_1 \ \Sigma_{11}] = N_2\left(\begin{bmatrix} \mu_2 \\ \mu_3 \end{bmatrix}, \begin{bmatrix} \sigma_{22} & \sigma_{23} \\ \sigma_{23} & \sigma_{33} \end{bmatrix}\right)$$

EXAM. 3

(a) If $\underset{(q_1 \ x_1)}{X_1}$ and $\underset{(q_2 \ x_1)}{X_2}$ are independent, then it is always true that cov = $(X_1, X_2) = 0$, a $q_1 x q_2$ matrix of zeros.

(b) If $\begin{bmatrix} x_1 \\ x_2 \end{bmatrix}$ is $Nq_1 + q_2\left(\begin{bmatrix} \mu_1 \\ \mu_2 \end{bmatrix}, \begin{bmatrix} \Sigma_{11} & 0 \\ 0 & \Sigma_{22} \end{bmatrix}\right)$, then x_1 and x_2 are independent if and only if $\Sigma_{12} = 0$.

(c) If X_1 and X_2 are independent and are distributed as $Nq_1(\mu_1, \ \Sigma_{11})$ and $Nq_2(\mu_2, \ \Sigma_{22})$, respectively, then $\begin{bmatrix} X_1 \\ X_2 \end{bmatrix}$ is multivariate normal

$$Nq_1 + q_2\left(\begin{bmatrix} \mu_1 \\ \mu_2 \end{bmatrix}, \begin{bmatrix} \Sigma_{11} & 0 \\ 0 & \Sigma_{22} \end{bmatrix}\right)$$

EXAM. 4: Let X be N_3 (μ, Σ) with

$$\Sigma = \begin{bmatrix} 5 & 2 & 0 \\ 2 & 4 & 0 \\ 0 & 0 & 3 \end{bmatrix} \text{ Let } X_1 \text{ be } \begin{bmatrix} x_1 \\ x_2 \end{bmatrix} = \begin{bmatrix} 5 & 2 \\ 2 & 4 \\ 0 & 0 \end{bmatrix} \text{ and } X_2 = X_3 = \begin{bmatrix} 0 \\ 0 \\ 3 \end{bmatrix}$$

Are X_1 and X_2 independent? What about (X_1, X_2) and X_3?

Since X_1 and X_2 have covariance $\sigma_{12} = 2$, they are not independent. However partitioning X and Σ as

$$x = \begin{bmatrix} x_1 \\ x_2 \\ x_3 \end{bmatrix} \text{ and } \Sigma = \begin{bmatrix} S_{11} & S_{12} \\ 2x2 & (2x1) \\ \\ S_{21} & S_{12} \\ (4x2 & 1x1 \end{bmatrix}$$

We see that $X_1 = \begin{bmatrix} X_1 \\ X_2 \end{bmatrix}$ and X_3 have covariance matrix $\Sigma_{12} = \begin{bmatrix} 0 \\ 0 \end{bmatrix}$. Therefore $(X_1,$ $X_2)$ and (X_3) are independent. This implies X_3 is independent of X_1 and also of X_2.

8.3 THE CONDITIONAL DENSITY OF A BIVARIATE NORMAL DISTRIBUTION

The conditional density of X1 given $X_2 = x_2$, for any bivariate distribution, is defined by
$f(x_1 \mid x_2) = \{$conditional density of X_1 given $X_2 = x_2\}$

$$= \frac{f(x_1, x_2)}{f(x_2)}$$

where $f(x_2)$ is the marginal distribution of X_2. If $f(x_1, x_2)$ is the bivariate normal density, we will show that $f(x_1 \mid x_2)$ is

$$N\left(\mu_1 + \frac{S_{12}}{S_{22}}(x_2 - \mu_2), s_{11} - \frac{S_{12}^2}{S_{22}} \right)$$

Here $\sigma_{11} - \sigma_{12}^2/\sigma_{22} = \sigma_{11}(1 - \rho_{12}^2)$. The two terms involving $x_1 - \mu_1$ in the exponent of

the bivariate normal density become, apart from the multiplicative constant $-\dfrac{1}{2}(1 - \rho_{12}^2)$,

$$\frac{(x_1 - \mu_1)^2}{s_{11}} - 2?_{12}\frac{(x_1 - \mu_1)(x_2 - \mu_2)}{\sqrt{s_{11}}\sqrt{s_{22}}}$$

$$= \frac{1}{\sigma_{11}}\left[x_1 - \mu_1 - \rho_{12}\frac{\sqrt{\sigma_{11}}}{\sqrt{\sigma_{22}}}(x_2 - \mu_2)\right]^2 - \frac{\rho_{12}^2}{\sigma_{22}}(x_2 - \mu_2)^2$$

Because $\quad \rho_{12} = \dfrac{\sigma_{12}}{\sqrt{\sigma_{11}}\sqrt{\sigma_{22}}}\quad$ or

$$\rho_{12} = \sqrt{\sigma_{11}}\cdot\sqrt{\sigma_{22}} = \frac{\sigma_{12}}{\sigma_{22}},\quad \text{the complete exponent is}$$

$$\frac{-1}{2\,(1-?_{12}^2)}\left[\frac{(x_1-\mu_1)^2}{s_{11}} - 2?_{12}\frac{(x_1-\mu_1)\,(x_2-\mu_2)}{\sqrt{s_{11}}\,\sqrt{s_{22}}} + \frac{(x_2-\mu_2)^2}{s_{22}}\right]$$

$$= \frac{-1}{2\sigma_{11}(1-\rho_{12}^2)}\left(x_1 - \mu_1 - \rho_{12}^2\frac{\sqrt{\sigma_{11}}}{\sqrt{\sigma_{22}}}(x_2-\mu_2)\right)^2 - \frac{1}{2(1-\rho_{12}^2)}\left(\frac{1}{\sigma_{22}} - \frac{\rho_{12}^2}{\sigma_{22}}\right)(x_2-\mu_2)^2$$

$$= \frac{-1}{2\sigma_{11}(1-\rho_{12}^2)}\left(x_1 - \mu_1 - \frac{\sigma_{12}}{\sigma_{22}}(x_2-\mu_2)\right)^2 - \frac{1}{2}\frac{(x_2-\mu_2)^2}{\sigma_{22}}$$

The constant term $2\pi\sqrt{\sigma_{11}\sigma_{22}(1-\rho_{12}^2)}$

also factors as $\sqrt{2\pi}\sqrt{\sigma_{22}}\times\sqrt{2\pi}\sqrt{\sigma_{11}(1-\rho_{12}^2)}$

Dividing the joint density x_1 and x_2 by the marginal density

$$f(x_2) = \frac{1}{\sqrt{2\pi}\sqrt{\sigma_{22}}}e^{-(x_2-\mu_2)^2/2\sigma_{22}}$$

and canceling terms yields the conditional density

$$f(x_1/x_2) = \frac{f(x_1,x_2)}{f(x_2)}$$

$$= \frac{1}{\sqrt{2\pi\sigma_{11}(1-\rho_{12}^2)}}\exp\left\{\frac{-\left[(x_1-\mu_1)-\dfrac{\sigma_{12}}{\sigma_{22}}(x_2-\mu_2)\right]^2}{2\sigma_{11}(1-\rho_{12}^2)}\right\}$$

Thus with our customary notation, the conditional distribution of x_1 given $X_2=x_2$

is $N\left(\mu_1 + (\sigma_{12}/\sigma_{22})(x_2-\mu_2),\ \sigma_{11}(1-\rho_{12}^2)\right)$

Now $\sum_{11} - \sum_{12} \sum_{22}^{-1} \sum_{21} = \sigma_{11} - \sigma_{12}^2 / \sigma_{22} = \sigma_{11}(1 - \rho_{12}^2)$

and $\sum_{12} \sum_{22}^{-1} = \sigma_{12} / \sigma_{22}$ agreeing with result (6) which we obtained by an indirect method.

[Note :
1. All conditional distributions are multivariate normal
2. The conditional mean is of the from

$$\mu_1 + \beta_{1,q+1}\left(x_{q+1} - \mu_{q+1}\right) + \cdots + \beta_{1,p}\left(x_1 - \mu_p\right)$$

$$\vdots \quad \vdots$$

$$\mu_q + \beta_{q,q+1}\left(x_{q+1} - \mu_{q+1}\right) + \cdots + \beta_{q,p}\left(x_p - \mu_p\right),$$ where the b's are defined by

$$\sum_{12} \sum_{22}^{-1} = \begin{bmatrix} \beta_{1,q+1} & \beta_{1,q+2} & \cdots & \beta_{1,p} \\ \beta_{2,q+1} & \beta_{2,q+2} & \cdots & \beta_{2,p} \\ \vdots & \vdots & \vdots & \vdots \\ \beta_{q,q+1} & \beta_{q,q+2} & \cdots & \beta_{q,p} \end{bmatrix} \quad \ldots(9)$$

3. The conditional covariance $\sum_{11} - \sum_{12} \cdot \sum_{22}^{-1} \sum_{12}$ does not depend upon the value(s) of the conditioning variable(s)

The following are some of the results that con be established.

1. Let **X** be distributed as $N_p(\mu, \Sigma)$ with $|\Sigma| > 0$. Then

   (a) $(\mathbf{x} - \mu)^T \Sigma^{-1}(\mathbf{x} - \mu)$ is distributed as χ^2 where χ^2 denotes the chi-square distribution with p degrees of freedom
   (b) The $N_p(\mu, \Sigma)$ distribution assigns probability $(1 - \alpha)$ to the solid ellipsoid
   (c) $\left\{X : (\mathbf{x} - \mu)^T \Sigma^{-1}(\mathbf{x} - \mu) \leq \chi_p^2(\alpha)\right\}$, where $\chi_p^2(\alpha)$ denotes the upper (100α) the percentile of the χ^2 distribution. $\qquad \ldots(10)$

2. Let $X_1, X_2 ..., X_n$ be mutually independent with X distributed as $N_p(\mu, \Sigma)$ (Note each X_j has the same covariance matrix S). Then
   $$\mathbf{V}_1 = c_1 X_1 + c_2 X_2 + ... + c_n X_n$$

   is distributed as $N_p\left(\sum_{j=1}^{n} c_j \mu_j, \left(\sum_{j=1}^{n} c_j^2\right) \Sigma\right)$. Moreover $\mathbf{V}_1$ and $\mathbf{V}_2 = b_1 X_1 + b_2$

   $X_2 + ... + b_n X_n$ are jointly multivariate normal with covariance matrix

$$
\begin{bmatrix}
\left(\displaystyle\sum_{j=1}^{n} c_j^2 \right) S & \left(b^T c \right) S \\[3ex]
\left(b^T c \right) S & \left(\displaystyle\sum_{j=1}^{n} b_j^2 \right) S
\end{bmatrix}
$$

Consequently, V_1 and V_2 are independent if $b^T c = \displaystyle\sum_{j=1}^{n} c_j b_j = 0$.

Joint density function of all observations, when $X_1, X_2, ..., X_n$ are mutually independent and each has a distribution $N_p(\mu, \Sigma)$ p being the dimension of the normal distribution, is given by

$$
\begin{Bmatrix} \text{Joint} \\ \text{density of} \\ X_1, X_2, ... X_n \end{Bmatrix} = \prod_{j=1}^{n} \left\{ \frac{1}{(2\pi)^{p/2} |\Sigma|^{1/2}} \exp \left\{ -\frac{1}{2} (x_j - \mu)^T \Sigma^{-1} (x_j - \mu) \right\} \right.
$$

$$
= \frac{1}{2p^{np/2}} \frac{1}{|S|^{n/2}} \exp \left\{ -\frac{1}{2} \sum_{j=1}^{n} (x_j - \mu)^T \Sigma^{-1} (x_j - \mu) \right\} \qquad ...(11)
$$

Likelihood:

When the numerical values of the observations become available, these may be substituted for the x_j in equation (11). The resulting expression, now considered as a function of μ and Σ for the fixed set of observations $X_1, X_2, ..., X_n$ is called the likelihood.

The population parameters are the "best" to explain the observed data. By selecting the parameter values that maximize the joint density evaluated at the observations. This technique is called maximum likelihood estimation and the maximizing parameter values are called maximum likelihood estimates.

[Note : The trace of a square matrix is the sum of its diagonal elements]

Using the concept of trace, we can write the joint density of a random sample from a multivariate normal population as

$$
\begin{Bmatrix} \text{Joint} \\ \text{density of} \\ X_1, X_2, ... X_n \end{Bmatrix} = (2\pi)^{-np/2} |\Sigma|^{-n/2} \exp \left\{ - \mathrm{tr} \left[\Sigma^{-1} \left(\sum_{j=1}^{n} (x_j - \bar{x})(x_j - \bar{x})^T \right. \right. \right.
$$

$$
\left. \left. + n(\bar{x} - \mu)(\bar{x} - \mu)^T \right] / 2 \right\} \qquad ...(13)
$$

Substituting the observed values $X_1, X_2, ..., X_n$ into the joint density is yields the likelihood function. This is denoted by $L(\mu, \Sigma)$ to stress the fact that it is a function of the (unknown) population parameters μ and Σ. Thus we have

$$L(\mu, S) = \frac{1}{(2p)^{np/2} |S|^{n/2}} \exp\left\{-\operatorname{tr}\left[S^{-1}\sum_{j=1}^{n}\left(x_j - \overline{x}\right)\left(x_j - \overline{x}\right) + n\left(\overline{x} - \mu\right)\left(\overline{x} - \mu\right)^T /2\right]\right\}$$

8.4 MAXIMUM LIKELIHOOD ESTIMATION OF μ AND Σ

The maximum likelihood estimates of m and S are those values denoted by $\hat{\mu}$ and $\hat{\Sigma}$ - that maximize the function $L(\mu, \Sigma)$ in (13). The estimates $\hat{\mu}$ and $\hat{\Sigma}$ will depend on the observed values $X_1, X_2 ..., X_n$ through the summary statistics $\overline{x}$ and S

Let $X_1, X_2, ..., X_n$ be a random sample from a normal population with mean μ and covariance Σ respectively. Then $\hat{\mu} = \overline{x}$ and

$$\hat{S}: \frac{1}{n}\sum_{j=1}^{n}\left(x_j - \overline{x}\right)\left(x_j - \overline{x}\right)^T = \frac{n-1}{n}S \qquad ...(14)$$

are called the maximum likelihood estimates of μ and Σ.

Sufficient Statistics

Let $X_1, X_2, ..., Xn$ be a random sample from a multivariate normal population with mean m and covariance Σ. Then $\overline{x}$ and S are sufficient statistics

$$\left[\text{Here}: \sum_{j=1}^{n}\left(\overline{x}_j - \overline{x}\right)\left(\overline{x}_j - \overline{x}\right)^T = (n-1)S\right] \qquad ...(15)$$

The sample distribution of $\overline{x}$ and S

The tentative assumption that the columns of

$$X = \begin{bmatrix} x_{11} & x_{12} & \cdots & x_{1n} \\ x_{21} & x_{22} & \cdots & x_{2n} \\ \vdots & \vdots & \vdots & \vdots \\ x_{p1} & x_{p2} & \cdots & x_{pn} \end{bmatrix} = \left[X_1, X_2, ... X_n\right]$$

constitute a random sample from a normal population with mean μ and covariance Σ completely determines the sampling distributions of X and S.

The sampling distributions of $\overline{x}$ and S are presented by drawing a parallel with the familiar univariate conclusions.

In the univariate case (p = 1), we know that $\overline{x}$ is normal with mean μ = (population mean) and variance

$$\frac{1}{n}s^2 = \frac{\text{Population variance}}{\text{Sample size}}$$

The sample variance, is given by

$$(n-1)s^2 = \sum_{j=1}^{n}\left(x_j - \overline{x}\right)^2 \text{ and is}$$

distributed as σ^2 times a chi-square variable having $(n-1)$ degrees of freedom. That is $(n-1)s^2$ is distributed as

$$\sigma^2(Z_1^2 + Z_2^2 + ... + Z_{n-1}^2) = (\sigma Z_1)^2 + (\sigma Z_2)^2 + ... + (\sigma Z_{n-1})^2$$

The individual (σZ_j) are independently distributed as $N(0, \sigma^2)$.

The sampling distribution of the sample covariance matrix is called the **Wishart Distribution** after its discoverer. It is defined as the sum of independent products of multivariate normal random vectors. Specifically

$$W_m(.| \Sigma) = \text{Wishart distribution with m d.f}$$

$$= \text{distribution of } \sum_{j=1}^{m} Z_j Z_j^T \quad ...(15)$$

where the Z_j are each independently distributed as $N_p(0, \Sigma)$.

We summarize the sampling distribution results below

Let $X_1, X_2, ..., X_n$ be a random sample of size n from a p-variate normal distribution with mean m and covariance matrix Σ. Then :

(1) $\overline{x}$ is distributed as $N_p\left(\mu, \dfrac{1}{n} S\right)$

(2) $(n-1)S$ is distributed as a wishart random matrix with $(n-1)$ d f.   ...(16)

(3) $\overline{x}$ and S are independent.

8.5 Principal Components

The principal component analysis explains the variance-covariance structure through a few linear combinations of the original variables and achieves data reduction and interpretation. By replacing the original p variables with a smaller of k of the principal components, the original data consisting of n measurements on p variables is reduced to one consisting of n measurements on k principal components.

Algebraically, principal components are particular linear combinations of the p random variables $X_1, X_2, ..., X_p$. Geometrically, these linear combinations represent the selection of a new coordinate system obtained by rotating the original system with $X_1, X_2, ..., X_p$ as coordinate axes. The new axes represent the directions with maximum variability and provide simple comprehensible covariance structure.

The development of the principal components depend solely on the covariance matrix Σ (or the correlation matrix ρ) of $X_1, X_2, ..., X_p$ and does not require a multivariate normal assumption. But the principal components derived for multivariate normal populations have useful interpretations in terms of the constant density ellipsoids. In addition, inferences can be made from the sample components when the population is multivariate normal.

Let the random vector $\mathbf{x}^T = [x_1, x_2, ..., x_P]$ have the covariance matrix Σ with eigen values $?_1 \geq ?_2 \geq ... \geq ?_p \geq 0$. Consider the combinations,

$$Y_1 = Y\ell_1^T \underline{x} = \ell_{11}x_1 + \ell_{21}x_2 + ... + \ell_{p1}x_p$$

$$Y_1 = Y\ell_2^T \underline{x} = \ell_{12}x_1 + \ell_{22}x_2 + ... + \ell_{p2}x_p$$

$$...\qquad ....\ ...$$

$$Y_p = Y\ell_p^T \underline{x} = \ell_{1p}x_1 + \ell_{2p}x_2 + ... + \ell_{pp}x_p$$

Here Var $(Y_i) = \ell_i^T \Sigma \ell_i, i = 1, 2, ..., p$ and cov $(Y_i, Y_k) = \ell_i^T \Sigma \ell_k, i, k = 1, 2 ..., p$. The principal are those **un-correlated** linear combinations of $Y_1, Y_2, ..., Y_p$ whose variances are as large as possible.

The first principal component is the linear combination with maximum variances. That is, it maximizes var $(Y_1) = \ell_1^T \underline{\Sigma}\ell_1$. It can be increase by multiplying any ℓ_1 by some constant. It is convenient to restrict the coefficient vectors to unit length. Consequently we define;

First principal component = linear combination of $\ell_1^T X$ that maximizes var $(\ell_1^T X)$

subject to $\ell_1^T \ell_1 = 1$

Second principal component = linear combination $\ell_2^T x$ that maximizes Var $(\ell_2^T x)$ subject to $\ell_2^T \ell_2 = 1$ and cov $(\ell_1^T x, \ell_2^2 x) = 0$.

At the i th step,

i the principal component = linear combination $\ell_i^T x$ that maximizes var $(\ell_i^T x)$ subject to $\ell_i^T \ell_i = 1$ and cov $(\ell_i^T x, \ell_i^T x) = \mathbf{0}$ for $k < i$

[Note: 1. Let Σ have the eigen value-eigenvector pairs $(\lambda_1, e_1), (\lambda_2, e_2), ..., (\lambda_p, e_p)$ where $\lambda_1 \geq \lambda_2,^3 ... \geq \lambda_p \geq 0$. The **i the principal** component is given by

$$Y_i = e_i^T x = e_{1i}x_1 + Y_{2i}x_2 + ... + e_{pi}x_p, i = 1, 2, ..., p \text{ and var } (Y_i) = Y_i^T \sum Y_i = \lambda_i, i = 1,$$

$2, ..., p \text{ cov } (Y_i Y_k) = e_i^T \sum e_k = 0, i \neq k$. If some λ_i are equal, the choices of the corresponding coefficient vectors e_i and hence Y_i are not unique

1. $\sigma_{11} + \sigma_{22} + ... + \sigma_{pp} = \displaystyle\sum_{i=1}^{p} \text{var}(x_i) = \lambda_1 + \lambda_2 + ... \lambda_p$

$$= \sum_{i=1}^{p} \text{var}(Y_i)$$

Hence the total population variance

$$= \sigma_{11} + \sigma_{22} + ... + \sigma_{pp} = \lambda_1 + \lambda_2 + ... + \lambda_p$$

Hence the proportion of total variance due to the k th principal component is

$$= \frac{\lambda_k}{\lambda_1 + \lambda_2 + ... + \lambda_p}, k = 1, 2, ..., p$$

(3) The correlation coefficient between the components Y_i and the variables X_k is given by

$$\rho_{y_i, x_k} = \frac{e_{ki} \sqrt{\lambda_i}}{\sqrt{\sigma_{kk}}}, i, k = 1, 2, ... p$$

8.6 PRINCIPAL COMPONENTS OBTAINED FROM STANDARDIZED VARIABLES

The principal components may also be obtained for the standardized variables

$$Z_1 = \frac{(X_i - \mu_i)}{\sqrt{\sigma_{ii}}}, i = 1, 2, ..., p$$

In matrix notation,

$$Z = \left(V^{\frac{1}{2}}\right)^{-1}(X - \mu)$$

where the diagonal standard deviation matrix $V^{\frac{1}{2}}$ is defined by

$$V^{\frac{1}{2}} = \begin{bmatrix} \sqrt{\sigma_{11}} & 0 & \cdots & 0 \\ 0 & \sqrt{\sigma_{22}} & \cdots & 0 \\ \vdots & \vdots & & \vdots \\ 0 & 0 & \cdots & \sqrt{s_{pp}} \end{bmatrix}$$

Clearly E(Z) = 0 and

$$\text{Cov}(Z) = \left(V^{\frac{1}{2}}\right)^{-1} \Sigma \left(V^{\frac{1}{2}}\right)^{-1} = \rho$$

The principal components of Z may be obtained from the eigenvectors of the correlation matrix $\int$ of ρ.

We shall use the notation Y_i to refer to the s the principal component and (λ_i, e_i) for the eigen value – eigen victor pair. However, these quantities derived from Σ are, in general not the same as the ones derived from ρ.

[Note 1. The i th principal component of the standardized variables $z^T = \left[Z_1, Z_2, ..., Z_p\right]$ with cov (z) = ρ is given by

$$Y_i = e_i^t Z = e_i^T \left(V^{\frac{1}{2}}\right)^{-1} (X - \mu), i = 1, 2, ..., p$$

Also $\displaystyle\sum_{i=1}^{p} \text{var}(Y_i) = \sum_{i=1}^{p} \text{var}(z_i) = p$

and $\quad ? \ y_i \ z_k = e_{ki} \sqrt{?_i}, i, k = 1, 2, ..., p$

In this case, $(\lambda_1, e_1), (\lambda_2, e_2), ...(\lambda_p, e_p)$ are the eigen value-eigenvector pairs for $\underline{\rho}$ with $\lambda_1 \geq \lambda_2 \geq ... \lambda_p \geq$

2. The proportion of (standardized) population variance due to k th principal component

$= \dfrac{?_k}{p}$, k = 1, 2, ...p where the $?_k^i$s are the eigen values of $\underline{\rho}$.

Assume that the vector $X^T = (X_1, X_2, ... X_p)$ be normally distributed with expectation $E(X) = \mu^T = \left[\mu_1, \ \mu_2, ... \mu_p\right]$.

Let Σ be the covariance matrix. Consider the case where Σ is singular (i.e., positive semi definite) and Σ has multiple roots. Let l be a p component column vector such that $\ell^T \ell = 1$. The variance of $\ell^T x$ is

$$E\left[(\ell^T X)^2\right] = E\left[\ell^T X X \ell^T\right] = \ell^T \sum \ell \qquad \qquad ...(1)$$

In order to determine the normalized linear combination $\ell^T X$ with maximum variance, a vector ℓ must be found satisfying $\ell^T \ell = 1$ which maximize $\ell^T \Sigma \ell$

Let $\phi = \ell^T \Sigma \ell - \lambda(\ell^T \ell - 1)$

$$= \sum_{i\,k=1}^{p} \ell_i \, \sigma_{ik} \ell_k - \lambda \left(\sum_{i=1}^{p} \ell_i^2 - 1 \right) \qquad \ldots(2)$$

where λ is a Lagrange multiplies. The vector of partial derivatives $\left(\dfrac{\partial \phi}{\partial \ell_i} \right)$ is

$$\frac{\partial \phi}{\partial \ell} = 2 \sum \ell - 2\lambda \ell \qquad \ldots(3)$$

A vector ℓ maximizing $\ell^T \sum \ell$ must satisfy the expression (3) equated to zero; that is

$$\left(\sum - \lambda I \right) \ell \; = \; 0 \qquad \ldots(4)$$

Hence $\left(\sum - \lambda I \right)$ must be singular., in other words, λ must satisfy

$$\left| \sum - \lambda I \right| \; = \; 0 \qquad \ldots(5)$$

(5) is the characteristic polynomial of the matrix Σ in λ of degree p. The roots are arranged as $\lambda_1 \ge \lambda_2 \ge \ldots \ge \lambda_p$

By multiplying (4) on the lift by ℓ^T, we obtain

$$\ell^T \Sigma \ell \; = \; \lambda \ell^T \ell = \lambda \quad \ldots(6)$$

This shows that if ℓ satisfies (4) (and $\ell^T \ell = 1$), then the variance of $\ell^T x$ is λ. Thus for the maximum variance, we should use in (4), the largest root λ_1.

Let $\ell^{(1)}$ be the normalized solution of $\left(\sum - \lambda I \right) \ell = 0$. Then $V_1 = \left[\ell^{(1)} \right]^T X \qquad \ldots(7)$ is a normalized linear combination with maximum variance.

Let us find a normalized linear combination $\ell^T X$ that has maximum variance of all linear combinations uncorrelated with V_1. Lack of correlation means

$$\begin{aligned} 0 &= \; E\left[\ell^T X V_1 \right] \\ &= \; E\left[\ell^T X X^T \ell^{(1)} \right] \\ &= \; \ell^T \sum \ell^{(1)} \\ &= \; \lambda_1 \ell^T \ell^{(1)} \qquad \ldots(8) \end{aligned}$$

since $\quad \sum \ell^{(1)} \; = \; \lambda_1 \ell^{(1)}$

Thus $\ell^T x$ is orthogonal to V in both the statistical sense (of lack of correlation) and the geometric sense (of the inner product of vectors ℓ and ℓ^T being Zero).

[Note : $\lambda_1 \ell^T \ell^{(1)} = 0$ only if $\ell^T \ell^{(1)} = 0$ when $\lambda_1 \neq 0$ and $\lambda_1 \neq 0$ if $\sum \neq 0$.

8.7 THEOREMS

Theorem 1

Let the ρ-component random vector X have $E[X] = D$ and covariance matrix $E[XX^T]$ $= \sum$. Then there exists an orthogonal linear transformation

$$U = \beta^T X \qquad \qquad ...(9)$$

such that the covariance matrix of U is $E[UU^T] = \wedge$ and

$$\wedge = \begin{bmatrix} \lambda_1 & 0 & 0 & \cdot & \cdot & \cdot & \cdot & \cdot & 0 \\ 0 & \lambda_2 & 0 & \cdot & \cdot & \cdot & \cdot & \cdot & 0 \\ \cdot & & & & & & & & \\ \cdot & & & & & & & & \\ \cdot & & & & & & & & \\ 0 & 0 & 0 & \cdot & \cdot & \cdot & \cdot & \cdot & \lambda_p \end{bmatrix} \qquad ...(10)$$

where $\lambda_1 \geq \lambda_2 \geq ... \geq \lambda_p \geq o$ are the roots of

$$(\sum - \lambda I) = 0.$$

The r-th the column of β, β^T satisfies $(\sum - \lambda_r I)\beta^r = 0$. The r-th component of U, $U_r = [\beta^{(r)}]^T X$ has maximum variance of all normalized linear combinations uncorrelated with $U_1, U_2, ..., U_{r-1}$.

The vector U is defined as the vector of principal components of X

8.7.1 *Maximum Likelihood Estimators of the Principal Components and their Variances*

Theorem 2

Let $x_1, x_2 ..., x_N$ N(> P) observations from $N(\mu, \Sigma)$, where Σ is a matrix with different characteristic roots. Then a set of maximum likelihood estimators of $\lambda_1, \lambda_2, ..., \lambda_p$ and $b^{(1)}$ $b^{(2)}, ..., b^{(p)}$ defined in Theorem (1) is the previous section consists of the roots $k_1 > ... > kp$ of

$$\left| \hat{\Sigma} - kI \right| = 0$$

and a set of corresponding vectors
$b^{(1)}, b^{(2)} \dots b^{(p)}$ satisfying

$$\left(\hat{\Sigma} - k_i I \right) b^{(i)} = 0$$

$\left[b^{(i)} \right] b^{(i)} = 1,$ Where $\hat{\Sigma}$ is the maximum likelihood estimate of Σ.

8.8 Multivariate Normal Distribution

The univariate normal density function can be written as

$$ke^{-\frac{1}{2}\alpha(x-\beta)^2} = ke^{\frac{1}{2}(x-\beta)\alpha(x-\beta)} \qquad \dots(1)$$

where α is positive and k is chosen so that the integral of (1) over the entire x-axis is unity.

The density function of a multivariate normal distribution of $X_1, X_2, \dots, X_p$ has an analogous form. The sealer variable x is replaced by a vector

$$x = \begin{bmatrix} x_1 \\ \cdot \\ \cdot \\ \cdot \\ x_p \end{bmatrix} \qquad \dots(2)$$

the sealer constant β is placed by a vector

$$\beta = \begin{bmatrix} b_1 \\ \cdot \\ \cdot \\ \cdot \\ b_p \end{bmatrix} \qquad \dots(3)$$

and the positive constant a is replaced by a positive definite (symmetrical) matrix

$$A = \begin{bmatrix} a_{11} & a_{12} & \cdot & \cdot & \cdot & a_{1p} \\ a_{21} & a_{22} & \cdot & \cdot & \cdot & a_{2p} \\ \cdot & & & & & \\ \cdot & & & & & \\ \cdot & & & & & \\ a_{p1} & a_{p2} & \cdot & \cdot & \cdot & a_{pp} \end{bmatrix}, a_{ij} = a_{ji} \qquad \qquad ...(4)$$

The square $\alpha(x - b)^2 = (x - b) \, \alpha(x - b)$ is replaced by the quadratic form

$$(x - b)^T A(x - b) = \sum_{i,\,j=1}^{p} a_{ij}(x_i - b_i)\,(x_j - b_j) \qquad \qquad ...(5)$$

Thus the density function of a p-variate normal distribution is

$$f(x_1, x_2, ... x_p) = k e^{-\frac{1}{2}(x_i - b_i)\, A(x_j - b_j)}, \qquad \qquad ...(6)$$

where k(> 0) is chosen so that the integral over the entire p-dimensional Euclidean space of $x_1, x_2, ...x_p$) is unity. Here $f(x_1, x_2, ... x_p)$ is non negative and bounded by k and A is positive definite.

Let
$$k^* = \int_{-\infty}^{\infty} ... \int_{-\infty}^{\infty} e^{-\frac{1}{2}(x - b)' A(x - b)} dx_p ... dx_1 \qquad \qquad ...(7)$$

Since A is position definite, there exists a non singular matrix C such that
$$C^T AC = I, \qquad \qquad ...(8)$$
where I denotes the identity matrix and C^T the transpose of C. Let
$$x - b = Cy, \text{ where} \qquad \qquad ...(9)$$

$$y = \begin{bmatrix} y_1 \\ y_2 \\ \vdots \\ y_p \end{bmatrix} \qquad \qquad ...(10)$$

Hence $(x - b)^T A\,(x - b) = y^T C^T ACy = y^T y$ \qquad ...(11)
The Jacobian of the transformation is
$$J = \text{mod } |C| \qquad \qquad ...(12)$$
where mod |C| indicates the absolute value of the determinant of C. Thus (7) becomes

$$K^* = \text{mod } |C| \int_{-\infty}^{\infty} ... \int_{-\infty}^{\infty} e^{-\frac{1}{2}y^T y} dy_p ... dy_1 \qquad \qquad ...(13)$$

$$= \text{mod } |C| \prod_{i=1}^{p} \left\{ \int_{-\infty}^{\infty} e^{\frac{1}{2}y_i^2} dy_i \right\}$$

$$= \text{mod } |C| \ \underset{i=1}{\overset{p}{\pi}} \ \left\{ \sqrt{2\pi} \right\}$$

$$= \text{mod } |C| \ (2\pi)^{\frac{1}{2}p} \qquad \qquad ...(14)$$

Corresponding to (8) is the determinantal equation

$$|C^T| \ |A| \ |C| \ = \ |I| \qquad \qquad ...(15)$$

Since $\qquad |C^T| \ = \ |C|$ and $|I| = 1$,

we deduce from (15) that

$$\text{mod } |C| \ = \ 1\Big/\sqrt{|A|} \qquad \qquad ...(16)$$

Thus $\qquad k \ = \ \dfrac{1}{k^*} = \sqrt{|A|}\,(2\pi)^{-\frac{1}{2}p} \qquad \qquad ...(17)$

The normal density function is

$$\frac{\sqrt{|A|}}{(2\pi)^{\frac{1}{2}p}} e^{-\frac{1}{2}(x-b)^T A(x-b)} \qquad \qquad ...(18)$$

Theorem 3

If the density of p-dimensional random vector X is

$$\frac{\sqrt{|A|}}{(2\pi)^{p/2}} \ e^{-(x-b)^T A(x-b)}, \qquad \qquad ...(1)$$

then the expected value of X is **b** and covariance matrix is A^{-1}.

Conversely, given a vector **m** and a positive definite matrix Σ, there is a multivariate normal density

$$(2\pi)^{-p/2}|\Sigma|^{-1/2}e^{-\frac{1}{2}(x-\mu)^T\Sigma^{-1}(x-\mu)} \qquad \qquad ...(2)$$

such that the expected values of the vector with this density is μ and the covariance matrix is Σ

8.9 Correlation Coefficient Between X_1 and X_2

It is defined as

$$\rho_{ij} \ = \ \frac{s_{ij}}{\sqrt{s_{ii}}\ \sqrt{s_{jj}}} = \frac{s_{ij}}{s_i s_j} \qquad \qquad ...(3)$$

This measure of association is by metric in X_i and X_j : $?_{ij} = ?_{ji}$ since

$$\begin{bmatrix} s_{ii} & s_{ij} \\ s_{ji} & s_{jj} \end{bmatrix} = \begin{bmatrix} s_i^2 & s_i s_j \; ?_{ij} \\ s_i s_j \; ?_{ij} & s_j^2 \end{bmatrix}$$

is position definite because the determinant

$$\begin{vmatrix} s_i^2 & s_i \; s_j \; ?_{ij} \\ s_i s_j \; ?_{ij} & s_j^2 \end{vmatrix} = s_i^2 s_j^2 (1 - ?_{ij}^2)$$

is positive. Therefore $-1 < ?_{ij} < 1$ ...(4)

• For a bivariate normal distribution, the mean vector is

$$E \begin{bmatrix} X_1 \\ X_2 \end{bmatrix} = \begin{bmatrix} \mu_1 \\ \mu_2 \end{bmatrix}$$ ...(5)

and the covariance matrix

$$\Sigma = E \begin{bmatrix} (X_1 - \mu_1)^2 & (X_1 - \mu_1)(X_2 - \mu_2) \\ (X_2 - \mu_2)(X_1 - \mu_1) & (X_2 - \mu_2)^2 \end{bmatrix}$$

$$= \begin{bmatrix} \sigma_{11} & \sigma_{12} \\ \sigma_{21} & \sigma_{22} \end{bmatrix} = \begin{bmatrix} \sigma_1^2 & \sigma_1\sigma_2 \; \rho \\ \sigma_1\sigma_2 \; \rho & \sigma_2^2 \end{bmatrix},$$ ...(6)

where σ_1^2 is the variance X_1, σ_2^2 is the variance of X_2 and ρ is the correlation between X_1 and X_2. The inverse of (6) is

$$\Sigma^{-1} = \frac{1}{1 - \rho^2} \begin{bmatrix} \dfrac{1}{\sigma_1^2} & -\dfrac{\rho}{\sigma_1\sigma_2} \\ \dfrac{-\rho}{\sigma_1\sigma_2} & \dfrac{1}{\sigma_1^2} \end{bmatrix}$$ ...(7)

The density function of X_1 and X_2 is

$$\frac{1}{2\pi\sigma_1\sigma_2\sqrt{1-\rho^2}} \exp\left\{ -\frac{1}{2(1-\rho^2)} \left[\frac{(x_1 - \mu_1)^2}{\sigma_1^2} \right] - 2\rho\frac{(x_1 - \mu_1)(x_2 - \mu_2)}{\sigma_1\sigma_2} + \frac{(x_2 - \mu_2)^2}{\sigma_2^2} \right\}$$ ...(8)

8.10 THEOREMS

Theorem 1

Let **X** (with p components) be distributed according to $N(\mathbf{\mu}, \mathbf{\Sigma})$. Then **Y** = **CX** is distributed according to $N(C\mu, C\Sigma C^T)$ for C non singular

Theorem 2

If $X_1, X_2, ..., X_p$ have a joint normal distribution, a necessary and sufficient condition for one set of random variables and the subset consisting of the remaining variables to be independent is that each covariance of a variable from one set and a variable from the other set is zero.

Theorem 3

If X is distributed according to $N(\mu, \Sigma)$ the marginal distribution of any set of components of X is multivariate normal with means, variances and covariances by taking the corresponding components of μ and Σ, respectively

Theorem 4

If X is distributed according to $N(\mu, \Sigma)$, then $Z = DX$ is distributed according to $N(D\mu, D\Sigma D^T)$ where D is a q x p matrix of rank $q \leq p$.

8.10.1 Relationships Among the Covariance, Standard Deviation and the Correlation Matrices

Let the means and covariances of the (px1) random vector X can be written as matrices.

The mean vector is $\mu = E(X)$ and the p-variances σ_{ii} and the $\dfrac{p(p-1)}{2}$ distinct covariance $\sigma_{ik}(i < k)$ are contained is the symmetric variance – covariance matrix

$$\Sigma = E\left[(X - \mu)(X - \mu)^T \right]$$

$$= \begin{bmatrix} s_{11} & s_{12} & \cdot & \cdot & \cdot & s_{1p} \\ s_{12} & s_{22} & \cdot & \cdot & \cdot & s_{2p} \\ \cdot & \cdot & & & & \cdot \\ \cdot & \cdot & \cdot & & & \cdot \\ \cdot & \cdot & & \cdot & & \cdot \\ s_{1p} & s_{2p} & \cdot & \cdot & \cdot & s_{pp} \end{bmatrix}$$

The correlation coefficient ρ_{ik} is defined in terms of the covariance σ_{ik} and variances σ_{ii} and σ_{kk}

as $$?_{ik} = \frac{s_{ik}}{\sqrt{s_{ii}}\sqrt{s_{kk}}}$$

The correlation coefficient measures the amount of linear association between the random variables X_i and X_k. The population correlation matrix is (pxp) symmetrical matrix ρ.

where

$$\rho = \begin{bmatrix} \dfrac{s_{11}}{\sqrt{s_{11}}\sqrt{s_{11}}} & \dfrac{s_{12}}{\sqrt{s_{11}}\sqrt{s_{22}}} & \cdots & \cdots & \dfrac{s_{1p}}{\sqrt{s_{11}s_{pp}}} \\[2em] \dfrac{s_{12}}{\sqrt{s_{11}}\sqrt{s_{22}}} & \dfrac{s_{22}}{\sqrt{s_{22}}\sqrt{s_{22}}} & \cdots & \cdots & \dfrac{s_{2p}}{\sqrt{s_{11}}\sqrt{s_{pp}}} \\[2em] \vdots & \vdots & & & \\[1em] \dfrac{s_{1p}}{\sqrt{s_{11}}\sqrt{s_{pp}}} & \dfrac{\sqrt{s_{2p}}}{\sqrt{s_{22}}\sqrt{s_{pp}}} & \cdots & \cdots & \dfrac{s_{pp}}{\sqrt{s_{pp}}\sqrt{s_{pp}}} \end{bmatrix}$$

$$= \begin{bmatrix} 1 & ?_{12} & \cdots & \cdots & ?_{1p} \\ ?_{12} & 1 & \cdots & \cdots & ?_{2p} \\ \vdots & \vdots & & & \\ ?_{1p} & ?_{2p} & \cdots & \cdots & 1 \end{bmatrix}$$

and the (p x p) standard deviation matrix, $V^{\frac{1}{2}}$ is

$$V^{1/2} = \begin{bmatrix} \sqrt{s_{11}} & 0 & \cdots & \cdots & 0 \\ 0 & \sqrt{s_{22}} & \cdots & \cdots & 0 \\ \vdots & \vdots & & & \vdots \\ \vdots & \vdots & & & \vdots \\ 0 & 0 & & & \sqrt{s_{pp}} \end{bmatrix}$$

It can be verified that

$$V^{\frac{1}{2}}?\,V^{\frac{1}{2}} = \Sigma$$

and $\qquad \rho = \left(V^{\frac{1}{2}}\right)^{-1} S \left(V^{\frac{1}{2}}\right)^{-1}$

That is, Σ can be obtained from $V^{\frac{1}{2}}$ and ρ where as ρ can be obtained from Σ.

• If $z_1, z_2, \ldots z_n$ are standardized observations with covariance matrix R, the i-th sample principle component is

$$\hat{y}_i = [\hat{e}_i]^T = \hat{e}_{1i}z_1 + \hat{e}_{2i}z_2 + \ldots + \hat{e}_{pi}z_p, i = 1, 2, \ldots, p$$

where $(\hat{\lambda}_i, \hat{e}_i)$ is the i-th eigen value eigen vector pair of **R** with $\hat{\lambda}_1 \ge \hat{\lambda}_2 \ge \ldots \ge \hat{\lambda}_p \ge 0$.

Also, Sample variance $(\hat{y}_i) = \hat{?}_i, i = 1, 2, \ldots p$, sample covariance $(\hat{y}_i, \hat{y}_k) = 0, i \ne k$. In addition,

Total (standardized) sample variance

$$= \text{tr}(R) = p = \hat{?}_1 + \hat{?}_2 + \ldots + \hat{?}_p$$

and $r_{\hat{y}_i, z_k} = \hat{e}_{ki}\sqrt{\hat{\lambda}_i}, i, k = 1, 2, \ldots, p$

$$\left.\begin{array}{l}\text{Proportion of (standardized)}\\ \text{sample variance due to the}\\ \text{i-th sample prinicipal}\\ \text{component}\end{array}\right\} = \frac{\hat{?}_i}{p}, i = 1, 2, \ldots, p.$$

8.11 Results Concerning Sample Principal Components

If $S = \{s_{ik}\}$ is the p x p sample covariance matrix with eigen value eigen vector pairs $(\hat{\lambda}_1, \hat{e}_1), (\hat{\lambda}_2, \hat{e}_2), \ldots (\hat{\lambda}_p, \hat{e}_p)$, the i the sample principal component is given by

- $\hat{y}_i = (\hat{e}_i)^T X = \hat{e}_{1i}x_1 + \hat{e}_{2i}x_2 + \ldots + \hat{e}_{pi}x_p, \quad i = 1, 2, \ldots, p,$
- where $\hat{\lambda}_1 \ge \hat{\lambda}_2 \ge \ldots \ge, \hat{\lambda}_p \ge 0$ and **X** is any observation on the variables $X_1, X_2, \ldots X_p$. Also
- Sample variance $(\hat{y}_k) = \hat{?}_k, k = 1, 2, \ldots, p$
- Sample covariance $(\hat{y}_i, \hat{y}_k) = 0, \quad i \ne k.$
- In addition,

- Total sample variance $\sum_{i=1}^{p} s_{ii} = \hat{?}_1 + \hat{?}_2 + ... + \hat{?}_p$ and

- $r_{\hat{y}_i}, r_k = \dfrac{\hat{e}_{ki} \sqrt{\hat{?}_i}}{\sqrt{s_{kk}}}, i, k = 1, 2, ... p$

  - **R** is the sample correlation matrix
  - The sample mean of each principal component is zero.

8.12 PRINCIPAL COMPONENTS BY GRAPHING

Plots of the principal components can reveal suspect observations and also provide checks on the assumptions of normality. As the principal components are linear combinations of the original variables, they may be taken as nearly normal. By verifying the first few principal components whether they are approximately normally distributed while they are used as the input data for additional analyses.

In order to specifically precise of suspecting the observations, the last principal component is taken up for discussion. Each observation $\mathbf{X}_j$ can be expressed as a linear combination

$$\mathbf{x}_j = (\mathbf{x}_j^T \hat{\mathbf{e}}_1)\hat{\mathbf{e}}_1 + (\mathbf{x}_j^T \hat{\mathbf{e}}_2)\hat{\mathbf{e}}_2 + ... + (\mathbf{x}_j^T \hat{\mathbf{e}}_p)\hat{\mathbf{e}}_p$$

$$= \hat{y}_{1j}\hat{\mathbf{e}}_1 + \hat{y}_{2j}\hat{\mathbf{e}}_2 + ... + \hat{y}_{pj}\hat{\mathbf{e}}_p$$

of the complete set of eigenvectors $\hat{\mathbf{e}}_1, \hat{\mathbf{e}}_2, ... \hat{\mathbf{e}}_p$ of **S**. The magnitudes of the last principal components can determine how well the first few fit the observations that fit the data. This means, that $\hat{y}_{1j}\hat{\mathbf{e}}_1 + \hat{y}_{2j}\hat{\mathbf{e}}_2 + ... + \hat{y}_{q-1,j}\hat{\mathbf{e}}_j$ differs from $\mathbf{X}_j$ by $\hat{y}_{qj}\hat{\mathbf{e}}_q + \hat{y}_{q+1j}\hat{\mathbf{e}}_{q+1,j} + ... + \hat{y}_{pj}\hat{\mathbf{e}}_p$ whose squared length is $(\hat{y}_{qj})^2 + ... + (\hat{y}_{pj})^2$. Suspect observations will often be such that at least one of the coordinates $(\hat{y}_{qj}), ... (\hat{y}_{pj})$ contributing to this squared length will be large.

Note :

- The normal assumption can be checked by constructing the scatter diagrams for pairs of the first few principal components. Develop Q-Q plots from the sample values generated by each principal components.
- Construct scatter diagrams and Q-Q plots for the last few principal components. These will help to identify suspect observation.
- We can find a 95 percent confidence interval for λ_1, the variance of the first population component, using the given data. The data represents independent drawings from an $N_p(\mu, \Sigma)$ population where Σ is position definite with distinct eigen values $\lambda_1 > \lambda_2 > ... > \lambda_p > 0$. If $n = 100$, we use

$$\frac{\lambda_i}{1+z(\alpha/2)\sqrt{2/n}} \le \lambda_i \le \frac{\lambda_i}{\left(1-z(\alpha/2)\right)\sqrt{2/n}}$$

where $Z(\alpha/2)$ is the upper $100(\alpha/2)$th percentile of a standard normal distribution, with $i = 1$, to construct a 95 percent confidence interval for λ_1. As an example if $\lambda_1 = 0.0016$ and, is addition, $Z(0.025) = 1.96$, the 95 percent confidence, becomes

$$\frac{0.0016}{1+1.96\sqrt{\dfrac{2}{100}}} \le \lambda_1 \le \frac{0.0016}{1-1.96\sqrt{\dfrac{2}{100}}}$$

or $0.0012 \le \lambda_1 \le 0.0022$

8.12.1 Examples

EXAM. 1: Suppose the random variables X_1 and X_2 have covariance matrix

$$\Sigma = \begin{bmatrix} 9 & -2 \\ -2 & 6 \end{bmatrix}$$

(i) Find the eigen values and the corresponding eigen vectors
The eigen values are given by

$$\begin{vmatrix} 9-\lambda & -2 \\ -2 & 6-\lambda \end{vmatrix} = 0 \ (\text{or}) \ \lambda^2 - 15\lambda + 50 = 0 \ (\text{or}) \ \lambda_1 = 10, \ \lambda_2 = 5$$

For $\qquad \lambda_1 = 10$, We have $\begin{bmatrix} 9 & -2 \\ -2 & 6 \end{bmatrix}\begin{bmatrix} x_1 \\ x_2 \end{bmatrix} = 10\begin{bmatrix} x_1 \\ x_2 \end{bmatrix}$ and

this gives $x_1 + 2x_2$ If $x_1 = 1$, then $x_2 = \dfrac{1}{2}$

$$\therefore \ \mathbf{X_1} = \begin{bmatrix} 1 \\ -1/2 \end{bmatrix} \text{ or } \mathbf{e_1} = \begin{bmatrix} 2/\sqrt{5} \\ -1/\sqrt{5} \end{bmatrix}$$

For $\qquad \lambda_2 = 5$, we have $\begin{bmatrix} 9 & -2 \\ -2 & 6 \end{bmatrix}\begin{bmatrix} x_1 \\ x_2 \end{bmatrix}$ and

this gives $x_1 + 2x_2 = 0$ If $x_1 = 1$, then $x_2 = \dfrac{1}{2}$

$$\therefore \qquad \mathbf{X_2} = \begin{bmatrix} 1/2 \\ 1 \end{bmatrix} \text{ or } \mathbf{e_2} = \begin{bmatrix} 1/\sqrt{5} \\ 2/\sqrt{5} \end{bmatrix}$$

(ii) Find the principal components Y_1 and Y_2

We have $\quad \mathbf{y}_1 = \mathbf{e}_1^T \mathbf{x} = \dfrac{2}{\sqrt{5}} x_1 - \dfrac{2}{\sqrt{5}} x_2$

And $\quad \mathbf{y}_1 = \mathbf{e}_1^T \mathbf{x} = \dfrac{2}{\sqrt{5}} x_1 - \dfrac{2}{\sqrt{5}} x_2$

$$\text{Var}(y_1) = \text{var}\left(\frac{2}{\sqrt{5}} x_1 - \frac{1}{\sqrt{5}}\right) x_2$$

$$= \frac{4}{5}.9 + \frac{1}{5}.6 - \frac{2.2}{5}(-2) = 10 \quad (= \lambda_1)$$

$$\text{Var}(y_2) = \text{var}\left(\frac{x_1}{\sqrt{5}} + \frac{2}{\sqrt{5}} x_2\right)$$

$$= \frac{1}{5}\text{var}(x_1) + \frac{4}{5}\text{var}(x_2) + 2\left(\frac{1}{\sqrt{5}}\right)\left(\frac{2}{\sqrt{5}}\right)(-2)$$

$$= \frac{1}{5}(9) + \frac{4}{5}(6) - \frac{8}{5} = 5 \quad (= \lambda_2)$$

$$\text{Cov}\quad (y_1, y_2) = \text{cov}\left(\frac{2}{\sqrt{5}} x_1 - \frac{1}{\sqrt{5}} x_2, \frac{1}{\sqrt{5}} x_1 + \frac{2}{\sqrt{5}} x_2\right)$$

$$= \begin{bmatrix} \dfrac{2}{\sqrt{5}} & -\dfrac{1}{\sqrt{5}} \end{bmatrix}\begin{bmatrix} 9 & -2 \\ -2 & 6 \end{bmatrix}\begin{bmatrix} 1/\sqrt{5} \\ 2/\sqrt{5} \end{bmatrix}$$

$$= \begin{bmatrix} \dfrac{2}{\sqrt{5}} & -\dfrac{1}{\sqrt{5}} \end{bmatrix}\begin{bmatrix} \sqrt{5} \\ 2/\sqrt{5} \end{bmatrix} = 2 - 2 = 0$$

(iii) Find the total population variance

∴ Total population variance is

$$\sigma_{11} + \sigma_{22} = 9 + 6 = 15 = \lambda_1 + \lambda_2 = 10 + 5$$

(iv) Find the proportion of total population variance due to (a) the first principal component and (b) the second principal component.

(a) $\dfrac{\lambda_1}{\lambda_1 + \lambda_2} = \dfrac{9}{15} = \dfrac{3}{5}$ (b) $\dfrac{\lambda_2}{\lambda_1 + \lambda_2} = \dfrac{6}{15} = \dfrac{2}{5}$

(v) Find the correlation coefficient between y_i and x_k, i = 1, 2, k = 1, 2

Using the covariance matrix Σ, we have

$$\rho_{y_i, x_k} = \frac{\mathbf{e}_{ki}}{\sqrt{\sigma_{kk}}} \sqrt{\lambda_i}, \text{ we have}$$

$$\rho_{y_1, x_i} = \frac{\mathbf{e}_{11}}{\sqrt{\sigma_{11}}} \sqrt{\lambda_1} = \frac{2/\sqrt{5} \sqrt{10}}{3} = \frac{2\sqrt{2}}{3}$$

$$\rho_{y_1, x_2} = \frac{\mathbf{e}_{21}}{\sqrt{\sigma_{22}}} \sqrt{\lambda_2} = \frac{\frac{-1}{5}\sqrt{6}}{\sqrt{6}} = -\frac{1}{\sqrt{5}} = -\frac{1}{\sqrt{5}}$$

$$?_{y_2, x_1} = \frac{\mathbf{e}_{21}\sqrt{\lambda_1}}{\sqrt{\sigma_{11}}} = \frac{\frac{1}{5} - \sqrt{10}}{3} = \frac{\sqrt{2}}{3}$$

$$?_{y_2, x_2} = \frac{\mathbf{e}_{22}\sqrt{\lambda_2}}{\sqrt{\sigma_{22}}} = \frac{\frac{2}{\sqrt{5}}\sqrt{5}}{\sqrt{6}} = \frac{2}{\sqrt{6}}$$

(vi) Convert the covariance matrix Σ to a correlation matrix.

We have $\quad \Sigma = \begin{bmatrix} 9 & -2 \\ -2 & 6 \end{bmatrix}$

The correlation coefficient ρ_{ik} is defined interns of the covariance σ_{ik} and variances σ_{ii} and σ_{kk} as

$$\rho_{ik} = \frac{s_{ik}}{\sqrt{s_{ii}} \sqrt{s_{kk}}}, i, k = 1, 2, \ldots$$

$$\therefore \quad \rho_{11} = \frac{\sigma_{11}}{\sqrt{\sigma_{11}} \sqrt{\sigma_{11}}} = 1; \quad \rho_{12} = \frac{\sigma_{12}}{\sqrt{\sigma_{11}} \sqrt{\sigma_{22}}} = -\frac{2}{\sqrt{9} \sqrt{6}} = \frac{-\sqrt{2}}{3\sqrt{3}}$$

$$\therefore \quad \rho_{22} = \frac{\sigma_{22}}{\sqrt{\sigma_{22}} \sqrt{\sigma_{22}}} = 1;$$

Hence

$$\rho = \begin{bmatrix} \dfrac{\sigma_{11}}{\sqrt{\sigma_{11}} \sqrt{\sigma_{11}}} & \dfrac{\sigma_{12}}{\sqrt{\sigma_{11}} \sqrt{\sigma_{22}}} \\ \dfrac{\sigma_{12}}{\sqrt{\sigma_{11}} \sqrt{\sigma_{22}}} & \dfrac{\sigma_{22}}{\sqrt{\sigma_{22}} \sqrt{\sigma_{22}}} \end{bmatrix}$$

$$= \begin{bmatrix} 1 & -\dfrac{1}{3}\sqrt{\dfrac{2}{3}} \\ -\dfrac{1}{3}\dfrac{\sqrt{2}}{3} & 1 \end{bmatrix}$$

(vii) Find the eigenvalues and the corresponding eigen vectors of ρ.

The eigen values are given has

$$\begin{vmatrix} 1-\lambda & -\dfrac{1}{3}\sqrt{\dfrac{2}{3}} \\ -\dfrac{1}{3}\sqrt{\dfrac{2}{3}} & 1-\lambda \end{vmatrix} = 0$$

or $\qquad (1 - \lambda)^2 = \dfrac{2}{27}$

$$1 - \lambda = \pm\dfrac{\sqrt{2}}{3\sqrt{3}}$$

$\therefore \qquad 1 - \lambda_1 = \dfrac{-1}{3}\sqrt{\dfrac{2}{3}}$ or $\lambda_1 = 1 + \dfrac{\sqrt{2}}{3\sqrt{3}}$

$$1 - \lambda_2 = +\dfrac{1}{3}\dfrac{\sqrt{2}}{\sqrt{3}} \text{ or } \lambda_2 = 1 - \dfrac{\sqrt{2}}{3\sqrt{3}}$$

For $\qquad \lambda_1 = 1 + \dfrac{1}{3}\dfrac{\sqrt{2}}{\sqrt{3}}, \mathbf{x}_1 = \begin{bmatrix} 1 \\ 1 \end{bmatrix}; \mathbf{e}_1 = \begin{bmatrix} \dfrac{1}{\sqrt{2}} \\ \dfrac{1}{\sqrt{2}} \end{bmatrix}$

For $\qquad \lambda_1 = 1 - \dfrac{\sqrt{2}}{3\sqrt{3}}, \mathbf{x}_2 = \begin{bmatrix} 1 \\ -1 \end{bmatrix}; \mathbf{e}_2 = \begin{bmatrix} \dfrac{1}{\sqrt{2}} \\ \dfrac{-1}{\sqrt{2}} \end{bmatrix}$

In terms of the correlation matrix ρ the principal components become

$$Y_1 = \dfrac{1}{\sqrt{2}}Z_1 + \dfrac{1}{\sqrt{2}}Z_2 = \dfrac{1}{\sqrt{2}}\left(\dfrac{X_1 - \mu_1}{\sqrt{9}}\right) + \dfrac{1}{\sqrt{2}}\left(\dfrac{X_2 - \mu_2}{\sqrt{6}}\right)$$

$$= \frac{1}{3\sqrt{2}}(X_1 - \mu_1) + \frac{1}{2\sqrt{3}}(X_2 - \mu_2)$$

and

$$Y_2 = \frac{1}{\sqrt{2}}Z_1 - \frac{1}{\sqrt{2}}Z_2$$

$$= \frac{1}{\sqrt{2}}\left(\frac{X_1 - \mu_1}{3}\right) - \frac{1}{\sqrt{2}}\left(\frac{x_2 - \mu_2}{\sqrt{b}}\right)$$

$$= \frac{1}{3\sqrt{2}}(x_1 - \mu_1) - \frac{1}{2}\frac{1}{\sqrt{3}}(x_2 - \mu_2)$$

When the variable X_1 and X_2 are standardized, we have

$$?_{y_1, z_1} = e_{11}\sqrt{\lambda_1} = \frac{1}{\sqrt{2}}\sqrt{1 + \frac{1}{3}\sqrt{\frac{2}{3}}} = 0.0798$$

$$?_{y_1, z_2} = e_{21}\sqrt{\lambda_1} = \frac{1}{\sqrt{2}}\sqrt{1 + \frac{1}{3}\sqrt{\frac{2}{3}}} = 0.0798$$

Similarly

$$?_{y_2, z_1} = e_{21}\sqrt{\lambda_2} = \frac{1}{\sqrt{2}}\sqrt{1 + \frac{1}{3}\sqrt{\frac{2}{3}}} = 0.6039$$

$$?_{y_2, z_2} = e_{22}\sqrt{\lambda_2} = -\frac{1}{\sqrt{2}}\sqrt{1 - \frac{1}{3}\sqrt{\frac{2}{3}}} = -0.6039$$

EXAM. 2: Suppose the random variables X_1, X_2, X_3 have the covariance matrix

$$\Sigma = \begin{bmatrix} 6 & -2 & 2 \\ -2 & 3 & -1 \\ 2 & -1 & 3 \end{bmatrix}$$

(i) Find the eigen values and the corresponding eigen vectors of Σ
The eigen values are given by

$$\begin{bmatrix} 6-\lambda & -2 & 2 \\ -2 & 3-\lambda & -1 \\ 2 & -1 & 3-\lambda \end{bmatrix} = O$$

$$(6 - \lambda)(\lambda^2 - 6\lambda + 8) + 2(2\lambda - 4) + 2(2\lambda - 4) = 0$$
$$(6 - \lambda)(\lambda^2 - 6\lambda + 8) + 8(\lambda - 2) = 0$$

$(6 - \lambda)(\lambda - 2)(\lambda - 4) + 8(\lambda - 2) = 0$

or $(\lambda - 2)[(6 - \lambda)(\lambda - 4) + 8] = 0$

$(\lambda - 2)(\lambda^2 - 10\lambda + 16) = 0$

$(\lambda - 2)(\lambda - 2)(\lambda - 8) = 0$

$\lambda_1 = 8, \lambda_2 = 2, \lambda_3 = 2.$

For $\lambda_1 = 8$, where $\begin{bmatrix} 6 - \lambda_1 & -2 & 2 \\ -2 & 3 - \lambda_1 & -1 \\ 2 & -1 & 3 - \lambda \end{bmatrix} \begin{bmatrix} x_1 \\ x_2 \\ x_3 \end{bmatrix} = O$

$$-2x_1 - 2x_2 + 2x_3 = 0 \qquad\qquad x_1 + x_2 - x_3 = 0$$
$$-2x_1 - 5x_2 - x_3 = 0 \quad \text{or} \quad 2x_1 + 5x_2 + x_3 = 0$$
$$2x_1 - x_2 - 5x_3 = 0 \qquad\qquad 2x_1 + x_2 - 5x_3 = 0$$

The above equations are linearly dependent and hence it is sufficient to solve any two of the above equation.

$\therefore$ We take $2x_1 + 5x_1 + x_3 = 0$

$2x_1 - x_2 - 5x_3 = 0$

or $\dfrac{x_1}{-24} = \dfrac{x_2}{12} = \dfrac{x_3}{-12}$

or $\dfrac{x_1}{2} = \dfrac{x_2}{-1} = \dfrac{x_3}{1}$

or $\therefore \quad X_1^T = (2, -1, 1) \therefore e_1^T = \left(\dfrac{2}{\sqrt{6}}, \dfrac{-1}{\sqrt{6}}, \dfrac{1}{\sqrt{6}} \right)$

For $\lambda_2 = 2$, we have

$$4x_1 - 2x_2 + 2x_3 = 0 \qquad\qquad 2x_1 - x_2 + x_3 = 0$$
$$-2x_1 + x_2 - x_3 = 0 \qquad\qquad 2x_1 - x_2 + x_3 = 0$$
and $\qquad 2x_1 - x_2 + 3x = 0 \qquad\qquad 2x_1 - x_2 + x_3 = 0$

There is only one equation and it is given by

$2x_1 - x_2 + x_3 = 0$

Choose $x_1 = 1$, $x_2 = 0$. we get $x_3 = -2$

$\therefore \quad X_3^T = [1, 2, 0] \therefore e_3^T = \left(\dfrac{1}{\sqrt{5}}, \dfrac{2}{\sqrt{5}}, 0 \right)$

and choose $x_1 = 1, x_3 = 0$, we get $2x_1 - x_2 = 0$

$2 - x_2 = 0$ or $x_2 = 2$

$\therefore \quad X_3^T = [1, 2, 0] \therefore e_3^T = \left(\dfrac{1}{\sqrt{5}}, \dfrac{2}{\sqrt{5}}, 0 \right)$

(ii) Find the principal components,

The principle components become

$$Y_1 = e_1^T X = \frac{2}{\sqrt{6}} X_1 - \frac{1}{\sqrt{6}} X_2 + \frac{1}{\sqrt{6}} X_3$$

$$Y_2 = e_2^T X = \frac{1}{\sqrt{5}} X_1 - \frac{2}{\sqrt{5}} X_3$$

and

$$Y_3 = e_3^T X = \frac{1}{\sqrt{5}} X_1 + \frac{2}{\sqrt{5}} X_2$$

(iii) Find the variances and covariance

we have var $(Y_2) = e_i^T \Sigma e_i = \lambda_i \quad i = 1, 2, 3$

$\text{cov} (y_i, y_k) = e_i^T \Sigma e_k = 0, i \neq k$

EXAM. 3: Let $= X_{(3x1)}$ be $N_3(\mu, S)$ with

$$\Sigma = \begin{bmatrix} 4 & 1 & 0 \\ 1 & 9 & 0 \\ 0 & 0 & 25 \end{bmatrix}$$

(i) Are X_1 and X_2 independent? What about (X_1, X_2) and X_3
SOLUTION
Since X_1 and X_2 have covariance $\sigma_{12} = 1$, they are not independent. However, partitioning
X and **E** as

$$X = \begin{bmatrix} X_1 \\ X_2 \\ \cdot \\ X_3 \end{bmatrix}, S = \begin{bmatrix} 4 & 1 & \cdot & 0 \\ 1 & 9 & \cdot & 0 \\ \cdot & \cdot & \cdot & \cdot \\ 0 & 0 & \cdot & 25 \end{bmatrix} = \begin{bmatrix} \Sigma_{11} & \cdot & \Sigma_{12} \\ \cdot & \cdot & \cdot \\ \Sigma_{12} & \cdot & \Sigma_{22} \end{bmatrix}$$

We see that

$X_1 = \begin{bmatrix} X_1 \\ X_2 \end{bmatrix}$ and X_3 have the covariance matrix $S_{12} = \begin{bmatrix} 0 \\ 0 \end{bmatrix}$, therefore, (X_1, X_2) and X_3

are Independent. This implies that X_3 is independent of X_1 and also of X_2.

8.13 EXERCISES

In multivariate normal distributions
[*Note:*

- $F(x, y) = P\{X \le x, Y \le y\}$ for all $(x, y) Î R^2$

- $\dfrac{\partial^2 F}{\partial x \partial y}(x, y) = f(x, y)$

- $F(x, y) = \displaystyle\int_{-\infty}^{y} \int_{-\infty}^{x} f(u, v)\, du\, dv$

- $P\{x \leq X \leq x + \Delta x,\ y \leq Y \leq y + \Delta y\}$

- $\displaystyle\int_{y}^{y+\Delta y} \int_{x}^{x+\Delta x} f(u, v),\ \Delta x > 0,\ \Delta y > 0$

- $P\{(x, y) \in E\} = \displaystyle\iint_{E} f(x_1 y)\, dxdy$

- $F(x) = \displaystyle\int_{-\infty}^{x} \int_{-\infty}^{\infty} f(u, v)\, dudv = \int_{-\infty}^{\mu} f(u)\, du$

- $G(y) = \displaystyle\int_{-\infty}^{y} \int_{-\infty}^{\infty} f(u, v)\, dudv = \int_{-\infty}^{y} g(v)dv$

- $f(x, y) = \dfrac{\partial^2 F}{\partial x \partial y}(x, y) = \dfrac{\partial^2}{\partial x \partial y}\{F(x)G(y)\}$

- $\dfrac{d}{dx} F(x) \dfrac{d}{dy} G(y) = f(x)g(y)$ for the two random variables X and Y to be statistically independent. Similarly

$$F(x, y) \quad = \int_{-\infty}^{y} \int_{-\infty}^{x} f(u, v)\, dudv$$

$$\int_{-\infty}^{x} f(u)\, du \int_{-\infty}^{y} g(v)\, dv = F(x)G(y)$$

The above results can be extended to the random vector

$$X^T = \left[X_1, X_2, ..., X_p \right]$$

Conditional distributions

$$P\{x_1 \leq x \leq x_2 \mid y_1 \leq y \leq y_2\}$$

$$= \frac{\int\limits_{x_1}^{x_2} \int\limits_{y_1}^{y_2} f(u, v)\, du\, dv}{\int\limits_{x_1}^{x_2} g(v)\, dv}$$

$$P\{y_1 \le y \le y_2 \,|\, x_1 \le x \le x_2\} = \frac{\int\limits_{x_1}^{x_2} \int\limits_{y_1}^{y_2} f(u, v)\, du\, dv}{\int\limits_{x_1}^{x_2} f(u)\, du}$$

EXAM. 4:

(i) Let $f(x, y) = \begin{cases} 1, & 0 \le x \le 1, \; 0 \le y \le 1 \\ 0, & \text{other wise} \end{cases}$

(ii) $f(x, y) = \begin{cases} 2, & 0 \le y \le x \le 1 \\ 0, & \text{other wise} \end{cases}$

Find the following for both the above density functions
(a) $F(x,y)$ (b) $F(x)$ (c) $f(x)$ (d) $G(y)$
(e) $g(y)$ (f) $(x \,|\, y)$ (g) $f(y|x)$
(h) Test for independence of the random variables X and Y

EXAM. 5: Let X_1, X_2, X_3 and X_4 be independent $N_p\,(\mu, \Sigma)$ random vectors

(a) Find the marginal distributions of the random vectors

$$V_1 = \frac{1}{6}(X_1 + X_2 - X_3 - X_4) \text{ and } \qquad V_2 = \frac{1}{6}(X_1 + X_2 + X_3 - X_4)$$

(b) Find the joint density of the random vectors V_1 and V_2 defined in (a)

EXAM. 6: Consider a bivariate normal population with $\mu_1 = 0$, $\mu_2 = 3$, $\sigma_{11} = 4$, $\sigma_{22} = 9$ and $\sigma_{12} = 0.5$

(a) Write out the bivariate normal density function

(b) Write out the squared generalized distance expression $(X - \mu)^T \Sigma (X - \mu)$ as a function of x_1 and x_2.

EXAM. 7: Let $x = N_3\,(\mu, S)$ with $\mu^T = (-2, 5, 5)$ and $S = \begin{bmatrix} 2 & -3 & 0 \\ -3 & 6 & 0 \\ 0 & 0 & 3 \end{bmatrix}$.

Which of the following random variables are independent? Explain.

(a) X_1 and X_2 (b) X_2 and X_3 (c) (X_1, X_2) and X_3

(d) $\dfrac{X_1 + X_2}{2}, X_3$ (e) X_1 and $X_2 - \dfrac{3}{2}X_1 - X_3$

(f) Find the distribution of $4X_1 - 3X_2 + 2X_3$

EXAM. 8 Given any symmetric matrix **B** then there exists an other orthogonal matrix **C** such that

$$C^T B C = D = \begin{bmatrix} d_1 & 0 & . & . & . & . & 0 \\ 0 & d_2 & . & . & . & . & 0 \\ . & & & & & & . \\ . & & . & . & . & . & . \\ 0 & 0 & . & . & . & . & d_p \end{bmatrix} \text{ if } \mathbf{B} \text{ is positive definite with } d_i > 0, i = 1, 2, ..., p$$

$$0 = C^T \left| B - \lambda I \right| \left| C \right| = \left| C^T (B - \lambda I) C \right| = \left| C^T B C - \lambda I \right| = \left| D - \lambda I \right|$$

$$= \begin{vmatrix} d_1 - \lambda & 0 & 0 & 0 & . & . & . & . & . & 0 \\ 0 & d_2 - \lambda & 0 & 0 & 0 & . & . & . & 0 \\ . & & . & & & & & & . \\ 0 & 0 & 0 & 0 & 0 & . & . & . & d_p - \lambda \end{vmatrix}$$

Thus the characteristic to are the diagonal elements of the transformed matrix D

Let $A = \begin{bmatrix} 7 & 3 & 2 \\ 3 & 4 & 1 \\ 2 & 1 & 2 \end{bmatrix}$ Find the density function and Σ

$$\left. \begin{array}{l} \text{Hint:} \\ \text{Density} \\ \text{Function} \end{array} \right\} = \frac{\sqrt{A}}{(2\pi)^{p/2}} e^{-\frac{1}{2}(X^T A X)}$$

EXAM. 9: Determine the following pertaining to the covariance matrix Σ of the random variables

$X^T = [X_1, X_2 ..., X_P]$

(a) The eigen values and the corresponding eigen vectors

(b) The principal components

$$Y_1 = e_i^T X = e_{1i1i}X_1 + e_{2i} + X_2 + ... + e_{pi} X_p, i = 1, 2, ..., p$$

(c) $\text{Var}\left[(Y_i) = e_i^T S e_i = \lambda_i, \quad i = 1, 2, ... p\right.$

(d) $\text{cov}\left[(Y_i Y_k) = e_i^T S e_k = 0, \quad i \neq k\right.$

(e) (e) $\displaystyle\sum_{i=1}^{p} \text{var}(X_i) = \lambda_1 + \lambda_2 + ... + \lambda_p = \sum_{i=1}^{p} \text{var}(Y_i)$

(f) Proportion of total population variance due to kth principal component

$$= \frac{\lambda_k}{\lambda_1 + \lambda_2 + ... + \lambda_p}, \quad k = 1, 2 ... p$$

(g) $\boldsymbol{\rho}_{Y_i}, X_k, = \dfrac{e_{ki} \sqrt{\lambda_i}}{\sqrt{\sigma_{kk}}}, \ i, k = 1, 2, ... p$ (the correlation coefficients)

(h) Population correlation matrix $\mathbf{r}$; standard deviation matrix, $\mathbf{V}^{\frac{1}{2}}$ and the relationships

$$\mathbf{V}^{\frac{1}{2}} \boldsymbol{\Sigma} \mathbf{V}^{\frac{1}{2}} = \mathbf{r} \text{ and } \boldsymbol{\rho} = (\mathbf{V}^{1/2})^{-1} \boldsymbol{\Sigma} (\mathbf{V}^{1/2})^{-1}$$

(i) Find Principal components of the standardized variables

$$\mathbf{Z} = \left(\mathbf{V}^{\frac{1}{2}} \right)^{-1} (\mathbf{X} - \boldsymbol{\mu}) \text{ where } Z_i = \frac{X_i - \mu_i}{\sqrt{\sigma_{ii}}}, \ i = 1, 2, ... p$$

(j) Verify that $E(\mathbf{Z}) = 0$ and

$$\text{Cov}(\mathbf{Z}) = \left(\mathbf{V}^{\frac{1}{2}} \right)^{-1} \boldsymbol{\Sigma} \left(\mathbf{V}^{\frac{1}{2}} \right)^{-1} = \boldsymbol{\rho}$$

EXAM. 10:

(i) $S = \begin{bmatrix} 5 & 2 \\ 2 & 2 \end{bmatrix}, \begin{bmatrix} 2 & 4 \\ 4 & 1 \end{bmatrix}, \begin{bmatrix} 1 & 4 \\ 4 & 100 \end{bmatrix}, \begin{bmatrix} 5 & 4 \\ 4 & 5 \end{bmatrix}$

(ii) $\Sigma = \begin{bmatrix} 4 & 1 & 2 \\ 1 & 9 & -3 \\ 2 & -3 & 25 \end{bmatrix}, \begin{bmatrix} 13 & -4 & 2 \\ -4 & 13 & -2 \\ 2 & -2 & 10 \end{bmatrix},$

$\begin{bmatrix} 2 & 0 & 1 \\ 0 & 2 & 0 \\ 0 & 0 & 2 \end{bmatrix}, \begin{bmatrix} 6 & -2 & 2 \\ -2 & 3 & -1 \\ 2 & -1 & 3 \end{bmatrix}, \begin{bmatrix} 8 & -6 & 2 \\ -6 & 7 & -4 \\ 2 & -4 & 3 \end{bmatrix}$

$\begin{bmatrix} 4 & 0 & 0 \\ 0 & 9 & 0 \\ 0 & 0 & 1 \end{bmatrix}, \begin{bmatrix} 3 & 1 & 2 \\ 1 & 2 & 3 \\ 2 & 3 & 1 \end{bmatrix}, \begin{bmatrix} 2 & -1 & 3 \\ -1 & 5 & -3 \\ 3 & -3 & 5 \end{bmatrix}$

CHAPTER 9

THEORY OF ESTIMATION

9.1 INTRODUCTION

Statistical constants of the population considered, like the mean (μ), variance (σ^2), etc, are referred to as **Parameters.** On the other hand statistical measures computed from the sample observations of a Random variable (X) alone, like mean ($\bar{x}$), variance (σ^2), etc are termed by Professor R.A. Fisher as **Statistic**. While population parameters are not exactly known, their estimates based on the sample values are generally used. Thus the statistic $\bar{x}$, obtained from a sample is a function of the sample values only. As there are multiple choices of the samples that can be drawn from a population the statistic $\bar{x}$ varies from sample to sample. This variation is attributed to chance fluctuations of sampling.

Around 1930, Prof. R.A. Fisher founded the theory of estimation.

Consider a random sample $x_1, x_2, \ldots x_n$ of size n from a population concerning a random variable X whose distribution is given by the probability function $f(X, \theta_1, \theta_2, \ldots, \theta_k)$, where θ_1, $\theta_2, \ldots, \theta_k$ are the population parameters. The Probability density function

$$f(X, \mu, \sigma) = \frac{1}{\sigma\sqrt{2\pi}} \exp\left\{-(X-\mu)\frac{2}{2\sigma^2}\right\}, -\infty < X < \infty$$

with μ and σ are population parameters, is an example. There will then be always an infinite number of functions of sample values, called statistic, which may be regarded as estimates of one or more of the parameters

The **best estimate** would be one that falls nearest to the true value of the parameter to be estimated.

9.2 Basic Problem of Estimation

It is desirable to determine the functions of the sample observations $\hat{\theta}_1(x_1, x_2 ... x_n)$, $\hat{\theta}_2(x_1, x_2 ... x_n),..., \hat{\theta}_k(x_1, x_2 ... x_n), ...,$ such that their distribution is concentrated, as closely as possible near the true value of the parameter. The estimating functions are called **estimaters**.

9.2.1 *Characteristics of Estimators*

A good estimator should satisfy the following Criteria:
(i) Consistency; (ii) Unbiasedness ; (iii) Efficiency; and (iv)Sufficiency.
Each are of these terms are explained briefly as follows:

• *Consistency*

Let $\hat{\theta}_n$ be an estimator of θ based on a sample of size n. Then $\{\hat{\theta}_n\}$ is a consistent sequence of estimators of θ, if

$$\hat{\theta}_n \xrightarrow{P} \theta \text{ as } n \to \infty$$

i.e, for every $\epsilon > 0$, $\lim_{n \to \infty} P\left(\left|\hat{\theta}_n - \theta\right| > \epsilon\right) = 0$

or equivalently for every $\epsilon > 0, \eta > 0$,

$$P\left(\left|\hat{\theta}_n - \theta\right| > \epsilon\right) > 1 - \eta, n \geq N$$

where N is some very large value of n.
Consistency is a property concerning the behavior of an estimator for indefinitely large values of the sample size and not for finite n.

• *Unbiasedness*

Unbiasedness is a property associated with finite n. A statistic $\hat{\theta}_n = \hat{\theta}_n(x_1, x_2 ..., x_n)$ is said to be an unbiased estimate of parameter θ, if $E(\hat{\theta}_n) = \theta$.

Thus in sampling from a population with mean μ and variance σ^2, we have $E(\bar{x}) = \mu$;

and $E(\sigma^2) \neq \sigma^2$, but $E(\sigma^2) = \sigma^2$, where $\sigma^2 = \dfrac{1}{n-1} \sum_{i=1}^{n}(x_i - \bar{x})^2$, since one degree of freedom

is last while using $\bar{x}$.

The amount of biasedness is given by $b(\theta) = E(\hat{\theta}) - \theta$.

• *Efficiency*

While confining to unbiased estimates, there will in general exist more than one consistent estimator of a parameter. If in a class of consistent estimators for a parameter, there exists one whose sampling variance is less than that of any such estimator, it is called the most efficient estimator. Whenever such an estimator exists, it provides a criterion for measurement of efficiency of other estimators.

If $\hat{\theta}_1$ is the most efficient estimator with variance v_1 and $\hat{\theta}_2$ is any other estimator with variance v_2, then the efficiency E of $\hat{\theta}_2$ is defined as $E = \dfrac{v_1}{v_2}$.

[Note: E cannot exceed unity]

• *Sufficiency*

An estimator is said to be sufficient for a parameter, if it contains all the information in the sample regarding the parameter. More precisely, if $T = t(x_1, x_2, \ldots x_n)$ is an estimator of a parameter q based an a sample $x_1, x_2, \ldots x_n$ of size n from the population with density $f(x, \theta)$ such that the conditional distribution of $x_1, x_2, \ldots x_n$ given T is independent of θ, then t is sufficient estimator for θ

9.3 MINIMUM VARIANCE UNBIASED (MVU) ESTIMATORS

If a statistic $t = t(x_1, x_2, \ldots x_n)$ based on a sample of size n is such that

(i) t is unbiased and
(ii) it has the smallest variances among the class of all unbiased estimators of q, then t is called minimum variance unbiased estimator(MVUE) of q

9.4 METHODS OF ESTIMATION

The commonly used methods are

(i) Method of Maximum Likelihood Estimator
(ii) Method of Minimum Variance
(iii) Method of Moments
(iv) Method of least—squares
(v) Method of Minimum Chi-square
(vi) Method of Inverse Probability.

Each one is briefly discussed below.

9.4.1 *Method of Maximum Likelihood Estimation*

Likelihood Function

Let $x_1, x_2, \ldots x_n$ be a random sample of size n from a population with density function $f(X, \theta)$. Then the likelihood function of the sample values $x_1, x_2, \ldots x_n$ denoted by L, is the

joint density function given by $L = f(x_1, \theta) f(x_2, \theta) \ldots f(x_n, q) = \prod_{i=1}^{n} f(x_i, \theta)$

For a given sample $x_1, x, \ldots x_n$, L is a function of the single variable θ, the parameter.

The principle of maximum likelihood consists in finding an estimator of the parameter which maximizes L for variations in the parameter. Thus, if there exists a function $\hat{\theta} = \hat{\theta}(x_1, x_2 \ldots x_n)$ of the sample values which maximizes L for variations in θ, then $\hat{\theta}$ is to be taken as an estimator of θ. $\hat{\theta}$ is usually called Maximum Likelihood Estimator (MLE). Thus $\hat{\theta}$ is the solution of

$$\frac{\partial L}{\partial \theta} = 0 \text{ and } \frac{\partial^2 L}{\partial \theta^2} < 0$$

Since $L > o$, so is log L and hence L and log L attain their extreme values (maxima or Minima) at the same value of $\hat{\theta}$. The first of the above two equation can be rewritten as

$$\frac{1}{L}\frac{\partial L}{\partial \theta} = 0 \text{ or } \frac{\partial}{\partial \theta}(\log L) = 0$$

The above is referred to as the Likelihood Equation.

• Some Properties of Maximum Likelihood Estimators

(i) It is assumed that $\frac{\partial}{\partial \theta}(\log L)$ and $\frac{\partial^2}{\partial \theta^2}(\log L)$ exist and are continuous functions of θ. In a range R including the true value θ_0 of the parameter for almost all x. For every θ in R, $\frac{\partial}{\partial \theta}(\log L) < F_1(x)$ and $\left|\frac{\partial^2}{\partial \theta^2}(\log L)\right| < F_2(x)$ where $F_1(x)$ and $F_2(x)$ are integrable functions over $(-\infty, \infty)$.

(ii) The third order derivative $\frac{\partial^3}{\partial \theta^3}(\log L)$ exists such that $\left|\frac{\partial^3}{\partial \theta^3}(\log L)\right| < M(x)$, where $E\{M(x)\} < k$, a positive quantity.

(iii) For every θ in R,

$$E\left\{-\frac{\partial^2}{\partial\theta^2}(\log L)\right\} = \int\limits_{-\infty}^{\infty}\int\limits_{-\infty}^{\infty}\cdots\int\limits_{-\infty}^{\infty}\left\{\frac{-\partial^2}{\partial\theta^2}(\log L)\right\}.L.dx_1, dx_2, \ldots dx_n = I(\theta),$$

the information on θ supplied by the sample $(x_1, x_2, \ldots x_n)$, if finite and non zero.

(iv) The range of integration is independent of θ, that is $\left|\frac{\partial}{\partial\theta}\left\{\int\limits_a^b f_1(x)\,dx\right\}\right| = \int\limits_a^b \frac{\partial f}{\partial ?}\,dx,$

if $f(a, \theta) = 0 = f(b, \theta)$

Under the above assumptions, the MLE possesses a number of important properties. These are stated below in the form of theorems.

• *Cramer–Rao Theorem*

With probability approaching unity as n tends to infinity, the likelihood equation $\frac{\partial}{\partial\theta}(\log L) = 0$ has a solution which converges in probability to the true value θ_0. This means that MLEs are consistent.

• *Hazoor Bazar's Theorem*

Any consistent solution of the likelihood equation provides a maximum of the likelihood with probability tending to unity as the sample size, n tends to infinity.

• A consistent solution of the likelihood equation is asymptotically normally distributed about the true value θ_0. Thus $\hat\theta$ is asymptotically $N\left(\theta_0, \frac{1}{I(\theta_0)}\right)$

• If MLE exists, it is most efficient in the class of such estimates.
• If a sufficient estimator exists, it is a function of the Maximum Likelihood Estimator.

9.4.2 *Method of Minimum Variance*

(Minimum Variance Unbiased Estimator - MVUE). Here the importance is given to find out for estimates which are unbiased and have minimum variance.

If $L = \prod\limits_{i=1}^{n} f(x_i, \theta)$ is the likelihood function of a random sample of n observations $x_1, x_2,$ … x_n from a population with Probability function $f(x, \theta)$, then the problem is to find a statistic $t = t(x_1, x_2, \ldots x_n)$, such that

$$E\{t\} = \int t\,L\,d\hat{x} = \theta \Rightarrow \int\limits_{-\infty}^{\infty} (t - \theta)\,L\,dx \qquad \ldots(1)$$

and $\quad \text{Var }\{t\} = \int\limits_{-\infty}^{\infty} \{t.E(t)\}^2 L\, dx = \int\limits_{-\infty}^{\infty} (t-\theta)^2 L\, dx$ is minimum, where $\quad$ ...(2)

$\int_{-\infty}^{\infty} dx$ represents the n-fold integration $\int_{-\infty}^{\infty}\int_{-\infty}^{\infty} .... \int_{-\infty}^{\infty} = \int_{-\infty}^{\infty} dx_1\, dx_2 .... dx_n$

Differentiating (1) partially with aspect to θ, yields $\int_{-\infty}^{\infty} t\dfrac{\partial L}{\partial \theta} dx = 1,$ ...(3)

provided $f(x, \theta)$ is independent of θ or $f(x, \theta)$ vanishes at the extremes dependent on θ.

also $\int_{-\infty}^{\infty} L\, dx = 1 \Rightarrow \int_{-\infty}^{\infty} \dfrac{\partial L}{\partial x} dx = 0$ $\qquad\qquad$ ...(4)

(3) θ times, (4) gives

$$\int\limits_{-\infty}^{\infty} (t-\theta)\frac{\partial L}{\partial \theta} dx = 1 \qquad\qquad ...(5)$$

$$\Rightarrow \int\limits_{-\infty}^{\infty} (t-\theta)\frac{\partial}{\partial \theta}(\log L)\, L dx = 1$$

$$\Rightarrow \int\limits_{-\infty}^{\infty} \left[(t-\theta)\sqrt{L}\right]\sqrt{L\frac{\partial}{\partial \theta}(\log L)}\, dx = 1 \qquad ...(6)$$

Applying Schwartz inequality extended to integrals viz,

$$\int\limits_{-\infty}^{\infty} \left[f(x)\right]^2 dx,\quad \int\limits_{-\infty}^{\infty} [g(x)]^2 dx,\quad \int\limits_{-\infty}^{\infty} [f(x)\, g(x)\, dx]^2,$$

The sign of equality holding if and only if

$\dfrac{f(x)}{g(x)} = k$, where k is a constant independent of x, using (5), we have

$$\int\limits_{-\infty}^{\infty} \left[(t-\theta)\sqrt{L}\right]^2 dx \int\limits_{-\infty}^{\infty} \left[\sqrt{L}\frac{\partial}{\partial \theta}(\log L)\right]^2 dx$$

$$\geq \int\limits_{-\infty}^{\infty} \left[(t-\theta)\sqrt{L}\sqrt{L}\frac{\partial}{\partial \theta}(\log L)\, dx\right]^2 = 1 \qquad\qquad ...(7)$$

$$\Rightarrow \geq \int\limits_{-\infty}^{\infty} \left[(t-\theta)^2 L\, dx \int\limits_{-\infty}^{\infty}\left[\frac{\partial}{\partial \theta}(\log L)^2 L\, dx \geq 1\right.\right.$$

$$\Rightarrow \text{Var}(t) \int_{-\infty}^{\infty} \left[\frac{\partial}{\partial \theta} (\log L) \right]^2 L dx \geq 1 \qquad \qquad ...(8)$$

$$\Rightarrow \text{Var}(t) \geq \frac{1}{\int_{-\infty}^{\infty} \left[\frac{\partial}{\partial \theta} (\log L) \right]^2 L dx} \qquad \qquad ...(9)$$

The minimum variance is attained if there is sign of equality in (9).

Using $\dfrac{f(x)}{g(x)} = \Theta$ the sign of equality holds in (9), iff

$$\frac{(t-\theta)\sqrt{L}}{\sqrt{L \dfrac{\partial}{\partial \theta} [\text{Log } L]}} = \lambda,$$

where λ is constant independent of x and may be a function of θ only. This implies

$$t - \theta = \lambda . \frac{\partial}{\partial ?} (\log L)$$

$$\text{or } \frac{\partial}{\partial \theta} (\log L) = \frac{t - \theta}{\lambda}, \qquad \qquad ...(10)$$

where $t = t(x_1, x_2, ... x_n)$ is a function of the sample observations only and λ may be a function of parameter only but independent of x.

Thus the necessary and sufficient condition for the existence of Minimum Variance Unbiased Estimator –MVUE is that the likelihood function can be expressed in the from (10).

• *Variance of MVUE*

Now $V(t) = \text{Var}\{t\} = \displaystyle\int_{-\infty}^{\infty} (t-\theta)^2 L dx$

$$= \int_{-\infty}^{\infty} \left[\lambda \frac{\partial}{\partial \theta} (\log L) \right]^2 L dx$$

$$= \lambda^2 \int_{-\infty}^{\infty} \left[\frac{\partial}{\partial \theta} (\log L) \right]^2 L dx \text{ being a constant independent of } x.$$

$$= -\lambda^2 \int_{-\infty}^{\infty} \frac{\partial^2}{\partial \theta^2} (\log L) L dx$$

$$= -\lambda^2 \int_{-\infty}^{\infty} \frac{\partial}{\partial\theta}\left(\frac{t-\theta}{\lambda}\right) L\,dx$$

$$= -\lambda^2 \int_{-\infty}^{\infty} \left[(t-\theta)\frac{\partial}{\partial\theta}\left(\frac{1}{\lambda}\right) - \left(\frac{1}{\lambda}\right)\right] L\,dx$$

$$= -\lambda^2 \frac{\partial}{\partial\theta}\left(\frac{1}{?}\right)\int_{-\infty}^{\infty} (t-\theta)\,L\,dx + \lambda \int_{-\infty}^{\infty} L\,dx$$

$$\therefore \qquad V(t) = 0 + \lambda.1 = \lambda \text{ or } V(t) = \lambda \text{ using (1)} \qquad\qquad ...(11)$$

9.4.3 *Method of Moments*

This method was studied by Karl Pearson.

Let $\phi(\theta_1, \theta_2, ..., \theta_n)$ be the density function of the parent population with k parameters $\theta_1, \theta_2, ..., \theta_k$, if μ'_r denotes the r th moment about the origin, then

$$\mu'_r = \int_{-\infty}^{\infty} x^r f(x, \theta_1, \theta_2, ..., \theta_k)\,dx\ (r = 1, 2, ...\,k), \qquad\qquad ...(12)$$

In General, $\mu'_1, \mu'_2 ... \mu'_k$ will be function of the parameters $\theta_1, \theta_2, ..., \theta_k$. $\qquad ...(13)$

Let x_i, $i = 1, 2, ...\,n$ be a random sample of size n from the given population.

The method of moments consists in solving the k-equations (13) for $\theta_1, \theta_2, ..., \theta_k$, in

terms of $\mu'_1, \mu'_2 ... \mu'_k$, μ and then replacing these moments μ'_r, $r = 1, 2, ...\,k$ by sample moments as follows:

$$\hat{\theta}_i = \theta_i(\hat{\mu}'_1, \hat{\mu}'_2 ......\hat{\mu}'_k)$$

$$= \theta_i(m'_1, m'_2, ... m'_k), i = 1, 2... k$$

where m'_i is the i-th moment about the origin in the sample.

Then by the method of moments, $\hat{\theta}_1, \hat{\theta}_2, ...\hat{\theta}_k$ are the required estimators of $\theta_1, \theta_2, ..., \theta_k$ respectively.

Note: 1. Let $(x_1, x_2, ... x_n)$ be a random sample of size n from a population with p.d.f. $f(x, \theta)$. Then X_i, $(i = 1, 2, ..., n)$ are independent identically distributed random variables, implying X_i^r, $(i = 1, 2, ...\,n)$ also independent identically distributed.

Hence if $E[X_i^r]$ exists, then by Weak Law of Large Numbers, we get

$$\frac{1}{n}\sum_{i=1}^{n} x_i^r \xrightarrow{P} E[X_i^r]$$

$$\Rightarrow m'_r \xrightarrow{\quad P \quad} \mu'_r \qquad \qquad ...(14)$$

Hence sample moments are consistent estimators of the corresponding population moments.

2. The estimators obtained by the method of moments are asymptotically normal, but not, is general efficient.

3. Generally, the method of moments yields less efficient estimators than those obtained from the principle of maximum likelihood. However, they are identical if the probability mass function or probability density function is of the form.

$$f(x, \theta) = \exp[b_0 + b_1 x + b_2 x^2 + ...] \qquad \qquad ...(15)$$

where b's are independent of x but may depend on $\theta = (\theta_1, \theta_2 ...)$. (15) implies that

$$L(x_1, x_2, ... x_n, ? = \exp[n\, b_0 + b_1 S x_i + b_2 S x_i^2 + ...]$$

$$\Rightarrow \frac{\partial}{\partial \theta_j} \text{Log } L = [b_0 + b_i S x_i + b_2 S x_i^2 + ...]$$

Thus both methods yield identical estimators, if MLE'S are obtained as linear functions of the moments.

9.4.4 *Method of Least Squares*

The principle of least squares is used to fit a curve of the form

$$y = f(x, a_0, a_1, ..., a_n) \qquad \qquad ...(16)$$

where a_i's are unknown parameters to a set of n sample observations (x_i, y_i), $i = 1, 2, ..., n$ from a bivariate population. The principle consists in minimizing the sum of squares of residuals, viz,

$$E = \sum_{i=1}^{n} [yi - f(x_i, a_0, a_1, ..., a_n)]^2 \qquad \qquad ...(17)$$

subject to variations in $a_0, a_1, ... a_n$. The procedure for estimating $a_0, a_1, ..., a_n$ consists in forming the normal equations given by

$$\frac{\partial E}{\partial a_i} = 0, i = 0, 1, 2, ... n \qquad \qquad ...(18)$$

Solving the above (n + 1) equations, we get the best estimators of $a_0, a_1, ..., a_n$

The method of least squares is used to fit the following types of curves

(a) Fitting a straight line

Let (x_i, y_i), I = 1, 2, ... n be a set of n points and let

$$y = a + bx \text{ be the line of best fit.} \qquad \qquad ...(19)$$

Taking $P_i(x_i, y_i)$ as the set of points in the scatter diagram, the error of estimate or residual for y_i, is

$$E_i = y_i - (a + bx_i)$$

According to the principle of least squares, we have to determine a and b so that

$$E = \Sigma E_i^2 = \sum_{i=1}^{n} (y_i - a - bx_i)^2 \text{ minimum. This means that}$$

$$\frac{\partial E}{\partial a} = 0 = -2\sum_{i=1}^{n}(y_i - a - bx_i)$$

and $$\frac{\partial E}{\partial b} = 0 = -2\sum_{i=1}^{n}(y_i - a - bx_i)x_i$$

The above equation become

$$\sum_{i=1}^{\hat{n}} y_i = na + b\sum_{i=1}^{n} x_i \qquad \qquad ...(20)$$

and $$\sum_{i=1}^{\hat{n}} x_i y_i = a\sum_{i=1}^{n} x_i + b\sum_{i=1}^{n} x_i^2 \qquad ...(21)$$

(20) and (21) are called the **normal equations.**

The quantities $\sum x_i, \sum x_i^2, \sum y_i$ and $\sum x_i y_i$ can be obtained from the given set of points (x_i, y_i), i = 1, 2, ..., n. With these values, a and b are obtained by solving the two normal equations and the line of best fit is obtained in the form y = a + bx.

9.5 EXAMPLES

EXAM. 1: If $x_1, x_2, ..., x_n$ is a random sample from a Poisson population with parameter θ, then $\hat{\theta} = \frac{1}{n}(x_1 + x_2 + ... + x_n) = \bar{x}_n$ is an estimator of θ. Show that $\hat{\theta}$ is a consistent estimator of θ.

SOLUTION:

Since x_i are independent and identically distributed random variables, $P(\theta)$ we have

$E[x_i] = \theta$ and $B_n = \text{Var}[x_i] = \theta < \infty$, for $n > n_0$ $\{\epsilon, \eta\}$ and $P\left\{\left|\hat{\theta} - \theta\right| < \epsilon\right\} > 1 - \eta$ with

$$\lim_{n \to \infty} \frac{\theta}{n^2} \to 0 \text{ for small } \epsilon \text{ and } \eta$$

Hence by the weak law of large numbers,

$$\hat{\theta} = \bar{x}_n \xrightarrow{P} \theta$$

$\therefore \hat{\theta}$ is a consistent estimator of θ.

EXAM. 2: If an element of the sample covariance matrix Σ is

$$s_{ij} = \frac{1}{N-1} \sum_{\alpha=1}^{N} (x_{i\alpha} - \mu_i)(x_{j\alpha} - \mu_i) - \frac{N}{N-1}(\overline{x}_i - \mu_i)(\overline{x}_j - \mu_j),$$

then

$$\overline{x} = \frac{1}{N} \sum_{\alpha=1}^{N} x_\alpha = [\overline{x}_1, \overline{x}_2, \ldots \overline{x}_N]^T$$

With b as any vector equal to the mean vector μ, show that **S**, the sample covariance matrix is a consistent estimator of Σ.

SOLUTION:

The second term of the covariance matrix **S**, $= [s_{ij}]$, the probability limit of the second term in s_{ij} is zero.

The probability limit of the first term is σ_{ij} if $x_1, x_2, \ldots$ are independently and identically distributed with mean μ and covariance matrix Σ.

Hence **S** is a consistent estimator of Σ.

EXAM. 3: Show that the sample covariance matrix S is an unbiased estimator of Σ.

SOLUTION:

Unbiasedness is a property associated with finite N:

Since

$$E[\overline{X}] = \frac{1}{N} E\left[\sum_{\infty=1}^{N} X_\alpha\right] = \mu,$$

the sample mean is an unbiased estimator of the population mean

Let

$$Z_\alpha = \sum_{\beta=1}^{N} b_{\alpha\beta} X_\beta$$

and

$$Z_N = \sum_{\beta=1}^{N} b_{N\beta} X_\beta = \sum_{\beta=1}^{N} \frac{1}{\sqrt{N}} X_\beta = \sqrt{N}\overline{X}.$$

Since Z_N is independent of $Z_1, Z_2, \ldots Z_{N-1}$.

$$E[Z_N] = \sum_{\beta=1}^{N} b_{N\beta} E[X_\beta] = \sum_{\beta=1}^{N} \frac{1}{N} \mu = \sqrt{N} \mu$$

Z_N is distributed according to $N(\sqrt{N}, \mu, \Sigma)$ and $\overline{X} = \left(\frac{1}{\sqrt{N}}\right) Z_N$ is distributed according

to $N\left(\mu \frac{1}{N} \Sigma\right)$

However

$$E[\hat{\Sigma}] = \sum_{\alpha=1}^{N} X_{\alpha}X_{\alpha}^{T} - N\bar{X}\bar{X}^{T}$$

$$= \sum_{\alpha=1}^{N} Z_{\alpha}Z_{\alpha}^{T} - N_{N}Z_{N}^{T}$$

$$= \sum_{\alpha=1}^{N-1} Z_{\alpha}Z_{\alpha}^{T} = \frac{N-1}{N}\Sigma$$

Thus $\hat{\Sigma}$ is a biased estimator of Σ.

$$\therefore S = \frac{1}{N-1}\sum_{\alpha=1}^{N-1} Z_{\alpha}Z_{\alpha}^{T},$$ the sample covariance matrix is an unbiased estimator of Σ.

EXAM. 4: If $x_1, x_2, \ldots x_n$ is a random sample from a normal population $N(\mu, 1)$. Show

that $\dfrac{1}{n}\sum_{i=1}^{n} x_i^2$ is an unbiased estimator of $\mu^2 + 1$.

SOLUTION:
Now $E[x_i] = m$, var $(x_i) = 1$ for every $i = 1, 2, \ldots, n$.
Now $E[x_i^2] = $ var $(x_i) + \{E[x_i]\}^2 = 1 + \mu^2$

$$\therefore E\left[\frac{1}{n}\sum_{i=1}^{n} x_i^2\right] = \frac{1}{n}\sum_{i=1}^{n}(1+\mu^2) = \frac{1}{n}n(1+\mu^2) = 1+\mu^2$$

EXAM. 5: If $x_1, x_2, \ldots x_n$ are random observations on a Bernoulli variable X taking the value 1 with probability p and value 0 with probability $1 - p$, show that an unbiased estimator

of p^2, where $t = \displaystyle\sum_{i=1}^{n} x_i$

SOLUTION:
$$E[x_i] = 1.p + 0(i-p) = p$$
$$E[x_i^2] = 1^2.p + 0^2.(1-p) = p \text{ for } i = 1, 2 \ldots n$$
$$V[x_i] = \text{Var}[x_i] = E[x_i^2] - \{E(x_i)\}^2 = p - p^2 = p(1-p)$$

$$E[t] = E\left[\sum_{i=1}^{n} x_i\right] = \sum_{i=1}^{n} E[x_i] = np, \ V[t] = \text{Var}[t] = V[x_1 + x_2 + \ldots + x_n]$$

$$= V[x_1] + V[x_2] + \ldots + V[x_n],$$

since covariance terms vanish, due to $x_1, x_2 \ldots x_n$ are independent

$$\therefore \qquad V(t) = \sum_{i=1}^{n} p(1-p) = np(1-p)$$

New $\qquad E[t^2] = V[t] + \{E[t]\}^2$
$$= np(1-p) + n^2 p^2$$
$$= np\,[1 - p + np]$$

$$E\left[\frac{t(t-1)}{n(n-1)}\right] = \frac{1}{n(n-1)}\left\{E(t^2) - E(t)\right\}$$

$$= \frac{1}{n(n-1)}\left\{np(1-p+np) - np\right\}$$

$$= \frac{1}{n(n-1)}\left\{np(1-p+np-1)\right\}$$

$$= \frac{1}{(n-1)}\left\{p(n-1)p\right\}$$

$$= \frac{1}{(n-1)}\,(p)(n-1)p = p^2$$

EXAM. 6: A random sample (x_1, x_2, x_3, x_4) of size 4 is drawn from a normal population with unknown mean μ. Consider the following estimators to estimate μ.

(i) $t_1 = \dfrac{x_1 + x_2 + x_3 + x_4}{4}$

(ii) $t_2 = \dfrac{x_1 + x_2 + x_3}{3} + x_4$

(iii) $t_3 = \dfrac{x_1 + 2x_2 + ?x_3}{3}$,

where λ is such that t_3 is an unbiased estimator. Find λ. Are t_1 and t_2 unbiased? State giving reasons, the estimator which is best among t_1, t_2 and t_3

SOLUTION:

We have $E(x_i) = \mu$, Var $(x_i) = \sigma^2$ (say);
Cov $(x_i, x_j) = 0$, $i \neq j = 1, 2, 3, 4 \ldots n(1)$

(i) $E(t_1) = \dfrac{1}{4}\sum_{i=1}^{4} E(x_i) = \dfrac{1}{4}.\sum_{i=1}^{4}\mu = \dfrac{1}{4}.4\mu = \mu$

This implies that t_1 is an unbiased estimator of μ

(ii) $E(t_2) = E\dfrac{(x_1 + x_2 + x_3)}{3} + E(x_4)$

$= \dfrac{1}{3}.3\mu + \mu = 2\mu$

This implies that t_2 in not an unbiased estimator of μ.

(iii) $E(t_3) = \mu$, implies, $\dfrac{1}{3}E(x_1 + 2x_2 + ?x_3) = \mu$

i.e., $E(x_1) + 2E(x_2) + ?E(x_3) = 3\mu$

$\mu + 2\mu + \lambda\mu = 3\mu$. or $\lambda\,\mu = 0$ or $\lambda = 0$ using (1), we get

$V(t_1) = \dfrac{1}{16}[\text{Var}(x_1) + \text{Var}(x_2) + \text{Var}(x_3) + \text{Var}(x_4)] = \dfrac{1}{16}; 4\sigma^2 = \dfrac{1}{4}\sigma^2$

$V(t_2) = \dfrac{1}{9}(\sigma^2 + \sigma^2 + \sigma^2) + \sigma^2 = \dfrac{4}{3}\sigma^2$

$V(t_3) = \dfrac{1}{9}(\sigma^2 + 4\sigma^2) = \dfrac{5}{9}\sigma^2$

Since $V(t_1)$ is least, t_1 is the best estimate in the sense of least variance.

EXAM. 7: Let x_1, x_2, x_3, and x_4 be a random sample of size 4 from a population with mean value μ and variance σ^2 Let T_1, T_2, T_3, T_4 be the estimators used to estimate mean value μ where

$T_1 = x_1 + x_2 + x_3 - 2x_4$
$T_2 = x_1 + 2x_2 + 3x_3 - 5x_4$
$T_3 = \dfrac{kx_1 + x_2 + x_3 + x_4}{4}$

(i) Find whether T1 and T2 are unbiased estimators?
(ii) Find the value of k such that T2 is unbiased estimator for m.
(iii) With this values of k is T3 a consistent estimator?
(iv) Which is the best estimator?

SOLUTION:

Since x_1, x_2, x_3, x_4 is a random sample from a population with mean m and variance σ^2, $E(x_i) = \mu$, $\text{Var}(x_i) = \sigma^2$ $\text{Cov}(x_i, x_j) = 0$, $(i \neq j = 1, 2, 3, \dots n)$ ...(1)

(i) $E(T_1) = 3\mu - 2\mu = \mu$ which implies that T_1 is a unbiased estimator of μ,
$E(T_2) = \mu + 2\mu + 3\mu - 5\mu = \mu$
which implies that T_2 is also an unbiased estimator of μ.

(ii) we are given $E(T_3) = \mu$, this implies $\dfrac{1}{2}(k\mu + 3\mu) = \mu$ $k + 3 = 4$ or $k = 1$.

(iii) with $\dot{k} = 1$, $T_3 = \dfrac{1}{4}(x_1, x_2, + x_3, + x_4) = \bar{x}$

Since sample mean is a consistent estimator of population μ, by the Week Law of large numbers, T_3 is a consistent estimator of μ.

(iv) Var $(T_1) = V(x_1) + V(x_2) + V(x_3) + 4.$ Var $(x_4) = 3\sigma^2 + 4\sigma^2 = 7\sigma^2$

Var $(T_2) = \sigma^2 + 4\sigma^2 + 9\sigma^2 + 25\sigma^2 = 39\sigma^2$

$$\text{Var } (T_3) = \frac{1}{16}\left[\sigma^2 + \sigma^2 + \sigma^2 + \sigma^2\right] = \frac{\sigma^2}{4}$$

Since Var (T_3) is minimum, T_3 is the best estimator of μ in the sense minimum variance.

9.6 Examples in Maximum Likelihood Estimation (MLE)

EXAM. 1: If $(x_1, x_2, ..., x_n)$ is a random sample from a normal population $N(\mu, \sigma^2)$, find the maximum likelihood estimators for

(i) μ when σ^2 is known
(ii) σ^2 when μ is known
(iii) The simultaneous estimation of μ and σ^2

SOLUTION:

Since $X \sim N(\mu, \sigma^2)$, we have

$$L = \prod_{i=1}^{n}\left[\frac{1}{s\sqrt{2p}} \exp\left\{-\frac{1}{2s^2}(x_i - \mu)^2\right\}\right]$$

$$= \left(\frac{1}{\sigma\sqrt{2\pi}}\right)^n \exp\left\{-\frac{1}{2\sigma^2}\sum_{1}^{n}(x_i - \mu)^2\right\}$$

Taking log on both sides, we got

$$\text{Log } L = -\frac{n}{2}\log(2\pi) - \frac{n}{2}\log\sigma^2 - \frac{1}{2\sigma^2}\sum(x_i - \mu)^2$$

Case (i) when σ^2 is known

$$\frac{\partial}{\partial\mu}\log L = 0 \text{ yields} -\frac{1}{2\sigma^2}\sum_{i=1}^{n}2(x_i - \mu)(1) = 0, \quad \sum(x_i - \mu) = 0 \text{ or } \sum_{i=1}^{n}x_i - n\mu = 0$$

$$\hat{\mu} = \frac{1}{n}\sum_{i=1}^{n}x_i = \bar{x}$$

Hence M.L.E for m is the sample mean.

Case (ii) when μ is known.

Here the likelihood equation for estimating σ^2 is

$$\frac{\partial^2}{\partial\sigma^2} \log L = 0; \text{ this implies } -\frac{n}{2}\frac{1}{\sigma^2} + \frac{1}{2\sigma^4}\sum_{i=1}^{n}(x_1 - \mu)^2 = 0$$

This means that $n - \dfrac{1}{\sigma^2}\displaystyle\sum_{i=1}^{n}(x_1 - \mu)^2 = 0$ or $\hat{\sigma}^2 = \dfrac{1}{n}\displaystyle\sum_{i=1}^{n}(x_i - \mu)^2$

Case (iii)

The likelihood equations for the simultaneous estimation of μ and σ^2 are

$$\frac{\partial}{\partial\mu} \log L = 0 \text{ and } \frac{\partial}{\partial\sigma^2} \log L = 0, \text{ giving}$$

$\hat{\mu} = \bar{x}$ and $\hat{\sigma}^2 = \dfrac{1}{n}\displaystyle\sum_{i=1}^{n}(x_i - \hat{\mu}) = \dfrac{1}{n}\displaystyle\sum_{i=1}^{n}(x_i - \bar{x})^2 = s^2$ the sample variance

Note :

(a) Though $E[\hat{\mu}] = E[\bar{x}] = \mu$ and $E[\hat{s}] = E[s^2] = s^2$, the M.L.E s need not necessarily be **unbiased**

(b) Since M.L.E is most efficient, we conclude that $\bar{x}$ is the most **efficient** estimator for the population mean μ

EXAM. 2: Obtain the maximum likelihood estimator of θ for the binomial distribution with p.d.f given by

$$f(x, \theta) = \binom{n}{x}\theta^x(1-\theta)^{n-x}, \quad n = 0, 1, 2, ..., n$$

SOLUTION:

Here $\qquad L = L(\theta, x_0, x_1, x_2, ..., x_n)$

$$= \mathop{p}_{i=0}^{n} f(x_i, ?)$$

$$\mathop{p}_{i=0}^{n}\left[\binom{n}{x_i}\theta^{x_i}(1-\theta)^{n-x_i}\right]$$

$$\therefore \log L = \sum_{i=0}^{n}\left[\log\left\{\left[\binom{n}{x_i}?^{x_i}(1-?)^{n-x_i}\right]\right\}\right]$$

$$\sum_{i=0}^{n}\log\binom{n}{x_i} + \sum_{i=0}^{n}x_i \log ? + \sum_{i=0}^{n}(n - n_i)\log(1-?)$$

The conditions are:

$$\frac{\partial L}{\partial \theta} = 0 \text{ which implies } \sum_{i=o}^{n} x_i \frac{1}{?} + \sum (n - x_i)\frac{(-1)}{(1-?)} = 0, \ \frac{1}{?}\sum_{i=0}^{n} x_i = \frac{1}{1-?}\sum_{i=0}^{n}(n - x_i)$$

$$\frac{1-?}{?}\sum_{i=0}^{n} x_i = \sum_{i=0}^{n} n - \sum_{i=0}^{n} x_i, \ \frac{1-?}{?}\sum x_i = -\sum_{i=0}^{n} x_i, \ \sum_{i=0}^{n} x_i\left[\frac{1-?}{?}+1\right] = n(n+1)$$

$$\sum_{i=0}^{n} x_i\left[\frac{1-\theta+\theta}{\theta}\right] = n(n+1), \ \frac{1}{?}\sum_{i=0}^{n} x_i = n(n+1)$$

$$\text{or } \theta = \frac{x_0 + x_1 + ... + x_n}{n(n+1)} = \frac{1}{n}\left\{\frac{1}{n+1}(x_0 + x_1 + ... + x_n)\right\} = \frac{1}{n}\overline{x} = \frac{\overline{x}}{n}$$

$$\therefore \hat{\theta} = \frac{\overline{x}}{n}$$

EXAM. 3: Find the maximum likelihood estimate for the parameter λ of a Poisson distribution on the basis of sample of size n. Also find its variance.

SOLUTION:

The probability density function of the Poisson distribution with parameter λ is given by

$$P(X = x) = P(x, \lambda) = \frac{e^{-\lambda}\lambda^x}{x!}, 0 \leq x \leq \infty$$

The likelihood function of the random sample x_1, ..., x_n of n observations from this population is

$$L = \underset{i=1}{\overset{n}{p}} P(x_i, ?) = \frac{e^{-n?}?^{(x_1 + x_2 + ... + x_n)}}{x_1! x_2! ... x_n!}$$

$$\therefore \log L = -n\lambda + \left(\sum_{i=1}^{n} x_i\right)\log \lambda - \sum_{i=1}^{n} \log(x_i!) = -n? + n\overline{x}\log ? - \sum_{i-1}^{n} \log(x_i!)$$

The likelihood equation for estimating λ is

$$\frac{\partial}{\partial \lambda}\log L = 0 \text{ yielding } -n + \frac{n\overline{x}}{?} = 0 \text{ or } \hat{?} = \overline{x}$$

Thus MLE for λ is the sample mean $\overline{x}$.

Since the variance of the poison distributor is also λ, the MLE of σ^2 is also the sample mean $\overline{x}$

9.7 Examples in Method of Moments

EXAM. 1: Let $(X_1, X_2, \ldots, X_n)$ be a random sample from the exponential distribution with probability density functions of X as

$$f(x, \theta) = \theta\, e^{-\theta x}, \ 0 < x < \infty, \ \theta > 0$$
$$= 0, \text{ elsewhere}$$

Estimate θ using the method of moments.

SOLUTION:

Here there is only one parameter θ to estimate so we must equate $E(X)$ to $\bar{X}$.

For the exponential distribution, $E(X) = \dfrac{1}{\theta}$

$\therefore E(X) = \bar{X}$ results in $\dfrac{1}{\theta} = \bar{X}$,

So $\hat{\theta} = \dfrac{1}{\bar{X}}$ is the moment estimation of θ.

EXAM. 2

The time to failure of an electronic component follows an exponential distribution with parameter λ. Eight units are randomly selected and tested resulting in the following failure time (in hours):

13.03, 6.07, 68.44, 17.11, 32.54, 8.77, 12.14, 23.42.

Find the moment estimate of λ.

SOLUTION:

Here $\bar{x} = \dfrac{1}{8}(13.03 + 6.07 + 68.44 + 77.11 + 32.54 + 8.77 + 12.14 + 23.42)$

$= \dfrac{1}{8}(181.52) = 22.69$

$\therefore$ the moment estimate of λ is

$\lambda = \dfrac{1}{\bar{x}} = \dfrac{1}{22.69} = 0.04407$

EXAM. 3

Suppose that $X_1, X_2, \ldots X_n$ is a random sample from a normal distribution $N(\mu, \sigma^2)$. Find the moment estimators of μ and σ^2.

SOLUTION:

For the normal distribution

$E(X) = \mu$ and $E(X^2) = \mu^2 + \sigma^2$. Equating $E(X)$ to $\bar{X}$ and $E(X^2)$ to $\dfrac{1}{n}\sum_{i=1}^{n} X_i^2$ gives

$$= \bar{X}, \quad \mu^2 + \sigma^2 = \frac{1}{n}\sum_{i=1}^{n} X_i^2$$

$$\therefore \qquad \hat{\mu} = \bar{X} \quad \text{and}$$

$$\hat{\sigma}^2 = \frac{\displaystyle\sum_{i=1}^{n} X_i^2 - \left(\frac{1}{n}\sum_{i=1}^{n} X_n\right)^2}{n}$$

$$= \frac{\sum (X_i - \bar{X})^2}{n}$$

EXAM. 4

$X_1, X_2, ..., X_n$ is a random sample from a gamma distribution with parameter r and λ, given by

$$f(x) = \frac{\lambda^r x^{r-1} e^{-\lambda x}}{\Gamma(r)}, \quad \text{for } x > 0$$

$= 0$, other wise.

Find the moment estimators of the parameters r and λ.

SOLUTION:

For the gamma distribution,

$$E(X) = \frac{r}{\lambda} \text{ and } E(X^2) = \frac{r(r+1)}{\lambda^2}.$$

The moment estimators are found by solving

$$\frac{r}{\lambda} = \bar{X}, \frac{r(r+1)}{\lambda^2} = \frac{1}{n}\sum_{i=1}^{n} X_i^2$$

i.e $r = 1$ $\bar{X}, \dfrac{r^2 + r}{\lambda^2} = \dfrac{1}{n}\sum_{i=1}^{n} X_i^2, \quad \dfrac{\lambda^2 \bar{X}^2 + \lambda\bar{X}}{\lambda^2} = \dfrac{1}{n}\sum X_i^2, \quad \bar{X}^2 + \dfrac{\bar{X}}{\lambda} = \dfrac{1}{n}\sum X_i^2$

or $\dfrac{\bar{X}}{\lambda} = \dfrac{1}{n}\sum X_i^2 - \bar{X}^2, \quad \lambda = \dfrac{\bar{X}}{\dfrac{1}{n}\sum_{i=1}^{n} X_i^2 - \bar{X}^2}$ or $\hat{\lambda} = \dfrac{\bar{X}}{\dfrac{1}{n}\sum X_i^2 - \bar{X}^2}$

Also $\hat{r} = \hat{?}.\bar{X} = \dfrac{\bar{X}^2}{\dfrac{1}{n}\displaystyle\sum_{i=1}^{n} X_i^2 - \bar{X}^2}$

EXAM. 5: In the example (Example 2), the squares of failure times are given by $(13.03)^2; (6.07)^2; (69.44)^2, (17.11)^2, (32.54)^2, (8.77)^2, (12.14)^2; (23.42)^2$

i.e. 169.70; 36.84; 4686.00; 292.70, 10549.00, 76.91, 147.30; 548.60

If x is a gamma variate with parameters λ and r, find the moment estimates of λ and r.

SOLUTION:

We have $\bar{x} = 22.69$ and $\displaystyle\sum_{i=1}^{d} X_i^2 = 7017.05$

Hence

$\hat{r} = \dfrac{(22.69)^2}{\dfrac{1}{8}(7017.05) - (22.69)^2} = \dfrac{514.7}{877.13 - 514.7}$ and $\hat{\lambda} = \dfrac{22.69}{\dfrac{1}{8}(7017.05) - (22.69)^2}$

$= \dfrac{22.69}{877.13 - 514.7}$ or $\hat{r} = \dfrac{514.7}{364.43} = 0.1413$ and $\hat{?} = \dfrac{22.69}{364.43} = 0.0637.$

9.8.1 Short Answer Examples

1. What is an Estimator?

   Estimator (or) point estimator is a procedure for producing an estimate of a parameter of interest. An estimator is usually a function of only sample data values, and when these data values are available, it results in an estimator of the parameter of interest.

2. Define a point estimator

   A point estimator of some population parameter θ is a single numerical values $\hat{\theta}$ of a statistic $\hat{\Theta}$. The statistic $\hat{\Theta}$ is called the point estimator.

3. What are the commonly used methods of point estimation?

   The following methods of estimation are commonly used for point estimation:
   1. Methods of maximum likelihood estimator
   2. Method of minimum variance
   3. Method of moments
   4. Method of least squares
   5. Method of minimum chi-square
   6. Method of inverse probability

4. What are the characteristics that should be satisfied by a good estimator?
   The following are some of the criteria that should be satisfied by a good estimator:
   (i) Consistency (ii) Unbiasedness (iii) Efficiency and (iv) Sufficiency

5. Define the minimum variance unbiased estimator (MVUE).
   If all the unbiased estimators of θ are considered the one with the smallest variance is called the minimum variance unbiased estimator

6. (a) Define a likelihood function
   Let $x_1, x_2 \ldots x_n$ be a random sample of size n from a population with density function $f(x, \theta)$. Then the likelihood function of the sample values $x_1, x_2 \ldots x_n$, usually denoted by L, is then joint density function, given by:

$$L = f(x_1, ?)\, f(x_1, ?) \ldots f(x_n, ?) = \prod_{i=1}^{n} f(x_i, ?)$$

   (b) Define the maximum likelihood estimator of θ.
   The maximum likelihood estimator of θ is the value of θ that maximizes the likelihood function

$$L(\theta) = \prod_{i=1}^{n} f(x_i, \theta)$$

7. Define the mean square error of an estimator $\hat{\theta}$ of the parameter θ.

   The mean square error of an estimator $\hat{\theta}$ is the expected squared difference between $\hat{\theta}$ and θ, given by

$$
\begin{aligned}
\mathrm{MSE}(\hat{\theta}) &= E[(\hat{\theta} - \theta)^2] \\
&= E[\hat{\theta} - E(\hat{\theta})]^2 + [\theta - E(\hat{\theta})]^2 \\
&= V(\hat{\theta}) + (\text{bias})^2
\end{aligned}
$$

   If $\hat{\theta}$ is an unbiased estimator of θ then the mean square error of $\hat{\theta}$ is equal to the variance of $\hat{\theta}$.

8. Mention the properties of the maximum likelihood estimator.
   Under very general and not restrictive conditions, when the sample size n is large and if $\hat{\theta}$ is the maximum likelihood estimator of the parameter θ,
   (i) $\hat{\theta}$ is an approximately unbiased estimator of θ, $[E(\hat{\theta}) \approx \theta]$.
   (ii) The variance of $\hat{\theta}$ is nearly as small as the variance that could be obtained with any other estimator.
   (iii) $\hat{\theta}$ has an approximate normal distribution.
   (iv) Invariance property

   If $\hat{\theta}_1, \hat{\theta}_2, \ldots \hat{\theta}_k$ are the maximum likelihood estimators of the parameters $\theta_1, \theta_2, \ldots \theta_k$, then the maximum likelihood estimator of any function $h(\theta_1, \theta_2, \ldots \theta_k)$ of these parameters is the same function $h(\hat{\theta}_1, \hat{\theta}_2, \hat{\theta}_3 \dots \hat{\theta}_k)$ of the estimators $\hat{\theta}_1, \hat{\theta}_2, \hat{\theta}_3 \dots \hat{\theta}_k$.

9. Write down the necessary and sufficient condition for the existence of Minimum Variance Unbiased Estimator (MVUE)

The n and s condition is

$$\frac{\partial}{\partial \theta} \log L = \frac{t - \theta}{\lambda}, \text{ where } t = t(x_1, x_2, \dots, x_n) \text{ is a function of the sample observations}$$

only and λ may be a function of parameter only but independent of x.

10. Define the population moment estimators.

Let $X_1, X_2, \dots X_n$ be a random sample from either a probability mass function or probability density function with m unknown parameters $\theta_1, \theta_2, \dots \theta_m$. The moment estimators $\hat{\theta}_1, \hat{\theta}_2, \dots \hat{\theta}_m$ are found by equating the first m population moments to the first m sample moments and solving the resulting equations for the unknown parameters.

11. Define minimum variance unbiased estimator. (MVUE).

If we consider all unbiased estimators of θ, the one with the smallest variance called the minimum variance unbiased estimator.

12. (a) If $X_1, X_2, \dots X_n$ is a random sample of size n from a normal distribution with mean μ and variance σ^2, the sample mean $\overline{X}$ is the MVUE for μ.

Consider the comparison of two possible estimators for m: the sample mean $\overline{X}$ and a single observation from the sample, say X_i. Since both the $\overline{X}$ and X_i are unbiased estimators of μ; for the sample mean we have

$$V(\overline{X}) = \frac{s^2}{n} \text{ and the variance of any observation is } V(X_i) = \sigma^2.$$

Since $V(\overline{X}) < V(X_i)$ for sample size $n \geq 2$, we conclude that the sample mean is a better estimator of μ than a single observation X_i.

(b) If a random sample of size 2n is taken from a population denoted by X and $E(X) = \mu$ and $V(X) = \sigma^2$, then find out which is a better estimator of μ, when

$$\overline{X}_1 = \frac{1}{2n} \sum_{i-1}^{2n} X_i \text{ and } \overline{X}_2 = \frac{1}{n} \sum_{i=1}^{n} X_i$$

Now $V(\overline{X}) = \frac{s^2}{2n}$ and $V(\overline{X}_2) = \frac{s^2}{n}$

Evidently $V(\overline{X}_1) < V(\overline{X}_2)$

Hence $\bar{X}_1$ is a better estimator of μ.

13. If X is distributed with $N(\mu, \sigma^2)$ for a sample size, find out by the methods of moments for the μ whether $\bar{X}$ or σ^2 is unbiased estimator for μ.

Since $m_1 = \dfrac{1}{n}\sum_{i=1}^{n} x_i$, and $m_2 = \dfrac{1}{n}\sum_{i=1}^{n} x_i^2$

$= E(X) = \bar{X} \quad E(X^2) = s^2 + \mu^2$

We solve for μ and σ^2, and get $m_1 = \bar{X}$ as the estimator for μ

Also $\dfrac{1}{n}\sum_{i=1}^{n} x_i^2 - \bar{X}^2 = \sum_{i=1}^{n}\dfrac{1}{n}(x_i - \bar{X})^2 = V^2$, the variance of the sample.

Therefore the estimators for μ and σ^2 are $\mu = \bar{X}$ and $s^2 = V$, Evidently $\mu = \bar{X}$ is unbiased, whereas $s^2 = V$ is biase

9.8.2 *Short Answer Questions*

1. When is a statistic called an unbiased estimator of a population parameter 2.
2. Show that (i) the sample mean $\bar{x}$ is an unbiased estimator of the population mean, μ (ii) the sample variance s^2 is not an unbiased estimator of the population variance σ^2.
3. When is a statistic $\hat{\theta}_1$ is said to be more efficient unbiased estimate of the parameter θ than the statistic $\hat{\theta}_2$.
4. Show that for random samples from normal populations, the mean is more efficient than the median as an estimate of μ.
5. Show that, if s^2 is the sample variance and n, the sample size then $\dfrac{ns^2}{(n-1)}$ is an unbiased estimate of the population variance.
6. If the time to failure of a component of a machine has the exponential distribution with parameter α, find the maximum likelihood estimate of the expected failure time and show that it is also unbiassed.
7. Write a note on point istimation and interval estimation, what is an unbiased estimator?.
8. State and explain the principle of maximum likelihood estimation of population parameter.
9. State some of its optimal properties.

10. Distinguish between the methods of moments and maximum likelihood.
11. Describe the important properties to be possessed by a good estimator.
12. Define Minimum variance unbiased estimator.
13. Describe the method of moments for estimating parameters.
14. (a) If $X_1, X_2, ..., X_n$ is a random sample from

$$F(x; a, b) = \frac{1}{b-a}, a < x < b$$

$= 0$, otherwise,

find estimates of a and b by the method of moments.

(b) Let X be uniformly distributed on the interval 0 to 9. The density functions as

$$f(x) = \frac{1}{a}, \text{ for } 0 \le x \le a = 9, \text{ otherwise.}$$

(c) A random sample of size n is taken from the distribution. Find the maximum likelihood estimator of a.

9.8.3 Exercises Theory of Estimation (Long answer questions)

1. Let $x_1, x_2 ..., x_n$ be uniformly distributed on the interval 0 to a. Show that the moment estimator of a is $\hat{a} = 2\bar{x}$. Is this an unbiased estimator 2.

2. Consider the probability density function $f(x) = c(1 + \theta x), -a \le x \le a$.
   (a) Find the value of the constant c.
   (b) What is the moment estimator for θ?
   (c) Find the maximum likelihood estimator for θ.

3. Let X be a geometric random variable with parameter p and with probability mass function given by

$$P(X = x) = (1 - p)^x p, \ x = 0, 1, 2,... \ \ 0 < p \le 1 = 0, \text{ otherwise}$$

   (a) Find the maximum likelihood estimator of p, based on a random sample of size n
   (b) Find also the moment estimator of p

4. (a) Estimate θ in the density function.
      $f(x, \theta) = (1 + \theta)x^\theta, 0 < x < 1, \theta > 0$
      by the method of (i) maximum likelihood estimate; (ii) moments
   (b) If 10 samples from the above distribution are given as:
      $0.50, 0.44; 0.65; 0.86, 0.67$
      $0.57; 0.76; 0.48, 0.63; 0.64$, find the estimate of θ by the above two methods.

5. (a) Define a maximum likelihood (ML) estimate. List its properties.
   (b) If the time to failure of a component of a machine has the exponential distribution with parameter α, find the ML estimate of the expected failure time and show that it is also unbiased.

6. Obtain the estimator for θ by
   (i) the method of moments and

(ii) the moments maximum likelihood, if

$$f(x; \theta) = \frac{1}{?} \exp\left(-\frac{x}{?}\right); 0 < x < \infty$$

$= 0$, otherwise

7. Consider the probability density function

$$f(x) = \frac{1}{?^2} x \exp\left(-\frac{x}{?}\right), 0 \le x < \infty, 0 < ? < \infty$$

Find the maximum likelihood estimator for θ.

8. Give one example of an unbiased consistent estimator, one example of a biased consistent estimator and one example of an unbiased inconsistent estimator.

9. Show that there are infinitely many unbiased estimators of the parameter θ of the Poisson distribution.

CHAPTER 10

TESTING OF HYPOTHESIS

10.1 INTRODUCTION

Sampling

Any finite or infinite collection of individual units or objects is called a **population.** Sometimes there is an actual **physical population** A **conceptual** or **hypothetical population** is one that does not exist physically. For any statistical investigation complete enumeration of the population is rather impracticable.

A finite subset of statistical individual in a population is called a **sample** and the number of individuals in a sample is called the **sampling size**.

There are two ways of drawing objects (one at a time) for obtaining a sample from a given set of objects, briefly referred to as **sampling from a population.**

- **Sampling with replacement** means that the object which was drawn at random is placed back to the given set and the set is mixed thoroughly. Then the next object is drawn at random.

- **Sampling without replacement** means that the object which was drawn is put aside **Sample space**.

The set of all possible outcomes of a random experiment is called the **sample space** of the experiment. The sample space is denoted as S. The sample space is often defined based on the objective of the analysis.

A sample space is **discrete** if it consists of finite or countable infinite set of outcomes. A samples space is **continuous** if it contains an interval (either finite or infinite) of real numbers.

An **event** is a subset of the sample space of a random experiment. Two events, denoted as E_1, as E_2 such that $E_1 \cap E_2 = \phi$ are said to be **mutually exclusive.**

Whenever a sample space consists of N possible outcomes that are equally likely, the probability of each outcome is $\dfrac{1}{N}$. For a discrete sample space, the probability of an event E, denoted by P(E), equals the sum of the probabilities of the outcomes in E.

Probability is a number that is assigned to each member of collection of events from a random experiment that satisfies the following axioms.

If S is a sample space and E is any event in a random experiment,

1. $P(S) = 1$; $2.0 \le P(E) \le 1$;
3. for two events E_1 and E_2 with $E_1 \cap E_2 = \phi$, $P(E_1 \cup E_2) = P(E_1) + P(E_2)$

10.2 Census And Sampling Methods

Census is concerned with examination of every item with regard to the attributes or characteristics possessed by each individual item. Enumeration of a population of a country is an example for census.

In order to understand the characteristics of a population, a certain number of items is selected from the population and this constitutes a **sample**. The procedure of obtaining the sample is called the **sampling method.**

Necessity for sampling

Except the population census, sampling methods are adopted to know about the population. There is an intuitive feeling that sampling items will reveal some thing about the population. Due to time and cost considerations and the impossibility of analyzing the characteristics of every one of the individual in the population, sampling methods are adopted to know about the population.

Objectives of sampling

With minimum effort, it is possible to obtain maximum amount of information about the population. The confidence reposed in the estimates from the samples, will strengthen to know about the entire population from which samplings are taken.

10.3 Types Of Sampling

Some of the commonly known and frequently used types sampling are :

- Purposive sampling
- Random sampling
- Systematic sampling
- Stratified sampling

- Multiple sampling
- Sequential sampling.

Purposive Sampling

Here the sample units are selected with definite purpose in view. In order to give the picture that the standard of living has increased in a particular locality, we may take individuals with sample from say middle class families by ignoring rich and low incomes groups. This sampling suffers from the drawback of favoritisms and nepotism and does not give a representative sample of the population.

Random Sampling

Here the sample units are selected at random by giving each unit of the population an equal chance of being included in it. Hence a sample is said to be random if it is selected in such a way so that every possible sample item has the same probability of being selected.

Fairly good random samples can be obtained by the use of Tippet's random number tables or by throwing of an unbiased die or draw a lottery or a random number generator.

Systematic Sampling

All the houses in a city have fixed door numbers. In order to obtain a sample, take any house with the specific door number and from it select at equal number of intervals, the houses from the streets in the city. This gives a systematic sample.

Stratified Sampling

Here the entire heterogeneous population is divided into a number of homogenous groups, usually termed as **strata**, which differ from one another but each of these groups is homogeneous with in itself. Then units are sampled at random from each of these stratum, the sample size in each stratum varies according to the relative importance of the stratum in the population. The sample, which is the aggregate of the sampled units of each of the stratum, is termed as **stratified sample** and the technique of drawing this sample is known as **stratified sampling.**

Multiple Sampling

Sometimes while conducting an experiment on the basic element or a production item, there is the possibility of involvement of expenses as will as destruction of the units. On such occasions, it is desirable to conduct an experiment of a certain number of units constituting a Single sampling. Afterwards yet another sampling of items can be subjected to experiment. This sampling is called double sampling.

Using the same principle, multiple number of samplings can be undertaken to test each one of the samples in the experiment. Such a technique yields a multiple sampling.

Sequential Sampling

Without fixing in advance the maximum number of items to be included in the sample for experimentation, each time items are included one by one till the required number of items are included in the sample, we get insequence a sampling called **sequential sampling.** In such a process, it is possible to reduce the number of items for experimentation.

Box and Whisker Plot

The **box plot** is a graphical display describing at the same time features like a data set which indicates the central tendency, spread, unsymmetry and identification of unusual observations outside the range under consideration. Also it displays the three quartiles – left or lower quartile Q_1, right or upper quartile Q_3 and median or second quartile, Q_2 positioned at the 50th percentile. **A line or wishker** extends from each end of the box. The lower and upper wishker (or extension) cover as an extension to the left of the lower quartile and to the right of the upper quartile upto 1.5 inter quartile ranges from the box edge are called outliers. A point more than 3 interquartile ranges from the box edge is called an **extreme outliers.**

Box plots are very useful in graphical comparisons among data sets to convey visual impact and easy to understand

Example

For the sample data given below (a) calculate the sample mean, median, the quartiles, the variance and the standard derivation; (b) construct a box plot—of the data and comment on the information in the display

Class	43.5	44.5	45.5	46.5	47.5	48.5	49.5	50.5	51.5	52.5	
											Total
Interval	44.5	45.5	46.5	47.5	48.5	49.5	50.5	51.5	52.5	53.5	
Frequency	1	2	2	3	1	2	4	4	3	2	24

(a)
$$N = 24 \frac{N}{4} = 6 \frac{3N}{4} = 18$$

$$Q_1 = 46.5 + \frac{6-5}{3} = 46.88; \; Q_3 = 50.5 + \frac{18-15}{4} = 51.25$$

$$\text{Median} = 49.5 + \frac{12-11}{4} = 49.75$$

Quartile range $= Q_3 - Q_1 = 51.25 - 46.88 = 4.37$
On the left side whisker extends

$$\text{from the first quartile upto} = Q_1 + \frac{3}{2} \times 4.37$$

$$= 46.88 - 6.585$$

$$\simeq 40.38$$

$$= 46.88 - 6.585$$

In the right side whisker extends

$$\text{from the third quartzite} = 40.38$$

$$= Q_3 + 6.585$$

(b)   Description of a box plot $\simeq 57.83$

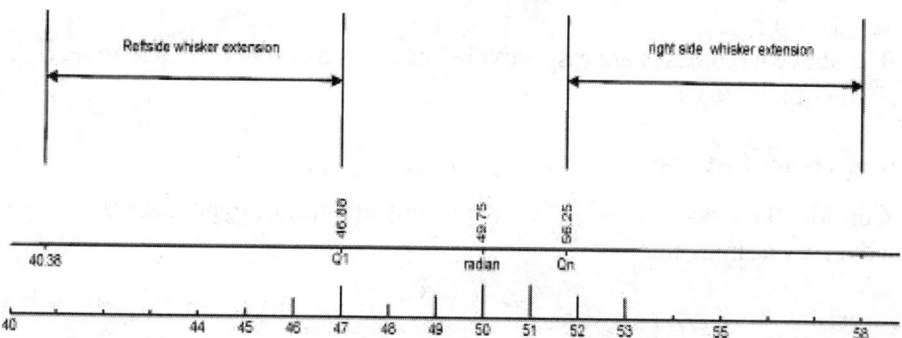

Frequency distribution—line diagram there is no outliers and extreme outliers.

10.4 Statistical Hypotheses

Many problems in engineering require a decision whether to accept or reject a statement about a parameter. The statement is called a **hypothesis** and the decision making procedure about the hypothesis is called **hypothesis testing.** This is one of the most useful aspects of statistical inference. There is a close connection between hypothesis testing and confidence intervals.

• Null Hypothesis

The technique of randomization used for selection of sample units makes the test of significance valid for us. For applying the tests of significance, we first set up a hypothesis—a definite statement about the population parameter. Such a hypothesis, which is usually a hypothesis of no difference, is called Null hypothesis and is usually denoted by H_0. According to R.A.FISHER, **Null hypothesis is the hypothesis which is tested for possible rejection under the assumption that it is true.**

Having set up the null hypothesis we compute the probability P that the deviation between the observed sample statistic and the hypothetical parameter value might have occurred due to fluctuations of sampling. If the deviation comes out of to be significant, the null hypothesis is refuted or rejected at the particular level of significance adopted and if the deviation is not significant, null hypothesis may be retained at that level. (The null hypothsis determines the probability of TypeI error for the test procedure).

• **Alternative Hypothesis**

Any hypothesis which is complementary to the null hypothesis is called an alternative hypothesis, usually denoted by H_1.

To test the null hypothesis that the population has a specified mean μ_0(say), i.e., $H_0 : \mu = \mu_0$, then the alternative hypothesis could be

(i) $H_1 : \mu \neq \mu_0$ (i.e., $\mu > \mu_0$ or $\mu < \mu_0$)

(ii) $H_1 : \mu > \mu_0$

(iii) $H_1 : \mu < \mu_0$

The above hypotheses are respectively known as a two tailed alternative, right tailed and left tiled alternatives.

Types of errors in tests

Consider the risks of making false decisions in a test of hypothesis $\theta = \theta_0$ against an alternative, say a single number θ_1.

Type I error

In hypothesis testing, an error incurred by rejecting a null hypothesis when it is actually true is called a **TypeI error** we write also as:

$P\{$ reject H_0 when it is true$\} = P\{$Reject $H_0/H\} = \alpha$

Type II error

In hypothesis testing an error incurred by failing to reject a null hypothesis when it is actually false is called a Type II error.

We write also as

$P\{$Accept H_0 when it is wrong$\} = P\{$Accept $H_0/H_1\} = \beta$

α and β are called the sizes of Type I error and Type II error, respectively.

10.5 Sampling Distributions

Statistical inference is concerned with making **decisions** about a population based on the information contained in a random sample from that population.

It we draw a sample of size n from a given finite population of size N, then the total number of possible samples is:

$$\binom{N}{n} = \frac{N!}{n!(N-n)!} = k \text{ (say)}$$

For each of these the samples, we can compute some statistic $t = t(x_1, x_2, \ldots, x_n)$, in particular, the mean $\bar{x}$, the variance s^2 etc as given below

Sample number

$$
\text{Statistic} \begin{cases} t & t_1 & t_2 & t_3 & \cdots & \cdots & \cdots & t_k \\ \bar{x} & x_1 & x_2 & x_3 & \cdots & \cdots & \cdots & x_k \\ s^2 & s_1^2 & s_2^2 & s_3^2 & \cdots & \cdots & \cdots & s_k^2 \end{cases}
$$

with column headings 1 2 3 . . . . . k

The set of the values of the statistic so obtained, one for each sample, constitutes what is called the **sampling distribution** of the statistic. Each has a probability distribution. Hence the probability distribution of a statistic is called a sampling distribution.

Sampling distributions of means

Suppose that a random sample of size n is taken from a normal population with mean μ and variance σ^2. Now each observation in this sample, say $x_1, x_2, \ldots x_n$ is a normally and independently distributed random variable with mean μ and variance σ^2. Then by the reproductive property of the normal distribution, we conclude that the sample mean

$$
\bar{X} = \frac{1}{n}(x_1 + x_2 + \ldots + x_n)
$$

has normal distribution with mean

$$
\mu_{\bar{x}} = \frac{1}{n}(\mu + \mu + \ldots + \mu) = \mu
$$

and variance

$$
\sigma_{\bar{x}}^2 = \frac{1}{n^2}(\sigma^2 + \sigma^2 + \ldots \sigma^2) = \frac{s^2}{n}
$$

If we are sampling from a population that has an unknown probability distribution, the sampling distribution of the sample mean will still be approximately normal with mean μ and variance σ^2, if n is large. This is one of the most useful theorems caller the **control limit theorem**.

If $x_1, x_2, \ldots x_n$ is a random sample of size n taken from a population with mean μ and finite variance σ^2 and if $\bar{X}$ is the sample mean, the limiting form of the distribution of

$$
Z = \frac{\bar{X} - \mu}{\sigma/\sqrt{n}}
$$

as $n \to$ is the standard normal distribution.

It two independent populations with means μ_1 and μ_2 and variance σ_1^2 and σ_2^2 and if $\bar{X}_1$ and $\bar{X}_2$ are the sample means of two independent random samples of sizes n_1 and n_2 from these populations, then the sampling distribution of

$$Z = \frac{\bar{X}_1 - \bar{X}_2 - (\mu_1 - \mu_2)}{\sqrt{s_1^2/n_1 + s_2^2/n_2}}$$

is approximately standard normal, if the conditions of the central limit apply: If two populations are normal, the sampling distribution of Z is exactly standard normal.

Confidence interval

A confidence interval estimate for m is an interval of the form $l \leq \mu \leq u$, where the end points l and u are computed from the sample data. Because different samples will produce different values of l and u, these end-points are values of random variables L and U, respectively. If L and U values are determined such that the following probability statement is true:

$P\{L \leq \mu \leq \cup\} = 1 - \alpha$, where $0 \leq \alpha \leq 1$)

This means that there is a probability of $(1 - \alpha)$ of selecting a sample for which the confidence interval will contain the true value of μ. Once a sample is selected and the values of l and u are computed the resulting **confidence interval for m** is $l \leq \mu \leq u$.

The end points or bounds l and u are called the **lower** and **upper confidence limits** respectively, and $(1 - \alpha)$ is called the **confidence coefficient**.

Because $Z = (\bar{X} - \mu)/(\sigma/\sqrt{n})$

has a standard normal distribution, we may write

$$P\left\{-Z_{\alpha/2} \leq \frac{\bar{X} - \mu}{\sigma/\sqrt{n}} \leq Z_{\alpha/2}\right\} = 1 - \alpha$$

or $P\left\{\bar{X} - Z_{\alpha/2}\frac{s}{\sqrt{n}} \leq \mu \leq \bar{X} + Z_{\alpha/2}\frac{s}{\sqrt{n}}\right\} = 1 - \alpha$

or $P(L \leq \mu \leq U\} = 1 - \alpha$, and

L and U are called the lower and upper confidence limits ($s/\sqrt{n}$ is called the standard error of $\bar{X}$).

If $\alpha = 0.05$, then $Z_{\alpha/2} = Z_{0.025} = 1.96$.

$$\therefore L = \bar{X} - 1.96\frac{s}{\sqrt{n}} \text{ and } U = \bar{X} + 1.96\frac{s}{\sqrt{n}}$$

Similarly the 99% of confidence limits are $L = \bar{X} - 2.58\frac{s}{\sqrt{n}}$ and $U = \bar{X} + 2.58\bar{X}\frac{s}{\sqrt{n}}$.

The following table gives for large samples the standarderror for some of the well known statistic.

Statistic : $\bar{x}$ p   s $\bar{x}_1 - \bar{x}_2$ $s_1 - s_2$ $p_1 - p_2$

Standard

error : $\dfrac{\sigma}{\sqrt{N}}$ $\sqrt{\dfrac{pq}{n}}$ $\dfrac{s}{\sqrt{2n}}$ $\sqrt{\dfrac{s_1^2}{2n_1} + \dfrac{s_2^2}{2n_2}}$ $\sqrt{\dfrac{s_1^2}{2n_1} + \dfrac{s_2^2}{2n_2}}$ $\sqrt{\dfrac{p_1q_1}{n_1} + \dfrac{p_2q_2}{n_2}}$

(If the variance of the population σ^2 is not known, it has to be estimated from the sample variance s^2).

The confidence interval for the correlation coefficient r is.

$$\tanh\left\{\text{arc tanh } u - \frac{Z_{a/2}}{\sqrt{n-3}}\right\} \leq ? \leq \tanh\left\{\text{arc tanh } r + \frac{Z_{a/2}}{\sqrt{n-3}}\right\}$$

where $\tanh u = \dfrac{e^u - e^{-u}}{e^u + e^{-u}}$

Tests of Significance

A very import aspect of the sampling theory is the study of the tests of significance, which enables us to decide on the basis of sample results, if

    (i) The deviation between the observed statistic and the hypothetical parameter value

        or

   (ii) The deviation between two sample statistics, are taken for analysis.

Tests based on normal distribution for large samples

The following steps are used for tests based upon the normal distribution for large values of n.

   (1) Compute the test statistic Z(under H_0 hypothesis)

   (2) If $|Z| > 3$, H_0 is always refuted

   (3) If $|Z| < 3$, we test its significance at certain level of significance usually at 5% and sometimes at 1% level of significance. Thus for a two tail if $|Z| > 1.96$, H_0 is rejected at 5% level of significance and if $|Z| < 1.96$, H_0 is retained at 5% level of significance.

   (4) Similarly, if $|Z| > 2.58$, H_0 is contradicted at 1% level of significance and if $|Z| < 2.58$, H_0 may be accepted at 1% level of significance.

       For a Single tail test (right tail or left tail), we compare the computed value of $|Z|$ with 1.645 at 5% level and 2.33 at 1% level and accept or reject H_0 accordingly. For sampling of attributes, we compute the probability of proportion p for success in n trials

10.6 TESTS OF SIGNIFICANCE FOR SMALL SAMPLES

The sample size n is usually small in most engineering problems.

10.6.1. *The t-Distribution*

For $n < 30$, when the population standard deviation σ is known, the standard error of estimate of the mean m is defined as $s_m = \dfrac{s}{\sqrt{n}}$. When the population standard deviation σ is not known, σ is estimated from the sample standard deviation s, and the standard error of the unbiased estimate for σ_m is taken as $\dfrac{s}{\sqrt{n-1}}$. In this case

$$t = \frac{\overline{x} - m}{s/\sqrt{n-1}}.$$

In the above, the sample (arithmetic) mean $\overline{x}$, sample standard deviation, s, together with $(n-1)$ alone are used.

When $n < 30$, the t-distribution is based upon the small sample observations. Thet-distribution is given by the formula.

$$y = y_0\left(1 + \frac{t^2}{v}\right)^{-\frac{?+1}{2}}, ? = n-1$$

Here $y = y(t)$ is measured vertically above the t-axis on either side of the origin O, based on the probability measure. y_0 is the value of y when $t = 0$, and it is the maximum ordinate, $v = n - 1$ is the degree of freedom of the t statistic. y_0 is determined by taking

$$y_0 \int_{-\infty}^{+\infty}\left(1 + \frac{t^2}{v}\right)^{-\frac{v+1}{2}} dt = 1$$

10.6.2 *Attributes of the t-distribution*

(i) t-distribution is symmetrical about the vertical line at $t = 0$.

(ii) Even though t-distribution has similar humpback its tail areas have greater values when compared to the tail areas of the normal distribution.

(iii) As n tends to infinity, the t-distribution approaches the normal distribution.

(iv) The t-distribution curve does not touch the horizontal t-axis any where, in the range $\{-\infty, \infty\}$. In other words, the t-distribution is asymptotic to the t-axis.

(v) As the t-distribution curve is symmetrical about the vertical axis at $t = 0$, the total, area under the t curve is divided into two equal areas.

Whatever the value taken by the variable t in the interval $\{-\infty, \infty\}$, its probability measure is denoted by $P(-\infty \leq t \leq \infty) = \int_{-\infty}^{\infty} y(t) \, dt = 1$

$$\therefore P(-\infty \leq t £ 0) = P(0 \leq t \leq \infty) = \frac{1}{2}.$$

(vi) When t takes the value $\leq -t_0$ and $> + t_0$, the corresponding probability measure is given by

$$p = P(t \leq -t_0, t > t_0)$$
$$= P(-\infty \leq t \leq -t_0, t_0 < t < +\infty)$$
$$= P(|t| \leq t_0)$$
$$= 2\int_{t_0}^{\infty} y(t) \, dt$$

When $p = 0.05$, the corresponding t_0 values is determined from the table of areas of the t-distribution.

10.6.3 *Tests of significance for small samples (n < 30) based on t-distribution*

1. **Between the sample mean $\bar{x}$ and the population mean m, (standard deviation σ, not known).** Let s be the sample standard deviation.

Set $\qquad t = \dfrac{|\bar{x} - m|}{\dfrac{s}{\sqrt{n-1}}}$, $v = n - 1$ (degree of freedom)

From the t-table, for $v = n - 1$, find the t_0 value corresponding to the 5% level of probability.

If $t > t_0$, the difference is significant; if $t < t_0$, the difference is not significant.

The confidence limits for the population mean m is.

$$\left[\bar{x} - \frac{s\,t_0}{\sqrt{n-1}}, \bar{x} + \frac{s\,t_0}{\sqrt{n-1}} \right]$$

2. **Between the two Sample means $\bar{x}_1$ and $\bar{x}_2$** from two samples of sizes n_1 and n_2 taken from the population (μ, σ unknown). Let s_1 and s_2 be the sample standard deviations.

The estimate for the population standard deviation based on the standard deviations of the two samples s_1 and s_2 is given by

$$\sigma^2 = \frac{n_1 s_1^2 + n_2 s_2^2}{n_1 + n_2 - 2}$$

Set

$$t = \frac{\left|\overline{x}_1 - \overline{x}_2\right|}{\text{S.E of } (\overline{x}_1 \sim \overline{x}_2)} = \frac{\left|\overline{x}_1 - \overline{x}_2\right|}{s\sqrt{\dfrac{1}{n_1} + \dfrac{1}{n_2}}}$$

$$= \frac{\left|\overline{x}_1 - \overline{x}_2\right|}{\sqrt{\dfrac{n_1 s_1^2 + n_2 s_2^2}{n_1 + n_2 - 2}\left(\dfrac{1}{n_1} + \dfrac{1}{n_2}\right)}}, ? = n_1 + n_2 - 2$$

From the t table, for $v = df = n_1 + n_2 - 2$ at 5% level, the corresponding value of t_0 is found.

If $t \geq t_0$, the difference is significant, and if $t < t_0$, the difference is not significant.

If $n_1 = n_2 = n$, then we are given n pairs of observations taken from the population.

Case 1: If the n-pairs of observation are independent, then

$$t = \frac{\overline{x}_1 \sim \overline{x}_2}{\sqrt{\dfrac{n(s_1^2 + s_2^2)}{2n - 2}\left(\dfrac{1}{n} + \dfrac{1}{n}\right)}} \quad v = n + n - 2 = 2n - 2.$$

$$= \frac{\overline{x}_1 \sim \overline{x}_2}{\sqrt{\dfrac{s_1^2 + s_2^2}{n - 1}}} \quad v = 2(n - 1)$$

From the t-table find the value of t_0 corresponding to the degree of freedom $2n - 2$ at 5% probability level of significance.

If $t \geq t_0$, the difference is significant and if $t < t_0$ the different is not significant.

Case 2 : If the n-pairs of observations are related to each other, then take the n-pairs of observations as $[x_r, y_r]$, $r = 1, 2, \ldots n$.

Let $z_r = x_r - y_r$, $\overline{z} = \overline{x} - \overline{y}$

and $s_2^2 = s_2^2 = \dfrac{1}{n}\Sigma z_r^2 - \overline{z}^2$

Set $t = \dfrac{\overline{z} - 0}{\dfrac{s_z}{\sqrt{n-1}}}$, $v = n - 2$

Find the t_0 value from the t table corresponding to the d.f, $v = n - 2$ at 5% probability level.

If $t \geq t_0$, the difference is significant, and if $t < t_0$, the difference is not significant.

10.6.4 *Test of significance of the observed correlation coefficient (r)*

for n < 30 pairs of related observations.

Assume the hypothesis H_0 that there is no correlation between the n pairs of observations. Then set

$$t = \frac{r - 0}{\sqrt{1 - r^2}/\sqrt{n-2}} = \frac{r\sqrt{n-2}}{\sqrt{1-r^2}} \text{ for } n = n-2.$$

Find the value of t_0, corresponding to the d.f $v = n - 2$ at 5% level of significance. If $t \geq t_0$, the value of r is significant and if $t < t_0$, the value of r is not significant.

10.7 CHI-SQUARE χ^2 DISTRIBUTION

Definition

Let x_1, x_2, ..., x_n be a random sample from a normal distribution with mean μ and variance σ^2 and let S^2 be the sample variance. Then the random variable

$$\chi^2 = \frac{(n-1)S^2}{\sigma^2} \text{ has a chi-square } (\chi^2) \text{ distribution with } (n-1) \text{ degrees of freedom.}$$

The probability density function of a χ^2 random variable as

$$f(x) = \frac{1}{2^{k/2} \rightleftarrows \lceil (k/2)} x^{(k/2)-1} e^{-x/2}, x = \chi^2 > 0, \text{ where k is the number of degrees}$$

of freedom (The gamma function is defined by $\lceil (r) = \int_0^\infty x^{r-1} e^{-x} dx$, for $r > 0 = (r-1)!$ if r

is a positive integer.

Also $\lceil (1) = 0! = 1$ and $\lceil \left(\frac{1}{2} \right) = \sqrt{\pi}$

The mean and variance of the χ^2 distribution are k and 2k respectively

The chi-square distribution is nonnegative and that the probability distribution is skewed to the right. However as k increases, the distribution becomes more symmetric. As $k \to \infty$, the limiting form of the chi-square distribution is the normal distribution.

Define $\chi^2 \alpha$, k as the percentage point or the value of the chi-square random variable with k degrees of freedom such that the probability the χ^2 exceeds this value is α. That is,

$$P(\chi^2 > \chi^2_{\alpha,k}) = \int\limits_{\chi^2_{\alpha,k}}^{\infty} f(u)\, du = \alpha$$

In the chi square table, the areas α are the column headings and the degrees of freedom k are given in the left column.

Confidence Interval

If s^2 is the sample variance form a random sample of n observations from a normal distribution with unknown variance σ^2, then a $100\,(1-\alpha)\%$ confidence interval on σ^2 is

$$\frac{(n-1)s^2}{\chi^2_{\alpha/2,\,n-1}} \leq \sigma^2 \leq \frac{(n-1) \rightleftarrows s^2}{\chi^2_{1-\frac{\alpha}{2},\,n-1}}$$

where $\chi^2_{\alpha/2,\,n-1}$ and $\chi^2_{1-\frac{\alpha}{2},\,n-1}$ are the upper and lower $100\ \alpha/2$ percentage points of

the chi-square distribution with $(n-1)$ degrees of freedom, respectively.

10.8 HYPOTHESIS TEST ON THE VARIANCE OF A NORMAL POPULATION

Suppose we wish to test the hypothesis that the variance of a normal population σ^2 equals a specified value, say σ_0^2. Let $x_1, x_2, \dots x_n$ be a random sample of n observations from this population. To test

$$H_0 : \sigma^2 = \sigma_0^2$$

$$H_1 : \sigma^2 \neq \sigma_0^2$$

we will use the test statistic

$$\chi_0^2 = \frac{(n-1)\,S^2}{\sigma_0^2}$$

If the null hypothesis $H_0 : \sigma^2 = \sigma_0^2$ is true, the test statistic χ_0^2 and the null hypothesis $H_0 : \sigma^2 = \sigma_0^2$ would be rejected if

$$\chi_0^2 > \chi^2_{\alpha/2,\,n-1} \quad \text{or if} \quad \chi_0^2 < \chi^2_{1-\alpha/2},\, n-1$$

where $X^2_{\alpha/2,\,n-1}$ and $X^2_{1-\alpha/2,\,n-1}$ are the upper and lower $100\ \alpha/2$ percentage point of the chi-square distribution with $(n-1)$ degrees of freedom, respectively.

For one sided alternative hypothesis, $(H_1 : \sigma^2 > \sigma_0^2)$, we would reject H_0 if $\chi_0^2 > \chi_{\alpha, n-1}^2$; whereas the other one-sided hypothesis $(A_1 : \sigma^2 < \sigma_0^2)$, we would reject H_0 if $\chi_0^2 > \chi_{1-\alpha, n-1}^2$.

10.9 TEST OF INDEPENDENCE OF TWO ATTTRIBUTES IN A "X × Y" CONTINGENCY TABLE

Contingency Table

Consider N elements of a sample from a population classified according to two attributes A and B. It is required to test for independent of the two attributes A and B. Let the attribute A be divided into $A_1, A_2 ... A_x$, x classes arranged in x rows and the attribute B be divided into $B_1, B_2 ..., B_y$, y classes arranged in y columns forming an "x × y" contingency table. There are x.y cells in the contingency table. In each cell, the observed frequencies of second order are shown. For instance, in the rth row and sth column, the corresponding to the 'rs' th cell, the observed frequency is denoted by (A_r, B_s). Assuming that the two classifications A_r and B_s, r = 1, 2 ..., 1, 2, ... y are independent, the expected frequencies are defined by

$(A_r.B_s)_0 = \dfrac{(A_r).(B_s)}{N}$. Let δ_{rs} denote the difference between the observed frequency $(A_r B_s)$ and the expected frequencies $(A_r B_s)_0$.

$$\therefore \ \delta_{rs} = (A_r B_s) - (A_r B_s)_0$$

Since $\Sigma\delta_{rs} = 0$, we find δ_{rs}^2, the square of the difference between the observed and expected frequencies for each cell. Dividing this δ_{rs}^2 by the expected frequency $(A_r B_s)_0$ for each cell, we form the chi-square , χ^2-value given by

$$\chi^2 = \sum_{r=1}^{x} \sum_{s=1}^{y} \frac{\delta_{rs}^2}{(A_r B_s)_0}$$

The "x × y" contingency contains on the whole xy cells. Since the row sums and column sums are equal, one degree of freedom is lost.

Hence the total number of linear constraints become x + y − 1, consequently, the number of degrees of freedom for the "x × y" contingency table becomes.

$$\nu = xy - (x + y - 1) = xy - x - y + 1 = (x - 1)(y - 1)$$

From table of chi square, find for the d.f. $\nu = (x - 1)(y - 1)$, the appropriate value χ_0^2.

If $\chi^2 > \chi_0^2$, the null hypothesis H_0 : (the attributes A and B are independent) is rejected .

On the otherhand if $\chi^2 < \chi_0^2$, the null hypothesis H_0 is **accepted**, i.e., the two attributes are independent (i.e., there is no association between the two attributes).

10.9.1 Test for Homogenity in a Contingency Table

Using the two way contingency table to test independence between two variables (or attributes) of classification in a sample from a single population of interest is only one application of the contingency table methods. Another common situation occurs, when there are r (rows) populations of interest and each population is divided into the same c(columns) categories. A sample is then taken from the i-th population {i = 1, 2, ..., r), and counts are entered in the appropriate columns), (j = 1, 2, ... c) of the i-th row. In this situation, we want to investigate whether or not the proportions in the c categories are the same for all populations.

The null hypothesis of the problem states that the populations are **homogeneous** with respect to the categories.

For example, when there are only two categories, such as success and failure, defective and non-defective, and so on, the test for homogeneity is really a test of the equality of r binomial parameters.

Calculation of expected frequencies, determination of the degrees of freedom and the computation of the chi-square statistic for the test of homogeneity are identical to the test of independence.

10.9.2 Testing for Goodness of fit

When the underlying distribution of the population is not known, we wish to test the hypothesis that a particular distribution will be satisfactory as a population model.

Probability Plotting is a graphical method for determining whether sample data conform to a hypothetical distribution based on a subjective visual examination of the data. Probability plotting typically uses special graph paper, known as **probability paper**, that has been designed for the hypothesized distribution. Probability paper is widely available for the normal. Lognormal, Weibull and various chi-square and gamma distributions.

Goodness of Fit Test

There is a formal goodness-of-test procedure based on the chi-square distribution.

The test procedure requires a random sample of size n from the population whose probability distribution is unknown. These n observations are arranged in a frequency histogram, having class intervals. Let O_i be the observed frequency in the i-th class interval from the hypothetical probability distribution. We compute the expected frequency in the i-th class interval, denoted by E_i. The test statistic is

$$\chi_0^2 = \sum_{i=1}^{K} \frac{(0_i - E_i)^2}{E_i}$$

If the population follows the hypothetical distribution, χ_0^2 has approximately, a chi- square distribution with k-p-1 degrees of freedom, where p represents the number of parameters of the hypothetical distribution estimated by sample statistics. The approximation improves as n increases. We would reject the hypothesis that the distribution of the population is the hypothetical distribution, if the calculated value of the statistic $\chi_0^2 > \chi_{\alpha, k-p-1}^2$.

[Note: 1. In the population, if the probability P for the occurrence of the event is not given, then this has to be estimated from the observed data. As a consequence we have to reduce by one of the degrees of freedom.

2. If the frequencies at the beginning and ending of the classes of the frequency distribution are very small, it is necessary to combine these classes at the beginning and end of the class intervals. We find the number of class intervals after the combination and reduce this number by one to determine the number of degrees of freedom, while applying the procedure for the good ness of fit test.]

10.10 THE F DISTRIBUTION

Suppose that we consider two independent normal populations whose population means and variances, say μ_1, σ_1^2, μ_2 and σ_2^2 are unknown we wish to test hypothesis about the equality of the two variances, say $H_0 : \sigma_1^2 = \sigma_2^2$.

Assume that two random samples of size n_1, from population 1 and of size n_2 from population 2 are available and let σ_1^2 and σ_2^2 be the sample variances. We wish to test the hypotheses.

$$H_0 : \sigma_1^2 = \sigma_2^2$$

$$H_1 : \sigma_1^2 \neq \sigma_2^2$$

The development of a test procedure for these hypotheses requires a new probability distribution, F distribution.

The estimated variances of the two populations in terms of the sampling variances are given, respectively

$$u^2 = \frac{n_1 s_1^2}{n_1 - 1}, \quad v^2 = \frac{n_2 s_2^2}{n_2 - 1}$$

Here $n_1 = n_1 - 1$ and $n_2 = n_2 - 1$ are the degrees of freedom for u^2 and v^2.

• To test for the difference between the variances u^2 and v^2

We have to set up the following hypotheses

1. The difference between u^2 and v^2 is due to chance factors
2. The two samples have been taken from the same population
3. The two samples have been taken from two independent populations having the same variance.

We first form the F statistic which is defined to be the ratio of two independent chi-square random variables each divided by its number of degrees of freedom. That is

$$F = \frac{u^2}{v^2} = \frac{n_1 s_1^2 / n_1 - 1}{n_2 s_2^2 / n_2 - 1} \quad (>0)$$

The probability density function of the random variable F is given by

$$y = f(F) = K.\frac{F^{(v_1 - 2)/2}}{(v_1 F + v_2)^{(v_1 + v_2)/2}},$$

where n_1 and n_2 are the degrees of freedom of u^2 and v^2 respectively. The value of K is determined by taking the area between the frequency curve of the F-distribution and positive direction of the F axis to be equal to the total probability measure one. The value of K is given by $v_1/2$.

$$K = \frac{\left(\dfrac{v_1}{v_2}\right)^{v_1/2}}{B\left(\dfrac{v_1}{2} + \dfrac{v_2}{2}\right)}$$

$$K = K = \frac{\left(\dfrac{v_1}{v_2}\right)^{v_1/2}}{\Gamma\left(\dfrac{v_1}{2}\right)\Gamma\left(\dfrac{v_2}{2}\right)}\,\Gamma\left(\dfrac{v_1 + v_2}{2}\right), \quad \text{Where B and G are the beta and gamma function}$$

respectively.

$$\therefore y = f(F) = \frac{\left(\dfrac{v_1}{v_2}\right)^{v_1/2}\Gamma\left(\dfrac{v_1 + v_2}{2}\right)}{\Gamma\left(\dfrac{v_1}{2}\right)\Gamma\left(\dfrac{v_2}{2}\right)}\,\frac{F^{\frac{v_1}{2}-1}}{\left[1 + \dfrac{v_1}{v_2}F\right]^{\frac{v_1+v_2}{2}}}, \quad 0 \le F \le \infty$$

It is the probability function of Snedecor's F-distribution with (v_1, v_2) degrees of freedom and denoted by $F(v_1, v_2)$.

The mean and variance of the F distribution are

$$m = \frac{v_2}{v_2 - 2}f \text{ and } \sigma^2 = \frac{2v_2^2(v_2 + v_1 - 2)}{v_1(v_2 - 2)^2(v_2 - 4)}, \quad v_2 > 4$$

[Note: 1. The F random variable is non negative.]

The distribution is skewed to the right. The F distribution looks very similar to the chi-square distribution; however, the two parameters v_1 and v_2 provide extra flexibility regarding shape.

3. The Value of F satisfies $0 \le F < \infty$.

4. When the degrees freedoms v_1 and v_2 are equal, the F distribution is a symmetrical distribution. That is

$$P\{F(v_1, v_2) = c\} = P\left\{F(v_2, v_1) = \frac{1}{c}\right\}$$

5. If $P\{F(v_1, v_2) = c\} = \alpha$, then $P\left\{F(v_2, v_1) \geq \frac{1}{c}\right\} = 1 - \alpha$

4. Test Procedure: Hypothesis tests on the ratio of two variances.

A hypothesis-testing procedure for equality of variances is based on the following result.

$\{x_{11}, x_{12}, \ldots x_{1n_1}\}$ is a random sample from a normal population with mean μ_1 and variance σ_1^2.

$\{x_{21}, x_{22}, \ldots x_{2n_2}\}$ is a random sample from a second normal population with mean μ_2 and variance σ_2^2.

Assume that both are normal populations and are independent.

Let S_1^2 and S_2^2 be the sample variance. Then the ratio $F = \dfrac{S_1^2 / \sigma_1^2}{S_2^2 / \sigma_2^2}$.

Has an F distribution with $(n_1 - 1)$ numerator degrees of freedom and $(n_2 - 1)$ denominator degrees of freedom.

The basis of test procedure is as follows :

Null hypothesis : $H_0 : \sigma_1^2 = \sigma_2^2$; Test statistic $F_0 = \dfrac{S_1^2}{S_2^2}$

Alternate Hypotheses	Rejection criterion
$H_1 : \sigma_1^2 \neq \sigma_2^2$	$f_0 > f_{a/2, \, n_1 - 1, \, n_2 - 1}$ or $f_0 < f_{1 - \frac{a}{2}, \, n_1 - 1, \, n_2 - 1}$
$H_1 : \sigma_1^2 > \sigma_2^2$	$f_0 > f_{\alpha, \, n_1 - 1, \, n_2 - 1}$
$H_1 : \sigma_1^2 < \sigma_2^2$	$f_0 < f_{1 - \alpha, \, n_1 - 1, \, n_2 - 1}$

[The percentage points f_α, n_1, n_2 of the F-Distribution are given in the F-Table with degrees of freedom n_1 for the numerator and degrees of freedom n_2 for the denominator].

10.11 GENERAL PROCEDURE FOR HYPOTHESIS TESTS

1. The following sequences of steps will be used in applying hypothesis testing methodology.

2. From the problem context, identify the parameter of interest.
3. State the null hypothesis, H_0.
4. Specify an appropriate alternative hypothesis, H_1.
5. Choose a significance level α.
6. Determine an appropriate test statistic.
7. State the rejection region for the statistic.
8. Compute any necessary sample quantities, substitute these into the equation for the test statistic, and compute that value.
9. Decide whether or not H_0 should be rejected and report that in the problem context.

10.12 EXERCISES

Explain the Type I and Type II errors in the testing of hypotheses. Also explain the use of OC curves in the same context.

1. Describe any four methods of selecting a sample explaining the situation when they are preferred.

2. Distinguish between large sample and small sample test and briefly explain the ssampling distribution used in the two cases.

3. (a) How will you test the significance of the difference between the means of two small samples when
   (i) They are drawn from normal populations with the same known standard deviation.
   (ii) They are drawn from normal populations with the same but unknown standard deviation.

   (b) The following table shows the mean number of bacterial colonies per plot obtainable by four slightly different methods from soil samples taken at 4 PM and 8 PM respectively.

	Methods A	Methods B	Methods C	Methods D
4 PM	31.75	29.50	32.25	29.80
8 PM	41.20	42.60	32.20	42.40

   Are there significantly more bacteria at 8 PM than at 4 PM ?

4. If 50 measurements of the specific gravity of aluminium had a mean of 2.686 and a standard deviation of 0.042, construct a 0.99 confidence interval for the true standard deviation of such measurements.

5. A random sample of size 30 from a normal population has a mean = 20.8 and s.d = 0.5, construct a 95 percent confidence interval for the observed property.

6. A random sample of 40 measurements of a certain physical characteristic has a mean of 24.45 and s.d = 2.1. construct a 95 percent confidence interval for the standard deviation of the measurements.

7. Measurements of the diameter of a random sample of 400 ball bearings made by a certain machine during one week showed a mean of 0.825 cms and a standard deviation of 0.045 cms. Find 99 percent confidence limits for the mean diameter of all the bearings.

8. A manufacturer claims that a gallon can of his paint will cover on the average 44.5 square meters and government agency wants to test the validity of his claim. Suppose that the agency takes a random sample of 36 cans of paint and decides to reject the manufacturer's

claim if the mean area cover by the paint less than 43.3 square meters. Find the probability of TypeI error, given the standard deviation of this area covered to be 5.3 square meters.

$$\left[\text{use} : \int_{-\infty}^{1.25} \frac{1}{\sqrt{2\pi}} \exp\left(-\frac{1}{2} z^2 \right) d_z = 0.8944 \right]$$

9. If 25 measurements of the coefficient of thermal expansion of nickel have a mean of 12.81 and a standard deviation of 0.04, construct a 0.95 confidence interval for the actual coefficient of expansion. Assume that the 25 measurements may be looked upon as a random sample from a normal population.

10. What are the important tests of significance? Where are they used?

11. Define student's t-statistic and indicate its importance in testing statistical hypothesis.

12. Nine individuals are chosen at random from a population and their coded weights are found to be in units of 111, 116, 119, 121, 123, 126, 129 131, 140. In the light of these data, discuss the suggestion that the mean weight of the population is 120 units. For 8 degrees of freedom at 5 percent level of significance the value of t = 2.31.

    Distinguish between large sample and small sample tests and briefly describe the sampling distributions used the two can.

13. The Arithmetic mean and Standard deviation of a sample of 14 observations are 75 and 3.65 respectively. Test at 5 percent level of significance whether the sample can be regarded as belonging to a normal population with mean 72.5.

14. (a) It is given that x is normal and you are to test the hypothesis that $\mu = 15.5$ 'based on a sample of four independent observations, viz., x_1, x_2, x_3 and x_4. Indicate how you test the hypothesis (ii) when σ is given to be 2.4 and (ii) when s is not known.

    (c) Test the hypothesis in a (i) above when the values of x_i are 9.4, 12.6, 13.5 and 8.5 (use symmetric two-sided test at 5 percent significant level).

15. The specification for a certain kind of ribbon, call for a mean breaking strength of 70 kg. If six pieces of the ribbon (randomly selected from different rolls) have a common breaking strength of 63 kg with a standard deviation of 3.2 kg, test the null hypothesis

    $H_0 : \mu = 70$ kg against the alternate hypothesis

    $H_1 : \mu = < 70$ kg at the level of significance $\alpha = 0.05$

16. A manufacturer of steel wire claims that the average force required to break a given kind of wire is 500 kg. A sample of 25 lengths of this steel wire were stressed, and the mean and the standard deviation of the forces required to break these specimens were found to be

    $\bar{x} = 465$kg and $\sigma = 55$ kg. On the assumption that the breaking strengths may be looked upon as random sample from a normal population with $\mu = 500$, use the t-statistic to test the manufacturer's claim.

    [Given : For $v = d.f = 24$, $P_n = (t > 2.797) = 0.005$]

17. A random sample of boots worn by 50 soldiers is a desert region should an average life of 1.24 years with a standard deviation of 0.55 years. Under standard conditions, such boots are known to have an average life of 1.40 years. Is there reason to assert at a level of significance of 0.05 that use in the desert causes the average life of such boots to decrease? [use : $Z_{0.05} = 1.645$]

18. It is desired to test the hypothesis $\mu = 0$ against the alternate $\mu > 0$ on the basis of a random sample of size 9 from a normal population whose variance is $\sigma^2 = 1$. If the probability of Type

I error is to be 0.05, verify that the region of rejection is $\bar{x} > 0.548$.

19. A testing laboratory wants to check whether the average life time of certain kind of cutting tool is 2000 pieces against the alternative that it is less than 2000 pieces. What conclusion will they reach at a level of significance of 0.01, if tests showed tool lives of 2010, 1980,1920,2005,1975 and 1950 pieces?

20. The following are the down times of a computer (in hours) during each of the 5 consecutive months. 29, 16,20,31,24. Use the t-statistic to check the reasonableness of the claim that on the average the computer can be expected to he out of commission atmost 20 hours per month.

21. Test the null hypothesis s = 0.25 for the weights of certain lead balls, if in a random sample of size 10, the weights of the balls had a variance $S^2 = 0.01$. Test at the 0.01 level of significance.

22. If 25 measurements of the coefficient of thermal expansion of nickel have a mean 12.81 and s.d = 0.04, construct a 95 percent confidence interval for the actual coefficient of expansion, assuming normality of the population.

23. (a) Explain the student 't' test for testing the significant difference between the t w o sample means.

    (b) The marks obtained by a group of 9 regular course students and another group of 11 part-time course students in a test are given below

Regular	57	63	64	55	61	52	68	70	59		
Part-time	63	71	73	64	61	57	75	65	73	69	67

Examine whether the marks obtained by regular students and part-time students differ significantly at 5 percent lived of significant and at 1 percent level of significance.

24. The mean product of wheat in a sample of 100 fields come to 200 kgs with a standard deviation of 10 kgs. Another sample of 150 fields gives the mean at 220 kgs with standard deviation of 12 kgs. Assume the standard deviation of the yield at 11kgs for the universe, find out if there is a significant difference between the mean yields of the two samples.

25. A random sample of 25 executives spent on the average Rs. 17,536, with a standard deviation of Rs. 1694, entertaining visiting dignitaries. Using Chi-square distribution, find a 0.95 confidence interval for the true standard deviation of such expenses.

[Given $\begin{aligned} x^2_{0.975,\ 24} &= 12.401 \\ x^2_{0.025,\ 24} &= 39.3641 \end{aligned}$

26. Obtain the relationship of s^2 with the chi-square distribution where s^2 is the variance of sample of size n.

27. An optical firm purchases glass to be ground into lenses, and past experience has shown that the variance of the refractive index of this kind of glass is 1.26×10^{-4}. To grind the glass into lenses having a given focal length, it is important that variance pieces of glass have nearly the same index of refraction. Hence the firm decides to reject the shipment if the sample variance of 20 pieces selected at random exceeds 2.00×10^{-4}. Assuming that the sample values may be looked upon as coming from a normal population with $\sigma^2 = 1.26 \times 10^{-4}$, find the probability that a shipment will be erroneously rejected.

28. Define the chi-square statistic. Explain its use in test of hypothesis. Out of 1250 throws with an unbiased coin, 620 heads were obtained. Is this consistent with the hypothesis that the

probability of getting head is $\frac{1}{2}$?

29 (a) Explain chi-square test as a test for independence.

(b) Can vaccination be regarded as preventive measure of smallpox as evidenced by the following data. "Of 1497 persons exposed to small pose in a locality, 372 in all were attacked. Of these 1497 persons, 346 were vaccinated and of these only 35 were attacked." Given chi-square at 5 percent level of significance for one degree of freedom as 3.841.

30. Five hundred rounds of ammunition have been fired and the following frequency table gives the resulting barred pressures in thousands kilograms per square centimeter.

Barrel pressure :	48.0–49.9	50.0–51.9	52.0–53.9	54.0–55.9	56.0–57.9	58.0–59.9
Frequency :	21	94	132	155	82	16

Based on this information, is it reasonable to conclude that the standard deviation of such barred pressures exceeds 2.2 thousand kilograms per square cm? Use a 0.05 level of significance.

31. A large electronic firm which hires many handicapped workers wants to determine whether their handicaps affect such worker's performance. Use the level of significance $\alpha = 0.05$ to decide on the basis of the sample data shown in the following table whether it is reasonable to maintain that the handicaps have no effect on the worker's performance. Performance.

	Performance Above average	Average	Below average
Blind	23	70	19
Deaf	18	54	15
No handicap	32	102	31

32. In a recent diet survey the following results were found:

No. of families	Group A	Group B	Total
Taking tea	1236	164	1400
Not taking teas	564	36	600
Total	1800	200	2000

Discuss whether there is any significant difference between the two communities in the matter of tea taking.

33. Describe the various tests that can be performed with the help of the chi-square distribution.

34. According to a genetic model, the proportions in the three groups should be p^2, $2pq$ and q^2 where $p + q = 1$.
What conclusions would you draw if the observed frequencies were 9, 51 and 45?

35. The pedestrian guard nails were installed at a busy complex junction in a city centre. The number of accidents at this junction and in the rest of the city centre were recorded for one year before and one year after the change, as follows :

Location	Before	After
Grand-nail junction	32	12
Rest of the city	240	216

Is the improvement due to guard –rails significant?

36. The following table shows classification of 42 randomly chosen children according to the nature of teeth and the type of feeding

	Normal teeth	Maloccluded teeth
Breast – fed	4	16
Bottle – fed	1	21

37. (a) How is F-distribution related to the chi-square distribution? what is the mean and the variance of the F-distribution?

(b) Two random samples of size 13 each of students are taken and their height measurement is $N(\mu, \sigma^2)$, what is the probability that the ratio of the sample variance (first to second) is at least 2.69?

What conclusions can be drawn from these figures?

38. If S_1^2 and S_2^2 are the variances of independent random samples of size n_1 and n_2 respectively, taken from two normal populations having the same variance, then prove that $\dfrac{FS_1^2}{S_2^2}$ is a value of a random variable having the F-distribution with parameters $v_1 = n_1 - 1$ and $v_2 = n_2 - 1$.

39. Two different lighting techniques are compared by measuring the intensity of light at selected locations in areas lighted by the tow methods. If 15 measurements in the first area had a standard deviation of 85 cms candles and 21 measurements in the seemed area had a standard deviation of 130 cms candles, can it be concluded that the lighting in the second area is less uniform? Use a 0.01 level of significance. What assumptions must be made as to how the two samples area obtained?

40. Using the following data, test the hypothesis (using chi-square) that inoculation against small-pox has no effect on immunity from attack

	Attacked	Not attacked
Inoculated	15	300
Not inoculated	65	500

41. The number of items produced by three different operators on three different machines during a prescribed period is given in the table below.

Operation Machine	1	2	3
I	17	19	12
II	10	9	13
III	13	11	16

If there evidence that some of the operators perform more efficiently in some machines than others? Given that when P = 0.05 for 5 degrees of freedom, $\chi^2 = 11.07$ and for 4 degree of freedom $\chi^2 = 9.49$.

10.13 ADDITIONAL EXERCISES

1. A sample of 9 shafts are inspected from a production line. The following measurements are the diameters (in m.m) of the shafts: 45.010, 45.020, 45.021, 45.015, 45.019, 45.018, 45.020, 45.023 and 45.005.

   If the production line meets the specifications laid by the I.S.I, with a standard deviation 0.006 m.m., estimate the 95 per cent confidence interval within which the true diameter of the shaft lies.

2. The mean mileage obtained from a sample of 17 automobile tires of a taxi-cab company was 28,000 with standard deviation 1800. Find the 95% confidence interval for the mean life of tires used by the company.

3. A criminologist claims that a person convicted of counterfeting spends on the average 29 months in jail. A law student who feels that this figure is high, takes a random sample of 25 such cases from court files and gets a mean of 27 months spend in jail, with a standard deviation of 4.5. What can he conclude from these figures of the level of significance of 0.01?

4. The following table shows the number of accidents occurring during a particular period of time in each of 3 eight hour shifts in a factory; the working conditions and the number of people exposed to risk being assumed similar for all the shifts. Is there a significant difference in the number of accidents occurring among the shifts?

Shifts	First	Second	Third
No of accidents	1	7	7

   Test at 5 percent level of significance.

5. In a sample of pea plants the number of round peas is 336 and the number of angular peas is 101. Is this in agreement with the Mendalian hypothesis which says that on the average (in this case) the ratio of these two should be 3:1 ?

6. The mean life time of a sample of 100 fluorescent light bulbs produced by a company is computed to be 1570 hours with a standard deviation of 120 hours. If m is the mean life time of all the bulbs produced by the company, test the hypothesis $\mu = 1600$ hours against the alternative hypothesis $\mu \neq 1600$ hours, using a level of significant of 0.05.

7. A machine is expected to produce nails of length 7 cm. A random sample of 10 nails were found to measure: 7.2, 7.3, 7.1, 6.9, 6.8, 6.5, 6.8, 6.9, 7.1, 7.2 cms respectively. On the basis of this sample, what can you say about the reliability of the machine? Given that the 5 percent values of t for 9degree of freedom is 2.26.

8. Ten individuals are chosen at random from a normal population and their heights (in cms) are found to be 157.5, 157.5, 165.0, 167.5, 170.0, 172.5, 175.0, 175.0, 175.0, 177.5.

   Test if the sample belongs to the population whose mean height is 165.0 cms (given that the value of |t| significant at 5 per cent level of probability for 9 degrees of freedom is 2.62).

9. (a) Explain the logic of statistical tests of significance.
   (b) The following table furnishes the results of breaking load tests an two types of yarn

Types of Yarn sample	Mean breaking load in ounces	Standard deviation	Number in
X	6.84	1.25	18
Y	7.50	1.36	19

Do the yearns differ significantly in their mean values?

10. (a) The standard deviations of heights of 15 men and 7 women are given to be 6.00 cms and 6.75 cms respectively. Do you consider that these are significantly different?

   (b) The mean heights in cm of the above two groups are 171.5 cms and 163 cms respectively. Examine in the light of the given data if men are taller than women in general.

11. (a) We are given two independent samples. Here can we test if the means of the two samples are equal? In which way the above is to be modified if we are to test whether the mean of the first sample is greater than that of the second?

   (b) Illustrate your answer with reference to the following data:

	Sample size	Mean	Standard Deviation
First sample	9	12.5	1.5
Second sample	16	11.8	2.3

12. The following data represents the percentage of fat content in 5 samples each of two brands of ice-cream. Test the hypothesis whether the average percentages of fat content in the two brands of ice-cream are significantly different at a level of significance 0.05.

Brand A :	13.2	14.0	12.9	13.5	13.4
Brand B :	13.5	13.4	14.1	14.5	14.0

13. The following data pertain to the cruising speed (in knots) of aircrafts (9 in number) before painting and after painting.

Aircraft	1	2	3	4	5	6	7	8	9
Curising speed (knots) Before Painting	426	478	424	438	440	420	412	409	425
After Painting	416	400	420	431	430	404	398	405	422

Will the engineer's claim that painting reduces the cruising speed be justified?

14. Use the following data to test the hypothesis, by using chi-square test, that inoculation against smallpox has no effect on immunity from the attack

	Attacked	Not Attacked
Inoculated	15	300
Not Inoculated	65	500

15. The following data give the number of items produced by three different operators on three different machines during a prescribed period. Is there evidence that some of the operators perform more proficiently on some machines than others?

Machine	Operators		
	1	2	3
I	17	19	12
II	10	9	13
III	13	11	16

Given that when p = 0.05 for 5 degrees of freedom, $\chi^2 = 9.49$.

16. (a) Explain briefly the uses and limitation of chi-square test.
    (b) Two independent samples of 8 and 7 items respectively had the following values of the variable.

Sample I:	9,	11,	13,	11,	15,	9,	12,	14
Sample II:	10,	12,	10,	14,	9,	8,	10	

Do the estimates of the population variance differ significantly? Given that for 7 and 6 degrees of freedom the 5 percent values of F is 4.20.

17. The following table shows the number of recruits taking (i) a preliminary and (ii) a final test in car driving.

		Preliminary		
		Pass	Fail	Total
Final	Pass	605	135	740
	Fail	195	65	260
	Total	800	200	1000

Is there any association between the results of the preliminary test and those of the final test?

18. (a) What is a contingency table? Discuss its uses.
    (b) Use the following data to test whether "type of tenure" is dependent on "fertility of soil".

Fertility of soil (Grade)	Tenure Owned	Rented	Mixed
I	37	68	50
II	31	60	49
III	58	87	80

19. In experiments on breeding, Mendal obtained the following frequency of seeds.

Round and yellow	Wrinkled and yellow	Round and green	Wrinkled and green
315	101	108	32

The theory predicts that the frequencies should be 9:3:3:1.Examine the correspondence between theory and experiment.

χ^2 for 3 d.f at 5 per cent = 7.815.

20. In one sample of 8 observations, the sum of the squares of deviations of the sample values from the sample mean was 84.4 and in the other sample of 10 observations it was 102.6. Test whether the two samples could have been drawn from the same normal population given that the 5 percent point of F for $v_1 = 7$, and $v_1 = 9$ degrees of freedom is 3.29.

21. Samples of air were collected from a vegetated sanitary buffer zone and a terrain devoid of vegetation for analysis. The study revealed the concentration of 4 pollutants (in milligrams per cubic meter) as follows:

With out vegetation :	0.14	0.03	0.13	0.9
With vegetation :	0.08	0.025	0.07	0.7

Test the hypothesis that vegetation has no effect on atmospheric pollution.

22. In a random samples of 400 College students in a city, 168 said that they travel by bus. Using a confidence co-efficient of 0.99 make an interval estimate of the proportion of College students in the entire city who would travel by bus.

23. The variance in textile strength of a certain material is 36 units. A change in the manufacturing procedure is expected to reduce the variance. The following observations are recorded (coded data)

4, 9, 0, 6, 8, 1, 13, 2, 13, 9, 14, 7, 15, 16, 2.

Would you recommend the change in the manufacturing procedure ($\alpha = 0.05$)?

24. x_1 and x_2 represent the existing method and a new method for producing gasoline from a crude oil. All others things being equal, it was decided to abandon x_1 in favor of x_2, only if the average yield of the latter was substantially greater. The following are the percent yield of gasoline.

x_1:	23.2,	26.6,	24.4,	23.5,	22.6,	25.7,	25.5
x_2:	25.7,	27.7,	26.2,	27.9,	25.0,	21.4,	26.1

In the basics of the data, given your recommendations at 1 percent level of significance.

25. A particular brand of door-lining of car has a mean life of 972 days write a standard deviator of 140 days. A new supply of 36 pieces of door lining showed a mean life of 893 days. Can you conclude that the new supply has become inferior at 1 percent lever of significance?

10.14 SHORT ANSWER EXERCISES

1. What is meant by a statistical hypothesis?
2. What are simple and composite hypothesis?

3. Define a null and alternative hypothesis.
4. How is statistical hypothesis tested?
5. Explain the following terms:
    (a) Errors of the first and second kind
    (b) The best critical region
    (c) Power function of a test
    (d) Level of significance
    (e) Simple and composite hypotheses.
6. In each of the following situations, state whether it is a correctly stated hypothesis testing problem and why.
    (a) $H_0 : \mu = 20$, $H_1 : \mu \neq 20$
    (b) $H_0 : \sigma > 3$, $H_1 : \sigma = 3$
    (c) $H_0 : \bar{x} = 50$, $H_1 : \bar{x} \neq 50$,
    (d) $H_0 : P = 0.05$, $H_1 : P \neq 0.10$,
    (e) $H_0 : S = 25$ $H_1 : S > 25$
    (Ans : (a) yes. (b) No, since s is a population parameter and should be well determined. (c) No, because $\bar{x}$ is not a population parameter (d) No, because H_1 should pertain to $P = 0.05$ and not any other specific value (e) No, because S is note population parameter).
7. Identify the composite hypotheses in the following where μ is the mean and σ^2 is the variance of a distribution
    (a) $H_0 : \mu \leq 1, \sigma^2 = 3$
    (b) $H_0 : \mu = 0, \sigma^2 = 0$
    (c) $H_0 : \mu \leq 1, \sigma^2 =$ arbitrary
    (d) $H_0 : \sigma^2 = \sigma^2_0$ (a given value, m arbitrary)
    (Note : A simple statistical hypothesis will specify the population completely : other win it is called a composite statistical hypothesis.)
    (Ans : (a) composite : (b) simple (c) composite : (d) composite]

10.14.1 Exercises in Type I and Type II Errors

1. A textile fiber manufacturing company claims that the mean thread yarn elongation has mean 12 kilo grams with a standard deviation of 0.5 kilo grams. The company wants to test the   hypothesis
    $H_0 : \mu = 12$ against $H_1 : \mu < 12$ using a random sample of 16 specimens
    (a) What is the Type I error probability if the critical region is defined as $\bar{x} < 11.5$ kilograms?
    (b) Find β for the case where the true mean elongation is 11.25 kilograms?

$$[\text{Ans ; (a) } Z = \frac{11.5 - 12}{0.125} = -4 \quad \sigma = P(Z - 4) = 0$$

$$(b) Z = \frac{11.5 - 11.25}{0.125} = 2 \quad \therefore \beta = P(Z < 2) = 0.022750$$

(c) Find the boundary of the critical region if Type I error probability is specified to be 0.05.

$$[\text{Hint: } Z = \frac{\bar{x} - 12}{.5/5} = \frac{\bar{x} - 12}{0.125}, \quad \alpha = P\left(Z < -\frac{\bar{x} - 12}{0.125}\right) = 0.025 \, \bar{x} \, 11.5875]$$

2. For $\mu = 100, \sigma = 2, n = 9$, test $H_0 : \mu = 100$ versus $H1 : \mu \neq 100$ to find, (a) when the acceptance region is defined as $98.5 \leq \bar{x} \leq 101.5$ (b) find the Type I error probability (c) Find β for the case when the true mean is 103

(Ans : (a) $\alpha = P(Z < \dfrac{98.5 - 100}{\dfrac{2}{3}} = \dfrac{-1.523}{2} = -2.25)$

$+ P(z > 2.25) = 0.01224 + 0.01224 = 0.02448$

(b) $Z_1 = \dfrac{98.5 - 103}{\dfrac{2}{3}} = -6.75$, $Z_2 = \dfrac{101.5 - 103}{\dfrac{2}{3}} = -\dfrac{1.5 \times 3}{2}$

$\therefore \beta = P(-6.75 \leq Z \leq -2.25)$
$= P(Z \leq -2.25) - P(Z \leq -6.75)$
$= 0.01224 - 0 = 0.0122]$

3. The proportion of Ph.D Scholars in an University is estimated to be $p = 0.4$. To test the hypothesis, a random sample of 15 scholars is selected. If the number scholars is between 4 and 8, the hypothesis will be accepted, otherwise, it will be concluded that $P \neq 0.4$.
   (a) Find the Type I error probability for this procedure, assuming that $p = 0.4$
   (b) Find the probability of committing a Type II error if the true proportion is really $P = 0.2$

[Ans : (a) $Z_1 = \dfrac{4/15 - 0.4}{\dfrac{1}{15}\sqrt{15 \times 4 \times 6}} = -1.04$ $Z_2 = \dfrac{(8/15 - 0.4)}{\dfrac{1}{15}\sqrt{15 \times 4.8.6}} = 1.04$]

$\therefore \alpha = P(Z < -1.040 + P(Z > 1.04) = 0.141.859 + 0.146859 = .293718 = 0.29372$

(b) $Z_1 = \dfrac{\left(\dfrac{4}{15} - 0.2\right)15}{\sqrt{15 \times 2 \times .8}}$; $Z_2 = \dfrac{\left(\dfrac{8}{5} - 0.2\right)15}{\sqrt{15 \times .2 \times .8}} = 0.6445 = 3.22$

$\therefore \beta = (0.6445 \leq Z \leq 3.22) = 0.2572]$

4. A shampoo producing company is interested in foam height in millimeters. Assuming foam height is approximately normal with standard deviation of 25 millimeters. Using the results of $n = 16$ samples, the company desires to test $H_0 : \mu = 210$ m.m versus $H_1 : \mu > 210$ m.m.
   (a) Find the Type I error probability α if the critical region is $\bar{x} = 222$ mm
   (b) What is the probability of Type II error if the true mean foam height if. 234 m.m.?

5. If μ is the population mean equal to 0.9 and sample mean and standard deviation are $\bar{x} = 0.921$ and $s = 0.0270$, using one-sided alternative hypothesis as $H_0 : \mu = 0.9$, against $H_1 : \mu > 0.9$ at $\alpha = 0.05$, and $n = 9$.
   (Ans : $t_0 = 2.33$, $(t_{\alpha = 0.05} V = 8 = 1.860)$
   $t_0 = 2.33, > 1.860$. Hence H_0 is rejected and μ exceeds 0.9]

6. What the connection between hypothesis tests and confidence intervals?
   [Ans : Though they have equivalent procedure in inference about the mean μ, the confidence interval provides a range of likely values for μ at a stated confident level, while the hypothesis testing provides the risk levels in decision making].

7. How to determine the appropriable sample size in order to obtain a particular value of β for the given true mean as $\mu = \mu_0 + \delta$, $\delta > 0$ and α

10.15 GOODNESS OF FIT TESTS WHEN ALL PARAMETERS ARE SPECIFIED

Let $y_1, y_2, \ldots y_n$ be n independent random variables, each taking on one of the values 1, 2, ..., k. It is desirable to test the null hypothesis that $\{p_i, i = 1, 2, \ldots, k\}$ is the probability mass function of the y_j. Then the null hypothesis is:

$H_0 : P[y = i] = p_i$, $i = 1, 2, \ldots k$, where as the alternative hypothesis is

$H_1 : P\{y = i\} \neq p_i$, for the same $i = 1, 2, \ldots, k$. In order to test the null hypothesis let x_i $> i = 1, 2, \ldots, k$ denote the number of the y_j's that equal i. Then as each y_j will independently equal i, with probability $P\{y = i\}$, it follows that, under H_0, X_i is binomial with parameters n and p. Hence when H_0 is true,

$E[X_i] = np_i$ and so $\{X_i - np_i\}^2$ will be an indication as to how likely it appears that p_i indeed equals the probability that $y = i$. When this is large, say, in relationship to np_i, then it is an indication that H_0 is not correct.

Hence we consider the following test statistic

$$T = \sum_{i=1}^{K} \frac{(X_i - np_i)^2}{np_i}$$ and reject. The null hypothesis when T is large.

To determine the critical region we need first to specify a significance level a and then we must determine that critical value c such that $P_{H0}\{T \geq c\} = \alpha$. That is, we need to determine c so that the probability that the test statistic T is at least as larges c, when H_0 is true, is α. The test is to reject the hypothesis, at the α level of significance, when $T \geq c$ and to accept when $T < c$.

It remains to determine c. Hence, for n large, c can be taken to equal $\chi^2_{\alpha, k-1}$; and so the approximate α level test is

Reject H_0 if $T \geq c^2_{\alpha, \epsilon - 1}$:
Accept H_0 other wise.

If the observed value of T is T = t, then the preceding test is equivalent to rejecting H_0 if the significance level α is at least as large as the p-value given by

p-value $= P_{H0}\{ T \geq t\} \approx P\{c^2_{k-1} \geq t\}$, where χ^2_{k-1} is the chi-square random variable with $k - 1$ degree of freedom.

Remarks

(a) A computationally simpler formula for T can be obtained by expanding the square

in the above equation and using the results that $\sum_i p_i = 1$ and $\sum x_i = n$

[Note : since X_i, $i = 1, 2, \ldots, k$ denote the number of the Y_j, that equal i, we get

X_i = number of y_j's that equal 1 = 1 with prob. p_1

X_2 = number of y_j's that equal 1 = 1 with prob. p_2

X_k = number of y_j's that equal 1 = 1 with prob. p_k.

X_n = number of y_j's that equal 1 = 1 with prob. p_n

$$\therefore \quad \sum_{i=1}^{n} X_i = 1 + 1 + \ldots \text{ to n time} = n$$

Now consider

$$T = \sum_{i=1}^{k} \frac{\left(X_i - np_i\right)^2}{np_i}$$

$$= \sum_{i=1}^{k} \frac{X_i^2 - 2np_i\, X_i + n^2 p_i^2}{np_i}$$

$$= \sum_{i} \frac{X_i^2}{np_i} - 2\sum_{i} X_i + n\sum_{i} p_i$$

$$= \sum_{i} \frac{X^2}{np_i} - 2k + k$$

$$\therefore \quad T = \sum_{i=1}^{K} \frac{X_i^2}{np_i} - k$$

(b) The intuitive reason why T, which depends on the k values $X_1, X_2, \ldots, X_k$ has only $k - 1$ degrees of freedom is that 1 degree of freedom is lost because of the linear relationship

$$\sum_{i=1}^{n} X_i = n$$

(c) T has a chi-square distribution can be easily shown when $k = 2$. In this case $X_1 + X_2 = n$ and $p_1 + p_2 = 1$, we see that

$$T = \frac{\left(X_1 - np_1^2\right)}{np_1} + \frac{\left(X_2 - np_2\right)^2}{np_2}$$

$$= \frac{\left(X_1 - np_1\right)^2}{np_1} + \frac{\left[n - X_1 - n\left(1 - p_1\right)\right]^2}{n\left(1 - p_1\right)}$$

$$= \frac{(X_1 - np_1)^2}{np_1} + \frac{(X_1 - np_1)^2}{n(1 - p_1)}$$

$$= (X_1 - np_1)^2 \left[\frac{1}{np_1} + \frac{1}{n - np_1} \right]$$

$$= (X_1 - np_1)^2 \frac{(n - np_1 + np_1)}{np_1 \, n(1 - p_1)}$$

$$= (X_1 - np_1)^2 \frac{1}{np_1 \, (1 - p_1)}$$

However X_1 is a binomial random variable with mean np_1 and variance $np_1(1 - p_1)$ and these, by normal approximation to the binomial, it follows that $\dfrac{(X_1 - np_1)}{\sqrt{np_1(1 - p_1)}}$ has, for large n, approximately a unit standard distribution and so its square has approximately a chi-square distribution with 1 degree of freedom].

10.16 DETERMINING THE CRITICAL REGION BY SIMULATION

We have seen that T has approximately a chi-square distribution with $k - 1$ degrees of freedom. It is the only means available for determining the p-value of the goodness of fit. Due to recent advances in high speed inexpensive computer, use of simulation to obtain to a high level of accuracy the p-value of the test statistic is adopted.

The simulation approach is as follows:

• Firstly, the value of T is determined say $T = t$. We need to know the probability that T would be at least as large as t when Ho is true. To determine this probability that T would be at least as large as t when Ho is true. We simulate n independent random variables $Y_1^{(1)}, Y_2^{(2)}, ..., Y_n^{(n)}$, each having the probability mass function $\{p_i, i = 1, 2, ... k\}$ that is

$$P\{y_j^{(1)} = i\} = p_i, \quad i = 1, 2, 3, ... k, \quad j = 1, 2, ... n$$

$$X_i^{(1)} = \text{number } j : Y_i^{(1)} = 1 \text{ and set } T^{(1)} = \sum_{i=1}^{k} \frac{\left(X_i^{(1)} - np_i\right)^2}{np_i}$$

Now repeat this procedure by simulating a second set, independent of the first set, of n independent random variable $Y_1^{(2)}, Y_2^{(2)} ... Y_n^{(2)}$ each having the probability mass function $\{p_i, i = 1, 2 ... , k\}$ and then, as for the first set, determining $T_i^{(2)}$. Repeating this, a large number, say, r times, yields r Independent random variables $T^{(1)}, T^{(2)}, ..., T^{(r)}$, each of which has the same distribution as does the test statistic T when H_0 is true. Hence, by the law of large numbers, the proportion of the T_i that are as large as t will be very nearby equal to the probability that T is as large us t, when H_0 is true—that is

$$\frac{\text{number } 1 : T^{(1)} \geq t}{r} \approx P_{H_0} \{T \geq t\}$$

Letting r large enough, the above can be considered to be, with high probability, almost an equality. Hence, if that proportion is less than or equal to α, then p-value, equal to the probability of observing a T as large as t when H_0 is true, is less than α and so H_0 should be rejected.

Simulation procedure

To utilize the simulation approach to determine whether or not to accept H_0 when T is observed; we need to specify how one can simulate or generate, a random variable Y such that $P\{Y = i\} = p_i, i = 1, 2, ... k$. One way is as follows:

Step1 : Generate a random number U

Step2 : If $p_1 + p_2 + ... + p_{i-1} \leq U < p_1 + p_2 + ... + p_i$,

Set Y = i (where $p_1 + p_2 + ... + p_{i-1} = 0$, When i = 1 $U < p_1 + p_2 + ... + p_{i, ==} 0$ where i = 1)

That is

$U < p_1 \Rightarrow Y = 1$

$p_1 \leq U < p_1 + p_2 \Rightarrow Y = 2$

$P_1 + P_2 + ... P_{i-1} \leq U p_1 + p_2 \Rightarrow Y = i$

Since a random number is equivalent to a uniform (0, 1) random variable, we have that

$P\{a < U < b\} = b - a, 0 < a < b < 1$

And so $P\{Y = i\} = P\{p_1 + p_2 + ... + b_{i-1} < U < p_1 + p_2 + ... + p_i\} = p_i$

Simulation runs are necessary to continue so that the value r = 100 is usually sufficient as the conventional 5 per cent level of significance.

CHAPTER 11

DESIGN OF EXPERIMENTS

11.1 INTRODUCTION

Variations are an acceptable phenomenon. Since observed differences can always be due to causes than those postulated, it has become necessary to discuss some questions on experimental design so that, with reasonable assurance, statistically significant results can be attributed to particular causes. There are many pitfalls to which an experimenter is exposed. Some of the general principles of experimental design can be illustrated by furnishing designs most frequently in engineering and other applied research. It is assumed that systematic differences in measuring techniques and chance variability exist. The problem of determining whether the results produced by the laboratories are consistent by comparing the variability with an appropriate measure of chance variation. The problem of deciding consists of:

- The questions of sample size.
- The question of selection.
- The measuring techniques.
- The contribution of controlled variable. The purpose of randomization is to avoid confounding the variable under investigation with other uncontrolled variables.
- Inclusion of extraneous causes under the heading of chance variation.
- The complications that arise due controlled experimentation.
- Repeat the experiment in several blocks, where known sources of variability (due to extraneous variables) are held fixed in each block, but vary from block to block.

11.2 ANALYSIS OF VARIANCE

The **analysis of variance,** of more briefly **ANOVA,** refers broadly to a collection of experimental situations and statistical procedures for the analysis of quantitative responses from experimental units. The simplest ANOVA problem is referred to variously as a **single – factor, single-classification** or **one-way ANOVA** and involves the analysis either of data sampled from more than two numerical populations (distributions) or of data from experiments in which more than two treatments have been used. The characteristic that differentiates the treatments or populations from one another is called the **factor** under study, and the different treatments or populations are referred to as the **levels** of the factor.

11.3 ONE WAY CLASSIFICATION OR ONE-WAY ANALYSIS OF VARIANCE

Suppose the experimenter has available the results of k independent random samples, from k different populations (that is, data concerning k treatments, k groups, k methods of production etc.,); and the experimenter is concerned with testing the hypothesis that the means of these k populations are all equal.

In general, we denote the j-th observation in the i-th sample by y_{ij}, and the scheme for a one-way classification is as follows:

$$\text{Sample 1:} \quad y_{11}, \quad y_{12}, \quad \ldots \quad y_{1j} \quad \ldots \quad y_{1n_i} \quad \bar{y}_1 \quad \sum_{j=1}^{n_1}(y_{1j}-\bar{y}_1)^2$$

$$\text{Sample 2:} \quad y_{21}, \quad y_{22}, \quad \ldots \quad y_{2j} \quad \ldots \quad y_{2n_2} \quad \bar{y}_2 \quad \sum_{j=1}^{n=2}(y_{2j}-\bar{y}_2)^2$$

$$\text{Sample 3:} \quad \ldots \quad \ldots \quad \ldots \quad \ldots \quad \ldots \quad \ldots$$

$$\text{Sample i:} \quad y_{i1} \quad y_{i2} \quad \ldots \quad y_{ij} \quad \ldots \quad y_{in_i} \quad \bar{y}_i \quad \sum_{j=1}^{n_i}(y_{ij}-\bar{y}_i)^2$$

$$\text{Sample k} \quad y_{k_1} \quad y_{k_2} \quad \ldots \quad y_{kj} \quad \ldots \quad y_{kn_k} \quad \bar{y}_k \quad \sum_{j=1}^{n_k}(y_{kj}-\bar{y}_k)^2$$

[Note : The sample sizes n_i are unequal]

To simplify the calculations below, we use the notation T. for the sum of all the observations and N for the total sample size.

$$T = \sum_{i=1}^{k}\sum_{j=1}^{n_i} y_{ij}, \quad N = \sum_{i=1}^{k} n_i$$

The overall sample mean $\bar{y}$ is

$$\bar{y} = \frac{\sum\limits_{i=1}^{k}\sum\limits_{j=1}^{n_i} y_{ij}}{\sum\limits_{i=1}^{k} n_i} = \frac{\sum\limits_{i=1}^{k} n_i \bar{y}_i}{\sum\limits_{i=1}^{k} n_i} = \frac{T}{N}$$

[Note: The overall sample mean $\bar{y}$ is called the **grand mean**]

Theorem 1

$$\sum_{i=1}^{k}\sum_{j=1}^{n_i}(y_{ij}-\bar{y})^2 = \sum_{i=1}^{k}\sum_{j=1}^{n_i}(y_{ij}-\bar{y})^2 + \sum_{i=1}^{k}n_i(\bar{y}_i-\bar{y})^2$$

Proof

The proof of this theorem is based on the identity

$$y_{ij} - \bar{y} = (y_{ij} - \bar{y}_i) + (\bar{y}_i - \bar{y})$$

Squaring both sides and summing on i and j, we obtain

$$\sum_{i=1}^{k}\sum_{j=1}^{n_i}(y_{ij}-\bar{y})^2 = \sum_{i=1}^{k}\sum_{i=1}^{n_i}(y_{ij}-\bar{y}_i)^2 + \sum_{i=1}^{k}\sum_{j=1}^{n_i}(\bar{y}_i-\bar{y})^2 + 2\sum_{i=1}^{k}\sum_{j=1}^{n_i}(y_{ij}-\bar{y}_i)(\bar{y}_i-\bar{y})$$

We observe that

$$\sum_{i=1}^{k}\sum_{j=1}^{n_i}(y_{ij}-\bar{y}_i)(\bar{y}_i-\bar{y}) = \sum_{i=1}^{k}(\bar{y}_i-\bar{y})\sum_{i=1}^{n_i}(y_{ij}-\bar{y}_i) = 0$$

Since $\bar{y}_i$ is the mean of the i-th sample and hence,

$$\sum_{j=1}^{n_i}(y_{ij}-\bar{y}_i) = 0 \text{ for all i.}$$

To complete the proof of the Theorem 1, we have only to observe that the summand of the second sum of the right hand side of the above identity does not involve the subscript j and that, consequently,

$$\sum_{i=1}^{k}\sum_{j=1}^{n_i}(\bar{y}_i-\bar{y})^2 = \sum_{i=1}^{k}n_i(\bar{y}_i-\bar{y})^2$$

It is customary to denote the total sum of squares, the lefthand member of the identity of Theorem1, by SST. We refer to the first term on the righthand side as the **error sum of squares**, SSE. The term "error sum of squares" expresses the idea that the quantity estimates random (or chance) error. The second term on the right-hand side of the identity of Theorem 1, we refer as the **between samples sum of squares or treatment sum of squares**, SS (T$_r$). (Most of the early applications of this kind of analysis were in the field of agriculture, where the k-populations represented different treatments, such as fertilizers, applied to agricultural plots)

To be able to test the hypothesis that the samples were obtained from k population with equal means, we assume that we are dealing with **normal populations** having **equal variances**. But the development here is taken as fairly **robust**, that is, it is insensitive to violations of the assumption of normality as well as the assumption of equal variances.

If μ_i denotes the mean of the i th population and σ^2 denotes the common variance of k populations, we can express each observations y_{ij} as μ_i, plus the value of the random component j that is, we can write $y_{ij} = \mu_i + \in_{ij}$ for i = 1, 2, ... k, j = 1, 2, ... n$_i$.

We note here that $\in_{ij}$ are independent, normally distributed random variables with zero means and the common variance σ^2. It is customary to replace μ_i by $\mu + \alpha_i$ where μ is the mean

$$\sum_{i=1}^{k} n_i \mu_i \Big/ N$$ of the m$_i$ in the experiment and α_i is the **effect** of the i th treatment, hence

$$\sum_{i=1}^{k} n_i \alpha_i = 0$$

Using these new parameters, we can write the **model equation** for the one-way classification as

$$y_{ij} = \mu + \alpha_i + \in_{ij} \text{ for } i = 1, 2, ... k \ j = 1, 2, ... n_i$$

and the null hypothesis

$$H_0 : \mu_1 = \mu_2 = ... \mu_k$$

can be replaced by

$$H_0 : \alpha_1 = \alpha_2 = ... \alpha_k = 0$$

Each sum of squares is first converted to a **mean square**, so a test for the equality of treatment means can be performed.

$$\text{Mean Square} = \frac{\text{Sum of squares}}{\text{Degrees of freedom}}$$

When the population means are equal, both the

$$\text{treatment means square} = \frac{\sum_{i=1}^{k} n_i \left(\bar{y}_i - \bar{y} \right)^2}{k-1}$$

and the

$$\text{error means square} = \frac{\sum\limits_{i=1}^{k}\sum\limits_{j=1}^{n_i}(\bar{y}_i - \bar{y})^2}{N-k}$$

are estimates of σ^2. However, when the null hypothesis is **false**, the **treatment or between—sample mean square** can be expected to exceed the **error or within—sample mean square.** If the null hypothesis is true, it can be shown that the two mean squares are independent and that their ratio

$$F = \frac{\sum\limits_{i=1}^{k} n_i (\bar{y}_i - \bar{y})^2 \Big/ (k-1)}{\sum\limits_{i=1}^{k}\sum\limits_{j=1}^{n_i}(\bar{y}_{ij} - \bar{y})^2 \Big/ (N-k)} = \frac{SS(T_r)/(k-1)}{SSE/(N-k)}$$

has an F distribution with $(k - 1)$ and $(N - k)$ degree of freedom.

A large value of for F indicates large differences between the sample means. Therefore, the null hypothesis will be rejected, at level α, if the value of F exceeds F_α where F_α is obtained from the F distribution table $F v_1 = k - 1, v_2 = N - k$ where v_1 and v_2 are the degree of freedom.

The above results are summarized by means of the **analysis variance table**

Source of variation	Degrees of freedom	Sum of squares	Mean square	F
Treatments	$k - 1$	$SS(T_r)$	$MS(T_r) = SS(T_r)/(k-1)$	$\dfrac{MS(T_r)}{MSE}$
Error	$N - k$	SSE	$MSE = SSE/(N-k)$	
Total	$N-1$	SST		

where $N = \sum\limits_{i=1}^{k} n_i$

Short cut formulas

$$SST = \sum\limits_{i=1}^{k}\sum\limits_{j=1}^{n_i} y_{ij}^2 - C$$

$$SS(T_r) = \sum\limits_{i=1}^{k} \frac{T_i^2}{n_i} - C$$

where C is called the correction term for the mean and is given by

$$C\frac{T.^2}{N}$$

With $N = \sum_{i=1}^{k} n_i$, $T = \sum_{i=1}^{k} T_i$ and $T_i = \sum_{j=1}^{n_i} y_{ij}$

In this formulas T_i is the total of the n_i observations in the i-th sample. Where as T. is the grand total of all N observations. The error sum of squares, SSE is then obtained by subtraction, given by

$$SSE = SST - SS(T_r)$$

Evaluation of SST and SS(T_r) by formulating the tabular values as shown below.

Sample Number	Observations				Sum of each row	Square of the sum of each row	
1.	y_{11}	y_{12}	y_{1j}	y_{1n1}	T_1	T_1^2	T_1^2/n_1
2.	y_{21}	y_{22}	y_{2j}	y_{2n2}	T_2	T_2^2	T_2^2/n_2
i	y_{i1}	y_{i2}	y_{ij}	y_{inj}	T_i	T_i^2	T_i^2/n_i
K	y_{k1}	y_{k2}	y_{kj}	y_{knk}	T_k	T_k^2	T_k^2/n_k

Calculation

$$N = \sum_{i=1}^{k} n_i; \; T_i = \sum_{j=1}^{n_i} y_{ij}; \; T. = \sum_{i=1}^{k} T_i; \sum_{i=1}^{k} \frac{T_i^2}{n_i}$$

$$C = \frac{T.^2}{N};$$

$$SST = \sum_{i=1}^{k}\sum_{j=1}^{n_i} y_{ij}^2 - C$$

$$SS(T_r) = \sum_{i=1}^{k} \frac{T_i^2}{n_i} - C$$

and

$$SSE = SST - SS(T_r).$$

Example

EXAM.1: In an experiment to compute the tensile strengths of 5 different types of copper wires, 4 samples of each type were used. The between—samples and within samples of σ^2 were computed as MS(T_r) = 2570 and MSE = 1390, respectively. Use the F test-at level 0.05 to test $H_0 = \mu_1 = \mu_2 = \mu_3 = \mu_4 = \mu_5$ versus H_a : at least two μ_1's are un equal.

SOLUTION:

Here the types of copper wires is k = 5 and number of observations in each type
is $= n_1 = n_2 = n_3 = n_{4=} = n_5 = 4, \therefore N = 20$
MS(Tr) = 2570 $v_1 = k - 1 = 4$
MSE = 1390 $v_2 = N - k = 20 - 5 = 15$
and SST has N − 1 = 20 − 1 = 19 d.f

$$\therefore F_{v_1, v_2} = \frac{MS(Tr)}{MSE} = \frac{2570}{1390} = \frac{257}{139} = 1.8489 = f\,(say)$$

From the F table
$F_{0.5}, v_1 = 4, v_2 = 15 = 3.06$
$\therefore$ Here F = 1.8489 < 3.06 = $F_{0.5}, v_1, v_2$.
We do not reject the hypothesis i.e., we accept the hypothesis H_0.

EXAM. 2: To compare the effectiveness of three different types of phosphorescent coatings of airplane instrument dials, eight dials each are coated with the three types. Then the dials are illuminated by an ultraviolet light, and the following are the number of minutes each glowed after the light source was shut off:

Type 1: 52, 62, 57, 51, 59, 61, 60, 54
Type 2: 58, 55, 59, 63, 65, 61, 54, 57
Type 3: 71, 67, 63, 65, 75, 66, 73, 64

Test the null hypothesis that there is no difference in the effectiveness of three coatings at the 0.01 level of significance.

SOLUTION:

Here T. $= T_1 + T_2 + T_3 = 456 + 472 + 544 = 147.2$;
N = 8 + 8 + 8 = 24

$$\therefore C = \frac{T.^2}{N} = \frac{(1472)^2}{24} = 90290$$

$$SST = \sum_{i=1}^{k} \sum_{j=1}^{n_i} y^2_{ij} - C$$

$$= \sum_{i=1}^{3} \sum_{j=1}^{8} y^2_{ij} - C = 911196 - 90290 = 906$$

$$SS(T_r) = \sum_{i=1}^{K} \frac{T_r^2}{n_i} - C$$

$$= \frac{(456)^2}{8} + \frac{(472)^2}{8} + \frac{(544)^2}{8} - 90290 = 90962.5 - 90290 = 672.5$$

$$MS(T_r) = \frac{SS(T_r)}{k-1} = \frac{672.5}{2} = 336.2$$

$$SSE = SST - C = 906$$

$$MSE = \frac{SSE}{N-K} = \frac{906}{21} = 43.1$$

$$f = \frac{MS(T_r)}{MSE} = \frac{336.2}{43.1} = 7.8$$

From the F-Table we have

$F_{0.01}$, $v_1 = 2$, $v_2 = 21 = 5.78$

$\therefore$ Here $f = 7.8 > F_{0.01} = 5.78$

Hence, the difference in effectiveness are significant.

EXAM. 3: A laboratory test gives all the available bolts that connected the steel structure at 3 different positions on the roof. The forces required to shear each of these bolts (coded values) are as follows:

Position 1 :	90	82	77	99	83	91	
Position 2 :	103	89	93	102	90	94	87
Position 3 :	81	88	81	94			

Perform an analysis of variance to test at the 0.05 level of significance whether the difference among the sample means at the 3 positions are significant.

SOLUTION:

Step 1: Null hypothesis $\mu_1 = \mu_2 = \mu_3$

Alternative hypothesis μ_i's are not equal

Step 2: $\dfrac{\text{Level of Significance}}{\alpha = 0.05}$

Step 3: Criterion

Reject the null hypothesis if F (observed) > 3.74 the value of $F_{0.05}$, $v_1 = 3 - 1 = 2$, $v_2 = N - K =$ is $17 - 3 = 14$, where v_1 and v_2 are the degree of freedom, where F is to be determined by an analysis of variance.

Step 4: Calculations

Substituting $v_1 = 6$, $v_2 = 7$, $v_3 = 4$, $N = 17$, $T_1 = 522$, $T_2 = 658$, $T_3 = 344$ and T. $= 522 + 658 + 344 = 1524$ and

$$\sum_{i=1}^{3} \sum_{j=1}^{ni} y^2_{ij} = 137542$$

into the computing formulas for the sums of squares, we get

$$SST = 137542 - \frac{(1524)^2}{17} = 137542 - 136600 = 942$$

$$SS(T_r) = \frac{522^2}{6} + \frac{658^2}{7} + \frac{344^2}{4} - \frac{1524^2}{17} = 136840 - 136600 = 240$$

and SSE = 942 − 240 = 702

The remainder of the work is shown in the following analysis of variance table.

Source of Variation	Degrees of Freedom	Sum of Squares	Mean squares	F
Positions	2	240	120	2.39
Error	14	702	50.13	
Total	16	942		

Step 5: Decision: Since F = 2.39 does not exceed 3.74, the null hypothesis cannot be rejected. Hence we cannot conclude that there is a difference in the mean shear strengths of the bolts at the three different positions of the roof.

11.4 ANALYSIS OF TWO WAY CLASSIFICATION

There are **a** treatments distributed on **b** blocks. Let y_{ij} denote the observation pertaining to the i-th treatment and the j-th block ; let $\bar{y}_i$ denote the mean of the b observations for the i-th treatment; $\bar{y}_{.j}$ the mean of the a observations with j th block and $\bar{y}_{..}$, the grand mean of all the ab observations. The layout of the **two-way classification** is represented as follows:

Blocks

	B_1	B_2	....	B_j	...	B_b	Means
Treatment 1	y_{11}	y_{11}	...	y_{1j}	...	y_{1b}	$\bar{y}_{1.}$
Treatment 2	y_{21}	y_{22}	...	y_{2j}	...	y_{2b}	$\bar{y}_{2.}$
Treatment i	y_{i1}	y_{i2}	...	y_{ij}	...	y_{ib}	$\bar{y}_{i.}$
Treatment a	y_{a1}	y_{a2}	...	y_{aj}	...	y_{ab}	$\bar{y}_{a.}$
Means	$\bar{y}_{.1}$	$\bar{y}_{.2}$	...	$\bar{y}_{.j}$	...	$\bar{y}_{.b}$	$\bar{y}_{..}$

This kind of arrangement is also called a **randomized—block design**, provided the treatments are allocated at random **within** each block. Here it is to be noted that when a dot is used in place of a subscript, this means that the mean is obtained by summing over that subscript.

The underlying model considered here assumes for the analysis of this kind of experiment with one observation per cell (that is, there is one observation corresponding to each treatment within each block) is given by.

$$y_{ij} = \mu + \alpha_i + \beta_j + \in_{ij} \text{ for } i = 1, 2, \dots a; j = 1, 2 \dots b$$

Here μ is the grand mean, α_i is the effect of the i-th treatment, β_j is the effect of the j-th block, and the $\in_{ij}$ are **independent, normally distributed** random variables having **zero means** and the common variance σ^2. Here the parameters are restricted by imposing

the conditions that $\sum_{i=1}^{a} \alpha_i = 0$ and $\sum_{j=1}^{b} \beta_j = 0$.

In the analysis of two-way classification where each treatment is represented once in each block, the major objective is to test the significance of the difference among the $\overline{y}_{i.}$,

that is, to test the null hypothesis: $\alpha_1 = \alpha_2 = \dots \alpha_a = 0$

In addition, it may also be desirable to test whether the blocking has been effective, that is, whether the null hypothesis: $\beta_1 = \beta_2 = \dots = \beta_b = 0$, can be rejected. In either case, the alternative hypothesis is that at least one of the effects is different from zero.

These significance test on comparisons of estimates of σ^2-one based on the variation among treatments, one based on the variation among block, and one measuring the experimental error. Note that only the latter is an estimate of σ^2 when either (or both) of the null hypotheses do not hold. The required sums of squares are given by the there components into which the total sum of squares is partitioned by the following statement;

$$\sum_{i=1}^{a}\sum_{j=1}^{b}(y_{ij} - \overline{y}..)^2 = \sum_{i=1}^{a}\sum_{j=1}^{b}(y_{ij} - \overline{y}_{i.} - \overline{y}_{.j} + \overline{y}..)^2 + b\sum_{i=1}^{a}(\overline{y}_{i.} - \overline{y}..)^2 + a\sum_{j=1}^{b}(\overline{y}_{.j} - \overline{y}..)^2$$

The left hand side of this identity represents the total sum of squares SST, and the terms of the right hand side are, respectively, the error sum of squares, SSE, the treatment sum of squares, SS(T_r) and the block sum of squares SS(Bl). The statement can be established, by making use of the identity

$$y_{ij} - \overline{y}.. = \left((y_{ij} - \overline{y}_{i.} - \overline{y}_{.j} + \overline{y}.) + (\overline{y}_{.j} - \overline{y}..).\right)$$

The convenient formulas to calculate SST, SS(T_r) and SS(Bl) are given by

$$SST = \sum_{i=1}^{a}\sum_{j=1}^{b} y^2_{ij} - C$$

$$SS(T_r) = \frac{\sum_{i=1}^{a} T_i^2}{b} - C$$

$$SS(B1) = \frac{\sum\limits_{j=1}^{b} T_{\cdot j}^2}{a} - C$$

where C, the correction term, is given by

$$C = \frac{T_{\cdot\cdot}^2}{ab}$$

In these formulas, $T_{i\cdot}$ is the sum of the b observations for the i-th treatment, $T_{\cdot j}$ is the sum of the observations in the j-th block and $T_{\cdot\cdot}$ is the grand total of all the observations, in the respective totals, $T_{i\cdot}$ and $T_{\cdot j}$. The error sum of squares in then obtained by subtraction according the statement mentioned. We can write

$$SSE = SST - SS(T_r) - SS(Bl)$$

Using the sum of squares, we can reject the null hypothesis that the α_i are all equal to zero at the level of significance α if

$$\text{(F ratio for Treatments): } F_{Tr} = \frac{MS(T_r)}{MSE} = \frac{SS(T_r)/(a-1)}{SSE/(a-1)(b-1)}$$

exceeds F_α with $a - 1$ and $(a - 1)(b - 1)$ degrees of freedom.

The null hypothesis that the β_j are all equal to zero can be rejected at the level of significance α is

$$\text{(F ratio for blocks): } F_{Bl} = \frac{MS(Bl)}{MSE} = \frac{SS(Bl)/(b-1)}{SSE/(a-1)(b-1)}$$ exceeds F_α with $b - 1$ and $(a - 1)(b - 1)$ degrees of freedom.

The results obtained in this analysis are summarized in the following analysis of variance table:

Source of variation	Degrees of freedom	Sum of Squares	Mean square	F
Treatments	$a - 1$	$SS(T_r)$	$MS(T_r) = \dfrac{SS(T_r)}{a-1}$	$F_{Tr} = \dfrac{MS(T_r)}{MSE}$
Blocks	$b - 1$	$SS(Bl)$	$MS(Bl) = \dfrac{SS(Bl)}{b-1}$	$F_{Bl} = \dfrac{MS(Bl)}{MSE}$
ERROR	$(a - 1)(b - 1)$	SSE	$MSE = \dfrac{SSE}{(a-1)(b-1)}$	
Total	$ab - 1$	SST		

EXAM. 1: A Randomized block experiment is run with three treatments. The three treatment means are $\bar{y}_1. = 7; \bar{y}_2. = 8;$ and $\bar{y}_3. = 12$. The total (Corrected) sum of squares is

$$240 = \sum_{i=1}^{3} \sum_{j=1}^{b} (y_{ij} - \bar{y}..)^2.$$

The analysis of variance (ANOVA) table takes the form

Source of variation	Degrees of freedom	Sum of squares	Mean square	F
Treatments				
Blocks		144		
Error				
Total	11	240		

SOLUTION:

$a = 3; ab - 1 = 11; ab = 12; \therefore b = 4, \bar{y}_1 = 7; \bar{y}_2. = 8; \bar{y}_3. = 12; a = 3, b = 4$

$T_1. = 4 \times 7 = 28 ; T_2. = 4 \times 8 = 32 ; T_3. = 4 \times 12 = 48$

$T.. = T_1. + T_2. + T_3. = 28 + 32 + 48 = 108.$

$$C = \frac{T^2..}{ab} = \frac{108 \times 108}{12} = 972$$

$$SS(T_r) = \frac{\sum_{i=1}^{3} T_i^2}{b} - C = \frac{28^2 + 32^2 + 48^2}{4} - 972$$

$$= \frac{4(196 + 256 + 576)}{4} - 972$$

$$= 1028 - 972 = 56$$

$$\bar{y}.. = \frac{1}{3}\left[\bar{y}_1. + \bar{y}_2. + \bar{y}_3.\right] = \frac{7 + 8 + 12}{3} = \frac{27}{3} = 9$$

$$SST = \sum_{i=1}^{3} \sum_{j=1}^{4} (y_{ij} - \bar{y}..)^2$$

$$= \Sigma\Sigma y_{ij}^2 - 2\bar{y}..\Sigma\Sigma y_{ij} + 12\bar{y}^2.. = \sum_i \sum_j y_{ij}^2 - 2\bar{y}..12\bar{y}.. + 12 \times 27 \times 27$$

$$= \sum\sum y_{ij}^2 - 24\bar{y}.. + 12\bar{y}^2. = \sum\sum y_{ij}^2 - 12\bar{y}^2. = \sum\sum y_{ij}^2 - 12 \times 9^2$$

$$= \sum \sum y_{ij}^2 - 12 \times 81 = \sum \sum y_{ij}^2 - 972 \therefore \text{SST} = \sum_{i=1}^{3} \sum_{j=1}^{4} y_{ij}^2 - 972$$

$240 = \Sigma\Sigma y_{ij}^2 - 972 \quad \Sigma\Sigma y_{ij}^2 = 1212$

Now SSE = SST – SS(T_r) – SS(Bl)

= 240 – 56 – 144 = 240 – 200

= 40.

∴ The ANOVA table finally takes the form

Source of variation	Degrees of freedom	Sum of squares	Mean square	F
Treatments	2	56	28	$F_{\text{Tr}} = \dfrac{\text{MS}(T_r)}{\text{MSE}} = \dfrac{28.3}{20}$
Blocks	3	144	48	$F_{\text{Bl}} = \dfrac{\text{MS}(\text{Bl})}{\text{MSE}} = \dfrac{48.3}{20} = 7.2$
Error	6	40	$\dfrac{40}{6} = 6\dfrac{2}{3}$	
Total	11	240		

From the F-Table $F_{v1 = 2, v2 = 6} = 5.14$.

For treatment at 0.05 probability level. Here $F_{\text{Tr}} = 4.2 < 5.14$ Not significant.

For Blocks $F_{v1 = 3, v2 = 6} = 4.76$. Here $F_{\text{Bl}} = 7.2 > 4.72$ significant.

EXAM. 2: The number of kilometers of useful tread wear (1000's of k.m) was determined for tires of each of five different makes of subcompact car (factor A, with I = 5) in combination with each of four different brands of radial tires (factor B, with J = 4), resulting in IJ = 20 observations. The values of SSA = 35: SSB = 50 and SSE = 65 were then computed Assume that an additive model is appropriate.

(a) Test $H_o : \alpha_1 = \alpha_2 = \alpha_3 = \alpha_4 = \alpha_5 = 0$ (no differences in true average tire lifetime due to makes of cars) versus H_A : at least one $\alpha_i \neq 0$ using a level of 0.05 test.

(b) $H_o : \beta_1 = \beta_2 = \beta_3 = \beta_4 = 0$ (no differences in true average tire life time due to brands of tires versus H_B : at least one $\beta_j \neq 0$ using a level 0.05 test.

SOLUTION:

The ANOVA table is developed as follows: I – 5 , J = 4, IJ = 20

Source of variation	Degrees of freedom	Sum of squares	Mean square	F
A: Car makes	4	SSA = 35	$MS(A) = \dfrac{35}{4} = 8.75$	$F_A = \dfrac{8.75}{5.417} = 1.615$
B : Brands	3	SSB = 50	$MS(B) = \dfrac{50}{3} = 16.67$	$F_B = \dfrac{16.67}{5.417} = 3.077$
Error	12	SSE = 65	$MSE = \dfrac{65}{12} = 5.417$	
Total	19	SST = 150		

From the F-Table F_{v1} = numerator = 4, v_2 = denominator = 12,
For A : Car makes
$F_{0.05} = 3.26$ and hence $F_A = 1.615 < F_{0.05} = 3.26$
$\therefore$ We do not reject the hypothesis H_{0A}.
F_{v1} = numerator $v_1 = 3$, v_2 = denominator = 12
For B : Brands $F_{0.05} = 3.49$ and hence $F_B = 3.077 < F_{0.05} = 3.49$
$\therefore$ We do not reject the hypothesis H_{0B}.

Example 3

The following data represent the power consumption for five different brands of dehumidifier. Each brand is monitored at four different humidity levels. With in each level brands were assigned randomly to the five selected locations. The resulting amount of power consumption (annual kwh) is shown the table below.

Treatments (Brands)	Blocks (Humidity Level)				$T_.$	$\bar{y}_i.$
1	690	790	840	880	3200	800
2	720	810	890	960	3380	845
3	730	800	880	950	3360	840
4	810	890	950	1010	3660	915
5	830	920	980	1030	3760	940
$T_{.j}$	3780	4210	4540	4030	17360	$\bar{y}.. = \dfrac{1}{5}(4340) = 868$
$\bar{y}_{.j}$	756	842	908	966	- -	$= \dfrac{1}{4}(3472) = 868$

SOLUTION:

$$\sum_{i=1}^{5} \sum_{j=1}^{4} y_{ij}^2 = 100 \{4761 + 6241 + 7056 + 7744 + 5184 + 6261 + 7921 + 9216$$
$$+ 5329 + 6400 + 7744 + 9025 + 6561 + 7921 + 9025 + 10201$$
$$+ 6889 + 8464 + 9604 + 10609\}$$
$$= 100 \,(28724 + 35587 + 41350 + 46795) = 100 \,(152456) = 15245600$$

$$C = \frac{T..^2}{ab} = \frac{T..^2}{5 \times 4} = \frac{T..^2}{20}$$

$$T.. = T_{1.} + T_{2.} + T_{3.} + T_{4.} + T_{5.} = 17360$$

$$C = \frac{(17360)^2}{20} = \frac{(1736)^2 \times 100}{20} = (1736)^2 \times 5$$

$\therefore \quad$ $$SST = \sum\sum Y_{ij}^2 - C = 15245600 - 15070000 = 175600$$

$$SS(T_r) = \frac{1}{6} \sum_{i=1}^{a} T_{i.}^2 - C$$

$$= \frac{1}{4}\left[3200^2 + 3380^2 + 3360^2 + 3660^2 + 3760^2\right] - 15070000$$

$$= \frac{100 \times 100}{4}\left[32^2 + 33.8^2 + 33.6^2 + 37.6^2\right] - 15070000$$

$$= 2500 \,[6048] - 15070000 = 15120000 - 15070000 = 50000$$

$$SS(Bl) = \frac{1}{9} \sum_{j=1}^{b} T_{.j}^2 - C$$

$$= \frac{1}{5}\left[378^2 + 421^2 + 454^2 + 483^2\right] \times 100 - 15070000$$

$$= \frac{1}{5} \times 100 \times 100 \left[37.8^2 + 42.1^2 + 45.4^2 + 48.3^2\right] - 15070000$$

$$= 15192000 - 15070000 = 122000$$

$\therefore \quad$ $$SSE = SST - SS(T_r) - SS(Bl)$$
$$= 175600 - 50000 = 122000$$
$$= 175600 - 172000 = 3600$$

The ANOVA table is given by

Source of variation	Degrees of freedom	Sum of squares	Mean square	F
Treatments	4	50000	12500	$\dfrac{12500}{300} = 41.67$
Blocks	3	122000	40666	$\dfrac{40666}{300} = 135.53$
Error	12	3600	300	
Total	19			

Treatments

From F table for $v_1 = 4$, $v_2 = 12$, at 0.05 level $F_{v1, v2} = 3.26$
Hence $F_A = 41.67 > 3.26$ ∴ H_0 is rejected in favour of H_a.

Blocks

From F table for $v_1 = 3$, $v_2 = 12$, at 0.05 level $F_{v1, v2} = 3.49$.
Hence $F_B = 135.53 > 3.49$.

∴ H_0 is rejected in favour of H_a. The power consumption does depiction the brand of humidifier.

Example 4

There are 3 treatments and 4 blocks. The observations are given in the following table:

	1	2	3	4
Treatment 1	14	8	10	4
Treatment 2	7	7	4	2
Treatment 3	12	6	16	6

Set up the analysis of variance for the above randomized block experiment by decomposing each observation. Test the hypothesis of equal treatment effects.

Solution:

Each observation y_{ij} will be decomposed as

$$y_{ij} = \bar{y}.. + \left(\bar{y}_{i.} - \bar{y}..\right) + \left(\bar{y}_{.j} - \bar{y}..\right) + \left(y_{ij} - \bar{y}_{i.} - \bar{y}_{.j} + \bar{y}..\right)$$

Observation   Grand   Deviation   Deviation
                     mean   due to   due to
                     treatment   block   Error

Here the grand mean $\bar{y}.. = \dfrac{36 + 20 + 40}{12} = \dfrac{96}{12} = 8$

The respective treatment means are $\bar{y}_{.1} = 9; \bar{y}_{.2} = 5; \bar{y}_{.3} = 10; \bar{y}_{.4} = 4;$

And the respect block means are $\bar{y}_{.1} = 11; \bar{y}_{.2} = 7; \bar{y}_{.3} = 10; \bar{y}_{.4} = 4;$

For instance, $14 = 8 + (9 - 8) + (11 - 8) + (14 - 9 - 11 + 8) = 8 + 1 + 3 + 2$

Repeating this decomposition for each observation, we obtain the arrays.

Observation   Mean   Treatment

y_{ij} $\bar{y}..$ $\bar{y}_{i.} - y..$

$$\begin{bmatrix} 14 & 8 & 10 & 4 \\ 7 & 7 & 4 & 2 \\ 12 & 6 & 16 & 6 \end{bmatrix} = \begin{bmatrix} 8 & 8 & 8 & 8 \\ 8 & 8 & 8 & 8 \\ 8 & 8 & 8 & 8 \end{bmatrix} + \begin{bmatrix} 1 & 1 & 1 & 1 \\ -3 & -3 & -3 & -3 \\ 2 & 2 & 2 & 2 \end{bmatrix}$$

block   error

$\bar{y}_{ij} - \bar{y}.$ $y_{ij} - \bar{y}_{i.} \ \bar{y}_{.j} + \bar{y}..$

$$+ \begin{bmatrix} 3 & -1 & 2 & -4 \\ 3 & -1 & 2 & -4 \\ 3 & -1 & 2 & -4 \end{bmatrix} + \begin{bmatrix} 2 & 0 & -1 & -1 \\ -1 & 3 & -3 & 1 \\ -1 & -3 & 4 & 0 \end{bmatrix}$$

Taking the sum of squares for each array

Treatment sum of squares $= b \displaystyle\sum_{i=1}^{a} (\bar{y}_{i.} - \bar{y}..)^2 = 4(1)^2 + 4(-3)^2 + 4(2)^2 = 4 + 36 + 16$

$= 56$

Block sum of squares $= \displaystyle\sum_{i=1}^{a} \sum_{j=1}^{b} (y_{ij} - \bar{y}_{i.} - \bar{y}_{.j} + \bar{y}..)^2 = 3(3)^2 + 3(-1)^2 + 3(2)^2 + 3(-4)^2 = 90$

Error sum of squares $= \displaystyle\sum_{i=1}^{a} \sum_{j=1}^{b} (y_{ij} - \bar{y}_{i.} - \bar{y}_{.j} + \bar{y}..)^2$

$= (2^2 + 0^2 + 1 + 1) + (1 + 9 + 9 + 1) + (1 + 9 + 16 + 0) = 52.$

We obtain the body of the analysis of variance table. Their sum, $56 + 90 + 52 = 198$, the total sum of squares, is also equal to the sum of squares of the observations, $\{(196 + 64 + 100 + 16) + (49 + 49 + 16 + 4) + (144 + 36 + 256 + 36)\} = 966$, minus the sum of the squares $12 \times 8^2 = 768$ for grand mean array. The degrees of freedom of the treatments, blocks are $a - 1 = 2$ and $b - 1 = 3$ respectively. The degrees of freedoms of the error is equal to $(a - 1)(b - 1) = 6$. Hence the degrees of freedom can be decomposed as:

$$ab - 1 = (a - 1) + (b - 1) + (a - 1)(b - 1)$$

total   treatments   blocks   error

Hence the final analysis of variance (ANOVA) table is given by

Source of variation	Degrees of freedom	Sum of squares	Mean square	F
Treatments	2	56	28.00	$\dfrac{28}{8.67} = 3.23$
Blocks	3	90	30.00	$\dfrac{30}{8.67} = 3.46$
Error	6	52	8.67	
Total	11	198		

The values of $F_{0.05}$ with $v_1 = 2$, $v_2 = 6$ degrees of freedom is 5.14, Hence $F_{ob} = 3.23 < F_{0.05} = 5.14$ and the null hypothesis cannot be rejected. There is no significance difference between treatments.

11.5 LATIN—SQUARE DESIGN

In the randomized block design, a **balance** is achieved by assigning the same number of observations for each treatment in each block. Latin – square design is another kind of balanced design used to eliminate the effects of **two** extraneous sources of variability. For instance, it is desired to compare 3 treatments A, B and C in the presence of 2 other some of variability. Consider the case in which the 3 treatments are the **three methods** for soldering copper electrical leads and the two extraneous sources of variability are: the **three different operators** doing the soldering and the use of three **different solder fluxes.** When three operators and three fluxes are to be considered, the experiment can be arranged in the following pattern.

	Flux 1	Flux 2	Flux 3
Operator 1	A	B	C
Operator 2	C	A	B
Operator 3	B	C	A

Here A,B, and C are the three soldering method

Each soldering method is applied once by each operator in conjunction with each flux and if there are systematic effects due to differences among operators or differences among fluxes these effects are present equally for each treatment, that is, for each method of soldering.

An experimental arrangement like the one mentioned above is called a **Latin square.** An n x n Latin square is a square array of n district letters, with each letter, appearing once and once only in each row and in each column.

Note that in a Latin—square experiment involving n treatments, it is necessary to include n^2 observations, n for each treatment.

A Latin—square experiment without replication provides only $(n-1)(n-2)$ degrees of freedom for estimating the experimental error. When n is small, the Latin—square pattern is not repeated several times. If there is a total of n replicates, the entire Latin—square analysis of variance of data presumes the following model, where $y_{ij(k)l}$ is the observation in the i-th row and the j-th column of the *l*-th replicate and the subscript in parentheses, indicates that is pertaining to the k-th treatment.

The model equation for Latin—square is:

$y_{ij(k)l} = \mu + \alpha_i + \beta_j + \gamma_k + \rho_l + \epsilon_{ij(k)l}$, for i, j, k = 1, 2, ..., and l = 1, 2, ... n subject to the restrictions

$$\sum_{i=1}^{n} \alpha_i = 0; \sum_{j=1}^{n} \beta_j = 0; \sum_{k=1}^{n} \gamma_k = 0 \text{ and } \sum_{l=1}^{r} \rho_l = 0$$

Here μ is the grand mean, α_i is the effect of the i-th row, β_j is the effect of the j-th column, v_k is the effect of the k-th treatment, ρ_l is the effect of the l-th replicate, and the $\epsilon_{ij(k)l}$ are independent, normally distributed random variables with zero means and the common variance σ^2.

The effects of the rows and the effects of the columns represent the effects of the two extraneous variables, while the replication introduces the third extraneous variable k is automatically determined when i and j are known.

The main hypothesis is to test the null hypothesis $v_k = 0$ for all k, namely the null hypothesis that there is no difference in the effectiveness of the n treatments. However, we can also test whether the cross blocking of Latin- square design has been effective. That is, we can test the two null hypotheses $\alpha_i = 0$ for all I and $\beta_j = 0$ for all j (against suitable alternatives) to see whether the two extraneous variables actually have an effect on the phenomenon under consideration. In addition, we can test the null hypothesis $\rho_l = 0$ for all l against the alternative that the ρ_l are not all equal to zero. This test for effects of replicates may be important if the parts of the experiment representing the individual Latin—Square were performed on different days, by different technicians, at different temperatures, and so on.

The sums of squares required to perform these tests are usually obtained by means of the following short cut formulas, where $T_{i..}$ is the total of the r.n. observation of all the ith rows, $T_{.j.}$ is the total of the r.n observations in all of the j-th columns, T..l is the total of the n² observations with l-th replicate, T(k) is the total of all the r.n observations pertaining to the k-th treatment, and T... is the grand total of all the r.n² observations.

Sum of squares (Latin square)

$$C = \frac{(T...)^2}{r.n^2}$$

$$SS(T_r) = \frac{1}{r.n} \sum_{k=1}^{n} T_{(K)}^2 - C$$

$$SSR = \frac{1}{r.n} \sum_{i=1}^{n} T_{i..}^2 - C \text{ (for rows)}$$

$$SSC = \frac{1}{r.n} \sum_{j=1}^{n} T_{.j.}^2 - C \text{ (for columns)}$$

$$SS(Rep) = \frac{1}{n^2} \sum_{l=1}^{r} T_{..l}^2 - C \text{ (for replicates)}$$

$$SST = \sum_{i=1}^{n} \sum_{j=1}^{n} \sum_{l=1}^{r} y_{ij\,(k)l}^2 - C$$

$$SSE = SST - SS(T_r) - SSR - SSC - SS(Rep)$$

The Analysis of Variance Table:

Source of variation	Degrees of freedom	Sum of squares	Mean square	F
Treatments	$n-1$	$SS(T_r)$	$MS(T_r) = \dfrac{SS(T_r)}{n-1}$	$\dfrac{MS(T_r)}{MSE}$
Rows	$n-1$	SSR	$MSR = \dfrac{SSR}{n-1}$	$\dfrac{MSR}{MSE}$
Coloumns	$n-1$	SSC	$MSC = \dfrac{SSC}{n-1}$	$\dfrac{MSC}{MSE}$
Replicates	$r-1$	SSC(Rep)	$MS(Rep) = \dfrac{SS(Rep)}{r-1}$	$\dfrac{MS(Rep)}{MSE}$
Error	$(n-1)x$ $(rn+r-3)$	SSE	$MSE = \dfrac{SSE}{(n-1)(rn+r-3)}$	..
Total	$r\,n^2 - 1$	SST	..	..

EXAM. 1: There are 2 replicates; three methods for soldering copper electrical leads;

three different operators doing the soldering and three different solder fluxes are used. A,B,C, represent the three methods; The three operators are: operator 1, operator 2 and operator 3; the three flues are 1,2,3. The two replicates are: replicates are replicate 1 and replicate2.

The experiment uses the following arrangement

	Replicate I			Replicate II		
	flux			flux		
	1	2	3	1	2	3
Operator 1	A	B	C	C	B	A
Operator 2	C	A	B	A	C	B
Operator 3	B	C	A	B	A	C

The results showing the number of pounds of tensile force required to separate the soldered leads, were as follows.

Replicate I			Replicate II		
16	18	13	12	18	15
11	19	17	14	14	17
13	14	15	15	20	13

Analyze this experiment as a Latin Square and test at the 0.01 level of significance whether there are differences in the methods, the operators, the fluxes or the replicates.

SOLUTION:

1. **Null hypotheses:** $\alpha_1 = \alpha_2 = \alpha_3 = 0$; $\beta_1 = \beta_2 = \beta_3 = 0$; $\gamma_1 = \gamma_2 = \gamma_3 = 0$; $\rho_1 = \rho_2 = 0$
   Alternative hypotheses: The α's are not all equal to zero; the β's are not equal to zero; the γ's are not all equal to zero; the ρ's are not all equal to zero.
2. **Level of Significance:** $\alpha = 0.01$ for each test.
3. **Criteria:** For treatments, rows or columns reject the null hypothesis if $F > 7.56$, the values of $F_{0.01}$ for $v_1 = v - 1 = 3 - 1 = 2$ and $v_2 = (n - 1)(r\,n + r - 3) = (3 - 1)\,(2 \times 3 + 2 - 3) = 10$ degrees of freedom. For replicates, reject the null hypothesis if F

>10.04, the values of $F_{0.01}$ for $v_1 = n - 1 = 2 - 1 = 1$ and $v_2 = 10$ degrees of freedom. Here $n = 3$; $r = 2$.

4. **Calculations:**

	I			Row Total	II				
	Col 1	Col 2	Col 3		Col 1	Col 2	Col 3		
Row 1	16 A	18 B	13 C	47	12 C	18 B	15 A	45	$T_{1..}$
Row 2	11 C	19 A	17 B	47	14 A	14 C	17 B	45	$T_{2..}$
Row 3	13 B	14 C	15 A	42	15 B	20 A	13 C	48	$T_{3..}$
Column Total	40	51	45	136	41	52	45	138	

$$T_{1.} = 47 + 45 = 92; \ T_{2..} = 47 + 45 = 92; \ T_{3..} = 42 + 48 = 90$$
$$T_{.1} = 40 + 41 = 81; \ T_{.2} = 51 + 52 = 103; \ T_{.3.} = 45 + 45 = 90$$
$$T_{..1} = 136; \ T_{..2} = 138; \ T_{...} = 136 + 138 = 274$$
$$T_{(A)} = (16 + 19 + 15) + (15 + 14 + 20) = 50 + 49 = 99$$
$$T_{(B)} = 18 + 17 + 13) + (18 + 17 + 15) = 48 + 50 = 98$$
$$T_{(c)} = (13 + 11 + 14) + (12 + 14 = 13) = 38 + 39 = \mathbf{77}$$

$$\sum_{i=1}^{n}\sum_{j=11}^{n}\sum_{l=1}^{r} yi_{ij}^{2}{}_{(k)l} = \sum_{i=1}^{3}\sum_{j=11}^{3}\sum_{l=1}^{2} yi_{ij}^{2}{}_{(k)l}$$

$$= (16^2 + 18^2 + 13^2 + 11^2 + 1^2 + 17^2 + 13^2 + 14^2 + 15^2) + (12^2 + 18^2 + 15^2 + 14^2 + 17^2 + 15^2 + 20^2 + 13^2) = 2110 + 2168 = 4278$$

$$C = \frac{(T...)^2}{r.n^2} = \frac{(274)^2}{18} = \frac{75056}{18} = \frac{37538}{9} = 4170.88$$

$$SS(T_r) = \frac{1}{r.n}\sum_{k=1}^{n} T_{(k)}^2 - C = \frac{1}{6}\sum_{k=1}^{3} T_{(k)}^2 - C$$

$$= \frac{1}{6}(99^2 + 98^2 + 77^2) - 4170.88$$

$$= \frac{1}{6}(9801 + 9604 + 5929) - 4170.88 = \frac{25334}{6} - 4170.88$$

$$= 4222.33 - 4170.88 = 51.45$$

$$SSR = \frac{1}{r.n} \sum_{i=1}^{n} T_{i..}^2 - C$$

$$= \frac{1}{6}\sum_{i=1}^{3} T_{i..}^2 - C = \frac{1}{6}(92^2 + 92^2 + 90^2) - 4170.88 = \frac{1}{6}(8464 + 8464 + 8100) - 4170.88$$

$$= \frac{25028}{6} - 4170.88 = 4171.33 - 4170.88 = 0.45$$

$$SSC = \frac{1}{r.n} \sum_{j=1}^{n} T_{.j.}^2 - C$$

$$= \frac{1}{6}(81^2 + 103^2 + 90^2) - 4170.88 = \frac{1}{6}(6561 + 10609 + 8100) - 4170.88$$

$$= \frac{1}{6}(25270) - 4170.88 = 4216.66 - 4170.88 = 40.78$$

$$SS(Rep) = \frac{1}{n^2} \sum_{l=1}^{n} T_{..1}^2 - C = \frac{1}{9}(136^2 + 138^2) - 4170.88$$

$$= \frac{1}{9}(18496 + 19044) - 4170.88 = 4171.11 - 4170.88 = 0.23$$

$$SST = \sum_{i=1}^{n}\sum_{j=1l=1}^{n}\sum_{j=1l=1}^{n} Y_{ij\,(k)l}^2 - C = 4278 - 4170.88 = 107.12$$

Using the above calculations, the analysis of variance table becomes.

Source of variation	Degrees of freedom	Sum of squares	Mean square	F
Treatments (methods) = 7.56	2	51.45	25.725	18.1 $v_1 = 2, v_2 = 100, F_{0.01}$
Rows (Operators)	2	0.45	0/225	0.158 $F_{0.01} = 7.56$
Columns (Fluxes)	2	40.78	20.390	14.349 $F_{0.01} = 7.56$
Replicates	1	0.23	0.230	0.16 $F_{0.01} = 10.04$
Error	10	14.21	1.421	....
Total	17	107.12		

Decisions

For treatments (methods) and columns (fluxes) since F = 18.1 and F = 14.349 exceed 7.56, the corresponding null hypotheses must be rejected; for rows (operators) since F = 0.158 does not exceed 7.56 and for replicates, since F = 0.16 does not exceed 10.04, the corresponding null hypotheses can not be rejected.

In other words, we conclude that difference in methods and fluxes, but not differences in operators or replicates affect the solder strengths of the electrical leads,

EXAM. 2: The experiment consists of coating steel pipes with plastic. The study pertains to temperature profile, A, at three levels, type of plastic: B, at three levels and the speed of the rotating screw that forces the plastic through a to be forming die: C, at three levels. There are two replicates, at each combination of the level of the factors, yielding a total of 54 ($3 \times 3 \times 3 \times 2$) observations on output the sums of squares are:

SSA = 14,150, SSB = 5520, SSC = 244,700

SSAB = 1075, SSAC = 65; SSBC = 330

SSE = 3125 and SST = 270, 050.

(a) Construct the ANOVA table

(b) Use the appropriate F tests to show that none of the F ratios for two or three factor interactions is significant al level 0.05.

SOLUTION:

A. The ANOVA Table

Source	df	SS	MS	F	$F_{0.05}$
A	2	14150	7075	61.12	3.35
B	2	5520	2760	23.84	3.35
C	2	244700	122350	10.57	3.35
AB	4	1075	268.75	2.32	2.73
AC	4	65	16.25	0.14	2.73
BC	4	330	82.5	0.71	2.73
ABC	8	1085*	135.625	1.17	2.31
Error	27	3125	115.74		
Total	53	270050			

ABC = Total – (A + B + C + AB + AC + BC + Error) = 270050 – 268965 = 1085

EXAM. 3: Because of potential variability in aging due to different castings and segments on the castings, a Latin square design with n = 7 was used to investigate the effect of heat treatment ongoing. With A = Castings, B = Segments, C = heat treatments, the summary of statistics include:

SSA = 69

SSB = 51   ∴ SSE = 171 – (169 + 51 + 6)

SSC = 6   171 – 126 = 45

SST = 171   Set up the ANOVA Table

SOLUTION:

Source	df	SS	MS	F
A	6	69	11.5	$7.66 > F_{0.05,\,6,30} = 2.42$
B	6	51	8.5	$5.66 > F_{0.05,\,6,30} = 2.42$
C	6	6	1.0	$0.66 < F_C = 0.66 < 2.42$
Error	30	45	1.5	
Total	48	171	..	

∴ H_0 : C is NOT REJECTED

11.6 PERFORMANCE IMPROVEMENT

Statistically based experimental design techniques are useful for improving the performance of a manufacturing process, applying extensively in the development of new processes. Many processes can be described in terms of several if controllable variables, such as temperature, pressure feed rate etc. Using designed experiments, the engineer will be able to determine which subset of the process variables has the greatest influence on process variables has the greatest influence on process performance. The results of such an experiment will yield:

- Improved process yield.
- Reduction in process variability.
- Reduced design and development time.
- Reduced cost of operation.
- Evaluation and comparison of basis design configurations.
- Evaluation of different materials.
- Selection of design parameters to make the design robust.
- Determination of key product design parameters that affect product performance.

11.7.1 *Examples Involving Short Answers*

1. What do experiments convey?

   Experiments are a natural part of the engineering and scientific decision making process.

2. What does single factor convey?

   In civil engineering, for example, different curing methods on the mean compressive strength of concrete are investigated to find the effects of the cunning methods. If there are only two curing methods of interest, the experimenter can design and analyze them using the statistical hypothesis methods for two samples. Here, the experimenter has a single **factor** of interest, namely, curing methods and there are only 2 **levels** of the factor. He determines from the operating characteristic curves, the member of specimens can be determined in order to find the maximum compressive strength between the two means by using the t-test to decide if the two means differ.

3. What is the purpose of analysis of variance?

   When there are more than two levels of a single factor, the analysis of variance (abbreviated by ANOVA) can be used for comparing the means.

4. How does statistically based experimental design techniques are useful for the performance improvement in a manufacturing process?

   Almost all processes can be described interms of several controllable variables. By using designed experiments engineers can determine which subset of the process variables has the greatest influence on the process performance. The results of such an experiment will lead to:

   - Improved process yield.
   - Reduced variability in the process conforming to the target requirements
   - Reduced design and development time
   - Reduced cost of operation.

5. Mention some typical applications of statistical designed experiments in engineering design.

   Some of the applications are:

   1. Evaluation and comparison of basic design configuration
   2. Evaluation of different materials
   3. Station of a approficiate parameters so that the product will work will under different field conditions
   4. Determination of key product design parameters that affect product performance

6. What is completely randomized design?

   It is a type of experimental design in which the treatments or design factors are assigned to the experimental units in a random manner. In designed experiments, a completely randomized design results from running all of the treatment combinations in random order.

7. What does analysis of variance mean?

   It is a method of decomposing the total variability in a set of observations as measured by the sum of the squares of these observations from their average, into component sums of squares that are associated with specific defined sources of variation.

8. Explain fixed factor (or fixed effect).

   In analysis of variance, a factor or effect is considered fixed, if all the levels of interest for that factor are included in the experiment. Conclusions are then valid about this set of levels only, although when the factor is quantitative, it is customary to fit a model to the data for interpolating between these levels.

9. Whet is meant by random effects or components of variance model?

   The individual components of the total variance that are attributable to specific sources are called components of variance. It usually refers to the individual variance components arising from a random or mixed analysis of variance. While testing of hypotheses about the variability of the treatment effects, we try to estimate this variability and this in called the random effects of components of variance model.

10. Summarize in tabular form the analysis of variance (ANOVA) for a single-factor experiment, Fixed effects model.

Source of variation	Sum of squares	Degrees of Freedom	Mean square	F_0
Treatments	$SS_{treatments}$	$a - 1$	$MS_{treatments}$	$\dfrac{MS\ treatments}{MS_E}$
Error	SS_E	$a(n-1)$	MS_E	
Total	SS_T	$an - 1$		

11. What is an unbalanced design?

   In a single factor experiment, the number of observations taken under each treatment may be different. In such a case the design is called unbalanced. Let n_i observations be taken under treatment i (i = 1, 2, ... a) and let the total number of observation $N = \sum\limits_{i=1}^{a} n_i$.

   The sums of squares computing formulas for the ANOVA with unequal sample size n_i in each treatment are

   $$SS_T = \sum_{i=1}^{a} \sum_{j=1}^{n_i} y_{ij}^2 - \frac{y_{..}^2}{N}$$

$$SS_{treatments} = \sum_{i=1}^{a} \frac{y_{i.}^2}{n_i} - \frac{y_{..}^2}{N}$$

and $SS_E = SS_T - SS_{Treatments}$

12. Set up the procedure for a randomized complete block design.

The design consist of selecting b blocks and running a complete replicate of the experiment in each block. The data that result from running a randomized block design for investigating a single factor with a levels and b blocks are put in the table with treatments along the a rows and blocks in the b columns, so that there will be a observations (one per factor level) in each block and the order in which these observations are run randomly assigned within the block.

13. Summarize in tabular from the ANOVA for a Randomized Complete Block design.

Source of variation	Degrees of freedom	Sum of squares	Mean square	F
Treatments	$SS_{Treatments}$	$a - 1$	$\dfrac{SS_{Treatments}}{a-1}$	$\dfrac{MS_{Treatment}}{MS_E}$
Blocks	SS_{Blocks}	$b - 1$	$\dfrac{SS_{Treatments}}{b-1}$	
Error	SS_E(by subtraction	$(a-1)(b-1)$	$\dfrac{SS_E}{(a-1)(b-1)}$	
Total	SS_T	$ab - 1$		

14. Define a factorial experiment.

This is a type of experimental design in which every level of one factor is tested in combination with every level of another factor. In general, in a factorial experiment, all possible combinations of factor levels are tested.

15. Define a Latin square design.

An $n \times n$ Latin square is a square array of n distinct letters with each letter appearing once and once only in each row and in each column. In a Latin square experiment involving n treatments, it is necessary to include n^2 observations, n for each treatment.

11.7.2 *Short Answer Questions in Design of Experiments*

1. Explain the following basic principles of the design of experiments:
   (i) Randomization; (ii) Replication; (iii) Local Control
2. What do you mean by analysis of variance?
3. What is the aim of the design of experiments?
4. Explain briefly the following basic design of experiment:

(i) Completely Randomized Design

(ii) Randomized Block Design

(iii) Latin square Design

5. Compare the Randomized block Design and Latin square design.

6. What is the main advantage of Latin square Design over the Randomized Block Design?

II.8 ANALYSIS VARIANCE

Exercises

1. Discuss hour the means of several sets of measurements are compared by the method of Analysis of Variance. State the assumptions clearly and explain how the F-ratio can be used to make an appropriate test of significance. Give details of the setup of the analysis of variance Table.

2. In an agricultural experiment, six treatments (A, B, C, D, E and F) were each tried on four plots. The yields in ton) per acre shown below:

Treatments

A	B	C	D	E	F
16.4	15.6	11.1	17.2	18.3	14.3
17.4	14.2	14.5	15.2	15.0	13.1
13.2	16.3	12.9	15.1	17.1	14.9
15.8	15.9	14.3	15.7	16.8	14.5

Set up a suitable Analysis of variance table and test for treatment effect

3. Persons A,B,C,D, Scored the following marks in repetitive tests.

Set up the Analysis of variance Table and test for significance of the difference of scores between persons.

A	B	C	D
20	40	36	60
24	36	34	56
28	48	48	32
22	50	54	48
24	44	46	44

Exercises

One Way Randomized Designs

4. The following data represents 48 discs allotted randomly to 4 laboratories (A, B, C, D) equally pertaining to the actual pattern of tin . Coating thickness on the sheet of tin plate, cut into four strips and 12 discs are cut from each strip 1, 2, 3, 4 and sent to the form laboratories.

	Strip 1			Strip 2			Strip 3			Strip 4		
Lab A :	5	36	19	30	22	28	46	21	13	29	20	39
Lab B :	42	22	17	23	43	6	16	7	37	9	38	32
Lab C :	14	23	40	10	25	18	45	44	20	41	33	28
Lab D :	11	4	26	21	35	3	12	8	27	34	31	15

Test for the consistency of the four laboratories. Under chance variation.
Given : $F_{0.05 \ v1 = 3; \ v2 = 44;} = 2.82$

5. Considering one way classification, each laboratory measures the tin coating weights of 12 discs, the results are shown below.

Lab A:	0.27	0.25	0.24	0.28	0.29	0.26	0.30	0.26	0.29	0.23	0.23	0.26
Lab B:	0.20	0.26	0.23	0.21	0.23	0.22	0.25	0.21	0.22	0.24	0.27	0.28
Lab C:	0.21	0.23	0.25	0.26	0.20	0.28	0.26	0.26	0.27	0.22	0.23	0.21
Lab D:	0.25	0.28	0.26	0.26	0.26	0.32	0.22	0.20	0.26	0.26	0.24	0.23

Test the null hypothesis at 5 percent level. (Given $F_{0.05}$ with 3 and 44 df is 2.82)

6. The experiment consists in collapse of the roof building, by a testing laboratory which givers all the available bolts that connected the steel structure at 3 different positions on the roof. The forces required to shear each of these bolts (coded values) are as follows:

Position 1	87	85	82	95	86	88
Position 2	102	92	93	101	92	89
Position 3	86	92	83	91		

Perform an analysis of variance true test at the 0.05 level of significance, whether the differences among the sample means at the three positions are significant
(Given : $F_{0.05, \ v1 = 2, \ v2 - 14} = 3.74$)

7. The following data pertains to the three methods (A, B, C) of teaching JAVA language method A is a straight JAVA based instruction, method B involves personal attention by the instruction and some direct practical involvement method c involves personal attention but no work with JAVA language.
Test the level of significance at $\alpha = 0.01$ level whether the differences among the means obtained for the 3 samples are significant

Method A	67	71	67	71	69
Method B	85	81	81	85	83
Method C	72	71	73	75	74

8. The yield in grams per plot (plot size 1/2000 acre) for three varieties of cotton seed are given below.

Variety 1	76	70	62	83	95	81	88	101
Variety 2	109	106	137	79	133	78	125	99
Variety 3	46	70	71	65	60	40	47	73

(a) Write out the analysis of variance table.
(b) Test if the varieties differ significantly among themselves.
(c) If the result of (b) is affirmative, determine, which varieties differ.

9. The following figures relate to the production in kg of three varieties A,B, and C of wheat sown in 12 plots.

A:	14	16	18	
B:	14	73	15	22
C:	18	16	19	19	23

Is there any significant difference in the production of three varieties ?

10. The following table gives the yields on 15 sample plots under three varieties of seed :

A:	20	21	23	16	20
B:	78	20	17	15	25
C:	25	28	22	28	32

Find out if the average yields of land under different varieties of seed show significant differences.

Exercises In One Way Classification

11. (a) Explain the principle behind analysis of variance.
    (b) Three varieties of wheat A,B,C, were shown in 4 plots each and the following yields in quintals per hectare were obtained.

A:	9	5	7	8
B:	8	6	6	4 and(c)
C:	3	6	5	5

and another set

(c) A:	9	10	12	13
B :	7	9	11	5
C:	15	13	18	10

Test the significance of the difference between the yield of the varieties given that the 5% value of F for 2 and 4 degrees of freedom is 4.26

12. (a) Explain a random block design show how analysis of variance can be used to design an efficient sample.
    (b) The breaking strength of a fabric has found to be 100 kg with a standard deviation of 7.2 kg. A new manufacturing process seems to increase the strength of the fabric.
      (i) Find the criterion for rejecting the old process at 1 percent level of signify
      (ii) When 36 specimens of the fabrics are tested.
      (iii) Assuming that the standard deviation is 7.2 kg, find the probability of accepting the old process when the new process has improved the mean strength to 106 kg.

13. The following table gives data pertaining to the tensile strength of paper (psi) with 4 levels of concentration (%) having 6 observations at each level.

Hard wood concentration (%)	1	2	3	4	5	6	Total	Averages
5	8	9	16	12	10	11	66	11.00
10	13	18	14	19	20	16	100	16.67
15	15	19	20	18	17	19	108	18.00
20	20	26	23	24	19	21	133	22.17
							407	16.96

(a) Form the ANOVA table for the Tensile strength data.

(b) Test the hypothesis at $\alpha = 0.01$ level.

(c) $H_0 : \tau_1 = \tau_2 = \tau_3 = \tau_4 = 0$ against the alternative $H_1 : \tau_i \neq 0$ for at least one i.

14. A test was given to five students taken at random from fifth standard class of three schools of a town. The individual scores are: (our of 100).

School I :   100   80   70   60   90

School II :   80   50   60   50   60

School III :   70   60   70   80   70

Carry out the analysis of variance and state your conclusions.

15. The following are the number of mistakes made in 4 successive days for 4 pc typing persons in a Xerox organization:

Typist 1:   6   11   9   10

Typist 2:   11   9   10   10

Typist 3:   9   11   6   10

Typist 4:   8   7   4   9

Test at the level of significance a - 0.05 whether the difference among the 4 sample means can be attributed to chance

Two way classifications

16. The data pertaining to the weight losses of certain machine parts (in milligrams) due to friction, when 3 different lubricants (A, B, C) were used under controlled conditions with 4 blocks are given below :

SS(Tr) = 60 with 2 degrees of freedom;

SS(Bl) = 135 with 3 degrees of freedom;

SSE = 30 with 6 degrees of freedom;

SST = 225 with 11 degrees of freedom;

Given $F_{0.05} (v_1 = 2, v_2 = 6) = 5.79$ for T_r

and $F_{0.05} (v_1 = 3, v_2 = 6) = 4.76$ for Bl

17. In an experiment regarding the running temperature of a computer chip, 4 different types of coolingfans were used and tried on 5 different computers. The following observations on temperature, coded by subtracting the smallest values.

Blocks:

	1	2	3	4	5
Treatment 1:	13	11	10	5	11
Treatment 2:	16	15	15	11	13
Treatment 3:	14	12	15	14	15
Treatment 4:	17	10	12	10	11

(a) Decompose each observation y_{ij} as

$$y_{ij} = \bar{y}.. + (\bar{y}_{i.} - \bar{y}..) + (\bar{y}_{.j} - \bar{y}..) + (y_{ij} - \bar{y}_{i.} - \bar{y}_{.j} + \bar{y}..)$$

(b) Obtain the sums of squares and degrees of freedom for each component.

(c) Construct the analysis of variance table and test for differences among the treatments and blocks using $\alpha = 0.01$.

[Given $F_{0.01}$ $(v_1 = 3, v_2 = 12) = 3.49$ and $F_{0.01}$ $(v_1 = 4, v_2 = 12) = 3.26$]

18. The following table furnishes the data pertaining to the protection of buried metal pipe regarding the amount of corrosion due to either on the coating or on the type of the soil. Each piece is coated with one of the four coating and buried in one of the three types of soil for a fixed time period, after which the amount of corrosion (in 0.0001 in) is determined.

		Soil type B		
		1	2	3
Coating (A)	1	61	52	55
	2	51	53	58
	3	50	48	49
	4	54	55	50

(a) Obtain the sums of squares and degrees of freedom for each component.

(b) Construct the analysis of variance table and test for difference among coatings and difference among the soil types using $\alpha = 0.01$ level of significance.

Using $\alpha = 0.01$ level of significance.

[given $F_{0.01}$, $(v_1 = 3, v_2 = 6) = 9.78$ for coating type $F_{0.01}$, $(v_1 = 2, v_2 = 6) = 10.92$ for soil type]

19. Setup the analysis of variance table pertaining to the amount of coverage of light blue interior latex paint for testing the brand of paint used over the roller used. One gallon of each of the 4 brands of paint was applied using each of the three brands of roller. The data pertaining to the experiment is given below (number of square meters coved).

		Roller Brand		
		1	2	3
Paint Brand	1	151	148	154
	2	145	149	147
	3	146	147	148
	4	150	152	151

(a) Construct the ANOVA table (Hint: The computations can be expedited by subtracting 150 (or any other convenient number) from each observation. This does not affect the final results].

(b) State and test hypotheses appropriate for deciding whether the paint brand has any effect on coverage; use $\alpha = 0.05$.

(c) Repeat part (b) for brand of roller.

20. A particular corporation zone employs three assessors who are responsible for determining the value of residential property in that zone. To use whether these assessors differ systematically in their assessments, nine houses are randomly selected and each assessor is asked to determine the market value of each house. With the factor A denoting assessors (I =3) and factor B denoting houses (J = 9), suppose SSA = 13.5, SSB = 204.3 and SSE = 51.2.

(a) Test $H_0 : \alpha_1 = \alpha_2 = \alpha_3 = 0$ at level 0.05
(H_0 states that there are no systematic difference among assessors)

(b) Explain why a randomized block experiment with only 9 houses was used rather than a one way ANOVA experiment involving a total of 27 different houses with each assessor asked to assess 9 different house (a different group of 9 for each assessor).

21. In a behaviour research experiment two electro – shock conditions are given to find the rate of stuttering adaptation. The adaptation scores for three different treatments : (1) no shock, (2) shock following each stuttered word, and (3) shock during each moment of stuffering. These treatments were used on each of 15 stutters.

(a) Summary statistics includes

$x_{1.} = 750;$ $x_{2.} = 765;$ $x_{3.} = 780$

$x = 2295;$ $\sum_j x_{.j}^2 = 358, 580$ and $\Sigma \Sigma x_{ij}^2 = 139950$

Construct the ANOVA table and test at level 0.05 to see whether true average adaptation score depends on the treatment given.

(b) Judging from the Fraction for subjects (factor B), do you think that blocking on subjects was effective in this experiment? Explain.

22. (a) Show that a constant d can be added to (or subtracted from each y_{ij} without effecting any of the ANOVA sums of squares.

(b) Suppose that each y_{ij} is multiplied by a non zero constant c. How does this affect the ANOVA sums of squares? How does this affect the values of the F statistics F_{Tr} and F_{Bl} (TR—Treatment and Bl—Block)? What effect does "coding" the data by $Y_{ij} = cy_{ij} + d$ have on the conclusions resulting from the ANOVA procedures?
[Ans: (b) Each SS is multiplied by c^2 but f_{TR} and f_{Bl} are unchanged].

(c) Using the fact that $E(Y_{ij}) = \mu + \alpha_i + \beta_j$ with $\Sigma \alpha_i = \Sigma \beta_j = 0$, show that $E(\overline{Y}_{i.} - \overline{Y}..) = \alpha_i$ so

that $\hat{\alpha}_i = \overline{Y}_{i.} - \overline{Y}..$ is an unbiased estimator for a_i.

23. There are four detergents (A, B, C, D) and three engines (E, F, G). It is designed to determine whether there is interaction between the detergents and the engines, that is, whether one detergent (A) might perform better in engine (E), another might perform better in engine (E), another might perform better in engine (F) and so on. The experiment pertains to the study of the performances of the 4 different detergents for cleaning fuel infectors in the 3 engines. The data pertaining to the 2 replicates are given below:

Replicate I   **Replicate II**

Engine

	E	F	G		E	F	G
A	47	46	52	A	45	48	60
B	49	48	53	B	51	53	55
C	51	50	55	C	40	52	49
D	45	41	48	D	56	59	61

Detergent

Combining the data of the two replicates, test for significant interaction and discuss the results

(a) Given : For detergents, $v_1 = 3, v_2 = 6, F_{0.05} = 4.76$.

(b) For engines $v_1 = 2, v_2 = 6, F_{0.05} = 5.14$.

(c) For interaction $4 \times 3 = 12$ cells duplicated twice. Hence there is no gain in degrees of freedom for error.

24. (a) Describe a randomized block experiment in the context of an industrial problem and set up the analysis of variance table.

(b) Certain company had four salesmen A, B, C, D each of whom was sent for a week into three types of areas : Rural; (R), Sub-Urban (S) and City Area (C). The sales in units per week are given below.

Salesmen

Areas	A	B	C	D
R	40	80	40	40
S	90	60	50	80
C	110	70	90	90

Carryout analysis of variance and interpret the results stating the limitations under which your results are valid.

(c) Instead of there types of areas, you have 3 seasons: summer, winter and monsoon. Carryout an analysis of variance for the following data

Salesmen

Season	A	B	C	D
Summer	41	38	24	39
Winter	33	31	34	36
Monsoon	31	30	32	33

25. The following data relate to the five observations made by each tester with five different apparatuses.

Toster	Apparatus					T_i	T_i^2	$\sum_i X_{ij}^2$
	B_1	B_2	B_3	B_4	B_5			
A_1	15	3	4	6	7	35	1225	335
A_2	25	16	5	17	6	69	4761	1231
A_3	13	24	13	17	9	76	5776	1284
T.j	53	43	22	40	22	180	11762	2850
$T_{.j}^2$	2809	1849	484	1600	484	7226		

Set up the analysis variance table and test
(i)   The significance of the tester effect and
(ii) The significance of apparatus effect.

26. (a) Explain the rationale behind analysis of variance
    (b) The following data gives the water demand (as a percentage) among different consumer groups in two cities

	Cities	
Consumer groups	A	B
Residential	40	57
Industrial	20	12
Commercial	20	8
Public services and losses	20	23

Establish the analysis of variance whether there is significant difference between consumers and between cities.

27. Receivers of a certain type have been causing trouble because of malfunctioning. A series of tests, were performed to evaluate the variations of the signal generator output. These receivers had in their design tow different types of vacuum tubes, each of which had an influence on the signal generator output. A single receiver was chosen for the study and five vacuum tubes were chosen randomly from each type. The signal generator output in KUV data are as follows:

Vacuum tube type B	Vacuum tube – type A				
	1	2	3	4	5
1	19	22	26	29	19
2	22	19	15	23	23
3	24	22	22	24	25
4	38	20	22	21	25
5	18	14	21	16	24

Test for the effects of vacuum tubes at 1% level of significance.

28. Five scientists each test five formulas for a certain compound and observe the number of days each compound takes to develop. The results are as follows:
(Developing time in days)

Compounds

Scientists		1	2	3	4	5
	A	11	15	24	19	21
	B	12	16	25	18	22
	C	10	13	21	17	20
	D	9	14	18	18	21
	E	13	16	20	16	23

Discuss the difference between (i) scientists and (ii) compounds. Given from the tables the following values; $F_{0.05}(4, 16) = 3.01$; $F_{0.01}(4, 16) = 4.77$.

29. Work out the analysis of variance for two way classification
Yields in tons of 4 varieties of wheat in 3 blocks are given below.
Block:

Variety	1	2	3
A	11	10	9
B	8	8	7
C	9	6	5
D	6	5	5

In the difference between varieties significant ?

30. The following table shows the gains in weight of pigs in a comparative feeding trial. The 5 lots of pigs represent 5 different treatments and there are 5 pigs in each lot. make an analysis of variance for the data and test the significance of the treatment differences.
Gains in weight of pigs in lot

Replicate	Gains in weight of pigs in lot				
	I	II	III	IV	V
1	16	18	15	35	51
2	6	30	6	45	38
3	12	35	39	36	23
4	17	16	3	51	43
5	20	20	5	15	14

31. Set up the analysis of variance table for the following two way classification and offer the appropriate comment.
Treatment

Blocks	1	2	3	4	5	6
I	185	157	461	141	130	136
II	117	143	130	144	169	125
III	154	166	152	203	184	215
IV	165	186	125	157	165	180

32. In a sugar—beet experiment four blocks of four plots were used to compare the effects of form treatments (i.e., methods of cultivation). The yields in ton / acre are as follows:

	A	B	C	D
I	13.2	14.1	12.9	15.4
II	16.3	15.6	11.3	17.2
III	17.4	16.3	14.5	15.8
IV	15.8	15.9	14.0	15.1

Make a suitable analysis of variance, and test for the significance of :
(i) Treatment effect
(ii) Block effect

33. (a) The price of a certain commodity was ascertained in each of four towns, on each of 4 dates, one in each quarter of the year. Prices in rupee are shown in the table. Are the variation between the different localities, and between the different seasons, significant?

Quarters	Towers			
	A	B	C	D
I	7	6	7	6
II	6	5	7.5	6
III	5.5	4.5	5.5	6
IV	7.5	5.5.	7	8

(b) Four different drugs are used in three different hospitals and the results are given below in the table show the number of cases of recovery from the disease per 100 people who have taken the drugs.

	D_1	D_2	D_3	D_4
H_1	20	10	24	10
H_2	11	9	13	7
H_3	11	14	14	13

What conclusions can you draw based on analysis of variance?

34. (a) In a feeding experiment of cows, three rations R_1, R_2, R_3 were tried. The cows were put into three classes of three each according to their initial weight. The following

table gives the gain in body weight in kgms in a certain period

Ration class	Class I	Class II	Class III
R1	5	17	11
R2	15	19	20
R3	4	15	8

Analyze the data and state you conclusion. Has divisions into classes proved effective? For 2 and 4 degrees of freedom, F value, significant at 5 percent level of confidence is given as 6.94.

(b) Four different regions of a country were chosen and three different techniques of advertising were introduced in each region to promote sales of a consumer product. The sales figures in 1000 units are given below. Analyze the data and your comments.

	Techniques		
Regions	T_1	T_2	T_3
R_1	22	15	23
R_2	18	16	16
R_3	20	18	22
R_4	20	21	19

(c) The following table is based on a sample study of the relationship between race and blood type in a country.

	Techniques			
Race	O	A	B	AB
I	175	148	96	71
II	78	50	45	12
III	17	22	9	7

(i) Determine whether there is any relationship between race and blood type for the population for this country, taking the data as attributes in a contingency table.

(ii) Taking the data as a two way analysis of variance with interaction.

35. The experiment consists in distributing among 4 laboratories to find the systematic difference in tin-coating weight along the direction of rolling as well as across the rolling direction. In order to eliminate these two sources of variability, each of two sheets of tin plate is divided

into 16 parts, representing four positions across and four positions along the rolling direction. Then, four samples from each sheet are sent to each of the laboratories A, B, C and D, as shown and the resulting tin-coating weights are determined.

Replicate I					Replicate II			
C	A	D	B	Rolling	B	A	C	D
0.23	0.29	0.35	0.25	direction	0.28	0.25	0.20	0.27
B	C	A	D		D	B	A	C
0.25	0.21	0.27	0.23		0.27	0.23	0.21	0.25
D	B	C	A		C	D	B	A
0.20	0.24	0.24	0.28		0.30	0.24	0.26	0.32
A	D	B	C		A	C	D	B
0.24	0.26	0.30	0.24		0.31	0.20	0.25	0.24

Determine from these data whether the laboratories were obtaining consistent results. Also determine whether there are actual tin-coating weight difference across and along the rolling direction. Test the level at 0.05 significance.

(Given $F_{0.05}, (v_1 = 3, v_2 = 21) = 3.07$

$F_{0.05}, (v_1 = 1, v_2 = 21) = 4.32$)

36. There are 3 operators $0_1, 0_2, 0_3$; 3 devices (D_1, D_2, D_3) and 3 different methods are made available (A, B, C). The experiment is replicated twice. The results are given below.

		Replicate I			Replicate II		
		Devices					
Operators	1	A	B	C	C	B	A
		25	41	36	48	39	30
	2	B	C	A	A	C	B
		29	45	47	37	40	37
	3	C	A	B	B	A	C
		48	34	32	32	41	50

Using the 0.05 level of significance, determine whether there are significant differences attributable to the operators, the devices and the methods available.

37. Analyze the variance in the following Latin square of yields in kilograms of paddy, where A, B, C, D denote the different methods of cultivation.

D125   A124   C126   B125
B127   C126   A125   D128
A123   B122   D123   C124
C125   D126   B124   A125

Examine whether the different methods of cultivation have given significantly different yields. It is given that the 5% value of F for 3 and 6 degrees of freedom is 4.76.

38. Explain how the principle of randomization is useful in the design of experiments. What is meant by a completely randomized design? Randomized block design and Latin square design? Point out their comparative merits.

39. (a) Explain a Latin square design (b) An experiment was conducted to compare three burners winners A, B and C. a Latin square design was used as the tests were made on three engines and were spread over three days :

Day	Engine 1	Engine 2	Engine 3
1	A 17	B 18	C 21
2	B 17	C 22	A 16
3	E 16	A 13	B 14

Test the hypothesis that there is no difference between the burners.

40. (a) In a Latin square design explain how the total variance can be decomposed into variances due to row, columns and treatments and show that under the hypothesis of no difference between treatments, the ratio of the variance for treatments to the error variance follow the F distribution, set up the appropriate analysis of variance table.

(b) A randomized block experiment was laid out to 4 varieties of oats. The yield in bushals per acre are given below:

		Varieties		
Block	A	B	C	D
1	57	54	59	58
2	54	52	58	56
3	58	55	63	62
4	55	56	58	57

Test for the significance of the difference between varieties of oats.

41. Analyze the variance in the following Latin square of yields in kilograms of potatoes where A, B, C, D, denote the different methods of cultivation.

D23   A22   C24   B23
B25   C24   A23   D26
A21   B20   D21   C22
C23   D24   B22   A23

Examine whether the different methods of cultivation has given significantly different results. It is given that the 5% value of F for 3 and 6 degree of freedom is 4.76.

42. In an experimental station, the effect of two Manures M_1 and M_2 each applied at two different times, T_1 and T_2 were tested on a variety of wheat. The Latin square layout of the experiment together with the yields of each plot are given below. Test which of the four combinations gives the best result.

M_1T_1	M_1T_2	M_2T_2	M_2T_1
125	81	102	64
M_2T_1	M_2T_2	M_1T_2	M_1T_1
72	136	142	165
M_1T_2	M_1T_1	M_2T_1	M_2T_2
115	160	85	142
M_2T_2	M_2T_1	M_1T_1	M_1T_2
132	75	145	118

43. Analyze the variance in the following Latin square:

A9	C18	B10
C11	B19	A16
B13	A11	C16

44. An agricultural experiment on the Latin square plan gave the following results for the yield of what per acre, letters corresponding 15 varieties.

A17	B11	C12	D10	E10
E11	C10	A15	B13	D12
B16	D9	E9	C11	A19
D13	E7	B14	A14	C13
C14	A12	D11	E8	B15

Discuss the variation of yield with each of the following factors corresponding to the rows and columns.

45. The following is a Latin square design of five treatments. A, B, C, D, E

A15	B11	C23	D9	E8
D11	E10	A17	B9	C18
B13	C19	D10	E12	A19
E10	A17	B9	C12	D9
C13	D11	E10	A13	B17

Analyze the data and interpret the results.

CHAPTER 12

STATISTICAL QUALITY CONTROL ANALYSIS

12.1 A Brief Outline

Quality is concerned with characteristics pertaining to high degree of maintaining **Uniformity, Superior Quality Standard ,Tolerable variability and Fitness for use** Every manufacturing organization is interested in producing quality products conforming to high degree of uniformity, maintain appropriate standard and eliminate variability as far as possible. Ideal maintenance is not guaranteed in day to day environment; hence it is essential to devote continuously for **Quality improvement** Programs. **Quality** or **fitness for use** is determined through the interaction of **quality of design** and **quality of conformance**. The quality of design is concerned with the appropriate engineering and management decisions resulting in the different grades or levels of performance, reliability, service reliability, and function of execution. The quality of conformance concentrates on the systematic **reduction of variability** and **elimination of defects** until every unit produced is identical and free of defects. Employing designed experiments to improve the product in the design, production and assembly stages are rather given prominence to only inspect quality into the product after it is produced. Some of the special techniques of **Statistical Quality assurances** consist in the maintenance of quality control, establishment of tolerance limits and acceptance sampling. Quality refers quantitatively to some measurable or countable property of a product.

12.2 Quality Improvement Programs

A product is said to be a defective one if has any one of the observable defects. By plotting percent defectives per day over a period of say 4 weeks, as a scatter diagram, it is possible to comprehend the stable variation of the trend about the percentage defective pieces. This indicates that the process is predictable. It is possible to estimate the mean by the average over days and also the amount of variation. These may reveal that the process is stable .This reavealment does not make the process as good. It may turn out that the process has too many defective pieces. Improvement conveys systematic elimination of waste.

One realizes that the process needs improvement and hence action can be taken. Data collected on several possible sources of variation can be displayed in the PARETO diagram exhibiting the cumulative percentages of defects indicated by broken lines in the PARETO diagram of defects.

Based on these data, training programs can be imitiated to the operators. The new process after training, will also appear to be stable but with a lower mean. This process of improvement must be continued to gain progressively quality improvement in any manufacturing process. It is thus possible to reduce the amount of variation and the percentage of defectives. Since the process is stable, the effects of changes can be observed and factorial designs can be used to improve the process. The cost factor concerning the maintenance of high quality, tolerable production cost and customer satisfaction should also be taken into account by the companies competing globally while stressing quality improvement.

12.3 Deming 14 Points for Quality Improvement

W. Edwards Deming in his book on "A theory of Management for product or service improvement" has specified as recommendation, 14 points to the industrial organizations. The list is furnished below:

1. Create constancy of purpose towards product improvement and services in a competitive environment by providing employment opportunities.
2. Set the goal to achieve economic gain possessing competent leadership.
3. Avoid the dependence of mass scale inspection by building quality into the manufacturing product.
4. Give prominence to minimize the overall cost choosing reliable single supplier for every item based on loyalty and trust worthiness.
5. Improve the production and service systems to achieve quality standard articles leading to overall decreasing the cost.
6. Provide appropriate training programmes to update the knowledge on the jobs done by the employees in their hierarchy.
7. Inculcate leadership qualities among the supervisors utilizing the machines and gadgets for the individuals to perform their jobs excellently.
8. Prevent fear phobia for promoting work culture.

9. Effect coordination among departments. Create an environment in the organization among the people of various categories to work together as a team.
10. Try as much as possible to have zero defects in the articles produced and find the real causes for low quality of production on a joint endeavour.
11. Give prominence to leadership qualities among the management personnel.
12. Provide facilities for the promotion of workman culture so that everyone will feel a real satisfaction of the contribution to attain the objectives of the management.
13. Provide knowledge gaining and self-improvement for each workman.
14. Involve everyone to accomplish reliable and improved transformation to be adored by the society.

Quality improvement is based on statistical approach in every one of the aspects like equipment, components and men and materials that go into standardized production. The actual consumers has also a role for quality improvement .Design, production and marketing play a prominent role in promoting quality standards of the goods produced, The reactions of the buyers play a significant role in quality satisfaction.

Statistical methods of sampling will provide the realistic view of the existing production techniques to effect appropriate changes in design and production for the ample satisfaction of the buyers in the market.

[Source: OUT OF THE CRISIS-by W. Edwards Deming .M.I.T. Cambridge, 1986]

12.4. Procedure Involving a Quality Improvement Program

- Select the process which needs improvement.
- Form a committee involving managers, supervisors , engineering professionals, employees, cost accountants etc.
- Collect data about the types of defects, the frequency of the occurrence in a fixed period of each defect to yield the total number of defectives.
- Find the possible causes and effects and the factor that affect the major defects based on their occurrences.
- Use factorial design among two major suppliers.
- Form quality groups of the persons involved. Provide each group statistical training in experimental design.

12.5 Role of Statistical Methods in Quality Improvement:

Some Applications

- Statistical methods including designed experiments

(i) Can be used in design and development to compare different materials, components or ingredients so as to enable the tolerances determination, reduction in cost and time.

(i) To determine the capability of the manufacturing process and to reduce variability for improving the process.

(iii) To provide reliability in life testing of the product- leading to new and improved designs, longer useful lives and reduction in operating and maintenances costs.

12.6. Conduct Statistical experimental designs for quality

The design procedure is as follows:

- Select one input variable to minimize variation while another input variable holds the response on target.
- Create products that are not sensitive to variations in their components or environmental conditions.

For illustration proposes consider a 2^3 factorial design experiment three times to study the effects of three variables x_1, x_2, x_3 concerning the project undertaken considering x1,x2,x3 as Boolean variables, the terms of the Boolean functions are

$$x_1\,x_2\,x_3,\ x_1\,x_2\,x_3{}',\ x_1 x_2{}'\,x_3,\ x_1\,x_2{}'\ x_3{}',$$
$$x_1{}'\,x_2\,x_3,\ x_1{}'\,x_2\,x_3{}',\ x_1{}'\,x_2{}'\,x_3,\ x_1{}'\,x_2{}'\,x_3{}',$$

For each run the variance of the 3 responses is computed (s_i^2, i=1,2,3). Find the values of their (s_i^2) natural logarithms in order to determine if any of the factors have an influence on the variance.

For the variable x1 find the estimated effects on ln (variance)

using, for $\quad x_1 = \dfrac{1}{4}\sum_{i=1}^{8} \ln\!\left(s_i^{\,2}\right)$

Similarly, for x_2, x_3, $x_1 x_2$, $x_1 x_3$, $x_2 x_3$ and $x_1\,x_2\,x_3$ are estimated.

Plot the absolute values of the estimated effects ln (s_i^2); this will reveal, which factor is important.

12.7 Statistical Quality Control

Consider 2 apparently identical parts made under carefully controlled conditions, from the same batch of raw material, and only seconds a part by the same machine. These 2 parts will be different in many respects. This variation is of a random nature which cannot be completely eliminated.

It is assumed that a measurable characteristic is nearly normally distributed and its natural variation is within plus or minus 3 standard deviation (σ) of its mean. A typical base line assessment is to determine if this interval of length 6σ is within the specification limits.

The quantification of process capability is quantified as follows.

Let LSL be the lower specification (or control) limit and USL, the upper specification (control) limit for the process. The process capability is assessed by a process capability

index, $\overset{n}{C}_p = \dfrac{USL - LSL}{6s}$, where s is the standard deviation obtained by measuring from a sample of units chosen from the process population. When the process mean is not centered between the specification limits, the closest specification may be most important. An alternative

process-capability index having the estimated value given by $\hat{C}_{pk} = \dfrac{\min\left(\overline{x} - LSL, USL - \overline{x}\right)}{3s}$.

At this point the variability present in the production process is restricted to **chance variation** and the process is considered to be in a state of statistical control. Chance causes are an inherent part of the process. It is possible to have **assignable variation** due to poorly trained operators, the low quality of raw materials, faults pertaining to the machines used in the process etc. When assignable causes are present the process is said to be out of control. In this respect the control charts play a prominent role.

12.8 CONTROL CHARTS

Based upon the observations which are either concerned with measurements or count data, it is essential to have.

1. Control chants for measurements and
2. Control charts for attributes.

The reasons for popularity of charts for use in industry are:

- They are a proven technique for improving productivity.
- They are effective in defect prevention.
- They prevent unnecessary process adjustments.
- They provide diagnostic information.
- They provide information about process capability.

In both cases, a control chart consists of a **Central Line** corresponding to the average quality at which the process is to perform and lines corresponding to **the upper and lower control limits.**

When the sampling values fall between these two control limits, such occurrence of values are attributed to chance factors and the values fall beyond these two lines are interpreted as indicating a lack of control. Once the sample point falls beyond the control limits, it is essential to look for the real cause for such troubles. Corrective action should be taken. Experience and high sense of judgment play important roles in such situations.

Quality assurance is concerned with several aspects of engineering and management quality control specifications concerning a process.

12.9 CONTROL CHARTS FOR MEASUREMENTS

In the case of data pertaining to measurements, the usual prominence is given to control over the average quality of a process and then its **variability.**

The first goal is accomplished by plotting the means of periodic samples on a Control Chart for means, called an X-bar or $\bar{x}$ chart.

The second goal pertains to the variability. Variability is controlled by plotting the sample ranges or standard deviations, respectively, on an **R Chart** or a σ Chart based on the statistic used to estimate the population standard deviation.

Let μ and σ denote the **known process mean and standard deviation**. Treating the measurements as samples from a normal population, it is possible to assert with probability

$1-\alpha$, that the mean of a random sample of size n will fall between $\mu - z_{\alpha/2} \dfrac{\sigma}{\sqrt{n}}$ and

$\mu + z_{\alpha/2} \dfrac{\sigma}{\sqrt{n}}$. These two limits provide upper and lower controls limits and under the given assumptions, the quality control engineer is able to determine whether or not an adjustment is to be made in the process.

On many occasions, α and σ are not known and hence it is necessary to estimate their values from a large sample (or samples) taken while the process is in control. It is not possible to guarantee that the samples are from a normal population and hence the (1-α) 100% confidence level associated with the control limits is only approximate. For practical purposes the industry uses three-sigma limits obtained by substituting 3 for $Z_{\alpha/2}$, though high confidence is not possible to declare that the process is actually in control. If there exists a long history of the process is in good control, then and μ can σ be estimated from the past data. Thus, the central line of an chart is given by and the upper, lower three- sigma control limits are given by $\mu \pm A \alpha$

Where $A = \dfrac{3}{\sqrt{n}}$ and n is the size of each sample.

When the population parameters are unknown, it is necessary to estimate these parameters on the basis of preliminary samples. If k samples are used each of size n, and

$\bar{x}_i$ represents the mean of the i-th sample, then the grand mean $\bar{\bar{x}}$ is given by $\bar{\bar{x}} = \dfrac{1}{k} \sum_{i=1}^{k} \bar{x}_i$

The process variability can be estimated either from the standard deviations or the ranges of the k samples, Denoting the range of the ith sample by Ri , then the mean sample range $\bar{R}$ is given by $\bar{R} = \dfrac{1}{k}\sum_{i=1}^{k} R_1$

$\bar{x}$ is said to provide an unbiased estimate of the population mean μ, the central line for the chart is given by $\bar{x}$. However, the statistic $\bar{R}$ does not provide an unbiased estimate of σ, but multiplying $\bar{R}$ by the constant A_2 (specified for various values of n from table for this purpose), the central line and the upper and lower three-sigma control limits UCL and LCL for an $\bar{x}$ chart- are given by

$$\text{Central line} \quad = \bar{\bar{x}}$$
$$\text{UCL} \quad = \bar{\bar{x}} + A2$$
$$\text{LCL} \quad = \bar{\bar{x}} - A2$$

In addition to monitoring the population mean, when fluctuation of the $\bar{x}$'s are noticeable a more sensitive test of shits in process variability is provided by a separate control chart, called an R chart based on the sample ranges or a σ- Chart based on the sample standard deviations

The central line and control limits of an R chart are based on the distribution of the range of samples of size n from a normal population. The mean and the standard deviation of this sampling distribution are given by $d_2\,\sigma$ and $d_3\,\sigma$ respectively, when σ is known. Thus, three -sigma control limits for the range are given by $d_2\,\sigma \pm d_3\,\sigma$ and the complete set or control -chart values for an R chart-(with σ known) is given by:

$$\text{Central} \quad \text{line} = d_2\,\sigma$$
$$\text{UCL} = D_2\,\sigma$$
$$\text{LCL} = D_1\sigma$$

Here $D_1 = d_2 - 3d_3$ $D_2 = d_2 + 3d_3$ and the values of these constants are given in the special table for various values of n.

If σ is unknown, it is estimated from past data and the control chart values for R chart (with σ unknown) are given by:

Central line= $\bar{R}$

UCL = $D_4\,\bar{R}$

LUC = $D_3\,\bar{R}$

There $D_3 = \dfrac{D_1}{d_2}$ and $D_4 = \dfrac{D_2}{d_2}$

The values of these constants can be found in the special table for various values of n.

12.10 CONTROL CHARTS FOR ATTRIBUTES

It is often desirable to classify a product as either defective or non-defective on the basis of comparison with a standard. This classification brings economy and simplicity in the inspection operation without actual measurements made on a finished product. It is easier to check the "product" against specifications as an "attribute" or "go no-go" basis. Concerning attribute sampling two types of control charts are used, known as:

Fraction-defective chart or p chart and Number of defects chart or c chart

To clarify the distinction between the number of defective "number of defect", it has to be noted that a unit tested can have several defects, whereas on the otherhand, it is either defective if it is not. Consequently a unit is referred to as defective if it has at least one defect.

Control limits for fraction -defection chart are based on the sampling theory for proportions and on the normal curve approximate to the binominal distribution.

If p denotes the fraction defective, the central line is p and three-sigma control limits

for the fraction defective in random samples of size n are given by $p \pm 3\sqrt{\dfrac{p(1-p)}{n}}$.

In actual practice p is estimated from past data. If k samples are available, d_i is the number defectives in the i-th sample, and n_i is the number of observations in the i-th sample, then p is estimated as the proportion of defectives in the combined sample, as

$$\bar{p} = \frac{d_1 + d_2 + \ldots + d_k}{n_1 + n_2 + \ldots + n_k}$$

Hence the control-chart -values for a fraction defective consists of

Central line = $\bar{p}$

$\text{UCL} = \bar{p} + 3\sqrt{\dfrac{\bar{p}(1-\bar{p})}{n}}$

$\text{LCL} = \bar{p} - 3\sqrt{\dfrac{\bar{p}(1-\bar{p})}{n}}$

Whenever n and p are such that the underlying binomial distribution cannot be approximated by a normal curve, the Possion's approximation is used to the binomial distribution.

In the control chart for the number defectives in a sample of size n, one plots the number of defectives and the control chart values for this kind of chart are obtained by multiplying the above values for the control line and the control-limits by n. Thus, if p is estimated by $\bar{p}$, the control-chart values for a number-of- defectives charts are as follows:

Central line $= n\bar{p}$

$$UCL = n\overline{p} + 3\sqrt{n\overline{p}(1-\overline{p})}$$

$$LCL = n\overline{p} - 3\sqrt{n\overline{p}(1-\overline{p})}$$

In situations where it is necessary to control the number of defects (c) in a unit of product it is assumed that c is taken as a random variable having the Poisson distribution with parameter λ as mean and standard deviation as $\sqrt{\lambda}$.

If λ is unknown, then it is estimated, by taking k as the number of units product and c_i as the number of defects in the i-th unit. The estimated value of λ is given by $\overline{c} = \dfrac{1}{k}\sum_{i=1}^{k} c_i$

The control -chart values for the c chart are

Central line $= \overline{c}$

$$UCL = \overline{c} + 3\sqrt{\overline{c}}$$

$$LCL = \overline{c} - 3\sqrt{\overline{c}}$$

12.11 TOLERANCE LIMITS

In dealing with industrial quality control in a deeper sense the information about process capability or the performance of the process, when it is operating in control can be obtained by comparing some quality characteristic or measurement of a finished product against given specifications.

These specifications are called tolerance limits as stated by the customer or by the design engineer. Deviation from these limits will make the product unusable.

The tolerance chart and the histogram are the two graphical tools helping in assessing process capability.

When reliable information is available about the distribution of the measurement in question, the process is allowed to establish its own limits called the natural tolerance limits. The long experience concerning the product, it is assumed that a certain dimension is normally distributed with mean and standard deviation σ. It is easy then to construct limits between which one can find any given proportion of the population. For P = 0.90 the tolerance limits are $\mu \pm 1.645\,\sigma$ and for P = 0.95, the tolerance limits are $\mu \pm 1.96\,\sigma$ as obtained from the table of areas of the normal curve.

If the true values of μ and σ are not known, then the tolerance limits must be based on the mean $\overline{x}$ and standard deviation S of a random sample from the concerned population. The limits $\overline{x} \pm 1.965$ are themselves random variables at 95 % level and they may or may not include a given proportion of the population. However, it is possible to determine a constant k so that one can assert with $(1 - \alpha)$ 100% confidence that the proportion of the population contained between $\overline{x} - ks$ and $\overline{x} + ks$ is at least P. For various values of n special tables have been constructed for P = 0.90, 0.95 and 0.99 with 95% or 99% levels of confidence.

In this respect, it is essential to note the difference between **confidence limits** and **tolerance limits.**

The confidence limits are used to estimate a parameter of a population. The tolerance limits are used to indicate between what limits one can find a certain proportion of a population.

When n becomes large the length of a confidence internal approaches zero, while the tolerance limits will approach the corresponding values for the population. For large n, k approaches 1.96 for P = 0.95in the special table (available).

12.12 ACCEPTANCE SAMPLING

In sampling inspection items are selected from each lot prior to shipment or prior to acceptance by the consumer and a decision is made on the basis of this sample whether to accept or reject the lot. Acceptance of a lot generally implies that if can be shipped or accepted by the consumer, even though, it may contain some defective items. A rejected lot may be subjected to closer inspection with the aim of eliminating all defective items. Since inspection cost is not negligible, it is not always desirable to inspect each item in a lot. Hence, acceptance inspection usually involves **Sampling**

Acceptance Sampling bases a decision on a random sample selected from each lot The lot is accepted if the number of defectives found in the sample does not exceed given a **acceptance number**

The procedure is equivalent to a test of the null hypothesis that the proportion of defectives p in the lot equals some specified value p0 against the alternative that it equals p1where p1>p0 In acceptance sampling the value of p0 is called the acceptable quality level or AQL and p1 is called the lot tolerance percent defective or LTPD in association with given producer's and (or) consumer's risks. The probability of a Type I error, α can be interpreted as upper limit to the proportion of good lots (lots with $p \leq p0$) that will be rejects, called the producer's risk. The probability of a Type II error, β gives an upper bound to the proportion of bad lots (lots with $p \geq p_1$) that will be accepted, called the consumer's risk.

A given sampling plan is based on its operating Characteristic OC-curve, which gives the probability of acceptance for each value that can be assumed by the lot proportion of defectives, p. Thus, the OC-curve describes the degree of protection offered by the sampling plan against incoming lots of various quality.

If n represents the sample size taken from a let containing N units, and c is the acceptance number, then the probability of accepting a lot containing the proportion of defectives p (the lot contains N_p defectives) can be calculated by using the hyper geometric distribution.

$$L(p) = \sum_{x=0}^{c} h(x : n, Np, N))$$

A sampling plan can also be described by means of its average outgoing quality or

(AOQ) curve. This curve describes the degree of protection offered by the sampling plan. If the incoming lots are good quality, that is if their proportion defective is smaller than the AQL, very few lots will be rejected and the average outgoing quality or AOQ will be good.If the incoming lots are of poor quality, that is, if their proportion of defectives is larger than LTPD, most of then will be rejected. It is only when the average incoming quality lots. between the AOL and the LTPD, that the poorest quality of lots will be shipped.

In general, there will be a maximum AOQ over all values of incoming quality p, and this value is called the average outgoing quality limit(AOQL).

It the incoming quality is p the probability that a lot will be accepted is L(p) the average outgoing quality is given by $AOQ = p.L(p)$

Let N be the lot size and M denote the number of defectives in the lot. Let A be the event that a lot is accepted. The corresponding probability P (A) depends not only on the sample size n and c the specified number of defectives (c<n) for the lot to be accepted. c is called acceptance number. The producer and the consumer must agree on a certain sampling plan having the sampling size n and allowable number c. Such a plan is called a single sampling plan because it is based on a single sample. The Probability P(A) is given by

$$P(A) = P(X \le c) = \sum_{x=0}^{c} \binom{M}{x}\binom{N-M}{n-x} \Big/ \binom{N}{n}$$

If M = 0, (no defective in the lot), than X must assume the value 0.
Consequently

$$P(A) = \binom{0}{0}\binom{N}{n} \Big/ \binom{N}{n} = 1$$

if M=N, then

$$P(A) = P(X \le c) = 0 \text{, because } c < n.$$

The ratio $\theta = M/N$ is called the fraction defective in the lot
$\therefore M = N\theta$ and hence P(A)
becomes

$$P(A; ?) = \sum_{x=0}^{c} \binom{N?}{x}\binom{N-N?}{n-x} \Big/ \binom{N}{n}$$

Since θ can have one of the (N+1) values, $0, \dfrac{1}{N}, \dfrac{2}{N}, ..., \dfrac{N}{N}$, the probability P(A) is defined for these values only.

For fixed n and c, a smooth curve can be drawn called the operating characteristic curve (OC) of the sampling plan considered.

There are two types of possible errors in acceptance sampling and the related problem of choosing the sample size "n" and the number of defective items "c". In acceptance sampling the producer and the consumer have different interests. The problem may require the probability of rejecting a "good", or "acceptable", lot to be a small number, call it α. The consumer (buyer) may demand the probability of accepting a "bad", or "unacceptable", lot to be a small number β.

More precisely suppose that the two parties agree that a lot for which θ does not exceed a certain number θ_0 is an **acceptable lot** while a lot for which θ is greater than or equal to a certain number θ_1 is an **unacceptable lot**. Then is the probability of rejecting lot with $\theta \leq \theta_0$ and is called **producer's risk.** This corresponds to a Type I error in testing a hypothesis. β is the probability of accepting a lot with $\theta \geq \theta_1$ and is called **consumer's risk.** The corresponds to a. Type II error.

θ_0 is called the acceptable quality level (AQL) and θ_1, is called the **lot tolerance percent defective** (LTPD) of the rejectable quality level (RQL).

A lot-with $\theta_0 < \theta < \theta_1$ may be called an **indifferent lot.**

The points $(\theta_0, 1 - \alpha)$ and (θ_1, β) lie on the operating characteristic (OC) curve.

12.13 EXAMPLES

EXAM. 1: Twenty-five samples of size 5 are drawn from a process at one -hour intervals, and the following data are obtained.

$$\sum_{i=1}^{25} \overline{x}_i = 362.5, \ \sum_{i=1}^{25} r_i = 8.75, \ \sum_{i=1}^{25} s_i = 4.00$$

    (a) Find trial control limits for and R charts

    (b) Repeat part (a) for $\overline{x}$ and S charts.

SOLUTION

    For n = 5, $A_2 = 0.577$, $D_4 = 20115$, $D_3 = 0$, $C_4 = 0.94$

$$\overline{x} = \frac{362.5}{25} = 14.5; \ \overline{R} = \frac{8.75}{25} = 0.35; \ \overline{S} = \frac{4.00}{25} = 0.16$$

(a) For the $\overline{X}$ Chart, the trial control limits are

$$UCL = \overline{\overline{x}} + A_2\overline{r} = 14.5 + 0.577 \times .35 = 14.5 + 0.20195 = 14.7019 = 14.72$$

$$LCL = \overline{\overline{x}} - A_2\overline{r} = 14.5 - 0.20195 = 14.29805 = 14.30$$

For the R chart, the trial control limits are

$$UCL = D_4\overline{r} = (2.115) \times 0.35 = 0.74025 = 0.740$$

$$LCL = D_3\overline{r} = 0 \times 0.35 = 0$$

For the S chart, the trial control limits are

$$UCL = \overline{s} + \frac{3 \times 0.16}{c_4}\sqrt{1 - (c_4)^2} = 0.16 + \frac{3 \times 0.16}{0.94}\sqrt{1 - (0.94)2} = 0.16 + 0.06 = 0.22$$

$$LCL = \bar{\bar{S}} - \frac{3.\bar{S}}{c_4}\sqrt{1-(c_4)^2} = 0.16 - 0.06 = 0.10$$

(b) If S_i is the standard deviation of the i th sample, then

$$\bar{S} = \frac{1}{25}\sum_{i=1}^{25} S_i$$

Because $E\{\bar{S}\} = c_4\sigma$, an unbiased estimate of σ is $\bar{S}/c_4$, that is $\hat{s} = \dfrac{\bar{S}}{c_4}$

$\bar{X}$ Control Chart. from $\bar{S}$, the trial control limits are

$$UCL = \bar{\bar{x}} + \frac{3\bar{s}}{c_4\sqrt{n}}\sqrt{1-(c_4)^2} = 14.5 + \frac{3\times 0.16}{0.94\sqrt{5}}\sqrt{1-(0.94)^2}$$

$$LCL = \bar{\bar{x}} - \frac{3\bar{s}}{c_4\sqrt{n}}\sqrt{1-(c_4)^2} = 14.5 - \frac{3\times 0.16}{0.94\sqrt{5}}\sqrt{1-(0.94)^2}$$

EXAM. 2: A manufacturer of a certain bearing knows from a preliminary record of 25 hourly samples of size 4 that, for the diameters of these bearings, $x = 0.7925$ and $\bar{R} = 0.004$. When deviations are taken from the mean x in 0.0002 ems find the control limits for $\bar{x}$ chart (coded) and for R chart (coded), given that the central line for $\bar{x}$ is 2.0 and that the central line for R is 2.0, the values of $A_2 = 0.729$, $D_3 = 0$, $D_4 = 20282$ for samples of size are also given.

SOLUTION

$\bar{x}$ Chart (coded)   R Chart (coded)

Central line $\bar{\bar{x}}$	=	2.00	Central line $\bar{R} =$		2.00
$UCL = \bar{\bar{x}} + A_2\bar{R}$	=	2.00+0.729x2	$VCL = D_4\bar{R}$	=	2.282x2.00
	=	2.00+1.458			= 4.564
	=	3.5			= 4.6
$LCL = \bar{\bar{x}} - A_2\bar{R}$	=	2.00-1.45	LCL		= $D_3 \times \bar{R} = 0$
	=	0.542			
	=	0.5			

EXAM. 3: From a factory producing metal sheets, a sample of 5 sheets is taken at random every hour and the following data is obtained. Draw the control charts for mean and range and examine whether process is within control.

For n = 5 $A_2 = 0.577$, $D_4 = 2.115$, $D_3 = 0$

Sample No	1	2	3	4	5	6	7	8	9	10
Mean thickness of sheet	.025	.032	.042	.022	.028	.010	.026	.040	.026	.029
Sample Range	.025	.048	.012	.012	.019	.010	.006	.046	.010	.032

SOLUTION

$$\bar{\bar{x}} = \frac{\sum \bar{x}_i}{10} = \frac{0.280}{10} = 0.028 \quad \bar{R} = \frac{\sum R_i}{10} = \frac{.220}{10} = 0.022$$

For the $\bar{x}$ Chart, the trial control limits are

$$UCL = \bar{\bar{x}} - A_2.\bar{R} = 0.028 + 0.577 \times 0.022 = 0.028 + 0.012694 = 0.041$$

$$LCL = \bar{\bar{x}} - A_2.\bar{R} = 0.028 - 0.012694 = 0.015306 = 0.0153 = 0.02$$

For the R Chart, the trial control limits are

$$UCL = D_4 \bar{R} = 2.115 \times 0.022 \qquad\qquad = 0.046530 = 0.0465$$

$$LCL = D_3 \bar{R} = 0 \times 0.022 = 0$$

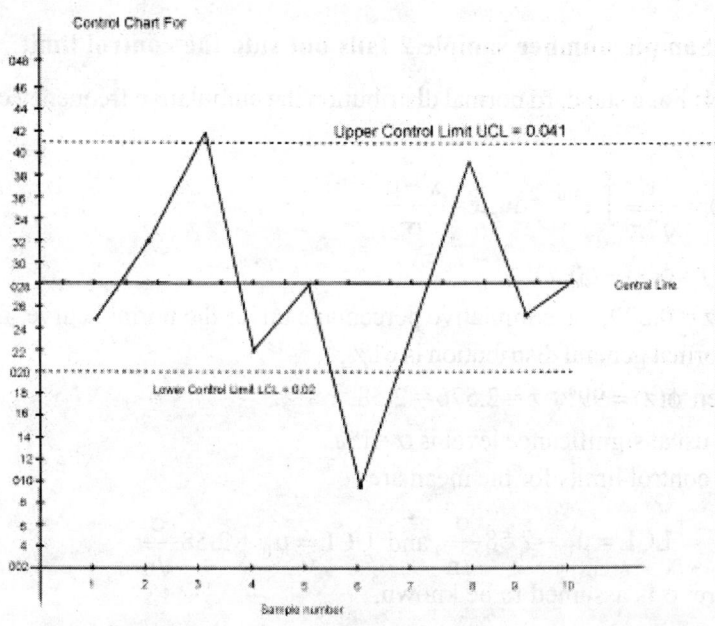

Control Chart For

Upper Control Limit UCL = 0.041

Central Line

Lower Control Limit LCL = 0.02

Sample number

Sample 3 and sample 6 fall out side the control limit

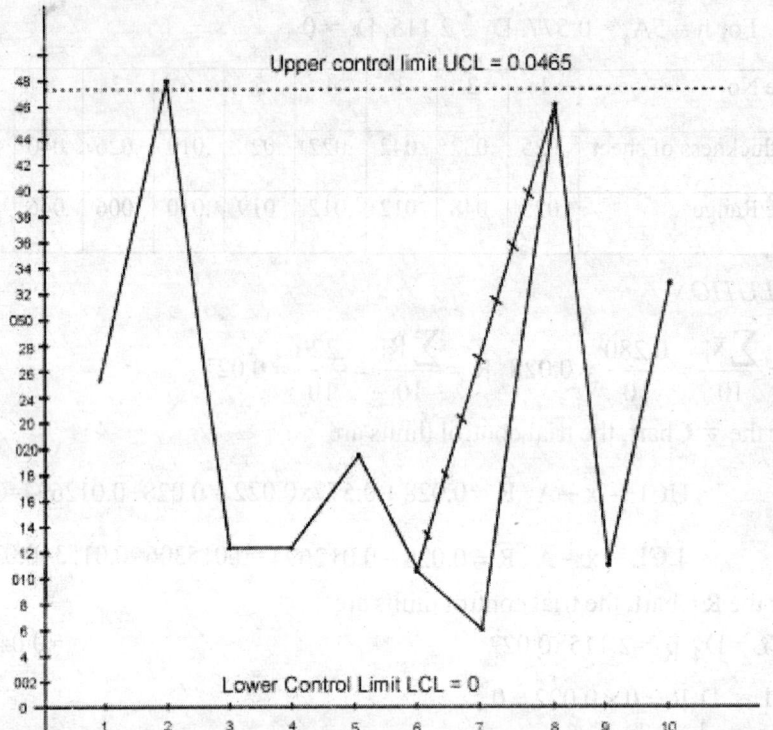

Sample number sample 2 falls out side the control limit

EXAM. 4: For a standard normal distribution the cumulative frequency curve is given by

$$\phi(z) = \frac{1}{\sqrt{2\pi}} \int_{-\alpha}^{z} e^{-u^2/2} du, u = \frac{x - \mu}{\sigma}$$

$D(2) = \phi(z) - \phi(-z)$

For $z = 0.279$, the cumulative percentage under the normal curve, from the table of normal general distribution is 61%.

When $\phi(z) = 99\%$ $z = 2.576 = 2.58$

The usual significance level is $\alpha = 1\%$

The control limits for the mean are

$$LCL = LCL = \mu_0 - 2.58\frac{\sigma}{\sqrt{n}} \text{ and } UCL = \mu_0 + 2.58\frac{\sigma}{\sqrt{n}}$$

There σ is assumed to be known.

EXAM. 5: A machine for filling cans with lubricating oil is set so that it follower a normal population with mean 1 gal and standard deviation 0.03 gal.

Find the LCL and UCL assuming that the sample size is 6.

SOLUTION

($\alpha = 1\%$ level)

$$\text{LCL} = 1 - \frac{2.58 \times 0.03}{\sqrt{6}} = 0.968 \text{ and UCL} = 1 + \frac{2.58 \times 0.03}{\sqrt{6}} = 1.032$$

EXAM. 6: What should be the sample size, if the mean is 1 gal and standard deviation is 0.03gall when UCL-LCL = 0.05.

SOLUTION

$$\text{UCL} = \mu_0 + \frac{2.58\sigma}{\sqrt{n}}, \text{LCL} = \mu_0 - \frac{2.58\sigma}{\sqrt{n}}$$

$$\therefore \text{UCL} - \text{LCL} = \frac{2 \times 2.58\sigma}{\sqrt{n}} = \frac{2 \times 2.58 \times .03}{\sqrt{n}}$$

$$0.05 = \frac{2 \times 2.58 \times 0.03}{\sqrt{n}} = 0.05 = \frac{2 \times 2.58 \times 0.03}{\sqrt{n}}$$

$$\sqrt{n} = \frac{2 \times 2.58 \times 0.03}{0.05} = \frac{6 \times 2.58}{5} = 1.2 \times 2.58 = 3096$$

or $n = (3.096)_2 = 10$

EXAM. 7: How should we change the sample size in controlling the mean of a normal population if we want UCL-LCL to decrease to half its original value.

SOLUTION

$$\text{UCL} - \text{LCL} = \frac{2 \times 2.58\sigma}{\sqrt{n}} \qquad \qquad ...(1)$$

For the sample size N, we have

$$\frac{\text{UCL} - \text{LCL}}{2} = \frac{2 \times 2.58\sigma}{\sqrt{N}} \qquad \qquad ...(2)$$

(1) $\div$ (2) yields $2 = \frac{\sqrt{N}}{\sqrt{n}}$ or N = 4n

The sample size should be chosen 4 times the original sample size.

EXAM. 8: Suppose the following fraction defective has been found in successive samples of size 100

										Total
0.10	0.08	0.15	0.03	0.05	0.11	0.12	0.05	0.14	0.13	0.96
0.13	0.14	0.09	0.06	0.07	0.14	0.06	0.09	0.12	0.08	0.96

(a) Using all the data, compute the trial control limits for a fraction-defective control chart, construct the chart- and plot the data.

(b) Determine whether the process is in statistical control. If not, assume assignable causes can be found and out of control point's eliminated. Revise the control limits.

$$\bar{p} = \frac{1.94}{20} = 0.097$$

$$UCL = \bar{p} + 3\frac{\sqrt{\bar{p}(1-\bar{p})}}{100} = 0.097 + \frac{3}{10}\sqrt{0.097 \times 0.903}$$

$$= 0.097 + 3x\frac{0.296}{10} = 0.097 + 0.0888 = 0.1858 = 0.19$$

$$LCL = \bar{p} - 3\frac{\sqrt{\bar{p}(1-\bar{p})}}{n} = 0.097. - 0.0888 = 0.0082 = 0.008$$

We have

UCL = 0.19and LCL = 0.008 ≅ 0.01

CONTROL CHART FOR FRACTION DEFECTIVE:P CHART

(b) Since all the points lie within the control limits, the process is in statistical control

EXAM. 9: A normally distributed process uses 66.666 % of the specification band. It is centered at normal dimension, located half way between upper and lower specification limits.

(a) Estimate PCR and PCR_k. Interpret these ratios.
(b) What fall out level (fraction defective) is produced?
Process capability ratio (PCR) is given by

$$PCR = \frac{USL - LSL}{6\sigma}$$

The numerator is the width of the specifications. The limits 3 on either side of the process mean are called natural tolerance limits. 6σ is often referred to as the width of the process. $\left(\frac{1}{PCR}\right)100$ is just the percentage of the specifications' width used by the process.

Here $\left(\frac{1}{PCR}\right)100 = 66.666... = \frac{2}{3} \times 100$

$$\therefore PCR = \frac{100 \times 3}{2 \times 100} = 1.5$$

The measure of actual capability PCR_k is defined as

$$\therefore PCR = Min\left(\frac{USL - \rho}{\sigma}, \frac{\mu - LSL}{\sigma}\right)$$

If the process is centered at the nominal dimension, then $PCR = PCR_k$.

$\therefore PCR_k = 1.5$

Since all items fall within the control limits, the fraction defective is p= o.

EXAM. 10: The number of defectives in samples of 100 ceramic substrates when 20 samples are taken in given below :

46,	50,	34,	52,	31,	48,	54,	46,	50,	31
32,	28,	40,	48,	32,	44,	43,	37,	54,	38

Find the control limits for fraction defectives.
SOLUTION

Here $\bar{p} = \frac{840}{2000} = 0.42$

Therefore, the trial parameters for the control chart are

$$UCL = 0.42 + 3 \frac{\sqrt{(0.42)(0.58)}}{100}$$

$$= 0.42 + 3 \times 0.046 = 0.42 + 0.1488 = 0.5688$$

$$LCL = 0.42 - 0.1488 = 0.2712$$

12.15 Examples in Acceptance Sampling

EXAM.1: There are tool bits packages 20 to a box. A sample of 3 tool bits is drawn and the box is accepted if and only if when all the 3 bits in the sample are good.

Here $N - 20$, $n = 3$, $c = 0$ and hence

$$P(A:\theta) = \binom{20\theta}{0}\binom{20-20\theta}{3}\binom{20}{3}$$

$$= \frac{(20-20\theta)(19-20\theta)(18-20\theta)1.2.3}{1.2.3 \qquad 20 \times 19 \times 18}$$

$$= \frac{20(1-\theta)(19-20\theta)2(9-10\theta)}{20 \times 19 \times 18}$$

The numerical values are:

θ	0.00	0.05	0.10	....
$P(a; \theta)$	1.00	0.85	0.716	....

The above tabular an values are used to draw the OC curve.

If n is such that $n\theta = \mu$ (say) is small, $P(A;\theta)$ can be approximations by the Poisson distribution. And is given by

$$P(A;\theta) \ e^{-\mu} \sum_{x=0}^{c} \frac{\mu^x}{x!}, (\mu = n\theta)$$

EXAM. 2: A sample of size 10 is taken. If contains not more than 1 defective the lot is accepted.

Here $\mu = n\theta = 10\theta$

$$\therefore \ P(A;\theta) = e^{-10\theta}\{1+10\theta\}$$

θ:	0	0.10	.20	.....
$P(A;\theta)$:	1	$e^{-1}(2)$	$e^{-2}(3)$	......

The OC curve can be drawn using the above tabular values.

EXAM. 3: A single sampling plan of sample size 40 and an acceptance number 1 is used to check the gaskets by a purchaser. Determine AOQL.

Compute the probability of acceptance of lots containing 1% of defective gaskets
SOLUTION

$$\theta = 1\% = \frac{1}{100}$$

$$\mu = n\theta = 40\left(\frac{1}{100}\right) = .4$$

$$P(A, \theta) = e^{-.4}\sum_{x=0}^{c}\frac{(.4)^x}{x!}$$

$$= e^{-.4}\left[1+\frac{.4}{1}\right] = e^{-.4}[1.4] = 0.6793 \times 1.4 = 0.9384$$

$$AOQL = \theta.P(A;\theta) = (1.0\%)(.9384) = 0.01 \times .9384 = .009384$$

EXAM. 4: Sampling of 3 fuses are drawn from lots and a lot is accepted if in the corresponding sample, we find no more than 1 defective fuse.

Draw the O.C. Curve

$$\mu = n\theta = 3\theta$$

$$P(A;\theta) = e^{-3\theta}\sum_{x=0}^{c}\frac{(3\theta)^x}{x!}$$

When c = 1,

$$P(A;3\theta) = e^{-3\theta}\left\{1+\frac{3\theta}{1!}\right\}$$

$$e^{-3\theta}[1+3\theta]$$

θ :	0	.1	.2	.2	.....
$P(A;\theta)$:	1	$e^{-3}(1.3)$	....	1.3	...
	1	96304	0.87808	0.77254	...

12.16 SHORT ANSWER EXAMPLES

1 **Explain the following terms :** (i) Quality; (ii) quality improvement; (iii) statistical quality control; (iv) statistical process control; (v) chance causes and assignable;

(i) **Quality** means **Fitness** for use. Quality or fitness for is determined through the interactive of quality of design and quality of performance. **Quality of design** pertains to the different grades or levels of performance, reliability. Serviceability and function that are the result of deliberate engineering and management decisions. **Quality of performance** is the systematic reduction of variability and elimination of defects until every unit produced is identical and defective free.

(ii) **Quality improvement** means the systematic elimination of waste. It is necessary to build quality into the product by concentrating on the equipment components, and materials that go into production. Statistical methods play a vital role in quality improvement pertaining to:

- Product design and development.
- Determine the capability of a manufacturing process.
- Investigate improvements in the process which can lead to higher yields and lower manufacturing costs.
- Provide reliability and other performance measures about the product.

(iii) **Statistical quality control** The term pertains to the statistical and engineering methods and to measure, monitor, control and improve quality.

(iv) **Statistical process control** is a set of problem- solving tools based on data that are used to improve a process. The major tools of statistical process control are Histogram; Pareto chart; cause and effect diagram; Defect-concentration diagram; Control chart; Scatter diagram; and check sheet. Its goal is the elimination of variability in the process.

(v) **Chance causes and assignable causes.**

In any production process a certain amount of inherent or natural variability will always exist. This natural variability is often called a "stable system of chance causes". The portion of variability in a set of observations that is due to only random forces and which can not be traced to specific sources, such as operators, materials or equipment. In general, it is also called a common cause.

The other type of variations that are not part of the chance cause pattern are termed as assignable causes. This portion of variability in a set of observations that can be traced to specific causes, such as operators, materials or equipment. These will produce adverse effects on the quality of items produced.

2. Mention the most important use of a control chart.

The most important use of a control chart is to improve the process.

In general most processes do not operate in a state of statistical control. Consequently, the routine and alternative use of control charts will identify assignable causes. If these causes can be eliminated from the process, variability will be reduced and the process will be improved. The control chart will only detect assignable causes. Management, operator and engineering action will be able to eliminate the assignable cause. It is essential to have an action plan for responding to control chart signals.

3. What is a control chart? Explain briefly.

Control chart is a graphical display to monitor a process. It usually consists of a horizontal central line corresponding to the in-control value of the parameter that is being monitored and lower and upper control limits. The control limits are determined by statistical criteria and are nor arbitrary nor are they related to specification limits. If sample points fall within the control limits, the process is said to be in control, or free from assignable causes. Points beyond the control limits indicate an out-of-control process; that is assignable causes are likely present. This signals the need to find and remove the assignable causes.

4. What are the two main types of control charts?

Controls charts are classified into two general types: variables control charts and attributes control charts.

Control charts for central tendency and variability are collectively called variables control charts.

The $\bar{X}$ chart is the most widely used chart for monitoring the central tendency where as charts based on either the sample range or the sample standard deviation are used to control process variability.

Variable control charts are for continuous random variable.

Many quality characteristics are not measured an a continues seale or even a quantitative scale. In these cases, we may judge each unit of product as either conforming or nonconforming on the basis of whether or not it possesses certain attributes or we may count the number of nonconformities(defects) appearing on a unit of product. Control charts for such quality characteristics are called attributes control charts.

5. Mention some of the main reasons for the popularity of control charts.

- The following are some of the main reasons for the popularity of control charts.
- Control charts are proven technique for improving productivity.
- Control charts are effective in defect prevention.
- Control charts prevent unnecessary process adjustments.
- Control charts provide diagnostic information.

- Control charts provide information about process capability.

6. Mention the four steps for continued product improvement.

The four steps essential for continued prcduct improvement are

- Design a product
- Make it
- Market it
- Find out the purchasers' reactions to the product.

7. Define the terms (i) Tolerance Interval and (iii) Tolerance limits.

Tolerance Interval is an interval that contains a specified proportion of a population with a stated level of confidence.

Tolerance limits indicate a set of limits between which some stated proportion of the values of a population must fall with a specified level of confidence.

12.17 Exercises in Quality Control

1. (a) Define quality cost and explain the broad categories of quality cost for an industry.
   (b) "Concentration on quality affects quantity". Is it true? Critically analyze this statement.
2. (a) Explain briefly the reporting system with regard to quality for an industry with which you are familiar.
   (b) "Quality is everybody's responsibility". Explain with illustrative examples.
3. Clearly and briefly explain each of the following concepts.
   (a) AQL and LTPD;
   (b) Producer's and consumer's risks
   (c) A.S.N and A.T.I
   (d) The O.C curve of a sampling plan;
   (e) Total quality control;
   (f) Type I and Type II errors
   (g) Quality assurance;
   (h) Interpretation of $\overline{X}$ and R charts
   (i) Rational subgroups;
   (j) Sampling frequency and sample sizes for control charts
4. A sample of 5 screws are drawn from a lot with fraction defective θ. The lot is accepted if the sample contains
   (a) no defective screws
   (b) at most one defective screws
   [use binomial distribution]

   [Ans (a) $(1-\theta)^5$ (b) $(1-\theta)^5 + 5\theta(1-\theta)^4$]

5. (a) If in a single sampling plan for large lots of spark plugs, the sample sizes 100 and it is desired the AQL to be 5% and producer's risk 3%, what acceptance number c should be chosen?

(b) What is the consumer's risk if RQL to be 10%.

6. Large lots of batteries for pocket calculators are inspected according to the following plan n = 30 batteries are randomly drawn from a lot and tested. If this sample contains at most c = 1 defective battery, the lot is accepted. Other wise it is rejected. Graph the OC curve of the plan, using Poisson approximation. Graph the AOQ and determine the AOQL

7. (a) Design a single sampling plan given the following data.

$\alpha = 10$ Percent, $\beta = 10$ percent, AQL = 1 percent and LTPD = 5. (b) Percent Design a sequential sampling plan for the same data given in (a).

8. (a) Give that the lot size is 1000units per lot and a double sampling plan of $n_1 = 50$, $n_2 = 90$, $C_1 = 1$ and $C_2 = 4$, find the probability of acceptance of the lot on the first sample and on the second sample, if the fraction defective of the incoming material is 4 percent.

(b) Assuming rectifying inspection programe with defectives replaced with good ones, find the values of AOQ (Average Outgoing Quality) and AI (Average number inspected).

9. For the NP Control chart, the central line is at and the control limits are:

$$UCL = n\bar{p} + 3\sqrt{n\bar{p}(1-\bar{p})}$$

$$LCL = n\bar{p} - 3\sqrt{n\bar{p}(1-\bar{p})}$$

and the number of defectives for each sample is plotted an the chart

(a) Verity that the control limit given above are correct

(b) Apply this for the $\bar{X}$ chart with samples of size 4, central line is at 100, and the upper and lower 3σ control limits are at 106 and 94 respectively.

(c) Will this chart always provide results that are equivalent to the usual p chart?

10. Consider the p chart with the usual 3 sigma control limits. Suppose that we define a new valuable:

$$Z_i = \frac{p_i - \bar{p}}{\sqrt{\dfrac{p - \bar{p}}{n}}}$$

as the quantity to plot on a control chart. It is proposed that this new chart will have a central line at o with the upper and lower control limits at ± 3. Verify that this standardized control chart will be equivalent to the original p chart.

11. Describe the construction and use of the $\bar{x}$ and R charts

12. Describe the important aspects of double sampling in acceptance inspection by attributes.

13. Explain any two types of control charts for variables and attributes for detection of lack of control in a continuous flow of manufactured products.

14. Control charts for $\bar{X}$ and R are to be setup for an important quality characteristic. The sample size is 5 and $\bar{x}$ and r are computed for each of 30 preliminary samples. The summary data are:

$$\sum_{i=1}^{30} \bar{x}_i = 6690, \quad \sum_{i=1}^{30} r_i = 1028$$

(a) Find the trial control limits for and R charts.

(b) Assuming that the process is in control estimate the process mean and standard deviation [given for n = 5, A_1 = 1.596, A_2 = 0.477, d_1 = 2.326, D_3 = 0, D_4 = 2.115, c_4 = 0.9400].

15. Control charts are to be constructed for samples of size 4 and $\overline{X}$ and s are computed for each of 25 preliminary samples as follows:

$$\sum_{i=1}^{25} \overline{x}_i = 5575 \ , \ \sum_{i=1}^{25} s_i = 339.5$$

   (a) Determine trial control limits for $\overline{X}$ and S charts.
   (b) Assuming the process is in control, estimate the process mean and standard deviation [given: for n=4, c_4=0.9213].

16. (a) The diameter of parts produced by a machine are measured by taking samples of 5 each. The data obtained for 20 samples is as follows : $\sum \overline{x}_i = 93.42$ and $\sum R_i = 0.83$. The specification limits are 4.5 0.05. Find the percentage of defective components produced by the process.
   (b) Calculate the control limits for mean and range.
   (c) What is the chance of catching the shift if the processes mean shifts to 4.82?

17. A manufacturer of a certain bearing knows from a preliminary record of 10 hourly samples of size 4 for the diameters of there bearings. The following tables given hourly coded sample values

Hour	Coded sample values				$\overline{X}$	R
1	1.9	2.4	2.1	1.4	1.950	1.00
2	1.2	1.6	1.3	1.5	1.400	0.40
3	1.6	2.6	3.0	2.9	2.500	1.40
4	2.0	2.2	1.3	0.3	1.450	1.90
5	1.3	1.4	3.3	1.8	1.95	2.00
6	2.5	2.3	2.3	1.9	2.25	0.60
7	2.1	1.9	2.0	1.6	1.90	0.50
8	1.5	2.2	4.1	1	2.20	3.10
9	1.9	3.8	1.2	1.7	2.15	2.60
10	1.8	0.8	1.2	1.0	1.20	1.00

Find the control chart for $\overline{X}$ and R. Comment on the statistical significance of the process based on the control charts.

18. (a) Units are being assembled, outputting at the rate of 2 per minute. Assembling is considered O.K if not more than 2% of the assemblies are defective. Suggest a control chart procedure to monitor the assembling process.

   (Checking an assembled unit for defectiveness requires about two minutes of testing time)
   (b) The following are the gross weights of packets in grams, taken in samples of size 3 in

half-hour intervals from the output of an automated packing system. Draw the appropriate control charts from these data and interpret the same.

Estimate the standard deviation of the packing system by use of the range charts values or otherwise.

Samples are : (270, 275, 265); (290, 280, 275); (275, 275; 275); (280; 275; 275); (280, 280, 295) (285, 280, 285); (275, 280, 285); (285, 285, 280); (290, 290, 280); (280,285,270);

19. Construct an $\overline{X}$ and R charts for the following data.

Sample number	Observations			
1	75	80	70	65
2	95	85	80	70
3	80	80	80	95
4	85	80	80	90
5	85	85	85	85
6	90	85	90	85
7	80	85	90	85
8	90	80	85	95
9	95	95	85	95
10	85	90	75	90

20. (a) What are the use of and R charts in industrial quality control?
    (b) From a factory producing metal sheets a sample of 4 sheets is taken every hour and the data obtained is as follows:

Sample No.	Tick ness of sheet(cm)				$\overline{X}$	R
1	0.25	0.24	0.22	0.25	0.24	0.03
2	0.32	0.33	0.34	0.29	0.32	0.05
3	0.42	0.39	0.42	0.41	0.41	0.03
4	0.28	0.35	0.30	0.27	0.30	0.08
5	0.30	0.31	0.29	0.26	0.29	0.05
6	0.40	0.35	0.39	0.38	0.38	0.05
7	0.26	0.33	0.31	0.30	0.30	0.07

Draw the control chart for the mean and range and examine whether the process is under control.

21. (a) Construct $\bar{X}$ and R charts using the following data on the diameter of cylindrical cavities in a component. Each sample is of size 4. Plot the points and check for the control.

Sample No.	1	2	3	4	5	6	7	8
$\bar{X}$	0.0975	0.1020	0.0977	0.1020	0.1007	0.1063	0.1022	0.1003
$\bar{R}$	0.011	0.009	0.015	0.014	0.028	0.009	0.013	0.026

22. (a) Compare $\bar{X} - R$ charts, p charts.
    (b) The diameter of parts produced by a machine are measured by taking samples of 5 each.

    The data obtained for 20 samples is as follows : $\Sigma \bar{X} = 93.42$, $\Sigma \bar{R} = 0.83$. The specification limits are 4.5 ± 0.05. Find the percentage of defective components produced by the process and also

    (i) Calculate the control limits for mean and range.
    (ii) What is the chance of catching the shift is the process shifts to 4.82.

23. (a) Explain the term O.C.C for a mean chart.
    (b) From a process 100 samples were tested. The percentage of defectives was found to be 5%.
        (i) Construct the control limits.
        (ii) If the process mean should shift to 5.6%, what is the probability of catching the shift in the first sample after the shift?
    (c) Define the OC function $L(\mu)$ and show that $L(\mu)$ is a strictly decreasing function of μ.
    (d) If the sample size n = 50 and the probability of Type I error is specified to be 0.05, determine C and $L(\mu)$.

24. The following are the gross weights of packets in grams, taken in samples of size 3 in half hour intervals from the output of an automated packing system. Draw the appropriate control charts from these data and interpret the same.

    Estimate the standard deviation of the packing system by use of the range chart values or otherwise.

Sample number	1	2	3	4	5	6	7	8	9	10
	270	290	275	280	280	285	275	285	290	280
Weights	275	280	275	275	280	280	280	275	290	285
	265	275	275	275	295	285	285	280	280	270

25. (a) What is implied when it is said that a manufacturing process is in a state of statistical control?

(b) What is process capability? Explain as to how this measure can be established for a certain operation on a machine tool.

(c) What are modified control limits? Indicate as to how they are established?

26. (a) A process is known to have a mean of 10.00 cm with a standard deviation of 0.01cm. A control chart using a sample size of 4 is to be established. Specify the control limits for $\overline{X}$, R charts.

(b) If the mean of the process in (a) above shifts to 10.005 cm and if the specification for the product turned out by the process is given as 10 ± 0.63, what proportion of the limits produced will be defective?

27. The following data relates to the inspection report in an electrical machine shop of an industry. The number of impellers inspected and rejected for five days are given below :

Data (in Dec 2008)	:	12th	14th	15th	16th	17th
Number inspected	:	1450	1530	1170	1680	1980
Number rejected	:	175	240	130	240	220

Calculate the control limits. Can you conclude that the process is quite satisfactory?

28. In manufacturing cars the following data were obtained:

Sample size	200	200	215	75	190	210	500	212	200
188									
No. of defectives	20	19	23	8	18	22	51	21	19
19									

Draw the control chart for defectives.

29. The average fraction defective in the number of castings produced in a formulry was 0.10. On one Monday, a sample of 100 was inspected and 18 were found to ben defective and on Tuesday, a sample of 150 units was inspected and 33 were found to be defective. It you are the foundry-in-change can you rust contented? If not, give your comments.

30. The following table gives the results of inspection of a certain instrument during a period of 12 weeks. Draw a p-chart-based on trial control limits and setup new control limits by discarding points falling outside the trial control limits.

Week No	1	2	3	4	5	6	7	8	9	10	11	12
No. of items inspected	110	120	30	40	60	160	25	50	35	180	180	25
No. of defectives	4	8	0	1	2	11	2	7	0	15	4	2

31. Ten samples of n=100 bottles of plastic liquid laundry detergent are inspected in time order of production and the fraction defective in each sample is reported as follows :

0.17, 0.20, 0.07, 0.11, 0, 15 0, 09, 0.11, 0.12, 0.08, 0.14

(a) Setup a p chart for this process. Is the process in statistical control?

(b) Suppose that instead if n=100, it is n=200.Use the data given to set up a P chart for this process. Revise the control limits if necessary.

(c) Compare your control limits for the p charts is parts (a) and (b). Explain why they differ. Also explain why your assessment about statistical control differs for two size of n.

32. A PC production company manufactures cover cases by injection mouldings. Samples of 5 cases are taken from the process periodically and the number of defects is noted. Ten samples save the number of defects as follows:

0, 1, 3, 4, 2, 0, 4, 5, 8, 3

(a) Using all the data, find the trial control of limits for this C chart for the process.

(b) Use the trial control limits from part (a) to identify out of control points. If necessary revise your control limits.

(c) Suppose that instead of samples of 5 cases, the sample was 10 cases. Repeat parts (a) and (b).

Explain how this changes alters your answers to parts (a) and (b)

33. It is required to conduct a single sampling plan with the following specifications :
AQL=0.10, L.T.P.D=0.06. Producer's risk $= \alpha = 0.05$, consumer's risk $= \beta = 0.10$.

(a) Set up the theoretical equations which enable the determination of the values of the sample size n and the acceptable number c and

(b) Using the Poisson approximation to the Binomial, determine the values of n and c by trial and error method.

34. Consider the single sampling plan N = 4000, n = 100, c = 2, in the usual notation.

(a) Construct the OC curve of the plan by plotting the points corresponding to lot quality p = 0.01, 0.02, 0.03, 0.04, 0.06 and 0.08. What is the in difference quality for this plan?

(b) For the above plan, calculate A.O.Q for lots with quality p=0.01, 0.02, 0.03, 0.04, 0.06 and 0.08. Draw a smooth AOQ curve and determine the A.O.Q.L. of the plan.

35. Out line the rational behind the acceptance sampling plan.

For the following plan :

N = 1000, n = 30, c = 2.

Does the operating characteristic and the average outgoing quality curves, assuming any reasonable appropriation (to be stated by you) to actual distribution of the number of defectives.

Indicate how these and other similar charts (curves) that you know of, can help you in chasing one from among many sampling inspection plan.

36. The following is a continuous sampling plan. Calculate the probabilities of acceptance at each stage sampling for an incoming material haring a fraction defective of 2 percent.

Cumulative sample size	20	40	60	80	100	120	140
Acceptance Number	-	-	0	0	1	1	2
Rejection Number	2	2	2	3	3	3	3

Indicate that acceptance is not possible at that stage.

37. It is required to construct a single sampling plan with the following specifications:
AQL=0.10, L.T.P.T=0.06,

Producer's risk $= \alpha = 0.05$,

Consumer's risk $= \beta = 0.10$,

Set up the theoretical equations which enable the determination of the values of the sample size n and the acceptable number c.

Using the Poisson Approximation to the Binomial, determine the values of n and c by trial and error method.

38. Consider the single sampling plan N= 4000, n = 100, c = 2, in the usual notation Construct the OC curve of the plan by plotting the points corresponding to lot quality for this plan?

For the above plan, calculate the AOQ for lots with quality p=0.01, 0.02, 0.03, 0.04, 0.06 and 0.08. Draw a smooth AOQ curve and determine the AOQL. of the plan.

39. (a) Explain is detail $\overline{X}$ and R charts. What are their advantages over p charts?

(b) The following 20 figures correspond to the number of defects in 20 sheets selected as 20 samples during a process of manufacturing metal sheets. Plot a suitable control chart.

29, 26, 20, 19, 14, 16, 22, 23, 14, 18

22, 24, 25, 18, 17, 15, 20, 16, 19, 23

40. The standard for a process producing tinplate back in a continuous strip is 6 defects in the from of pinholes per 30 meters. 25 observations are made giving the number of defects per 30 meters. They are 4, 3, 3 5, 5, 5, 7, 5, 2, 8. 6, 6, 5, 7, 7, 10, 6, 3, 7, 6, 12, 7, 7, 9, 3

Can it be concluded that the process is in control to this standard?

[Ans: Control line for c chart is 5.92, UCL = 13.22; LCL = 0. All the sampling items have the number of defects less than 13. Hence the process is in control].

41. Units are being assembled, out-putting at the rate of 2 per minute. Assembling is considered O.K, if more than 2 percent of the assemblies are defective. Suggest a control chart procedure to monitor the assembling process.

(Checking an assembled unit for defectives requires about two minutes of testing time).

CHAPTER 13

NUMERICAL METHODS AND APPLICATIONS

13.1 INTRODUCTION

Numerical methods were developed as early as 1930 in solving problems for which direct methods were either not suitable or not directly applicable. The branch of mathematics concerned with the development of such methods is called **Numerical Analysis.** The development of numerical analysis is influenced and determined by the advent of electronic computer in early 1960. The wide spread use of numerical methods in conjunction with computer helped scientists and engineers to solve several types of engineering problems. The solution of a problem involving numerical work consists of **modeling, choice of numerical methods, programming, operation** and **interpretation of results.**

13.2 DETERMINATION OF THE ROOTS OF AN EQUATION

Consider an equation of the form $f(x) = 0$. The function $f(x)$ may be given explicitly either by a polynomial $P_n(x)$ of degree n in x and

$f(x) = P_n(x) = x^n + a_1 x^{n-1} + a_2 x^{n-2} + \cdots + a_{n-1} x + a_n$ or as a transcendental function.

A number ξ is a solution of $f(x) = 0$ if $f(\xi) = 0$. Such a solution ξ is called a **root** or **zero** of $f(x) = 0$.

There are generally two types of methods used to find the roots of the equation $f(x) = 0$.

One is the direct method which gives the exact value of the roots in a finite number of steps. The other is the **iterative method** based on the idea of successive approximation, starting with an initial approximation to the root and getting a sequence of iterates which in the limit converges to the root.

Secant Method and Newton - Raphson Method are used in iteration methods to solve a first degree equation.

Iteration methods based on second degree equation are: The **Muller Method**, the **Chebyshev Method** and **Multipoint Iteration Method.**

The above iteration methods are special cases of the following

$$x_{k+1} = \phi(x_k), \quad k = 0, 1, 2, \cdots \tag{1}$$

The function $\phi(x)$ is known as an **iteration function**. The convergence of an iteration method depends on the suitable choice of the function $\phi(x)$ and x_0, a suitable initial approximation to the root.

Higher Order Methods

The iteration method (1) is said to be of the *p*-th order if

$$\phi'(\xi) = \phi''(\xi) = \cdots = \phi^{(p-1)}(\xi) = 0, \quad \phi^p(\xi) \neq 0, \text{ where } \xi \text{ is a solution of } x = \phi(x).$$

13.3 SYSTEMS OF LINEAR ALGEBRAIC EQUATIONS

Consider a system of n linear algebraic equations in n unknowns

$$a_{11}x_1 + a_{12}x_2 + \cdots + a_{1j}x_j + a_{1n}x_n = b_1$$

$$a_{21}x_1 + a_{22}x_2 + \cdots + a_{2j}x_j + a_{2n}x_n = b_2$$

$$\cdots \cdots \cdots \cdots \cdots \cdots \cdots \cdots \cdots \cdots \cdots \cdots$$

$$a_{i1}x_1 + a_{i2}x_2 + \cdots + a_{ij}x_j + a_{in}x_n = b_i$$

$$\cdots \cdots \cdots \cdots \cdots \cdots \cdots \cdots \cdots \cdots \cdots \cdots$$

$$a_{n1}x_1 + a_{n2}x_2 + \cdots + a_{nj}x_j + \cdots + a_{nn}x_n = b_n$$

or $\displaystyle\sum_{j=1}^{n} a_{ij}x_j = b_i, \quad i = 1, 2, \cdots n,$ \hfill (1)

where a_{ij}, $i, j = 1(1)\, n$ are the known coefficients, b_i, $i = 1(1)\ n$ are the known values and x_j, $(i = 1(1)n)$ are the unknowns to be determined.

The system (1) is called **homogeneous** if all b_i are zero; otherwise it is said to be **nonhomogeneous.**

The system (1) may be written as a single vector equation

$$\mathbf{Ax} = \mathbf{b} \qquad (2)$$

where the coefficient matrix $A = \left[a_{ij} \right]$ is the $n \times n$ matrix.

A **solution** of (1) is a set of numbers which satisfy all the n equations and a **solution vector** of (1) is a vector x whose components constitute a solution of (1).

13.4 DIRECT METHODS

The several direct methods can be classified as follows:

1. Cramer's rule or method of determinants
2. Method of leading coefficients
3. Gaussian elimination method
4. Gauss - Jordan elimination method
5. Crout's Method
6. Triangularization method
7. Cholesky method
8. Matrix inversion method
9. Eigin values and eigenvectors method

13.4.1 *Cramer's Rule Method*

In order to determine the unknown in the equations (1), we multiply the equations by the cofactors $A_{i1}, i = (1)n$ and add these together to get

$$\left(\sum_{i=1}^{n} a_{i1} A_{i1} \right) x_1 + \sum_{j=2}^{n} \left(\sum_{i=1}^{n} a_{ij} A_{i1} \right) x_j = \sum_{i=1}^{n} b_i A_{i1}$$

which gives

$$|A| \; x_1 = b_1 A_{11} + b_2 A_{21} + \cdots + b_n A_{n1}$$

Similarly, we may obtain

$$|A| \; x_2 = b_1 A_{12} + b_2 A_{22} + \cdots + b_n A_{n2}$$

$$|A| \; x_3 = b_1 A_{13} + b_2 A_{23} + \cdots + b_n A_{n3}$$

$$\vdots$$

$$|A| \; x_n = b_1 A_{1n} + b_2 A_{2n} + \cdots + b_n A_{nn}$$

Thus we have $\quad x_j = \left| \dfrac{\mathbf{B}_i}{\mathbf{A}} \right|, \quad i = 1(1)n,$

Where $|B_i|$ is the determinant of the matrix obtained by replacing the ith column of **A** by the right hand vector **b**. The process is known as the Cramer rule.

All the above equations may be written as

$$x = \frac{1}{|\mathbf{A}|}\begin{bmatrix} A_{11} & A_{21} & \cdots & A_{n1} \\ A_{12} & A_{22} & \cdots & A_{n2} \\ \vdots & & \cdots & \vdots \\ A_{1n} & A_{2n} & \cdots & A_{nn} \end{bmatrix}\begin{bmatrix} b_1 \\ b_2 \\ \vdots \\ b_n \end{bmatrix} = \frac{1}{|A|}\text{adj }(A)\mathbf{b} = \mathbf{A}^{-1}\mathbf{b}$$

Example

EXAM.1: Solve the equations: $2x_1 + x_2 + 4x_3 = 4$; $x_1 - 3x_2 - x_3 = -5$; $3x_1 - 2x_2 + 2x_3 = -1$ by Cramer's rule.

SOLUTION

(a) By Cramer's rule

$$|A| = \begin{vmatrix} 2 & 1 & 4 \\ 1 & -3 & -1 \\ 3 & -2 & 2 \end{vmatrix} = 7; \quad |B_1| = \begin{vmatrix} 4 & 1 & 4 \\ -5 & -3 & -1 \\ -1 & -2 & 2 \end{vmatrix} = 7; \quad |B_2| = \begin{vmatrix} 2 & 4 & 4 \\ 1 & -5 & -1 \\ 3 & -1 & 2 \end{vmatrix} = 14;$$

$$|B_3| = \begin{vmatrix} 2 & 1 & 4 \\ 1 & -3 & -5 \\ 3 & -2 & -1 \end{vmatrix} = 0$$

There fore $x_1 = \dfrac{|B_1|}{|A|} = \dfrac{7}{7} = 1;$ $\quad x_2 = \dfrac{|B_2|}{|A|} = \dfrac{14}{7} = 2$ and $x_3 = \dfrac{|B_3|}{|A|} = \dfrac{0}{7} = 0$

The solution is $x_1 = 1;$ $x_2 = 2;$ $x_3 = 0$

(b) By Matrix inversion

Also the cofactors of the elements $a_{ij}, i, j = 1(1)3$ are given by:

$$A_{11} = -8; \quad A_{12} = -5; \quad A_{13} = 7;$$

$$A_{21} = -10; \quad A_{22} = -8; \quad A_{23} = 7; \quad A_{31} = 11; \quad A_{32} = 6; \quad A_{33} = -7$$

$$\mathbf{x} = \frac{1}{7}\begin{bmatrix} -8 & -10 & 11 \\ -5 & -8 & 6 \\ 7 & 7 & -7 \end{bmatrix}\begin{bmatrix} 4 \\ -5 \\ -1 \end{bmatrix} = \begin{bmatrix} 1 \\ 2 \\ 0 \end{bmatrix}$$

$\therefore x_1 = 1;$ $x_2 = 2;$ $x_3 = 0$

13.4.2. *Method of Leading Coefficients*

In the given set of equations, the leading coefficients are rendered +1 or -1 in step 1. In step 2, the number of unknowns are reduced from n to $(n - 1)$. In step 3, the leading coefficients are again reduced to +1 or -1. In step 4, the number of unknowns are further reduced from $(n - 1)$ to $(n - 2)$.

The process is repeated until only one unknown is left and its value is determined. The other unknowns are evaluated by basic substitution.

The method is illustrated by the following example.

Example

Solve

$$2x - y + 3z = 8 \tag{1}$$

$$-x + 2y + z = 4 \tag{2}$$

$$3x + y - 4z = 0 \tag{3}$$

Step 1: Make the leading coefficients as +1 or -1 by dividing equation (1) by 2

$$x - 0.5y + 1.5z = 4 \tag{4}$$

$$-x + 2y + z = 4 \tag{5}$$

$$x + \frac{1}{3}y - \frac{4}{3}z = 0 \tag{6}$$

Step 2: Reduce the unknowns from n to $(n - 1)$

Here reduce the number of equations from 3 to 2

$$\text{Eq (4)} + \text{Eq (5)} \Rightarrow 1.5y + 2.5z = 8 \tag{7}$$

$$\text{Eq (5)} + \text{Eq (6)} \Rightarrow \frac{7}{3}y - \frac{1}{3}z = 4 \tag{8}$$

Step 3: Make the leading coefficient +1 or -1

$$[\text{Eq (7)} / 1.5] \Rightarrow y + \frac{5}{3}z = \frac{16}{3} \tag{9}$$

$$[\text{Eq (8)} / \ ] \Rightarrow y - \frac{1}{7}z = \frac{12}{7} \tag{10}$$

Step 4: Reduce the equations to one equation

$$y + \frac{5}{3}z = \frac{16}{3} \qquad \therefore \quad \frac{38}{21}z = \frac{76}{21} \quad \text{or} \ z = 2$$

$$-y + \frac{1}{7}z = -\frac{12}{7}$$

$\therefore$ From (10), $\quad y - \dfrac{2}{7} = \dfrac{12}{7}$ or $y = 2$

and from (6) $\quad x + \dfrac{2}{3} - \dfrac{8}{3} = 0$ or $x = 2$

13.4.3. *Gaussian Method of successive elimination*

One of the most important methods is the so called Gauss algorithm, which is a systematic elimination method. In this method for solving system of linear equations

$$\left.\begin{array}{l} a_{11}x_1 + a_{12}x_2 + \cdots + a_{1j}x_j + a_{1n}x_n = b_1 \\ a_{21}x_1 + a_{22}x_2 + \cdots + a_{2j}x_j + a_{2n}x_n = b_2 \\ \cdots\cdots\cdots\cdots\cdots\cdots\cdots\cdots\cdots\cdots \\ a_{n1}x_1 + a_{n2}x_2 + \cdots + a_{nj}x_j + \cdots + a_{nn}x_n = b_n \end{array}\right\} \quad (1)$$

we proceed as follows

Step 1:

Elimination of x_1 from the second, third ... n^{th} equation. We assume here that the order of the equations and the order of the unknown in each equation are such that $a_{11} \neq 0$. The variable x_1 can then be eliminated from the second .... equation by subtracting (a_{21}/a_{11}) times the first equation from the second equation, (a_{31}/a_{11}) times the first equation from the third equation, and so on. This new system as follows:

$$\left.\begin{array}{l} a_{11}x_1 + a_{12}x_2 + \cdots + a_{1n}x_n = b_1 \\ a_{22}^1 x_2 + \cdots + a_{2n}^1 x_n = b_2^1 \\ \cdots\cdots\cdots\cdots\cdots\cdots\cdots\cdots \\ a_{n2}^1 x_2 + \cdots + a_{nn}^1 x_n = b_n^1 \end{array}\right\} \quad (2)$$

Step 2:

Elimination of x_2 from the third, ..., n^{th} equation in (2)

If the coefficients $a_{22}^1, \cdots a_{nn}^1$ in (2) are not all zero, we may assume that the order of the equations and the unknowns is such that $a_{22}^1 \neq 0$. Then we may eliminate x_1 from the third, fourth, ... n^{th} equation of (2) by subtracting (a_{32}^1/a_{22}^1) time the second equation from the third equation, (a_{42}^1/a_{22}^1) times the second equation from, the fourth equation, etc. The further steps are now obvious. In the third step we eliminates and in the fourth step, we eliminate x_4, etc.

By successive elimination, we arrive at a single equation in the unknown x_n which can be solved and substituting this in the proceeding equation we obtain the value of x_{n-1}. In this manner we find x_1 when the elimination is completed. Also when the elimination is complete, the system takes the form

$$\left.\begin{array}{r}c_{11}x_1 + c_{12}x_2 + \cdots + c_{1n}x_n = d_1 \\ c_{22}x_2 \cdots + c_{2n}x_n = d_2 \\ \cdots\cdots\cdots\cdots\cdots\cdots\cdots \\ c_{nn}x_n = d_n\end{array}\right\} \qquad (3)$$

The above equations are in (upper) triangular form.

In this case there exists a unique solution. The new coefficient matrix is an upper triangular matrix; the diagonal elements e_{ii} are usually equal to 1.

EXAM. 1: Solve the following equations by Gaussian elimination method

$$3x_1 + x_2 + x_3 = 4 \qquad (1)$$

$$x_1 + 4x_2 - x_3 = 5 \qquad (2)$$

$$x_1 + x_2 - 6x_3 = -12 \qquad (3)$$

Step 1: Consider the equation having the largest coefficient of x_1

$$3x_1 + x_2 + x_3 = 4 \text{ or } 3x_1 = 4 - x_2 - x_3$$

$$x_1 = \frac{4}{3} - \frac{1}{3}x_2 - \frac{1}{3}x_3 \qquad (4)$$

Step 2: Substitute equation (4) in equations (2) and (3)

$$1.33 - 0.33x_2 - 0.33x_3 + 4x_2 - x_3 = -5 \text{ or}$$

$$3.67x_2 - 1.33x_3 = -6.33 \qquad (5)$$

$$1.33 - 0.33x_2 - 0.33x_3 + x_2 - 6x_3 = -12 \text{ or}$$

$$0.67x_2 - 6.33x_3 = -13.33 \qquad (6)$$

Step 3: Consider the equation (5) which has the largest coefficient of x_2

$$3.67x_2 - 1.33x_3 = -6.33$$

$$3.67x_2 = -6.33 + 1.33x_3$$

$$x_2 = -1.72 + 0.362x_3 \qquad (7)$$

Step 4: Substitute equation (7) in equation (6)

$$0.67 \; (-1.72 + 0.362x_3) - 6.33x_3 = -13.33$$

Therefore $x_3 = 2$

Step 5: By back substitution, evaluate x_1 and x_2.

From equation (7) $x_2 = -1.72 + 0.362 x_3$

$$= -1.72 + 0.363(2) = -1$$

From equation (4) $x_1 = [1.33 - 0.33(-1) - 0.33(2)] = 1$

Therefore $x_1 = 1$; $x_2 = -1$; and $x_3 = 2$

EXAM. 2: Solve the equations

$$10x_1 - x_2 + 2x_3 = 4$$

$$x_1 + 10x_2 - x_3 = 3$$

$$2x_1 + 3x_2 + 20x_3 = 7$$

Using the Gauss elimination method
SOLUTION:

The system is diagonally dominant and no pivoting is necessary. We have, after the first elimination state

$$10x_1 - x_2 + 2x_3 = 4$$

$$\frac{101}{10} x_2 - \frac{12}{10} x_3 = \frac{26}{10}$$

$$\frac{32}{10} x_2 + \frac{196}{10} x_3 = \frac{62}{10}$$

In the second elimination stage, we have

$$10x_1 - x_2 + 2x_3 = 4$$

$$\frac{101}{10} x_2 - \frac{12}{10} x_3 = \frac{26}{10}$$

$$\frac{20180}{1010} x_3 = \frac{5430}{1010}$$

Using back substitution, we get the solution

$$x_3 = 0.269 \; ; \quad x_2 = 0.2899 \text{ and } x_1 - 0.375$$

13.4.4 Gauss - Jordan Elimination Method

In the Gauss - Jordan method the given matrix is diagonalized by row operation so that the solution is directly obtained. This process of diagonalization eliminates the necessity of using back substitution process used in Gauss's method. In contrast to the Gaussian elimination, where the unknown is eliminated from all equation except the pivotal equation in the Gauss - Jordan method the unknown is eliminated from all the equations.

The diagonalization process is implemented using the following stepwise procedure:

1. In the first step, the element in the first column and first row is made 1 and subsequently the remaining elements in the first column are reduced to zero.
2. In the second step, the element in the second row and second column is reduced to 1 and the remaining elements in the second column are subsequently reduced to zero.
3. In the last step, the element in the third row and third column is reduced to unit followed by zeros in the third column.
4. In this way the diagonalization is done in a sequential manner and only row operation are performed

EXAM. 1: Apply Gauss - Jordan method to solve the following equations:

$$x + 2y + z = 8 \; ; \quad 2x + 3y + 4z = 20 \; ; \quad 4x + 3y + 2z = 16$$

SOLUTION:

The given equations are written in matrix form as

$$\begin{bmatrix} 1 & 2 & 1 \\ 2 & 3 & 4 \\ 4 & 3 & 2 \end{bmatrix} \begin{bmatrix} x \\ y \\ z \end{bmatrix} = \begin{bmatrix} 8 \\ 20 \\ 16 \end{bmatrix}$$

By successive row operations, the original matrix is reduced to a coefficient matrix by the procedure detailed below

Consider the augmented matrix

$$\begin{bmatrix} 1 & 2 & 7 & | & 8 \\ 2 & 3 & 4 & | & 20 \\ 4 & 3 & 2 & | & 16 \end{bmatrix} \begin{matrix} \\ r_2 - 2r_1 \\ r_3 - 4r_1 \end{matrix} \Rightarrow \begin{bmatrix} 1 & 2 & 1 & | & 8 \\ 0 & -1 & 2 & | & 4 \\ 0 & -5 & -2 & | & -16 \end{bmatrix} \Rightarrow \begin{bmatrix} 1 & 2 & 1 & | & 8 \\ 0 & 1 & -2 & | & -4 \\ 0 & 5 & 2 & | & 16 \end{bmatrix} \begin{matrix} r_1 - 2r_2 \\ \\ r_3 - 5r_2 \end{matrix}$$

$$\Rightarrow \begin{bmatrix} 1 & 0 & 5 & | & 16 \\ 0 & 1 & -2 & | & -4 \\ 0 & 0 & 12 & | & 36 \end{bmatrix} \Rightarrow \begin{bmatrix} 1 & 0 & 5 & | & 16 \\ 0 & 1 & -2 & | & -4 \\ 0 & 0 & 1 & | & 3 \end{bmatrix} \begin{matrix} r_1 - 5r_3 \\ r_2 + 2r_3 \end{matrix} \Rightarrow \begin{bmatrix} 1 & 0 & 0 & | & 1 \\ 0 & 1 & 0 & | & 2 \\ 0 & 0 & 1 & | & 3 \end{bmatrix}$$

$\therefore \quad x = 1, \quad y = 2, \quad z = 3$

The Gauss - Jordan method gives

$$[\mathbf{A} \mid \mathbf{b}] \xrightarrow[\text{Jordan}]{\text{Gauss}} [\mathbf{I} \mid \mathbf{d}]$$

Generally this method is not used for the solution of a system of equation as it is more expensive from the computation view - point than the Gauss elimination method. However, it gives a simple method for finding the inverse of a given matrix.

When the Gauss - Jordan procedure is completed, we obtain

$$[\mathbf{A}\,|\,\mathbf{I}]\xrightarrow[\text{Jordan}]{\text{Gauss}}[\mathbf{I}\,|\,\mathbf{A}^{-1}]$$

For example, find the inverse of the coefficient matrix of the system

$$\begin{bmatrix} 1 & 1 & 1 \\ 4 & 3 & -1 \\ 3 & 5 & 3 \end{bmatrix}\begin{bmatrix} x_1 \\ x_2 \\ x_3 \end{bmatrix}=\begin{bmatrix} 1 \\ 6 \\ 4 \end{bmatrix}$$

by the Gauss - Jordan method with partial pivoting and hence solve the system.

Using the augmented matrix $[\mathbf{A}\,|\,\mathbf{I}]$, we obtain

$$\begin{bmatrix} 1 & 1 & 1 & | & 1 & 0 & 0 \\ 4 & 3 & -1 & | & 0 & 1 & 0 \\ 3 & 5 & 3 & | & 0 & 0 & 1 \end{bmatrix}\sim\begin{bmatrix} 1 & \dfrac{3}{4} & \dfrac{1}{4} & | & 0 & \dfrac{1}{4} & 0 \\ 1 & 1 & 1 & | & 1 & 0 & 0 \\ 3 & 5 & 3 & | & 0 & 0 & 1 \end{bmatrix}\sim\begin{bmatrix} 1 & \dfrac{3}{4} & -\dfrac{1}{4} & | & 0 & \dfrac{1}{4} & 0 \\ 0 & \dfrac{1}{4} & \dfrac{5}{4} & | & 1 & -\dfrac{1}{4} & 0 \\ 0 & \dfrac{11}{4} & \dfrac{15}{4} & | & 0 & -\dfrac{3}{4} & 1 \end{bmatrix}$$

$$\sim\begin{bmatrix} 1 & \dfrac{3}{4} & -\dfrac{1}{4} & | & 0 & \dfrac{1}{4} & 0 \\ 0 & 1 & \dfrac{15}{11} & | & 0 & -\dfrac{3}{11} & \dfrac{4}{11} \\ 0 & \dfrac{1}{4} & \dfrac{5}{4} & | & 1 & -\dfrac{1}{4} & 0 \end{bmatrix}\sim\begin{bmatrix} 1 & 0 & -\dfrac{14}{11} & | & 0 & \dfrac{5}{11} & -\dfrac{3}{11} \\ 0 & 1 & \dfrac{15}{11} & | & 0 & -\dfrac{3}{11} & \dfrac{4}{11} \\ 0 & 0 & \dfrac{10}{11} & | & 1 & -\dfrac{2}{11} & -\dfrac{1}{11} \end{bmatrix}$$

$$\sim\begin{bmatrix} 1 & 0 & -\dfrac{14}{11} & | & 0 & \dfrac{5}{11} & -\dfrac{3}{11} \\ 0 & 1 & \dfrac{15}{11} & | & 0 & -\dfrac{3}{11} & \dfrac{4}{11} \\ 0 & 0 & 1 & | & \dfrac{11}{10} & -\dfrac{1}{5} & -\dfrac{1}{10} \end{bmatrix}\sim\begin{bmatrix} 1 & 0 & 0 & | & \dfrac{7}{5} & \dfrac{1}{5} & -\dfrac{2}{5} \\ 0 & 1 & 0 & | & -\dfrac{3}{2} & 0 & \dfrac{1}{2} \\ 0 & 0 & 1 & | & \dfrac{11}{10} & -\dfrac{1}{5} & -\dfrac{1}{10} \end{bmatrix}$$

Therefore the solution of the system is

$$\begin{bmatrix} x_1 \\ x_2 \\ x_3 \end{bmatrix} = \begin{bmatrix} \dfrac{7}{5} & \dfrac{1}{5} & -\dfrac{2}{5} \\ -\dfrac{3}{2} & 0 & \dfrac{1}{2} \\ \dfrac{11}{10} & -\dfrac{1}{5} & -\dfrac{1}{10} \end{bmatrix} \begin{bmatrix} 1 \\ 6 \\ 4 \end{bmatrix} = \begin{bmatrix} 1 \\ \dfrac{1}{2} \\ -\dfrac{1}{2} \end{bmatrix}$$

13.4.5 *Crout's Method*

This method has advantages over the Gauss method in the sense that it requires less computation. We shall explain this method by considering a system of three equations

Consider the system $\mathbf{Ax = b}$ (1)

That is

$$\begin{bmatrix} a_{11} & a_{12} & a_{13} \\ a_{21} & a_{22} & a_{23} \\ a_{31} & a_{32} & a_{33} \end{bmatrix} \begin{bmatrix} x_1 \\ x_2 \\ x_3 \end{bmatrix} = \begin{bmatrix} b_1 \\ b_2 \\ b_3 \end{bmatrix} \Rightarrow \begin{matrix} a_{11}x_1 + a_{12}x_2 + a_{13}x_3 = b_1 \\ a_{21}x_1 + a_{22}x_2 + a_{23}x_3 = b_2 \\ a_{31}x_1 + a_{32}x_2 + a_{33}x_3 = b_3 \end{matrix}$$

The augmented matrix of (1) is $\begin{bmatrix} a_{11} & a_{12} & a_{13} & b_1 \\ a_{21} & a_{22} & a_{23} & b_2 \\ a_{31} & a_{32} & a_{33} & b_3 \end{bmatrix}$ (2)

Then we can find a derived matrix

$$\begin{bmatrix} a_{11}^1 & a_{12}^1 & a_{13}^1 & b_1^1 \\ a_{21}^1 & a_{22}^1 & a_{23}^1 & b_2^1 \\ a_{31}^1 & a_{32}^1 & a_{33}^1 & b_3^1 \end{bmatrix}$$ (3)

which is obtained in the order given below

(i) To determine the first column

$a_{i1}^1 = a_{i1}$ for all i., i.e., the first column of the augmented matrix is the same as the first column of the coefficient matrix **A**.

(ii) To determine the first row

$$a_{1j}^1 = \frac{a_{1j}}{a_{11}}, \quad j = 2,3 \; ; \quad b_1^1 = \frac{b_1}{a_{11}}$$

(iii) To determine the elements of the second column except a_{12}^1

$$a_{i2}^1 = a_{i2} - a_{12}^1 a_{i1}^1 \; ; \quad i = 2,3$$

(iv) To determine the second row except a_{21}^1, a_{22}^1 and $a_{2j}^1 = \left(a_{2j} - a_1^1 \cdot a_{21}^1\right)/a_{22}^1$, $j = 3$

and $b_3^1 = (b_3 - b_1^1 a_{21}^1)/a_{22}^1$

(v) To determine the elements of the third column except a_{13}^1, a_{23}^1

$$a_{33}^1 = a_{33} - a_{23}^1 \cdot a_{32}^1 - a_{13}^1 \cdot a_{31}^1$$

(vi) To determine the remaining element of the third row

$$b_3^1 = (b_3 - b_2^1 \cdot a_{22}^1 - b_1^1 \cdot a_{31}^1)/a_{33}^1$$

The solution is given by

$$x_3 = b_3^1; \quad x_2 = b_2^{1'} - a_{22}^{1'} \cdot x_3; \quad x_1 = b_1^1 - a_{13}^1 x_3 - a_{12}^1 x_2$$

EXAM. 1 : Solve the problem $\begin{cases} x_1 + x_2 + x_3 = 1 \\ 3x_1 + x_2 - 3x_3 = 5 \\ x_1 - 2x_2 - 5x_3 = 10 \end{cases}$ by Crout's method

SOLUTION:

Coefficient matrix is given by:

$$\begin{bmatrix} a_{11} & a_{12} & a_{13} \\ a_{21} & a_{22} & a_{23} \\ a_{31} & a_{32} & a_{33} \end{bmatrix} = \begin{bmatrix} 1 & 1 & 1 \\ 3 & 1 & -3 \\ 1 & -2 & -5 \end{bmatrix}$$

The augmented matrix of the coefficient matrix is given by

$$\begin{bmatrix} a_{11}^1 & a_{12}^1 & a_{13}^1 & 1 \\ a_{21}^1 & a_{22}^1 & a_{23}^1 & 5 \\ a_{31}^1 & a_{32}^1 & a_{33}^1 & 10 \end{bmatrix}, \tag{1}$$

where $a_{i1}^1 = a_{i1}$, $i = 1, 2, 3$; $a_{11}^1 = 1$; $a_{21}^1 = -3$; and $a_{31}^1 = 1$

$$a_{12}^1 = \frac{a_{12}}{a_{11}} = \frac{1}{1} = 1 \; ; \quad a_{12}^1 = \frac{a_{13}}{a_{11}} = \frac{1}{1} = 1 \; ; \quad B_1^1 = \frac{b_1}{a_{11}} = \frac{1}{1} = 1 \tag{2}$$

$$\left. \begin{array}{l} a_{22}^1 - a_{22} - a_{12}^1 a_{21}^1 = 1 - 1 \times 3 = -2 \\ a_{32}^1 - a_{32} - a_{12}^1 a_{31}^1 = -2 - 1 \times 1 = -3 \end{array} \right\} \tag{3}$$

$$a_{23}^1 = (a_{23} - a_{13}^1 \cdot a_{21}^1)/a_{22}^1 = \frac{-3 - 1.3}{-2} = 3 \; ; \quad b_2^1 = \frac{5 - 1 \times 2}{-2} = -1 \tag{4}$$

$$a_{33}^1 = a_{33} - a_{23}^1 a_{32}^1 - a_{12}^1 a_{21}^1 = -5 - (3)(-3) - (1) \times (1) = 3 \tag{5}$$

$$b_3^1 = (b_3 - b_2^1 a_{32}^1 - b_1^1 a_{31}^1)/a_{33}^1 = 2 \qquad (6)$$

Hence the derived matrix is

$$\begin{bmatrix} 1 & 1 & 1 & 1 \\ 3 & -2 & 3 & -1 \\ 1 & -3 & 3 & 2 \end{bmatrix}$$

Thus $x_3 = b_3^1 = 2$; $x_2 = b_2^1 - a_{23}^1 x_3 = -1 - 3 \times 2 = -7$

and $x_1 = b_1^1 - a_{1x}^1 x_3 - a_{12}^1 x_2 = 1 - x_2 - x_3 = 1 - 2 + 7 = 6$

$\therefore$ The solution is $x_1 = 6$, $x_2 = -7$ and $x_3 = 2$

13.4.6 Triangularization Method

This method is also known as decomposition method. In this method, the coefficient matrix A of the system of equations $\mathbf{Ax = b}$ is decomposed or factorized into the product of a lower triangular matrix $\mathbf{L}$ and an upper triangular matrix $\mathbf{U}$. We write as

$$A = LU \text{ where } L = \begin{bmatrix} l_{11} & 0 & 0 & \cdots & 0 \\ l_{21} & l_{22} & 0 & \cdots & 0 \\ \vdots & & & & \\ l_{n1} & l_{n2} & l_{n3} & \cdots & l_{nn} \end{bmatrix}, \ U = \begin{bmatrix} u_{11} & u_{12} & u_{13} & \cdots & u_{1n} \\ 0 & u_{22} & u_{23} & \cdots & u_{2n} \\ 0 & 0 & u_{33} & \cdots & u_{3n} \\ \cdots & & & \cdots & \\ 0 & 0 & 0 & \cdots & u_{nn} \end{bmatrix}$$

Having determined the matrices $\mathbf{L}$ and $\mathbf{U}$, the system of equation $\mathbf{Ax = b}$ becomes $\mathbf{LUx = b}$

We write this equation $\mathbf{Ux = z}$ and $\mathbf{Lz = b}$

The unknowns $z_1, z_2, \cdots z_n$ are determined by forward substitution and the unknowns $x_1, x_2, \cdots, x_n$ are obtained by back substitution

Alternately, we can find $\mathbf{L}^{-1}$ and $\mathbf{U}^{-1}$ to get $\mathbf{z = L^{-1}b}$ and $\mathbf{x = U^{-1}z}$

The inverse of A can also be determined from $\mathbf{A^{-1} = U^{-1}L^{-1}}$

EXAM. 2: Consider the equations

$$x_1 + x_2 + x_3 = 1$$

$$3x_1 + 4x_2 - 2x_3 = 6$$

$$4x_1 + 3x_2 + 5x_3 = 3$$

Use the decomposition method to solve the system. We write

$$\begin{bmatrix} 1 & 1 & 1 \\ 3 & 4 & -2 \\ 4 & 3 & 5 \end{bmatrix} = \begin{bmatrix} l_{11} & 0 & 0 \\ l_{21} & l_{22} & 0 \\ l_{31} & l_{32} & l_{33} \end{bmatrix} \begin{bmatrix} 1 & u_{12} & u_{13} \\ 0 & 1 & u_{23} \\ 0 & 0 & 1 \end{bmatrix}$$

$$= \begin{bmatrix} l_{11} & l_{11}u_{12} & l_{11}u_{13} \\ l_{21} & l_{21}u_{12} + l_{22} & l_{21}u_{13} + l_{22}u_{23} \\ l_{31} & l_{31}u_{12} + l_{32} & l_{31}u_{13} + l_{32}u_{23} + l_{33} \end{bmatrix}$$

$$\therefore \ l_{11} = 1 \ l_{21} = 3, \ \ l_{31} = 4 \ \ 4 \cdot 1 + l_{32} = 3 \ \ 3 \times 1 + (+1)u_{23} = -2$$

$$u_{12} = 1 \qquad\qquad 3 \cdot 1 + l_{22} = 4 \ \therefore \ l_{32} = -1 \ \therefore \ u_{23} = -5$$

$$4(1) + (-1)(-5) + l_{35} = 5$$

$$l_{33} = -4$$

$$\mathbf{L} = \begin{bmatrix} 1 & 0 & 0 \\ 3 & 1 & 0 \\ 4 & -1 & -4 \end{bmatrix} \text{ and } \mathbf{U} = \begin{bmatrix} 1 & 1 & 1 \\ 0 & 1 & -5 \\ 0 & 0 & 1 \end{bmatrix}$$

13.4.7 *Cholesky Method*

Consider the system of simultaneous equations expressed in matrix form.

$[A][X] = [C]$ and that $(AX - C) = 0$

The above equation can be reduced as $(UX - K) = 0$, where U is the unit upper triangular matrix and hence $L(UX - K) = 0$

Therefore $L(UX - K) = AX - C$

where L is the lower triangular matrix. Hence we get

$L(UX - K) = AX - C$

$L \cdot K = C$

Hence the above equations can be written in the form

$$\begin{bmatrix} a_1 & a_2 & a_3 & c_1 \\ b_1 & b_2 & b_3 & c_2 \\ c_1 & c_2 & c_3 & c_3 \end{bmatrix} = \begin{bmatrix} l_{11} & 0 & 0 \\ l_{21} & l_{22} & 0 \\ l_{31} & l_{32} & l_{33} \end{bmatrix} \begin{bmatrix} 1 & u_{12} & u_{13} & K_1 \\ 0 & 1 & u_{23} & K_2 \\ 0 & 0 & 1 & K_3 \end{bmatrix}$$

Hence by multiplying and equating the corresponding coefficients, we get the unknowns. This method is efficient for real, symmetric and positive definite matrix.

EXAM. 3: Solve the following system of equations by Cholesky's method

$$3x + y + z = 2 \; ; \; x + 4y + z = 12 \; ; \; 2x + y + 2z = 10$$

The coefficient matrix (A) is expressed as

$$\begin{bmatrix} 3 & 1 & 1 & 2 \\ 1 & 4 & 1 & 12 \\ 2 & 1 & 2 & 10 \end{bmatrix} = \begin{bmatrix} l_{11} & 0 & 0 \\ l_{21} & l_{22} & 0 \\ l_{31} & l_{32} & l_{33} \end{bmatrix} \begin{bmatrix} 1 & u_{12} & u_{13} & K_1 \\ 0 & 1 & u_{23} & K_2 \\ 0 & 0 & 1 & K_3 \end{bmatrix}$$

Multiplying and equating the corresponding coefficients, we get

$$\begin{bmatrix} 3 & 0 & 0 \\ 1 & 3.67 & 0 \\ 2 & 0.34 & 1.28 \end{bmatrix} \begin{bmatrix} 1 & 0.33 & 0.33 & 0.67 \\ 0 & 1 & 0.18 & 3.08 \\ 0 & 0 & 1 & 5.953 \end{bmatrix}$$

The variables are determined as

$x_3 = 5.953 \; ; \; x_2 + 0.18x_3 = 3.08 \; \therefore \; x_2 = 2 \cdot 008$

and $x_1 + (0.33 \times 2.008) + (0.33 \times 5.953) = 0.67$ or $x_1 = -1.96$

Hence the values of the variable are

$x_1 = -1.96, \quad x_2 = 2.00$ and $x_3 = 5.95$

13.4.8 Matrix Inversion Method

In the matrix inversion method the given set of equations are expressed in the matrix form and the solution is determined by finding the inverse of the coefficient matrix. The procedure is outlined as follows: Given a set of simultaneous equations,

$$a_{11}x + a_{12}y + a_{13}z = k_1 \; ; \quad a_{21}x + a_{22}y + a_{23}z = k_2 \; ; \quad a_{31}x + a_{32}y + a_{33}z = k_3$$

We write these equations in matrix form given by

$$\begin{bmatrix} a_{11} & a_{12} & a_{13} \\ a_{21} & a_{22} & a_{23} \\ a_{31} & a_{32} & a_{33} \end{bmatrix} \begin{bmatrix} x \\ y \\ z \end{bmatrix} \quad \text{or} \quad [A][X] = [K]$$

Therefore $X = A^{-1}[K]$ where A^{-1} is called the inverse of the matrix A. Matrix inversion is defined so that $A^{-1} \cdot A = A \cdot A^{-1} = I = $ Identity matrix .

The inverse of the matrix is determined by finding the "adjoint" and the "determinant of the matrix. The inverse of the matrix is defined as

$$A^{-1} = \frac{\text{adj } A}{|A|}, \quad \text{adj } A = \begin{bmatrix} A_{11} & A_{12} & A_{13} \\ A_{21} & A_{22} & A_{23} \\ A_{31} & A_{32} & A_{33} \end{bmatrix}$$

= Transpose of the co-factors of elements of matrix A.

The determinant of A is expressed as $|A|$ and is evaluated as

$$|A| = a_{11}A_{11} + a_{21}A_{21} + a_{31}A_{31}.$$

Or in general for a square matrix of order n,

$$|A| = \sum_{j=1}^{n} a_{ij}A_{ij}, \quad i = 1, 2, \cdots n \text{ (rows)}$$

$$\sum_{i=1}^{n} a_{ij}A_{ij}, \quad j = 1, 2, \cdots n \text{ (columns)}$$

EXAM. 1: Solve the simultaneous equation by the matrix inversion method

$x + 2y - z = -3$; $4x - 3y + 4z = 1$;

SOLUTION:

In matrix notation the above equations are written as

$$\begin{bmatrix} 1 & 2 & -1 \\ 4 & -3 & 4 \\ 2 & -1 & 1 \end{bmatrix}\begin{bmatrix} x \\ y \\ z \end{bmatrix} = \begin{bmatrix} -3 \\ 1 \\ -2 \end{bmatrix} \quad \text{i.e. } AX = K \text{ or } X = A^{-1}K_j$$

where A^{-1} is the inverse of matrix A.

The cofactors of the matrix are determined as follows

$$A_{11} = (-1)^2 \begin{vmatrix} -3 & 4 \\ -1 & 1 \end{vmatrix} = -3 + 4 = 1, \cdots \text{ and } A_{33} = (-1)^6 \begin{vmatrix} 1 & 2 \\ 4 & -3 \end{vmatrix} = -11$$

$$\therefore \text{ Adj } A = \begin{bmatrix} A_{11} & A_{12} & A_{13} \\ A_{21} & A_{22} & A_{23} \\ A_{31} & A_{32} & A_{33} \end{bmatrix}^T = \begin{bmatrix} 1 & 4 & 2 \\ -1 & 3 & 5 \\ 5 & -8 & 11 \end{bmatrix}^T = \begin{bmatrix} 1 & -1 & 5 \\ 4 & 3 & -8 \\ 2 & 5 & 11 \end{bmatrix}$$

$$|A| = a_{11}A_{11} + a_{21}A_{21} + a_{31}A_{31} = T(1) + 4(-1) + 2(5) = 7$$

$$\therefore A^{-1} = \frac{\text{adj } A}{|A|} = \frac{1}{7}\begin{bmatrix} 1 & -1 & 5 \\ 4 & 3 & -8 \\ 2 & 5 & -11 \end{bmatrix}$$

Since $X = A^{-1} \cdot k$ we have

$$\begin{bmatrix} x \\ y \\ z \end{bmatrix} = \frac{1}{7} \begin{bmatrix} 1 & -1 & 5 \\ 4 & 3 & -8 \\ 2 & 5 & -11 \end{bmatrix} \begin{bmatrix} -3 \\ 1 \\ -2 \end{bmatrix} = \begin{bmatrix} -2 \\ 1 \\ 3 \end{bmatrix}$$

13.4.9 *Solution of Eigen Value Problems*

In the system of equations in matrix notation $\mathbf{Ax = b,}$ if all b_1 are zero then the system considered is said to be homogeneous. It has a unique solution if and only if the determinant of A, that is $|\mathbf{A}|$ is non zero and the solution is written as

$$\mathbf{X = A^{-1}b} \tag{1}$$

The homogeneous system $(b_i = 0, \quad i = 1(1)n)$ possesses only a trivial solution

$x_1 = x_2 = \cdots = x_n = 0$ if det $\mathbf{A} \neq 0$.

We therefore, consider the system in which a parameter λ occurs, and we determine values of λ, called **eigenvalues**, for which the homogeneous system has a non trivial solution. Such a solution is called an **eigenfunction** and the entire system is called an **eigenvalue problem** or **the characteristic value problem**. The system $\mathbf{AX = b,}$ (b = 0) may be written as $\mathbf{AX = ?X}$ or $(\mathbf{A} - \lambda \ \mathbf{I})\mathbf{X} = 0$.

This system will have a non trivial solution $X \neq 0$, if the determinant of the matrix $(\mathbf{A} - \lambda \ \mathbf{I})$ should be zero or det $(\mathbf{A} - \lambda \ \mathbf{I}) = 0$ \tag{2}

The equation (2) is called the **characteristic equation**. The n roots $\lambda_1, \lambda_2, \cdots, \lambda_n$ are called the eigenvalues of $\mathbf{A}$ and may be distinct or repeated, real or complex, The modulus of the largest eignenvalue is called the **spectral radious of A**. Corresponding to each eigenvalue λ_1 there exists an eigenvector $\mathbf{X}_i$ which is a non trivial solution of $(\mathbf{A} - \lambda_i\mathbf{I})\mathbf{X}_i = \mathbf{0}$.

If the n eigenvalues λ_i, $i = 1(1)n$ of A are different and X_i are the corresponding eigenvectors, then any vector y, in the eigenspace can be expressed as a linear combination of the $\mathbf{X}_i$,

$y = c_1X_1 + c_2X_2 + \cdots + c_nX_n$

EXAM. 1: Find the eigen value and the corresponding eigen vectors of the matrix A given by

$$A = \begin{bmatrix} 1 & -2 & 3 \\ 6 & -13 & 18 \\ 4 & -10 & 14 \end{bmatrix}$$

The characteristic equation of the matrix A is given by

$$\begin{vmatrix} 1-\lambda & -2 & 3 \\ 6 & -(\lambda+13) & 18 \\ 4 & -10 & 14-\lambda \end{vmatrix} = 0$$

$$(1-\lambda)[(\lambda+13)(\lambda+14)+180]+2[84-6\lambda-72]+3[-60+4\lambda+52]=0$$

$$(1-\lambda)(-\lambda-2+\lambda^2)-12(\lambda-2)+12(\lambda-2)=0$$

or $(1-\lambda)(\lambda-2)(\lambda+1)=0$

$\lambda_1 = 1, \quad \lambda_2 = -1, \quad \lambda_3 = 2$

For $\lambda_1 = 1$ we have $\begin{bmatrix} 0 & -2 & 3 \\ 6 & -14 & 18 \\ 4 & -10 & 13 \end{bmatrix} \begin{bmatrix} x_1 \\ x_2 \\ x_3 \end{bmatrix} = 0$

$-2x_2 + 3x_3 = 0, \ 6x_1 - 14x_2 + 18x_3 = 0 \ , \ 4x_1 - 10x_2 + 13x_3 = 0$

$x_2 = \dfrac{3}{2}x_3 \ , \ 6x_1 - 21x_3 + 18x_3 = 0 \qquad 6x_1 - 3x_3 = 0 \qquad x_1 = \dfrac{1}{2}x_3$

$4x_1 - 15x_3 + 13x_3 = 0 \qquad 4x_1 - 2x_3 = 0$

$\therefore x_1 = \dfrac{1}{2}x_3, x_2 = \dfrac{3}{2}x_3$ is arbitrary, let $x_3 = 2$ then $x_1 = 1, \ x_2 = 3$ and $x_3 = 2$

$\therefore$ For $\lambda_1 = 1$, the corresponding eigen vector is $\mathbf{x}_1 = [1,3,2]^T$.

Similarly for $\lambda_2 = -1$, $\mathbf{x}_2 = [0,3,2]^T$ and for $\lambda_3 = 2$, $\mathbf{x}_3 = [1,4,3]^T$

13.5 Iteration Methods

A general linear iterative method for the solution of $\mathbf{Ax} = \mathbf{b}$ may be defined in the form

$$\mathbf{x}^{(k+1)} = \mathbf{H} \ \mathbf{x}^{(k)} + \mathbf{c}, \quad k = 0,1,2,\cdots \tag{1}$$

where $x^{(k+1)}$ and $x^{(k)}$ are the approximations for x at the $(k+1)^{th}$ amd k_{th} iterations respectively. $\mathbf{H}$ is called the iteration matrix depending on $\mathbf{A}$ and c is a column vector. In the limiting case, when $k \to \infty$, $x^{(k)}$ converges to the exact solution

$$\mathbf{x} = \mathbf{A}^{-1}\mathbf{b} \tag{2}$$

and the iteration equation (1) becomes by substitution from (2),

$$\mathbf{A}^{-1}\mathbf{b} = \mathbf{H}\mathbf{A}^{-1}\mathbf{b} + \mathbf{c} \qquad \mathbf{c} = (\mathbf{I} - \mathbf{H})\mathbf{A}^{-1}\mathbf{b} \tag{3}$$

The determination of the iteration matrix $\mathbf{H}$ and the column vector $\mathbf{c}$ for a few well known iteration methods are given below.

13.5.1 *Jacobi Iteration Method*

Assuming that the quantities a_{ii} in the system of equations $\displaystyle\sum_{j=1}^{n} a_{ij}x_j = b_i, \quad i = 1, 2, \cdots n$ are pivot elements, we write

$$a_{11}x_1 = -(a_{12}x_2 + a_{13}x_3 + \cdots a_{1n}x_n) + b_1$$

$$a_{22}x_2 = -(a_{21}x_1 + a_{23}x_3 + \cdots a_{2n}x_n) + b_2 \qquad (4)$$

$$\vdots$$

$$a_{nn}x_n = -(a_{n1}x_1 + a_{n2}x_2 + \cdots a_{n,n-1}x_{n-1}) + b_n$$

The Jocobi iteration method may now be defined as

$$x_1^{(k+1)} = \frac{-1}{a_{11}}(a_{12}x_2^{(k)} + a_{13}x_3^{(k)} + \cdots + a_{1n}x_n^{(k)} - b_1)$$

$$x_2^{(k+1)} = \frac{-1}{a_{22}}(a_{21}x_1^{(k)} + a_{23}x_3^{(k)} + \cdots + a_{2n}x_n^{(k)} - b_2) \qquad (5)$$

$$x_n^{(k+1)} = -\frac{1}{a_{nn}}(a_{n1}x_1^{(k)} + a_{n2}x_2^{(k)} + \cdots + a_{n,n-1}x_{n-1}^{(k)} - b_n), \quad k = 0,1,2,\cdots$$

which in matrix notation becomes

$$\mathbf{x}^{(k+1)} = -\mathbf{D}^{-1}(\mathbf{L} + \mathbf{U})\mathbf{x}^{(k)} + \mathbf{D}^{-1}\mathbf{b}$$

$$= \mathbf{H}\mathbf{x}^{(k)} + \mathbf{c}, \quad k = 0,1,2,\cdots \qquad (6)$$

where $\mathbf{H} = -\mathbf{D}^{-1}(\mathbf{L} + \mathbf{U}), \quad \mathbf{c} = \mathbf{D}^{-1}\mathbf{b}$ and the matrices $\mathbf{L}$ and $\mathbf{U}$ are respectively lower and upper triangular matrices with zero diagonal entries the matrix $\mathbf{D}$ is the diagonal matrix, such $\mathbf{A} = \mathbf{L} + \mathbf{D} + \mathbf{U}$.

13.5.2 *Gauss - Seidel Iteration Method*

We now use on the right hand side of (5), all the available values from the present iteration we write the Gauss - Seidel method.

$$x_1^{(k+1)} = -\frac{1}{a_{11}}(a_{12}x_2^{(k)} + a_{13}x_3^{(k)} + \cdots + a_{1n}x_n^{(k)}) + \frac{b_1}{a_{11}}$$

$$x_2^{(k+1)} = -\frac{1}{a_{22}}(a_{21}x_1^{(k)} + a_{23}x_3^{(k)} + \cdots + a_{2n}x_n^{(k)}) + \frac{b_2}{a_{22}} \qquad (7)$$

$$\vdots$$

$$x_n^{(k+1)} = -\frac{1}{a_{nn}}(a_{n1}x_1^{(k+1)} + a_{n2}x_2^{(k+1)} + \cdots + a_{n,n-1}x_{n-1}^{(k+1)}) + \frac{b_n}{a_{nn}}$$

This may be rearranged in the form

$$a_{11}x_1^{(k+1)} = -\sum_{i=2}^{n} a_{1i}x_i^k + b_1$$

$$a_{21}x_1^{(k+1)} + a_{22}x_2^{(k+1)} = -\sum_{i=3}^{n} a_{2i}x_i^{(k)} + b_2 \tag{8}$$

$$\vdots$$

$$a_{n1}x_1^{(k+1)} + \cdots + a_{nn}x_n^{(k+1)} = b_n$$

In matrix notation, (8) becomes

$$(\mathbf{D}+\mathbf{L})\mathbf{X}^{(k+1)} = -\mathbf{U}\mathbf{X}^{(k)} + b$$

or $X^{(k+1)} = -(\mathbf{D}+\mathbf{L})^{-1}\mathbf{U}\mathbf{X}^{(t)} + (\mathbf{D}+\mathbf{L})^{-1}\mathbf{b}$ \tag{9}

$$= HX^{(k)} + C, \quad k = 0,1,2,\cdots$$

where $\mathbf{H} = -(\mathbf{D}+\mathbf{L})^{-1}\mathbf{U}$ and $c = (\mathbf{D}+\mathbf{L})^{-1}\mathbf{b}$

The Gauess - Seidal Iterative Method is an improvement over the Jocobi method in which the values of the variables are successively increased. In the Gauss - Seidal, the updated values of the variables are used instead of the values of the previous iteration. The limitation of this method is that it will converge to the correct solution only if the coefficient matrix has a strong leading diagonal.

EXAM. 1: Solve the following system of equations by Gauss - Seidal iteration method.

$$5x_1 + x_2 - x_3 = 4 \; ; \; x_1 + 4x_2 + 2x_3 = 15 \; ; \; x_1 - 2x_2 + 5x_3 = 12$$

SOLUTION:

The equation are tested for the diagonally dominance
$|5|>|1|+|-1|$; $|4|>|1|+|2|$; $|5|>|1|+|-2|$
Since the system is diagonally dominant, the iterative method can be used
The equation are rewritten in the form

$$x_1 = 0.8 - 0.2x_2 + 0.2x_3 \tag{1}$$

$$x_2 = 3.75 - 0.25x_1 - 0.5x_3 \tag{2}$$

$$x_3 = 2.4 - 0.2x_1 + 0.4x_2 \tag{3}$$

$x_i = 1$ for $i = 1, 2, 3$ will make

Eq (1) $x_2 = 1$, $x_3 = 1$ Therefore $x_1 = 0.8$

Eq (2) $x_1 = 0.8$, $x_3 = 1$ Therefore $x_2 = 3.05$

Eq (3) $x_1 = 0.8$, $x_2 = 3.05$ Therefore $x_3 = 3.46$

After the first iteration, the values the variables are $x_1 = 0.8$; $x_2 = 3.05$; $x_3 = 3.46$

In the second iteration, these values are used as initial values and a new set of values are obtained

Eq (1) $x_2 = 3.05$, $x_3 = 3.46$ Therefore $x_1 = 0.882$

Eq (2) $x_1 = 0.882$, $x_3 = 3.46$ Therefore $x_2 = 1.7995$

Eq (3) $x_1 = 0.882$, $x_2 = 1.7995$ Therefore $x_3 = 2.934$

The iterative procedure is repeated each time with updated values of the variables until the convergence is reached. The table below shows the number of iterations required for convergence and the final values.

Iteration	Gauss – Seidal Iteration		
Number	x_1	x_2	x_3
0	1.0	1.0	1.0
1	0.8	3.05	3.46
2	0.882	1.7995	2.934
3	1.0288	2.0271	3.0026
4	0.9963	1.9996	3.0005
5	1.0002	1.9997	2.9999
6	1.00	2.00	3.00

13.5.3 *Successive Over Relaxation (SOR) Method*

A matrix A is said to be Hermitian, denoted by $A*$ or A^H, if $A = (\bar{A})^T$, where $\bar{A}$ is the complex conjugate of $\mathbf{A}$. A matrix $\mathbf{A}$ is said to be unitary if $\mathbf{A}^{-1} = (\bar{\mathbf{A}})^T$ and a matrix $\mathbf{A}$ is said to be normal if $\mathbf{A} \cdot \mathbf{A}^* = \mathbf{A}^* \cdot \mathbf{A}$

The "property $\mathbf{A}$" is said to be possessed by a real matrix $\mathbf{B}$, iff there exists a permentation matrix $\mathbf{P}$ such that

$$\mathbf{PBP}^T = \begin{bmatrix} \mathbf{A}_{11} & \mathbf{A}_{12} \\ \mathbf{A}_{21} & \mathbf{A}_{22} \end{bmatrix}, \text{ where } \mathbf{A}_{11} \text{ and } \mathbf{A}_{22} \text{ are diagonal matrix.}$$

SOR method is a gerneralization of the Gauss - Seidel method. This method is often used when the coefficient matrix of the system of equations is symmetric and has "propery". We define an auxiliary vector $\tilde{\mathbf{X}}$ so that

$$\tilde{\mathbf{X}}^{(K+1)} = -\mathbf{D}^{-1}\mathbf{L}\mathbf{X}^{(K+1)} - \mathbf{D}^{-1}\mathbf{U}\mathbf{X}^{(K)} + \mathbf{D}^{-1}\mathbf{b} \tag{1}$$

The final solution is now written as

$$X^{(K+1)} = \mathbf{X}^{(K)} + \omega(\tilde{\mathbf{X}}^{(K)} - \mathbf{X}^{(K)})$$

or $X^{(K+1)} = (1-\omega)\mathbf{X}^{(K)} + \omega\tilde{X}^{(K+1)}$ \hfill (2)

Substituting from (1) into (2) and simplifying we have

$$X^{(K+1)} = (\mathbf{D} + w\mathbf{L})^{-1}[(1-w)\mathbf{D} - w\mathbf{U}]\mathbf{X}^{(k)} + w(\mathbf{D} + w\mathbf{L})^{-1}\mathbf{b}$$

$$= \mathbf{H}X^{(K)} + C, \quad k = 0,1,2,\cdots \tag{3}$$

where $\mathbf{H} = (\mathbf{D} + w\mathbf{L})^{-1}[(1-w)\mathbf{D} - w\mathbf{U}]$ $\mathbf{c} = w(\mathbf{D} + w\mathbf{L})^{-1}\mathbf{b}$ and

when $w = 1$, the equation (3) reduces to the Gauss - Seidel method. The quantity w is called the **relaxation parameter** and $X^{(K+1)}$ is a weighted mean of $\tilde{X}^{(K+1)}$ and $X^{(K)}$. The weights are non - negative for $0 \le w \le 1$. If $w > 1$, then the method is called **over relaxation method** and if $w < 1$, then it is called an **under relaxation method.**

13.5.4 *Exercises*

(A) Short Answer Questions

1. For solving the system $AX = b$, when there is no need pivoting in Gauss elimination method?
2. When the Gauss elimination is preferred instead of the Jordan method?

3. Setup the Jacobi iterative scheme for the following equations $\sum\limits_{j=1}^{n} a_{ij}x_j = b_i$, $i = 1,2,\cdots,n$

4. Show that the Gauss - Seidel method for solving $a_{11}x_1 + a_{12}x_2 = b_1$ and $a_{21}x_1 + a_{22}x_2 \ b_2$ will be converge if $|a_{11}a_{22}| > |a_{12}a_{21}|$.

5. State a sufficient condition for convergence of the Gauss - Seidel iteration method.
6. In what way the Gauss - Seidel method differs from Jacobi iteration method?
7. Explain the terms; block relaxation; over relaxation and under relaxation.
8. Why should care be exercised while using the relaxation technique?

(B) Solve, using determinants, the system of equations

1. $3x - y + z = 1;\quad x + 3y - 4z = 7;\quad 7x - y + 5z = 5$

2. $x + 2y + 3z = 6;\quad 2x + 4y + z = 7;\quad 3x + 2y + 9z = 14$

   Solve, using matrix methods, the system of equations:

1. $x + y + z = 2;\quad 2\ x - 2y + z = 6;\quad x + 2y + 3z = 5$

(C) Using Gauss elimination method solve the following system of equations

1. $3x - y + 2z = 12;\quad x + 2y + 3z = 11;\quad 2\ x - 2y - z = 33$ (Ans: $x = 3, y = 1, z = 2$)

2. $5x - y - 2z = 14;\quad x - 3y - z = -30;\quad 2x - y - 3z = 5$ (Ans: $x = 39.2,\ \ y = 16.7,\ \ z = 19$)

3. $2x + 4y + z = 3;\quad 3\ x + 2y - 2z = -2;\quad x - y + z = 6$ (Ans: $x = 2,\ \ y = -1,\ \ z = 3$)

4. $2x + y + 4z = 12;\quad 8x - 3y + 2z = 20;\quad 4x + 11y - z = 33$

5. $x_1 + 2x_2 + 4x_3 + 3x_4 = -2;\quad 3x_1 + x_2 + 3x_3 + 2x_4 = -10;\quad -x_1 + 3x_2 + x_3 + 4x_4 = -8$

6. $5x + 3y + 7z = 4;\quad x + 5y + 2z = 2;\quad 7x + 2y + 10z = 5$

7. $4x - y + 2z = 15;\quad -x + 2y + 3z = 5;\quad 5x - 7y + 9z = 8$ (Ans: $x = 4,\ \ y = 3,\ \ z = 1$)

8. $2x - 2y + 4z = -12;\quad 2x + 3y + 2z = 8;\quad -x + y - z = 7.5$

9. $4x + y + z = 4;\quad x + 4y - 2z = 4;\quad 3x + 2y - 4z = 6$

10. $5x - 2y + 3z = 18;\quad x + 7y - 3z = -22;\quad 2x - y + 6z = 22$

    Solve the following systems by Jordan method

1. $2x + y + 4z = 12;\quad 8x - 3y + 2z = 20;\quad 4\ x + 11y - z = 33$ (Ans: $x = 3, y = 2, z = 1$)

2. $x + 2y + z = 8;\quad 2x + 3y + 4z = 20;\quad 4x + 3y + 2z = 16$ (Ans: $x = 1,\ \ y = 2,\ \ z = 3$)

(D)

1. Given that $A = \begin{bmatrix} 1 & -2 & 0 \\ -2 & 1 & 1 \\ 0 & 1 & 1 \end{bmatrix}$; determine matrices **L** and **U** of the forms

   $L = \begin{bmatrix} l_{11} & 0 & 0 \\ l_{21} & l_{22} & 0 \\ l_{31} & l_{32} & l_{33} \end{bmatrix}$, $U = \begin{bmatrix} 1 & u_{12} & u_{13} \\ 0 & 1 & u_{23} \\ 0 & 0 & 1 \end{bmatrix}$, such that $LU = A$.

2. Factorize the matrix $\begin{bmatrix} 2 & 0 & 1 \\ 0 & 2 & 2 \\ 1 & 0 & 2 \end{bmatrix}$ as a product of lower triangular and a unit upper triangular

   matrix and hence find its inverse.

3. Factorize the matrix $A = \begin{bmatrix} 5 & -2 & 1 \\ 7 & 1 & -5 \\ 3 & 7 & 4 \end{bmatrix}$ into the form **LU** where **L** is the lower triangular

matrix and U an upper triangular matrix and hence solve the system of equations

$5x - 2y + z = 4$; $7x + y - 5z = 8$; $3x + 7y + 4z = 10$

Determine L^{-1} and U^{-1} and hence find A^{-1}

4. Solve the system of equation by cholestei's method:

$2x + y + w = 1$; $3x + 0.5y + z + w = 2$; $4x + 2y + 2z + w = -1$; $y + z + 2w = 0$

5. Solve by Crout's method the system of equations

$x + 2y - 3z = 4.5$; $4x + y - z = 6.3$; $3x - 7y + 5z = -8.7$

6. Solve by Crout's method:

$3x - y + 2z = 12$; $x + 2y + 3z = 11$; $2x - 2y - z = 2$

(Ans: $x = 3.15$, $y = 1.23$, $z = 1.82$)

7. Apply Crout's method to solve $10x - 7y + 3z + 5w = 6$; $= 6x + 8y - z - 4w = 5$; $3x + y + 4z + 11w = 2$; $5x - 9y - 2z + 4w = 7$ (Ans: $x = 5$, $y = 4$, $z = -7$, $w = 1$)

E. Solve the following equation by iteration

1. $27x + 6y - z = 85$; $6x + 15y + 2z = 72$; $x + y + 54z = 110$ (up to three stages)

2. $5x + y + 3z = 16$; $x + 4y + z + w = 11$; $-x + 2y + 6z - 2w = 23$; $x - y + z + 4w = -2$

3. $2x + 10y + z = 25$; $2x + 2y + 10z = 36$; $-10x + y + 5z = 7$ (apply Gauss - Seidel method)

4. $5x - y = 9$; $-x + 5y - z = 4$; $-y + 5z = -6$ (by Gauss Seidel method)

5. $8x - y + z - 18 = 0$; $2x + 5y - 2z - 3 = 0$; $x + y - 3z + 6 = 0$ (Tabulate the result upto the fifth iteration)

6. $x + 17y - 2z = 48$; $30x - 2y + 3z = 75$; $2x + 2y + 18z = 30$

7. $28x + 4y - z = 32$; $x + 3y + 10z = 24$; $2x + 17y + 4z = 35$ (by Gauss - Seidel method)

8. $x + 2y - z = 2$; $3x + 6y + z = 1$; $3x + 3y + 2z = 3$ (by Gauss - Seidel method, with $x = 0$, $y = 0$, $z = 0$ as initial approximate solution.)

9. $8x - 3y + 2z = 20$; $4x + 11y - z = 33$; $6x + 3y + 12z = 36$

10. Find to two significant figures the roots of the following system of equations by any iterative method

$$\begin{bmatrix} 2.5 & -3.0 & 4.6 \\ -3.5 & 2.6 & 1.5 \\ -6.5 & -3.5 & 7.3 \end{bmatrix} \begin{bmatrix} x \\ y \\ z \end{bmatrix} = \begin{bmatrix} -1.00 \\ -14.50 \\ -17.73 \end{bmatrix}$$

11. $27x + 6y - z = 85$; $6x + 15y + 2z = 72$; $x + y + 54z = 110$

(Ans: $x = 2.4254$; $y = 3.5730$; $z = 1.9259$)

F. Solve the following systems of equations by relaxation method

1. $x + 2y + z = 4$; $x - y + z = 5$; $2x + 3y - z = 1$

2. $2x - y + 3z = 8$; $-x + 2y + z = 4$; $3x + y - 2z = 4$

3. $y + z - 4x + 2 = 0$; $2x - 4y + \dfrac{10}{3} = 0$; $2x - 4z + \dfrac{2}{3} = 0$

4. $10x - 2y - 2z = 6$; $10y - 2z - x = 7$; $10z - x - y = 8$

5. $5x - y - z - 3 = 0$; $-x + 10y - 2z - 7 = 0$; $-x - y + 10z - 8 = 0$

6. $3x + 9y - 2z - 11 = 0$; $4x + 2y + 13z - 24 = 0$; $11x - 4y + 3z + 8 = 0$

7. $-2x + y + z = 0$; $2x - y + 4z = 15$; $3x - 4y + z = 5$

8. $3x - 3y + 5z = 4$; $x + 2y - 6z = 3$; $2x - y + 3z = 1$

9. Explain the block relaxation process and use it to solve the following system (to the nearest integer):

$-6x + 2y + 4z + 174 = 0$; $x + y - 5z + 917 = 0$; $x - 6y + 2z + 2198 = 0$

10. Set up Jacobi iterative scheme for the following equation and iterate three times starting with $\mathbf{X}^{(0)} = \mathbf{0}$.

$4x_1 + x_2 + 2x_3 = 4$; $7x_1 + 5x_2 + x_3 = 7$; $x_1 + x_2 + 3x_3 = 3$

[Ans: The first three iteration are $\left(1, \dfrac{7}{3}, 1\right)^T$; $\left(\dfrac{3}{10}, \dfrac{3}{5}, \dfrac{1}{5}\right)^T$; $\left(\dfrac{3}{4}, \dfrac{127}{100}, \dfrac{3}{4}\right)^T$]

11. Using Gauss - Seidel iterative scheme, solve the following systems:

(i) $x_1 + 8x_2 + x_3 = 10$; $x_1 + x_2 + 7x_3 = 9$; $9x_1 + x_2 + x_3 = 11$ (Ans: The successive iterations are:

$x_2 = x_3 = 0$, $x_1 = 1.222$, $x_2 = 1.097$, $x_3 = 0.953$)

(ii) $x_2 = 1.097$ $x_3 = 0.953$

$x_1 = 0.994$, $x_2 = 1.0066$, $x_3 = 0.9998$

(iii) $x_2 = 1.0066$, $x_3 = 9998$ $x_1 = x_2 = x_3 = 1$

12. $10x + y + z = 12$; $2x + 10y + z = 13$; $2x + 2y + 10z = 14$ (Ans: After 4 iterations, $x = y = z = 1$)

13. $20x + 2y + 6z + 28$; $x + 20y + 9z = -23$; $2x - 7y - 20z = -57$. Continue the iteration process until the maximum difference between successive values of x, y and z is less that 0.02

14. $10x - y - z = 13$; $x + 10y + z = 36$; $-x - y + 10z = 35$ (Ans: $x = 2$, $y = 3$, $z = 4$)

15. $27x + 6y - z = 85$; $6x + 15y + 2z = 72$; $x + y + 54z = 110$ upto three stages.

13.6 INTERPOLATION

13.6.1 *Introduction*

Interpolation is the art of reading between the lines of the tabulated values of a function $f(x)$ approximated by polynomials $P(x)$. The function is not given explicity and only the values of $f(x)$ and / or its certain order derivatives at a set of points, called **nodes, tabular**

points or **arguments** are known on a given interval $[a,b]$.

A polynomial $P(x)$ is called an interpolating polynomial if the values of $P(x)$ and / or its certain order derivatives coincide with those of $f(x)$ and / or its same order derivatives at one or more tabular points.

For example if the polynomial $P(x)$ is written as the Taylor's expansion, for the function $f(x)$ at a point x_0, $x_0 \in [a,b]$ in the form

$$P(x) = f(x_0) + (x - x_0)f'(x_0) + \frac{1}{2!}(x - x_0)^2 f''(x_0) + \cdots + \frac{1}{n!}(x - x_0)^n f^{(n)}(x_0) \tag{1}$$

then $P(x)$ may be regarded as an interpolating polynomial of degree n, satisfying the conditions

$$P^{(k)}(x_0) = f^{(k)}(x_0), \quad k = 0,1, \cdots n_{n+1} \tag{2}$$

The term $R_n = \frac{1}{(n+1)!}(x - x_0)^{n+1} f^{(n+1)}(\xi), \quad x_0 < \xi < x \tag{3}$

is called the **remainder** or **truncation error**.

In general, if there are $n+1$ distinct points $a \le x_0 < x_1 < x_2 \cdots < x_n = b$, then the problem of interpolation is to obtain $P(x)$ satisfying the conditions

$$\text{(i) } P(x_i) = f(x_i), i = 0,1,2, \cdots n$$

$$\tag{4}$$

or (ii) $P(x_i) = f(x_i), \quad P'(x_i) = f'(x_i), \quad i = 0,1,2, \cdots n \tag{5}$
and similar higher order derivatives

The condition (4) gives rise to Lagrange **interpolating polynomial** and (5) gives rise to **Hermite interpolating polynomial**.

13.6.2 *Lagrange and Newton Interpolations*

A function $f(x)$ defined in the interval $[a,b]$ is assumed to be continuous on $[a,b]$ and the values of $f(x)$ are known at the $n+1$ distinct points $a \le x_0 < x_1 < x_2 \cdots < x_n \le b$ of $[a,b]$. The polynomial $P(x) = a_0 + a_1 x + a_2 x^2 + \cdots + a_n x^n$ satisfy must the property (4), that is $P(x_i) = f(x_i)$, $i = 0,1,2, \cdots n$. The polynomial $P(x)$ so obtained is unique.

Depending on its form, the polynomial is called either **Lagrange** or the **Newton interpolation with divided differences**.

For $n = 1$, we want to determine the linear interpolation formula for $P(x) = a_1 x_1 + a_0$ with a_0 and a_1 being arbitrary constant, for a function $f(x)$, $x \in [x_0, x_1]$, such that

$f(x_0) = P(x_0) = a_1 x_0 + a_0$ and $f(x_1) = P(x_1) = a_1 x_1 + a_0$.

Eliminating a_0 and a_1 from the above equations, we get

$$\begin{vmatrix} P(x) & x & 1 \\ f(x_0) & x_0 & 1 \\ f(x_1) & x_1 & 1 \end{vmatrix} = 0 \tag{6}$$

or $P(x) = \dfrac{x - x_1}{x_0 - x_1} f(x_0) + \dfrac{x - x_0}{x_1 - x_0} f(x_1)$

$$= l_0(x) f(x_0) + l_1(x) f(x_1), \cdots \tag{7}$$

where $l_0(x) = \dfrac{x - x_1}{x_0 - x_1}$, $l_1(x) = \dfrac{x - x_0}{x_1 - x_0}$

The functions $l_0(x)$ and $l_1(x)$ are called the Lagrange fundamental polynomials and satisfy the conditions: $l_0(x) + l_1(x) = 1$, $l_0 x_0) = 1$, $l_0(x_1) = 0$

$$l_1(x_0) = 0, \quad l_1(x_1) = 0 \text{ or } l_i(x_j) = \delta_{ij} = \begin{cases} 1 \text{ if } i = j \\ 0 \text{ if } i \neq j \end{cases} \tag{8}$$

The degree of the polynomials $l_0(x)$ and $l_1(x)$ is one.

Newton's Divided Difference Interpolation

We expand the determinant (6) in terms of the first row and get

$$P(x) = f(x_0) + \frac{f(x_1) - f(x_0)}{x_1 - x_0} (x - x_0)$$

$$= f(x_0) + f[x_0, x_1](x - x_0) \tag{9}$$

where $\dfrac{f(x_1) - f(x_0)}{x_1 - x_0} = f[x_0, x_1]$ is called the first divided difference of $f(x)$ relative to x_0

and x_1.

We may write (9) as $\dfrac{P(x) - f(x_0)}{x - x_0} = f[x_0, x_1]$ \hfill (10)

The equation (9) or (10) is the linear **Newton** interpolating polynomial with divided differences

The polynomial $P(x)$ coincides with $f(x)$ at x_0 and x_1 and it deviates at all other points, in the interval (x_0, x_1). This deviation is called the truncation error and may be written as

$$E_1(f;x) = f(x) - P(x)$$

$$= \frac{1}{2}(x - x_0)(x - x_1)f''(\xi)$$

where $\min(x_0, x_1, x) < \xi < \max(x_0, x_1, x)$, using the mean value theorem

Higher Order Interpolation

The Lagrange fundamental polynomials of degree n based $n+1$ distinct points $a \le x_0 < x_1 < x_2 < \cdots < x_n = b$ and which satisfy (8) can be written in the form

$$l_i(x) = \frac{(x - x_0)(x - x_1)\cdots(x - x_{i-1})(x - x_{i+1})\cdots(x - x_n)}{(x_i - x_0)(x_i - x_1)\cdots(x_i - x_{i-1})(x_i - x_{i+1})\cdots(x_i - x_n)} \quad (11)$$

$$i = 0,1,2,\ldots n$$

An alternative form of (11) is given by

$$l_i(x) = \frac{w(x)}{(x - x_i)w'(x_i)},$$

where $w(x) = (x - x_0)(x - x_1)\cdots(x - x_n)$ and a prime represents differentiation with respect to x.

Thus the polynomial $P(x) = \sum_{i=0}^{n} l_i(x)f(x_i)$ (12)

where $l_i(x)$ are given by (11) is the Lagrange interpolating polynomial of degree n. The truncation error in the Lagrange interpolation is given by

$$E_n(f;x) = f(x) - P(x)$$

$$= \frac{w(x)}{(n+1)!}f^{(n+1)}(\xi) \quad (13)$$

The linear Newton divided difference interpolation (9) is easy to generalize. We define

$$f[x_0, x_1, x_2] = \frac{f[x_1, x_2] - f[x_0, x_1]}{x_2 - x_0}$$

$$f[x_0, x_1, x_2, \cdots, x_{k-1}, x_k] = \frac{f[x_1, x_2, \cdots, x_k] - f[x_0, x_1, \cdots, x_{k-1}]}{x_k - x_0}, \quad k = 3, 4, \cdots, n$$

In terms of function values, the nth divided difference can be written as

$$f[x_0, x_1, x_2, \cdots, x_n] = \sum_{i=0}^{n} \frac{f(x_i)}{\prod_{\substack{j=0 \\ i \ne j}}^{n}(x_i - x_j)} \quad (14)$$

The interpolating polynomial $P_n(x)$, interpolating at the $n+1$ distinct points $x_0, x_1, \cdots, x_n$ can be written as

$$P_n(x) = a_0 + (x - x_0)a_1 + (x - x_0)(x - x_1)a_2 + \cdots + (x - x_0)(x - x_1)\cdots(x - x_{n-1})a_n \quad \cdots(15)$$

Substituting, successively $x_0, x_1, \cdots, x_n$ in (15)

We have

$$a_0 = f[x_0], \quad a_1 = f[x_0, x_1], \quad a_2 = f[x_0, x_1, x_2], \quad \cdots \quad a_n = f[x_0, x_1 \cdots x_n] \qquad (16)$$

The divided difference interpolation polynomial becomes

$$P(x) = f(x_0) + (x - x_0)f(x, x_1) + \cdots (x - x_0)(x - x_1)\cdots(x - x_{n-1}) +$$

$$f[x_0, x_1, \cdots, x_n] \qquad (17)$$

13.6.3 *Finite Difference Operators*

If the tabular points $x_0, x_1, \cdots, x_n$ are equally spaced, that is $x_i = x_0 + ih, \ i = 0, 1, \cdots n,$ where h is called the step length, it is possible to define the following operators:

The shift operator: E $\qquad\qquad\qquad\qquad Ef(x_i) = f(x_i + h)$

The forward - difference operator: Δ $\qquad\quad \Delta f(x_i) = f(x_i + h) - f(x_i)$

The backward - difference operator: ∇ $\qquad \nabla f(x_i) = f(x_i) - f(x_i - h)$

The central - difference operator: δ $\qquad\quad \delta \ f(x_i) = f(x_i + \frac{h}{2}) - f(x_i - \frac{h}{2})$

The averaging - operator : μ $\qquad\qquad\quad \mu \ f(x_i) = \frac{1}{2}\left[f(x_i + \frac{h}{2}) + f(x_i - \frac{h}{2}) \right]$

By repeating the application of the operators, we get the higher order difference as follows $E^n f(x_i) = f(n_i + nh)$

$$\Delta^n f(x_i) = \Delta^{n-1} f_{i+1} - \Delta^{n-1} f_i = \sum_{k=0}^{n} (-1)^k \frac{n!}{k!(n-k)!} f_{i+n-k} \qquad (18)$$

$$\nabla^n f(x_i) = \nabla^{n-1} f_i - \nabla^{n-1} f_{i-1} = \sum_{k=0}^{n} (-1)^k \frac{n!}{k!(n-k)!} f_{i-k}$$

$$\delta^n f(x_i) = \delta^{n-1} f_{i+\frac{1}{2}} - \delta^{n-1} f_{i-\frac{1}{2}} = \sum_{k=0}^{n} (-1)^k \frac{n!}{k!(n-k)!} f_{i+\frac{n}{2}-k}$$

The various orders of the above operators can be successively calculated and represented in the tabular form.

The shift operator E can be used to represent the remaining other operators, since

$\Delta f_i = \nabla f_{i+1} = \delta f_{i+\frac{1}{2}}$, we have

$$\Delta = E - 1; \ \nabla = 1 - E^{-1}; \ \delta = E^{\frac{1}{2}} - E^{-\frac{1}{2}} \ \text{and} \ \mu = \frac{1}{2}[E^{\frac{1}{2}} + E^{-\frac{1}{2}}]$$

Similarly $\Delta, \nabla, \delta, \mu$ can be used successively to represent the other remaining operators.

Using these finite difference operators it is possible to obtain various forms of interpolation formulas.

Now the Newton divided differences in terms of forward, backward and central differences become

$$f[x_0, x_1] = \frac{f(x_1) - f(x_1)}{h} = \frac{1}{h} \Delta f_0$$

$$f[x_0, x_1, x_2] = \frac{f(x_1, x_0) - f(x_0, x_1)}{x_2 - x_0} = \frac{1}{2h}\left[\frac{1}{h}\Delta f_1 - \frac{1}{h}\Delta f_0\right]$$

$$= \frac{1}{2!h^2} \Delta^2 f_0 \qquad (19)$$

By induction, it can be shown that

$$f[x_0, x_1, \cdots, x_n] = \frac{1}{n!h^n} \Delta^n f_0$$

Similarly

$$f[x_0, x_1] = \frac{1}{h}\{f(x_1) - f(x_0)\} = \frac{1}{h}\nabla f_1$$

$$f[x_0, x_1, x_2] = \frac{1}{2h}\{f[x_1, x_2] - f[x_0, x_1]\} = \frac{1}{2h^2}(\nabla f_2 - \nabla f_1) = \frac{1}{2!h^2}\nabla^2 f_2$$

and

$$f[x_0, x_1, \cdots, x_n] = \frac{1}{nh}\{f[x_1, x_2, \cdots, x_n] - f[x_0, x_1, \cdots, x_{n-1}]\} = \frac{1}{n!h^n}\nabla^n f_n \qquad (20)$$

Also

$$f[x_0, x_1] = \frac{1}{h}\delta f_{\frac{1}{2}}$$

$$f[x_0, x_1, x_2] = \frac{1}{2!h^2}\delta^2 f_1$$

$$\cdots \cdots \cdots$$

and $\ f[x_0, x_1, \cdots, x_{2m}] = \frac{1}{(2m)!h^{2m}}\delta^{2m} f_m$

$$f[x_0, x_1, \cdots, x_{2m+1}] = \frac{1}{(2m+1)!h^{2m+1}}\delta^{2m+1} f_{m+\frac{1}{2}} \qquad (21)$$

13.6.4 *Interpolating Polynomials using finite differences*

1.Gregory - Newton Forward Difference Interpolation

Substituting the divided difference from (19) in terms of forward differences in the Newton divided difference interpolating polynomial (17), we get

$$P(x) = f_0 + \frac{x - x_0}{h} \Delta f_0 + \frac{(x - x_0)(x - x_1)}{2! h^2} \Delta^2 f_0 + \cdots$$

$$+ \frac{1}{n! h^n} \{(x - x_0)(x - x_1) \cdots (x - x_{n-1})\} \Delta^n f_0$$

(22)

If we put $\dfrac{(x - x_0)}{h} = u$, then (22) and the truncation error (13) becomes

$$P(x_0 + hu) = f_0 + u \Delta f_0 + \frac{u(u-1)}{2!} \Delta^2 f_0 + \cdots + \frac{u(u-1) \cdots (u - n + 1)}{n!} \Delta^n f_0$$

$$= \sum_{i=0}^{n} \binom{u}{i} \Delta^i f_0 \tag{23}$$

and $E_n(f; x) = \dfrac{u(u-1) \cdots (u - n)}{(n+1)!} h^{n+1} f^{(n+1)}(\xi)$ (24)

An alternate derivation of the interpolation polynomial (23) is given by

$$f(x) = f\left\{x_0 + \frac{x - x_0}{h} \cdot h\right\} = f(x_0 + uh) + E^u f | x_0)$$

$$= (1 + \Delta)^u f | x_0)$$

(25)

$$= f_0 + u \Delta f_0 + \frac{u(u-1)}{2!} \Delta^2 f_0 + \cdots + \frac{1}{n!} u(u-1) \cdots (u - n - 1) \Delta^n f_0$$

Neglating the difference $\Delta^{n+1} f_0$ and higher order differences, we get (23)

2. Gregory - Newton Backward Difference Interpolation

We write $f(x) = f(x_n + \dfrac{x - x_n}{h} \cdot h)$

$$= f(x_n + hu) = E^u f(x_n)$$

$$= (1 - \nabla)^{-u} f(x_n)$$

$$= f(x_n) + u\nabla f(x_n) + \frac{u(u+1)}{2}\nabla^2 f(x_n) + \cdots + \frac{1}{n!}\{u(u+1)\cdots(u+n-1)\}\nabla^n f(x_n) + \cdots$$

where $\dfrac{(x-x_n)}{h} = u$

Neglecting the difference $\nabla^{n+1} f(x_n)$ and higher order differences, we get the Gregory - Newton backward difference interpolation polynomial as

$$P(x_n + G_u) = f_n + u\nabla f_n + \frac{1}{2!}u(u+1)\nabla^2 f_n + \cdots + \frac{1}{n!}u(u+1)\cdots(u+n-1)\nabla^n f_n$$

$$= \sum_{i=0}^{n}(-1)^i \nabla^i f_n$$

(26)

And truncation error is

$$E_n(f;x) = \frac{u(u+1)\cdots(u+u)}{(n+1)!}h^{n+1}f^{(n+1)}(\xi)$$

(27)

3. Stirling and Bessel Interpolation Polynomials

Using central differences, for **never**, we assume that the nodal points are

$$x_{-p}, x_{-(p-1)}, x_{-(p-2)}, \cdots, x_{-1}, x_0, x_1, x_3, \cdots, x_{p-1}, x_p$$

The **Stirling** interpolation is given by

$$P(x) = f(x_0) + \frac{u}{2}\left[\delta f_{\frac{1}{2}} + \delta f_{-\frac{1}{2}}\right] + \frac{u^2}{2!}\delta^2 f_0 + \frac{u(u^2 - 1^2)}{3!}\frac{1}{2}\left[\delta f_{\frac{1}{2}} + \delta f_{-\frac{1}{2}}\right] + \cdots$$

$$+ \frac{u(u^2 - 1^2)(u^2 - 2^2)\cdots\{u^2 - (p-1)^2\}}{(2p-1)!}\frac{1}{2}\left[\delta f_{\frac{1}{2}} + \delta f_{-\frac{1}{2}}\right]$$

$$+ \frac{u^2(u^2 - 1^2)(u^2 - 2^2)\cdots\{u^2 - (p-1)^2\}}{(2p)!}\delta^{2p} f_0 \qquad (28)$$

where $u = \dfrac{(x-x_0)}{h}$

For n odd, we take the nodal points as $x_{-p}, x_{-(p-1)}, \cdots x_{-1}, x_0, x_1, x_2, \cdots x_{p+1}$ and write the **Bessel** interpolation as

$$P(x) = \frac{1}{2}[f_0 + f_1] + v\delta f_{\frac{1}{2}} + \frac{v^2 - \dfrac{1}{4}}{2!}\frac{1}{2}[\delta^2 f_0 + \delta^2 f_1]$$

$$+\frac{v(v^2-\frac{1}{4})}{3!}\delta^3 f_{\frac{1}{2}}+\cdots$$

$$+\frac{(v^2-\frac{1}{4})(v^2-\frac{9}{4})\cdots\{v^2-(2p-1)^2/4\}}{(2p)!}\frac{1}{2}[\delta^{2p}f_0+\delta^{2p}f_1]$$

$$+\frac{(v^2-\frac{1}{4})(v^2-\frac{9}{4})\cdots\{v^2-(2p-1)^2/4\}}{(2p+1)!}\delta^{2p+1}f_{\frac{1}{2}} \tag{29}$$

where $v=u-\frac{1}{2}$

Note: The Newton - Gregoly interpolations with forward and backward differences are used if the interpolation is used near the beginning and end respectively, of a table of values.

13.6.5 Examples

EXAM. 1: The values of x and y are given below

x:	3	6	7	9	10
y:	168	?	120	72	63

Find the best estimate for the value of the function   at , using Lagrange's formula
SOLUTION:

Here $x_0=3$, $x_1=7$, $x_2=9$, $x_3=10$

$$\therefore y_6=168\frac{(6-7)(6-9)(6-10)}{(3-7)(3-9)(3-10)}+120\frac{(6-3)(6-9)(6-10)}{(7-3)(7-9)(7-10)}$$

$$+72\frac{(6-3)(6-9)(6-10)}{(9-3)(9-7)(9-10)}+63\frac{(6-3)(6-7)(6-9)}{(10-3)(10-7)(10-9)}$$

$$=12+180-72+27-147$$

$\therefore$ The best estimate that can be given for the value of the function at the position by the independent variable is 147

EXAM. 2: Find the form of the function, given

x :	0	1	2	5
f(x):	2	3	12	147

Here $x_0 = 0,\ x_1 = 1,\ x_2 = 2,\ x_3 = 5$

Therefore, Lagrange's formula is

$$f(x) = \frac{(x-1)(x-2)(x-5)}{(0-1)(0-2)(0-5)} \times 2 + \frac{(x-0)(x-2)(x-5)}{(1-0)(1-2)(1-5)} \times 3$$

$$+ \frac{(x-0)(x-1)(x-5)}{(2-0)(2-1)(2-5)} \times 12 + \frac{(x-0)(x-1)(x-2)}{(5-0)(5-1)(5-2)} \times 147$$

$$= -\frac{1}{5}(x-1)(x-2)(x-5) + \frac{3}{4}x(x-2)(x-5) - 2x(x-1)(x-5) + \frac{49}{20}x(x-1)(x-2)$$

$$= -\frac{1}{5}(x^3 - 8x^2 + 17x - 10) + \frac{3}{4}(x^3 - 7x^2 + 10x) - 2(x^3 - 6x^2 + 5x) + \frac{49}{20}(x^3 - 3x^2 + 2x)$$

$$= \frac{1}{20}[-4x^3 + 32x^2 - 68x + 40 + 15x^3 - 105x^2 + 150x - 40x^3 + 240x^2$$

$$-200x + 49x^3 - 147x^2 + 98x]$$

$$= \frac{1}{20}[20x^3 + 20x^2 - 20x + 40]$$

$$= x^3 + x^2 - x + 2$$

$$\therefore\ f(x) = x^3 + x^2 - x + 2$$

13.6.6 *Exercises*

1. Establish the following

   (i) $hd = \sinh^{-1}(\mu\delta)$ where $D = \dfrac{d}{dx} = $ differential operator

   (ii) $\Delta - \nabla = \Delta\nabla$ (iii) $\Delta + \nabla = \dfrac{\Delta}{\nabla} - \dfrac{\nabla}{\Delta}$

   (iv) $\mu\delta = \dfrac{1}{2}\Delta E^{-1} + \dfrac{1}{2}\Delta$ (iv) $\Delta = \dfrac{1}{2}\delta^2 + \delta\sqrt{1 + \dfrac{1}{4}\delta^2}$

2. Form the divided difference table for the following data

x:	0	1	2	3	4	5	6
$f(x)$	3	4	13	36	79	148	249

3. Prove that

   (i) $\Delta\ f(x)g(x) = f(x)\ \Delta\ g(x) + g(x+h)\ \Delta\ f(x)$, (ii) $\Delta^n a^x = (a^h - 1)^n a^x$

4. Given 4 points $(1, 0)$, $(-2, 15)$, $(-1, 0)$ and $(2, 9)$. Write the Lagrangian form of the cubic that passes through them and express if in the form $ax^3 + bx^2 + cx + d$, using Lagrange's interpolation formula.

[Ans: $-\dfrac{x^3}{2}+4x^2+\dfrac{x}{2}-4$]

5. Find the cubic polynomial which takes the following values $y(0)=1$; $y(1)=1$; and $y(3)=10$. Hence or otherwise obtain

6. Using Lagrange's formula show that the form of the function $f(x)$ is equal to

$\dfrac{1}{72}(-x^3+29x^2+1602x+47448)$, given that

x :	0	2	3	6
$f(x)$:	659	705	729	804

7. Prove that the third divided differences with arguments a,b,c,d of the function $f(x)\dfrac{1}{x}$ is

$-\dfrac{1}{abcd}$

8. Show that for the function $f(x)=\dfrac{1}{x}$ the nth divided difference is $f(x_0,x_1,\cdots,x_n)=\dfrac{(-1)^n}{x_0\cdot x_1\cdots x_n}$

9. From the following table form the table of divided differences, and extend it to include the values of $f(x)$ for $x=3$ and $x=4$

x :	4	5	7	10	11	13
$f(x)$:	48	100	294	900	1210	2028

10. Prove that the divided differences of order n of a polynomial of the *n-th* degree are constant.

11. Find the unique polynomial of degree 2 or less such that $f(0)=1$, $f(1)=3$, $f(3)=55$ using
    (i) the Lagrange interpolation
    (ii) the Newton divided difference interpolation
    [Ans: $8x^2-6x+1$]

    Find the unique polynomial of degree 2 or less such that $P(1)=1$, $P(3)=27$ and $P(4)=64$ using
    (i) the Lagrange interpolation
    (ii) the Newton divided difference interpolation

12. If $f(x)=\dfrac{1}{x^2}$, find the divided differences: $f(a,b)$, $f(a,b,c)$, $f(a,b,c,d)$

13. Using Newton's divided difference formula for un equal intervals, find the form of the function $f(x)$, given

x :	0	1	4	5
$f(x)$:	8	11	68	123

14. Calculate the third differences of $f(51)$ from the following entries:

x :	51	52	53	54
$f(x)$:	132.651	140.608	148.877	157.464

15. By means of Newton's divided difference formula find the value of $f(8)$ and $f(15)$ from the following table

x :	4	5	7	10	11	13
$f(x)$:	48	100	294	900	1210	2028

[Ans: $f(8) = 448$, $f(15) = 3150$]

16. Find the interpolating polynomial of degree 3 that fits the following data

x :	1.0	2.7	3.2	4.8
$f(x)$:	14.2	17.8	22.0	38.3

17. Express the divided difference of the third order as a quotient of two determinants.

18. Use the Lagrange and Newton divided difference formulas to calculate $f(3)$ from the following data:

x :	0	1	2	4	5	6
$f(x)$:	1	14	15	5	6	19

19. The following data give the melting point of an alloy of lead with zinc; θ is the temperature in degrees centigrade and x is the percent of lead.

x :	40	50	60	70	80	90
θ :	184	204	226	250	276	304

Use Newton's Interpolation formula to find θ when $x = 43$ [Ans: 189.79]

20. The following table gives the value of θ

θ :	30°	31°	32°	33°	34°
$\sin\theta$:	0.5000	0.5150	0.5299	0.5446	0.5592

21. Write a polynomial passing through the following (x, y) points (2,23); (4,93); (6,259); (8,569); (10,1071); (12,1813); (14,2843) and hence find the value of y when $x = 4.2$.

22. Estimate the production of cotton in the year 2003 from the data given below

Year :	1999	2000	2001	2002	2003	2004	2005
Production (in billion of Lakhs):	171	130	140	96	-	124	182

[Ans: 124]

23. In the plus two examination the number of candidates who secured marks in mathematics between certain limits, in a urban school, were as follows:

Marks	0-19	20-39	40-59	60-79	80-99
No. of Candidates	41	62	65	50	17

Estimate the number of candidates getting marks less that 70. [Ans: 194]

24. Evaluate $f(1 \cdot 2)$ using the following table:

x:	0	1	2	3	4
$f(x)$:	1.0	1.5	2.2	3.1	4.3

25. Find $f(x)$ when $x = 9$ and $x = 15$ using Newton's formula.

x:	8	10	12	14	16
$f(x)$:	1000	1900	3250	5400	8950

[Ans: $f(9) = 1405.86$, $f(15) = 6952.725$]

13.6.7 *Hermite Interpolation*

The Hermite interpolating polynomials interpolates not only the function $f(x)$ but also its certain order derivatives at a given set of tabular points. The interpolating conditions are

$$P(x_i) = f(x_i) \quad \text{and} \quad P^1(x_i) = f^1(x_i), \quad i = 0, 1, 2, \cdots, n$$

Since there are $2n + 2$ conditions to be satisfied $P(x)$ must be a polynomial of degree $\leq 2n + 1$. The required polynomial may be written as

$$P(x) = \sum_{i=0}^{n} A_i(x) f(x_i) + \sum_{i=0}^{n} B_i(x) f'(x_i), \tag{1}$$

where $A_i(x)$ and $B_i(x)$ of polynomials of degree $\leq 2n + 1$ and satisfy

$$\text{(i)} \quad A_i(x_j) = \begin{cases} 0, & i \neq j \\ 1, & i = j \end{cases} ; \qquad \text{(ii)} \quad A'_i(x_j) = 0 \text{ for all } i \text{ and } j$$

$$\text{(iii)} \quad B_i(x_j) = 0 \text{ for all } i \text{ and } j ; \qquad \text{(iv)} \quad B'_i(x_j) = \begin{cases} 0, & i \neq j \\ 1, & i = j \end{cases} \tag{2}$$

Using the Lagrange fundamental polynomials $l_i(x)$, we write

$$A_i(x) = \gamma_i(x) l_i^2(x) \quad \text{and} \quad B_i(x) = \delta_i(x) l_i^2(x) \tag{3}$$

Since $l_i^2(x)$ is a polynomial of degree $2n$, $\gamma(x)$ and $\delta(x)$ must be linear polynomials.

Let $\gamma_i(x) = a_i x + b_i$ and $\delta_i(x) = c_i x + d_i$ $\tag{4}$

using the conditions (2), we obtain

$$\left. \begin{array}{l} a_i = -2l'_i(x_i), \quad b_i = 1 + 2x; \quad l'_i(x_i) \\ \quad c_i = 1 \text{ and } d_i = -x_i \end{array} \right\} \tag{5}$$

Substituting (5) in (4) and using (3), the equation (1) becomes

$$P(x) = \sum_{i=0}^{n} [1 - 2(x - x_i) l'_i(x_i)]^2 l_i^2(x) f(x_i) \sum_{i=0}^{n} (x - x_i) l_i^2(x) + f'(x_i) \tag{6}$$

(6) is called the **Hermite interpolating polynomial**. It can be shown that

$$l'_i(x_i) = \frac{w''(x_i)}{2w'(x_i)} \quad \text{and the truncation error can be written as}$$

$$E_{2n+1}(f; x) = \frac{w^2(x)}{(2n+2)!} f^{(2n+2)}(\xi), \quad x_0 < \xi < x_n \tag{7}$$

Example

EXAM. 1: Given the following values of $f(x)$ and $f'(x)$

$$x: \quad -1 \quad 0 \quad 1$$
$$f(x): \quad 1 \quad 1 \quad -5$$, find the Hermite interpolation
$$f'(x): \quad -9 \quad 2 \quad -5$$

SOLUTION:

Since there are 6 conditions $= 2n+2$

We have $n=2$

$\therefore$ The Hermite interpolation formula is given by

$$P(x) = \sum_{i=0}^{2} A_i(x) f(x_i) + \sum_{i=0}^{2} B_i(x) f'(x_i)$$

$$A_i(x) = (a_i x + b_i) l_i^2(x)$$

$$= \{-2l_i'(x_i)x + [1+2x_i l_i'(x_i)]\}l_i^2(x)$$

$$= [1 - 2(x-x_i) l_i'(x_i)] l_i^2(x)$$

$$B_i(x) = (x-x_i) l_i^2(x), \quad i = 0, \ 1, \ 2 \cdots$$

$$\therefore \ A_0(x) = [1 - 2(x-x_0) l_0'(x_0)] l_0^2(x)$$

$$A_1(x) = [1 - 2(x-x_1) l_1'(x_1)] l_1^2(x)$$

$$A_2(x) = [1 - 2(x-x_2) l_2'(x_2)] l_2^2(x)$$

$$B_0(x) = (x-x_0) l_0^2(x)$$

$$B_1(x) = (x-x_1) l_1^2(x)$$

$$B_2(x) = (x-x_2) l_2^2(x)$$

$$l_0(x) = \frac{(x-0)(x-1)}{(-1-0)(-1-1)} = \frac{x(x-1)}{2}; \quad l_0'(-1) = -\frac{3}{2}$$

$$l_1(x) = \frac{(x+1)(x-1)}{(0+1)(0-1)} = -(x^2-1); \quad l_1'(-0) = 0$$

$$l_2(x) = \frac{(x+1)(x-0)}{(1+1)(1-0)} = \frac{x(x+1)}{2}; \quad l_2'(1) = \frac{3}{2}$$

$$\therefore \ A_0(x) = [1+3(x+1)]\frac{x^2(x-1)^2}{4} = \frac{1}{4}[3x^5 - 2x^4 - 5x^3 + 42]$$

$$A_1(x) = [1 - 2(x-0)(0)](x^2-1)^2 = x^4 - 2x^2 + 1$$

$$A_2(x) = [1 - 3(x-1)]\frac{x^2(x+1)^2}{4} = \frac{1}{4}[-3x^5 - 2x^4 + 5x^3 + 4x^2]$$

$$B_0(x) = \frac{(x+1)x^2(x-1)^2}{4} = \frac{1}{4}[x^5 - x^4 - x^3 + x^2]$$

$$B_1(x) = x(x^2 - 1) + x^5 - 2x^3 + x$$

$$B_2(x) = \frac{(x-1)x^2(x+1)^2}{4} = \frac{1}{4}[x^5 + x^4 - x^3 - x^2]$$

$$\therefore P(x) = \frac{1}{4}(3x^5 - 2x^4 - 5)() + (x^4 - 2x^3 + 1) \tag{1}$$

$$+ \frac{1}{4}(-3x^5 - 2x^4 + 5x^3 + 4x^2)(-5)$$

$$+ \frac{1}{4}(x^5 - x^4 - x^3 + x^2)(-9) + (x^5 - 2x^3 + x) \tag{2}$$

$$+ \frac{1}{4}(x^5 + x^4 - x^3 - x^2)(-5)$$

$$= x^5\left[\frac{3}{4} + \frac{15}{4} - \frac{9}{4} + 2 - \frac{5}{4}\right] + x^4\left[-\frac{2}{4} + 1 + \frac{5}{4} + \frac{9}{4}\right]$$

$$+ x^3\left[-\frac{5}{4} - 2 - \frac{25}{4} + \frac{9}{4} - 4 + \frac{5}{4}\right]$$

$$+ x^2\left[1 - 5 - \frac{9}{4} + \frac{5}{4}\right] + x[2] + 1$$

$$= 3x^5 + 4x^4 - 3x^3 - 7x^2 + 2x + 1$$

EXAM. 2: Obtain the unique polynomial $P(x)$ of degree 5 or less corresponding to a function $f(x)$ when

$$f(x_0) = 1, \quad f'(x_0) = 2, \quad f''(x_0) = 1$$

$$f(x_1) = 3, \quad f'(x_1) = 0, \quad f''(x_1) = -2$$

Also find $P\left[\dfrac{x_0 + x_1}{2}\right]$

SOLUTION:

Let $f(x) = a_0 + a_1(x - x_0) + a_2(x - x_0)^2 + a_3(x - x_0)^3 + a_4(x - x_0)^4 + a_5(x - x_0)^5$

Now $f(x_0) = 1 = a_0$ $\therefore\ a_0 = 1$

$\qquad f'(x) = a_1 + 2a_2(x - x_0) + 3a_3(x - x_0)^2 + 4a_4(x - x_0)^3 + 5a_5(x - x_0)^4$

$\qquad\qquad \therefore\ f'(x_0) = 2 = a_1 \quad \therefore\ a_1 = 2$

$\qquad\qquad f''(x) = 2a_2 + 6a_3(x - x_0) + 12a_4(x - x_0)^2 + 20a_5(x - x_0)^3$

$\qquad \therefore\ f''(x_0) = 1 = 2a_2 \quad \therefore\ a_2 = \dfrac{1}{2}$

Let $\quad x_1 + x_0 + h$ or $h = x_1 - x_0$

Now $\quad f(x) = a_0 + a_1(x - x_0) + a_2(x - x_0)^2 + a_3(x - x_0)^3 + a_4(x - x_0)^4 + a_5(x - x_0)^5$

$\qquad \therefore\ f(x_1) = a_0 + a_1(x_1 - x_0) + a_2(x_1 - x_0)^2 + a_3(x_1 - x_0)^3 + a_4(x_1 - x_0)^4 + a_5(x_1 - x_0)^5$

$\qquad 3 = a_0 + a_1 h + a_2 h^2 + a_3 h^3 + a_4 h^4 + a_5 h^5$

$\qquad 3 = 1 + 2h + \dfrac{1}{2}h^2 + a_3 h^3 + a_4 h^4 + a_5 h^5$

or $\qquad 2 = 2h + \dfrac{1}{2}h^2 + a_3 h^3 + a_4 h^4 + a_5 h^5$ (1)

we have $f'(x_1) = a_1 + 2a_2 h + 3a_3 h^2 + 4a^4 h^3 + 5a_5 h^4$

or $\qquad 0 = 2 + h + 3a_3 h^2 + 4a_4 h^3 + 5a_5 h^4$

or $\qquad -2 = h + 3a_3 h^2 + 4a_4 h^3 + 5a_5 h^4$ (2)

Now $f''(x_1) = 2a_2 + 6a_3 h + 12a_4 h^2 + 20a_5 h^3$

or $\qquad -2 = 1 + 6a_3 h + 12a_4 h^2 + 20a_5 h^3$

or $\qquad -3 = 6a_3 h + 12ah^2 + 20a_5 h^3$ (3)

From (1), we have

$$a_3 h^3 + a_4 h^4 + a_5 h^5 = 2 - 2h - \frac{h^2}{2}$$

or $\qquad h^3(a_3 + a_4 h + a_5 h_2) = \dfrac{4 - 4h - h^2}{2}$

From (2), we have

$$h^2(3a_3 + 4a_4 h + 5a_5 h^2) = -2 - h$$

From (3), we have

$$h(6a_3 - 12a_4h + 20a_5h^2) = -3$$

or
$$6a_3 + 12a_4h + 20a_5h^2 = -\frac{3}{h} \tag{4}$$

$$3a_3 + 4a_4h + 5a_5h^2 = \frac{-2-h}{h^2} \tag{5}$$

$$a_3 + a_4h + a_5h^2 = \frac{4-4h-h^2}{2h^3} \tag{6}$$

(4) - 3(5) gives

$$-3a_3 + 5a_5h^2 = -\frac{3}{h} + \frac{3(2+h)}{h^2} = \frac{6}{h^2}$$

or
$$3a_3 - 5a_5h^2 = -\frac{6}{n^2} \tag{7}$$

(5) - 4(6) gives

$$-a_3 + a_5h^2 = \frac{-(2+h)}{h^2} - \frac{4-4h-h^2}{2h^3} \cdot 4$$

$$= \frac{1}{h^3}[-2h - h^2 - 8 + 8h + 2h^2]$$

$$= -\frac{(8-6h-h^2)}{43}$$

Or
$$a_3 - a_5h^2 = \frac{8-6h-h^2}{h^3} \tag{8}$$

(7) - 3(8) gives

$$-2a_5h^2 = -\frac{6}{h^2} - \frac{3(8-6h-h^2)}{h^3}$$

$$= \frac{1}{h^3}(-6h - 24 + 18h + 3h^2)$$

$$= \frac{3h^2 + 12h - 24}{h^3} = \frac{3(h^2 + 4h - 8)}{h^3}$$

Or
$$a_5 = \frac{3(8-4h-h^2)}{2h^5}$$

Using (9), (8) becomes

$$a_3 = \frac{8-6h-h^2}{h^3} + \frac{3}{2h^3}(8-4h-h^2)$$

$$= \frac{1}{2h^3}[16 - 12h - 2h^2 + 24 - 12h - 3h^2]$$

$$= \frac{2}{2h^3}[40 - 24h - 5h^2] \qquad (10)$$

Writing (6) as

$$a_4 \cdot h = \frac{4 - 4h - h^2}{2h^3} - a_3 - a_5 h^2$$

$$= \frac{4 - 4h - h^2}{2h^3} - \frac{1}{2h^3}(40 - 24h - 5h^2) - \frac{3}{2h^3}(8 - 4h - h^2)$$

$$= \frac{1}{2h^3}[4 - 4h - h^2 - 40 + 24h + 5h^2 - 24 + 12h + 3h^2]$$

$$\therefore a_4 = \frac{1}{2h^4}[7h^2 + 32h - 60] \qquad (11)$$

$$f(x) = a_0 + a_1(x - x_0) + a_2(x - x_0)^2 + a_3(x - x_0)^3 + a_4(x - x_0)^4 + a_5(x - x_0)^5$$

where $a_0 = 1$, $a_2 = 2$, $a_2 = \frac{1}{2}$, $a_3 = \frac{1}{2h^3}(40 - 24h - 5h)$

$$a_4 = \frac{1}{2h^4}(7h^2 + 32h - 60), \quad a_5 = \frac{3}{2h^3}(8 - 4h - h^2)$$

Let $\quad x = \dfrac{x_1 + x_0}{2} = \dfrac{x_0 + h + x_0}{2} = \dfrac{2x_0 + h}{2}$

$$\therefore \quad x - x_0 = \frac{2x_0 + h}{2} = x_0 = \frac{h}{2}$$

$$\therefore \quad P\left(\frac{x_1 + x_0}{2}\right) = P\left(\frac{h}{2}\right) = f\left(\frac{h}{2}\right)$$

$$= 1 + h + \frac{h^2}{2} + \frac{1}{16}(4 - 24h - 5h^2) + \frac{1}{32}(7h^2 + 32h - 60) + \frac{3}{64}(8 - 4h - h^2)$$

$$= \frac{1}{64}(128 + 20h - h^2) \quad \text{where } h = x_1 - x_0$$

13.6.8 Piecewise Interpolation

In order to keep the degree of interpolating polynomial small, we use **piecewise interpolation.** The given interval $[a,b]$ is subdivided into a number of subintervals

$[x_{i-1}, x_i]$, $i = 1, 2, \cdots, n$ and approximate the function by some lower degree polynomial in each sub interval, especially into a linear Lagrange interpolation, such that it agrees with the function $f(x)$ at $(n+1)$ nodal points. The such intervals are called **finite elements** in one space dimension and the nodal points are called **knots**.

Using the linear Lagrange interpolation, we have for $x \in [x_{i-1}, x_i]$, the piecewise linear interpolation is given by

$$P_{i,1}(x) = \frac{x - x_i}{x_{i-1} - x_i} f(x_{i-1}) + \frac{x - x_{i-1}}{x_i - x_{i-1}} f(x_i), \quad i = 1, 2, \cdots, n$$

Consequently, the interpolation polynomial becomes

$$P(x) = \sum_{i=0}^{n} P_{i,1}(x)$$, which agrees with $f(x)$ at $x_i, i = 0, 1, \cdots, n$ and also linear in each

sub interval $[x_{i-1}, x_i]$ and hence can be written as

$$P(x) = \sum_{i=0}^{n} N_i(x) f(x_i)$$, where

$$N_i(x) = \begin{cases} 0, & x \leq x_{i-1} \\ x - x_{i-1} / x_i - x_{i-1}, & x_{i-1} \leq x \leq x_i \\ (x_{i+1} - x)/(x_{i+1} - x_i), & x_i \leq x \leq x_{i-1} \end{cases}$$

The function $N_i(x)$ is called a **shape function.**

If in each subinterval $[x_{i-1}, x_i]$, $i = 1, 2, \cdots, n$ if the function $f(x)$ is approximated by a cubic polynomial $P_{i,3}(x)$, we obtain the **piecewise cubic interpolation.** The cubic polynomial on the interval $[x_{i-1}, x_i]$ can be obtained by using the Hermite type conditions

$$P_{i,3}(x_{i-1}) = f_{i-1}, P_{i,3}(x_i) = f_i \quad \text{and}$$

$$P_{i,3}(x_{i-1}) = f_{i-1}, \ P'_{i,3}(x_i) = f'_i$$

The polynomial thus obtained is called piecewise cubic Hermite Interpolation and we get

$$P_{i,3}(x) = A_{i-1}(x) f_{i-1} + A_i(x) f_i + B_{i-1}(x) f^1_{i-1} + B_i(x) f^1_i \ ,$$

where $A_{i-1}(x) = \dfrac{(x - x_i)^2}{(x_{i-1} - x_i)^2} \left[1 + \dfrac{2(x_{i-1} - x)}{(x_{i-1} - x_i)} \right]$

$$A_i(x) = \frac{(x - x_{i-1})^2}{(x_{i-1} - x_i)^2} \left[1 + \frac{2(x - x_i)}{(x_{i-1} - x_i)} \right]$$

$$B_{i-1}(x) = \frac{(x-x_{i-1})(x-x_i)^2}{(x_{i-1}-x_i)^2}$$

$$B_i(x) = \frac{(x-x_i)(x-x_{i-1})^2}{(x_{i-1}-x_i)^2}$$

The interpolation polynomial becomes

$$P_3(x) = \sum_{i=1}^{n} P_{i,3}(x) \ ,$$

which agrees with $f(x)$ and $f'(x)$ at x_i, $i = 0,1,\cdots n$ and is cubic in each subinterval $[x_{i-1}, x_i]$ can be written as

$$P_3(x) = \sum_{i=0}^{n} N_i(x)f(x_i) + \sum_{i=0}^{n} H_i(x)f'(x_i) \ ,$$

where

$$N_i(x) = \begin{cases} 0, & x \le x_{i-1} \\ \dfrac{(x-x_{i-1})^2}{(x-x_{i-1})^2}\left[1 + \dfrac{2(x_i-x)}{(x_{i-1}-x_i)}\right], & x_{i-1} \le x \le x_i \\ \dfrac{(x-x_{i+1})^2}{(x_{i+1}-x_i)^2}\left[1 + \dfrac{2(x-x_i)}{(x_{i-1}-x_i)}\right], & x_i \le x \le x_{i+1} \\ 0, & x \le x_{i-1} \end{cases}$$

and

$$H_i(x) = \begin{cases} 0, & x \le x_{i-1} \\ \dfrac{(x-x_{i-1})^2(x-x_i)}{(x_i-x_{i-1})^2}, & x_{i-1} \le x \le x_i \\ \dfrac{(x-x_{i+1})(x-x_i)^2}{(x_{i+1}-x_i)^2}, & x_i \le x \le x_{i+1} \\ 0, & x \ge x_{i+1} \end{cases}$$

It is to be noted that

$$P_{i-1,3}(x_i) + P_{i,3}(x_i) = f_i, \quad i = 1,2,\cdots,n$$

and $P'_{i-1,3}(x_i) + P'_{i,3}(x_i) = f'_i, \quad i = 1,2,\cdots,n$

The polynomial $P_3(x)$ is continuously differentiable on $[a,b]$

13.6.9 Spline Interpolation

The name **spline** is derived from thin rods, called splines, which engineers have used for a long time to fit curves through given points; spline approximation is piecewise polynomial approximation. This means that a function $f(x)$ is given on an interval $a \le x \le b$

and it is desired to approximate $f(x)$ by n polynomials. Approximating functions $g(x)$ are obtained and then functions are more suitable in many problems of **approximation** and **interpolation**.

A spline function $g(x)$ is of degree n with knots $x_i, i = 1, 2, \cdots n$ which satisfies the properties

(i) In each sub interval $[x_{i-1}, x_i]$, $1 \leq i \leq n$, $g(x)$ is a polynomial of degree n;

and (ii) $g(x)$ and its first $(n-1)$ derivatives are continuous on $[a, b]$

We shall consider **cubic splines**. By definition, a **cubic spline** $g(x)$ on $a \leq x \leq b$ corresponding to the partition.

$a = x_0 < x_1 < x_2 < \cdots < x_n = b$ is a continuous function $g(x)$ which has continuous first and second derivatives every where in that interval and in each subinterval of that partition, is represented by a polynomial of degree not exceeding three. Hence $g(x)$ consisits of cubic polynomials, one in each subinterval.

If $f(x)$ is given on $a \leq x \leq b$ and a partion of this interval is chosen, a cubic spline $g(x)$ can be uniquely obtained approximating $f(x)$ such that

$$g(x_0) = f(x_0); \quad g(x_1) = f(x_1); \quad \cdots g(x_n) = f(x_n), \quad \text{and} \quad g'(x_0) = k_0 \quad \text{and} \quad g'(x_n) = k_n$$

(k_0 and k_n) are given numbers.

This can be established as follows:

Let $I_j = x_j \leq x \leq x_{j+1}$ and the spline function $g(x)$ must agree with a cubic polynomial $p_j(x)$ such that $p_j(x_j) = f(x_j)$, $p_j(x_{j+1}) = f(x_{j+1})$

Let $c_j = \dfrac{1}{x_{j+1} - x_j}$ and $p_j^1(x_j) = k_j$, $p_j^1(x_{j+1}) = k_{j+1}$

where k_0 and k_n are given and $k_1, k_2, \cdots, k_{n-1}$ will have to be determined.

It can be verified that the unique polynomial $p_j(x)$ satisfying the above conditions is

$$p_j(x) = f(x_j)c_j^2(x - x_{j+1})^2[1 + 2c_j(x - x_j)]$$
$$+ f(x_{j+1})c_j^2(x - x_j)^2[1 + 2c_j(x - xj + 1)]$$
$$+ k_j c_j^2(x - x_j)(x - x_{j+1})^2 + k_{j+1}c_j^2(x - x_j)^2(x - x_{j+1}) \tag{1}$$

Differentiating (1) twice, we obtain

$$p_j^{\cdot}(x_j) = -6c_j^2 f(x_j) + 6c_j^2 f(x_{j+1}) - 4c_j k_j - 2c_j k_{j+1} \tag{2}$$

and $p_j^{\cdot}(x_{j+1}) = 6c_j^2 f(x_j) - 6c_j^2 f(x_{j+1}) + 2c_j k_j + 4c_j k_{j+1}$ $\tag{3}$

By definition, $g(x)$ has continuous second derivatives. This gives the conditions

$P''_{j-1}(x_j) = P''_j(x_j), \quad j = 1, 2, \cdots n-1$. By using (3) with j replaced by $j-1$, and (2), these $(n-1)$ equations become

$$c_{j-1}k_{j-1} + 2(c_{j-1} + c_j)k_j + c_j k_{j+1} = 3[c_{j-1}^2 \nabla f_j + c_j^2 \nabla f_{j+1}] \qquad (4)$$

where $\nabla f_j = f(x_j) - f(x_{j-1})$ and $\nabla f_{j+1} = f(x_{j+1}) - f(x_j)$, $j = 1, 2, \cdots n-1$.

This system of $(n-1)$ linear equations has a unique solution $k_1, k_2, \cdots, k_{n-1}$, because all the coefficients of the system are nonnegative and each element in the principal diagonal is greater than the sum of the other elements in the corresponding row, so that the coefficient determinant cannot be zero. Hence we are able to determine unique $k_1, k_2, \cdots, k_{n-1}$ of the first derivative at the nodes.

EXAM. 1: First the following points by the cubic splines

j:	0	1	2	3
x_j:	1	2	3	4
f_j	1	5	11	8

Use the end conditions $f''_0 = 0, \quad f''_3 = 0$

SOLUTION:

x_j	f_j	∇f_j	$\nabla^2 f_j$	$\nabla^3 f_j$
1	1	4	2	−11
2	5	6	−9	
3	11	−3		
4	8			

$$c_j = \frac{1}{x_{j+1} - x_j}$$

$p_j(x_j) = f(x_j)$

$p_j(x_{j+1}) = f(x_{j+1})$

$p'_j(x_j) = k_j; \quad p'_j(x_{j+1}) = k_{j+1}$

From the given conditions

$p'(x_0) = k_0 = f'(x_0) = f'(0) = f''(0) = 0$

$p'(x_3) = k_3 = f'(x_3) = f''(3) = 0; \quad k_0 = 0, \quad k_3 = 0$

The unique polynomial satisfying the above condition is

$$p_j(x) = f(x)c_j^2(x - x_{j+1})^2[1 + 2c_j(x - x_j)]$$

$$+ f(x_{j+1})c_j^2(x - x_j)^2[1 - 2c_j(x - x_{j+1})]$$

$$+ k_j c_j^2(x - x_j)(x - x_{j+1})^2 + k_{j+1}c_j^2(x - x_j)^2(x - x_{j+1})$$

Here the $k_j, j = 1, 2, \cdots, (n-1)$ are the unknowns.

The system of $(n-1)$ linear equations are given by

$$c_{j-1}k_{j-1} + 2(c_{j-1} + c_j)k_j + c_j k_{j+1} = 3[c_{j-1}^2 \nabla f_j + c_j^2 \nabla f_{j+1}]$$

where $\nabla f_j = f(x_j) - f(x_{j-1})$ and $\nabla f_{j+1} = f(x_{j+1}) - f(x_j)$, $j = 1, 2, \cdots, n$

For $j = 1$, we have

$$c_0 = c_1 = c_2 = 1, \quad k_0 = 0 \quad k_3 = 0$$

Hence we have $4k_1 + k_2 = 30$

For $j = 2$, we have

$$c_1 k_1 + 2(c_1 + c_2)k_2 + c_2 k_3 = 3[c_1^2 \nabla f_2 + c_2^2 \nabla f_3]$$

$$c_1 k_1 + 2(c_1 + c_2)k_2 + c_2 k_3 = 3[6c_1^2 - 3c_2^2]$$

or $\quad\quad\quad k_1 + 4k_2 + k_3 = 9$

or $\quad k_1 + 4k_2 = 9 \quad$ since $k_3 = 0$

Solving for k_1 and k_2, we get

$$k_1 = \frac{37}{5} \quad \text{and} \quad k_2 = \frac{2}{5}$$

The cubic spline polynomials are

$$p_0(x) = f(0)c_0^2(x - x_1)^2[1 + 2c_0(x - x_0)]$$

$$+ f(x_1)c_0^2(x - x_0)^2[1 - 2c_0(x - x_1)]$$

$$+ k_0 c_0^2(x - x_0)(x - x_1)^2 + k_1 c_0^2(x - x_0)^2(x - x_1)$$

$$= \frac{63}{5}x^3 + \frac{43}{5}x^2 - 4x + 1 \quad [\, p_0(0) = f(0) = 1 \text{ is verified}]$$

$$p_1(x) = f(x_1)c_1^2(x - x_2)^2[1 + 2c_1(x - x_1)]$$

$$+ f(x_2)c_1^2(x - x_1)^2[1 - 2c_1(x - x_2)]$$

$$+ k_1 c_1^2(x - x_1)(x - x_2)^2 + k_2 c_1^2(x - x_1)^2(x - x_2)$$

$$= 5(x - 2)^2[1 + 2(x - 1) + 11(x - 1)^2[1 - 2(x - 2)]]$$

$$+k_1(x-1)(x-2)^2 + k_2(x-1)^2(x-2)$$

$$p_1(x) = \frac{1}{5}[-21x^3 + 77x^2 - 54x + 23] \quad [\, p_1(1) = f(1) = 5 \text{ is verified}]$$

$$p_2(x) = f(x_2)c_2^2(x-x_3)^2[1+2c_2(x-x_2)]$$

$$+f(x_3)c_2^2(x-x_2)^2[1-2c_2(x-x_3)]$$

$$+k_2c_2^2(x-x_2)(x-x_3)^2 + k_3c_2^2(x-x_2)^2(x-x_3) \quad [k_3=0]$$

$$=11(x-4)^2[1+2(x-3)]+8(x-3)^2[1-2(x-4)]+k_2(x-3)(x-4)^2$$

$$=\frac{1}{5}[32x^3 - 337x^2 + 1160x - 1256] \quad [\, p(2) = f_2 = 11 \text{ is verified}]$$

The cubic spline polynomial

$$p(x) = \begin{cases} \dfrac{63}{5}x^3 + \dfrac{43}{5}x^2 - 4x + 1, & 0 \le x < 1 \\[2mm] \dfrac{1}{5}(-21x^3 + 77x^2 - 54x + 23), & 1 \le x < 2 \\[2mm] \dfrac{1}{5}(32x^3 - 334x^2 + 1160x - 1256), & 2 \le x < 3 \end{cases}$$

(A) Determination of cubic spline $g(x)$ for equidistant nodes:

$$x_0, x_1 = x_0 + h, \cdots, x_n = x_0 + nh$$

for the given function $f(x)$ such that

$$g(x_0) = f(x_0), \quad g(x_1) = f(x_1), \cdots g(x_n) = f(x_n) = f(x_0 + nh)$$

In addition we require that $g'(x_0) = k_0;\ g'(x_n) = k_n$, the numbers k_0 and k_n are given numbers, prescribing the tangential direction of $g(x)$ of the end points. A unique cubic spline can be determined. (The data points need not be evenly spaced)

$g'(x_0) = f'(x_0)$ and $g'(x_n) = f'(x_n)$ are called the **clamped conditions.**

Other conditions of practical interests are the free or natural conditions

$$f''(x_0) = 0, \quad g''(x_n) = 0 \quad \text{(Zero curvature at the ends)}$$

Now $c_j = \dfrac{1}{x_{j+1} - x_j} = \dfrac{1}{h}$ for all $j = 0,1,2,\cdots n$.

$\therefore$ The conditions

$$c_{j-1}k_{j-1} + 2(c_{j-1} + c_j)k_j + c_j k_{j+1} = 3[c_{j-1}^2 \nabla f_j + c_j^2 \nabla f_{j+1}]$$

where $\nabla f_j = f(x_j) - f(x_{j-1})$ and $\nabla f_{j+1} = f(x_{j+1}) - f(x_j)$, $j = 1, 2, \cdots, n-1$, become

$$k_{j-1} + 4k_j + k_{j+1} = \frac{3}{h}(f_{j+1} - f_{j-1}), \quad j = 1, 2, \cdots, n-1$$

given $k_0 = f'(a)$, $k_n = f'(b)$ or any other known values. The other $k_1, k_2, \cdots, k_n$ are determined by solving the linear system of the $(n-1)$ equations

In order to determine the coefficients of the spline $g(x)$, we consider the interval $x_j \le x \le x_{j+1} = x_j + h$, let

$$p_j(x) = a_{j0} + a_{j1}(x - x_j) + a_{j2}(x - x_j)^2 + a_{j3}(x - x_j)^3, \quad j = 0, 1, 2, \cdots, n-1$$

Using Taylor's formula, we obtain

$$a_{j0} = p_j(x_j) = f_j; \quad a_{j1} = p_j'(x_g) = k_j$$

$$a_{j2} = \frac{1}{2} p_j''(x_j) = \frac{3}{h^2}(f_{j+1} - f_j) - \frac{1}{h}(k_{j+1} + 2k_j)$$

and $a_{j3} = \frac{1}{6} p_j'''(x_j) = \frac{2}{h^3}(f_j - f_{j+1}) + \frac{1}{h^2}(k_{j+1} - k_j)$

As a consequence,

$$p_j''(x_{j+1}) = 2a_{j2} + 6a_{j3}h = \frac{6}{h^2}(f_j - f_{j+1}) + \frac{2}{h}(k_j + 2k_{j+1})$$

EXAM. 1. Find the cubic spline $g(x)$ to the data given below

$$f_0 = f(0) = 0; \quad f_1 = f(1) = 1; \quad f_2 = f(2) = 6; \quad f_3 = 10, \quad k_0 = 0, \quad k_3 = 0$$

x	f(x)	$\Delta f(x)$	$\Delta^2 f(x)$
0	0	1	4
1	1	5	-1
2	6	4	
3	10		

Here $n = 3$, and $h = 1$, hence

$$k_{j-1} + 4k_j + k_{j+1} = \frac{3}{h}(f_{j+1} - f_{j-1}), \quad j = 1, 2$$

$$4k_1 + k_2 = 18$$
$$k_1 + 4k_2 = 27$$

For $j = 1$, $j = 2$ $\quad \therefore \quad \dfrac{16k_1 + 4k_2 = 72}{-15k_1 \quad\quad = -45}$

$$k_0 + 4k_1 + k_2 = 3(f_2 - f_0) = 3(6 - 0) = 18$$

$$k_1 + 4k_2 + k_3 = 3(f_3 - f_1) = 3(10 - 1) = 27$$

$$\therefore \quad k_1 = 3, \quad 3 + 4k_2 = 27, \quad k_2 = 6$$

$$k_0 = 0, \quad k_1 = 3, \quad k_2 = 6, \quad k_3 = 0$$

with $\quad j = 0$

$$a_{00} = f_0 = 0, \quad a_{01} = k_0 = 0, \quad a_{02} = 3(1) - 1(3 + 0) = 0$$

$$a_{03} = 2(0 - 1) + 1(3) = 1$$

$$\therefore \; P_0(x) = a_{00} + a_{01}(x - x_0) + a_{02}(x - x_0)^2 + a_{03}(x - x_0)^3$$

$$= x^3$$

with $\quad j = 1,$

$$P_1(x) = a_{10} + a_{11}(x - x_1) + a_{12}(x - x_1)^2 + a_{1n}(x - x_1)^3$$

$$= a_{10} + a_{11}(x - 1) + a_{12}(x - 1)^2 + a_{13}(x - 1)^3$$

$$a_{10} = f_1 = 1; \quad a_{11} = k_1 = 4; \quad a_{12} = 3(6 - 1) - (6 + 6) = 3$$

$$a_{13} = 2(1 - 6) + 1(6 + 3) = -10 + 9 = -1$$

$$\therefore \; P_1(x) = 1 + 4(x - 1) + 3(x - 1)^2 - (x - 1)^3$$

with $\quad j = 2$

$$P_2(x) = a_{20} + a_{21}(x - 2) + a_{22}(x - 2)^2 + a_{23}(x - 2)^3$$

$$a_{20} = k_2 = 6; \quad a_{21} = k_2 = 6; \quad a_{22} = 3(f_3 - f_2) - (k_3 + 2k_2)$$

$$a_{23} = 2(f_2 - f_3) + (k_3 + k_2) \qquad = 3(10 - 6) - (0 + 12)$$

$$= 2(6 - 10) + (0 + 6) = -2 \qquad = +12 - 12 = 0$$

$$\therefore \; P_2(x) = 6 + 6(x - 2) - 2(x - 2)^3$$

$$P_2(x) = 6 + 6(x - 2) - 2(x - 2)^3$$

There are four alternative choices to specify the end conditions.

1. $g''(x_0) = 0, \; g''(x_n) = 0$ the curvature at the end points is zero. This condition is called "free" or natural conditions and the fitted spline is called natural spline.

2. The slopes at each end is assumed to have specified values. $f'(x_0) = A, \; f'(x_n) = B$. We use the divided differences in our derivation $f'(x_1) = f'(x_0) = A = k_0$) and $f'(b) = f'(x_n) = B = k_n$

3. Take $g''(x_0) = g''(x_1)$ and $g''(x_{n-1}) = g''(x_n)$. This is equivalent to assuming that the end cubes approach parabolas at their extremities.

4. Take $f''(x_0)$ as a linear extrapolation from $f''(x_1)$ and $f''(x_2)$ and $f''(x_n)$ as a

linear extrapolation from $f''(x_{n-1})$ and $f''(x_{n-2})$. Only this condition gives cubic spline curves that match exactly to $f(x)$ when $f(x)$ is itself a cubic.

(A) For equally spaced points $x_0, x_1 = x_0 + h, \cdots, x_n = x_0 + nh$

Condition 1: $g''(x_0 = 0); g''(x_n) = 0$

$$k_0 = 0 \qquad\qquad k_n = 0 \qquad \text{known}$$

The other unknowns are $k_1, k_2, \cdots, k_{n-1}$ can be solved, with the coefficient matrix as:

$$\begin{bmatrix} 4h & h & & & \\ h & 4h & h & & \\ & h & 4h & h & \\ & & \vdots & & \\ & & & h & 4h \end{bmatrix}$$

Condition 2: $g'(x_0) = A; \quad g'(x_n) = B$

The corresponding coefficient matrix is

$$\begin{bmatrix} 2h & h & & & \\ h & 4h & h & & \\ & h & 4h & h & \\ \vdots & \vdots & & \vdots & \\ & & & h & 2h \end{bmatrix}$$

Condition 3: $k_0 = k_1; \quad k_n = k_{n-1}$ (the end cubic are parabolas)

$$\begin{bmatrix} 5h & h & & & \\ h & 4h & h & & \\ & h & 4h & & \\ \vdots & \vdots & \vdots & & \\ & & h & . & 5h \end{bmatrix}$$

Condition 4: $f''(x_0)$ and $f''(x_n)$ are linear extrapolations from k_1 and k_2 and k_n as a linear extrapolation from k_{n-1} and k_{n-2}

$$\begin{bmatrix} 6h & 0 & & & & \\ h & 4h & h & & & \\ & h & 4h & h & & \\ \vdots & \vdots & \vdots & \vdots & & \\ & & & 0 & 6h \end{bmatrix}$$

Determination of cubic spline $g(x)$ for un equal step length when the step length is h_i for these interval $[x_i, x_{i+1}]$; we have $h_i = x_{i+1} - x_i$, $i = 0,1,2,\cdots,n-1$

When the end conditions corresponding to x_0 as $k_0 = f'(x_0) = 0$ and $k_n = f'(x_n) = 0$, the equation of $k_1, k_2, \cdots, k_{n-1}$ in matrix form, we get

Row

$$\begin{matrix} 0 \\ 1 \\ 2 \\ \\ n-1 \\ n \end{matrix} \begin{bmatrix} h_0 & 2(h_0 + h_1) & h_1 & & & \\ & h_1 & 2(h_1 + h_2) & h_2 & & \\ & & h_2 & 2(h_2 + h_3)h_3 & & \\ & & & \vdots & & \\ & & & & \vdots & \\ & & & h_{n-2} & 2(h_{n-2} + h_{n-1}) & h_{n-1} \end{bmatrix} \begin{bmatrix} k_0 \\ k_1 \\ k_2 \\ \\ k_{n-1} \\ k_n = 0 \end{bmatrix}$$

$$= 6 \begin{bmatrix} f[x_1,x_2] - f[x_0,x_1] \\ f[x_2,x_3] - f[x_1,x_2] \\ f[x_3,x_4] - f[x_2,x_3] \\ \vdots \\ f[x_{n-1} - x_n] - f[x_{n-2} - x_{n-1}] \end{bmatrix}$$

Note : $f[x_0, x_1] = \dfrac{f(x_1) - f(x_0)}{x_1 - x_0} = \dfrac{f(x_0) - f(x_1)}{x_0 - x_1} = f[x_1, x_0]$

In general, $f[x_s, x_t] = \dfrac{f_t - f_s}{x_t - x_s} = \dfrac{f_s - f_t}{x_s - x_t} = f[x_t, x_s]$

Also $f[x_0, x_1, x_2] = \dfrac{f[x_1, x_2] - f[x_0, x_1]}{x_2 - x_0}$

$f[x_0, x_1, x_2] = f[x_2, x_1, x_0] = f[x_1, x_2, x_0]$

Now

$f[x_0, x_1, x_2] = \dfrac{f[x_1, x_2] - f[x_0, x_1]}{x_2 - x_0}$

$$= \frac{f(x_2)}{(x_2 - x_0)(x_2 - x_1)} + \frac{f(x_1)}{(x_1 - x_0)(x_1 - x_2)} + \frac{f(x_0)}{(x_0 - x_1)(x_0 - x_2)}$$

$$= \sum \frac{f(x_0)}{(x_0 - x_1)(x_0 - x_1)}$$

Which is symmetric function $f(x)$ of arguments x_0, x_1 and x_2

2. Find the cubic splines for the following table of values:

$$x: \quad 1 \quad 2 \quad 3$$
$$y: \quad -6 \quad -1 \quad 16 \qquad W_0 = 0, \quad W_2 = 0$$

We have

$$W_{j-1} + 4W_j + W_{j+1} = 6(y_{j-1} - 2y_j + y_{j+1})$$

$$j = 1$$

$$W_0 + 4W_1 + W_2 = 6(y_0 - 2y_1 + y_2)$$

$$4W_1 = 6(-6 + 2 + 16) = 72 \quad \therefore \quad W_1 = 18$$

The spline polynomial in the interval (x_j, x_{j+1}) is

$$g(x) = \frac{1}{6}[(x_j - x)^3 W_{j-1} + (x - x_{j-1})W_j] + (x_j - x)[f_{j-1} - \frac{1}{6}W_{j-1}]$$

$$+ (x - x_{j-1})(t_j - \frac{1}{6}W_j)$$

$j = 1$, the interval $(x_0, x_1) = (1, 2)$

$$g(x) = \frac{1}{6}[(x_1 - x)^3 W_0 + (x - x_0)W_1] + (x_1 - x)[f_0 - \frac{1}{6}W_0] + (x - x_0)[f_0 - \frac{1}{6}W_1]$$

$$= \frac{1}{6}[(2 - x)^3 W_0 + (x - 1)W_1] + (2 - x)[-6 - \frac{1}{6} \times 0] + (n - 1)[-1 - \frac{1}{6} \cdot 78]$$

$$= 3(x - 1)^3 + (2 - x)(-6) + (x - 1)(-1 - 3)$$

$$= 3(x - 1)^3 + (2 - x)(-6) + (n - 1)(-1 - 3)$$

$$= 3(x - 1)^3 - 12 + 6x + 4 - 4x$$

$$= 3(x - 1)^3 + 2x - 8$$

$$= 3(x^3 - 3x^2 + 3x - 1) + 2x - 8$$

$$= 3x^3 - 9x^2 + 9x - 3 + 2x - 8$$

$$= 3x^3 - 9x^2 + 11x - 11 \qquad (1 \le x \le 2)$$

The spline function in the interval (2,3) is given by

$$g(x) = \frac{1}{6}[(x_2 - x)^3 W_1 + (x - x_1)^3 W_2] + (x_2 - x)(f_1 - \frac{1}{6}W_1) + (x - x_1)(f_2 - \frac{1}{6}W_2)$$

$$= \frac{1}{6}[(3 - x)^3 \cdot 18] + (3 - x)(-1 - \frac{1}{6}(18)) + (x - 2)(16)$$

$$= 3(3 - x)^3 - 4(3 - x) + 16(x - 2)$$

$$= 3[27 - 27x + 9x^2 - x^3] - 12 + 4x + 16x - 32$$

$$= 81 - 81x + 27x^2 - 3x^3 - 12 + 20x - 32$$

$$= 37 - 61x + 27x^2 - 3x^3$$

$$= -3x^3 + 27x^2 - 61x + 37 \qquad (2 \leq x \leq 3)$$

3. Fit a natural Cubic spline curve for the data given below

	x	f(x)		
x_0 :	1	22	$f[x_1, x_0] = 8$	$f[x_2, x_1] - f[x_1, x_0] = 18$
x_1 :	2	30	$f[x_2, x_1] = 26$	$f[x_3, x_2] - f[x_2, x_1] = -18$
x_2 :	4	82	$f[x_3, x_2] = 8$	
x_3 :	7	106		

$$h_0 = 1, \quad h_1 = 2, \quad h_2 = 3$$

For the natural cubic spline we use the end condition 1: $k_0 = 0, \quad k_3 = 0$

The matrix equation involving k_1 and k_2 is

$$\begin{bmatrix} h_0 & 2(h_0 + h_1) & h_1 & 0 & 0 \\ 0 & h_1 & 2(h_1 + h_2) & h_2 & \\ 0 & 0 & h_2 & 2(h_2 + h_3) & h_3 \\ \cdots & \cdots & \cdots & \cdots & \cdots \end{bmatrix} \begin{bmatrix} k_0 = 0 \\ k_1 \\ k_2 \\ k_3 = 0 \end{bmatrix}$$

$$\begin{bmatrix} 2(h_0 + h_1) & h_1 = 2 \\ h_1 = 2 & 2(h_1 + h_2) \end{bmatrix} \begin{bmatrix} k_1 \\ k_2 \end{bmatrix} = 6 \begin{bmatrix} 18 \\ -18 \end{bmatrix}$$

Solving we get $k_1 = 23.14, \quad k_2 = -75.42$

$$p_j(x) = a_{j0} + a_{j1}(x - x_j) + a_{j2}(x - x_j)^2 + a_{j3}(x - x_j)^3, j = 0,1,2,\cdots$$

The spline function in the interval $[x_0, x_1] = [1, 2]$ is

$$p_0(x) = a_{00} + a_{01}(x - x_0) + a_{02}(x - x_0)^2 + a_{03}(x - x_0)^3$$

We obtain

$$a_{00} = p_0(x_0) = f_0 = 1, \quad a_{01} = p_{01}^1(x_0) = k_0 = 0$$

$$a_{02} = \frac{1}{2}p_0^{''}(x_0) = 24; \quad a_{03} = \frac{1}{6}p_0^{'''}(x_0) = 7.143$$

$$\therefore \ p_0(x) = 1 + 24x^2 + 7.143x^3$$

$$j = 1, \quad p_1(x) = a_{10} + a_{11}(x - x_1) + a_{12}(x - x_1)^2 + a_{13}(x - x_1)$$

$$= a_{10} + a_{11}(x - 2) + a_{12}(x - 2)^2 + a_{13}(x - 2)^3$$

$$a_{10} = f_1 = 22, \quad a_{11} = k_1 = 23.14, \quad a_{12} = \frac{2}{4}(f_2 - f_3) - \frac{1}{2}(k_2 + 2k_1)$$

$$= 27.57$$

$$a_{13} = \frac{2}{8}(f_1 - f_2) + \frac{1}{4}(k_2 - k_1) = 11.64$$

$$\therefore \ p_1(x) = 22 + 23.14(x - 2) + 27.57(x - 2)^2 + 11.645(x - 2)^3$$

$$j = 2 \quad p_2(x) = a_{20} + a_{21}(x - 4) + a_{22}(x - 4)^2 + a_{23}(x - 4)^3$$

$$a_{20} = f_2 = 82; \quad a_{21} = k_2 = -75.42$$

$$a_{22} = \frac{3}{16}(f_3 - f_2) - \frac{1}{4}(k_3 + 2k_2) = -34.24$$

$$a_{23} = \frac{1}{64}(f_2 - f_3) + \frac{1}{16}(k_3 - k_1) = 5.085$$

$$\therefore \ p_2(x) = 8.2 - 75.42(x - 4) - 34.24(x - 4)^2 + 5.085(x - 4)^3$$

A practical technique of "spline-fitting" for interpolating between the data points $(x_0, y_0), (x_1, y_1), \cdots, (x_n, y_n)$ by means of cubic splines.

Let the cubic spline $g(x)$ to be determined for the given function $y(x) = f(x)$ whose values are known at the n equidistance nodes

$$x_1 = x_0 + ih, \quad i = 1, 2, \cdots, n \quad \text{such that}$$

$$g(x_0) = f(x_0) = y_0, \cdots g(x_n) = f(x_n) = f(x_0 + nh) = y_n$$

for $n = 0, 1, 2, \cdots, n$; h being the step length.

In addition we require that $g'(x_0) = W_0$, $g'(x_n) = W_n$, the numbers W_0 and W_n are known indicating the tangential direction of $g(x)$ at the end points. A unique cubic spline can be determined in each subinterval (x_j, x_{j+1}) $j = 0, 1, 2, \cdots, n$, such that

i. $g(x)$ is a linear polynomial outside the interval (x_0, x_n)

ii. $g(x)$ is a cubic polynomial in each of the subintervals

iii. $g'(x)$ and $g''(x)$ are continuous at each point

Also $g'(x) = f'(x_0)$ and $g'(x) = f'(x_n)$ and are called clamped conditions.

Other conditions of practical interests are the free or natural conditions $g''(x_0) = 0$, $g''(x_n) = 0$ (zero curvature at the ends)

Since $g(x)$ is cubic in each subintervals $g''(x)$ shall be linear.

Consider the *j-th* subinterval (x_j, x_{j+1}), so that $x_{j+1} - x_j = h$; we can write

$$g''(x) = \frac{1}{h}[(x_{j+1} - x)g''(x_j) + (x - x_j)g''(x_{j+1})]$$

Integrating twice, we have

$$g(x) = \frac{1}{h}\left[\frac{(x_{j+1} - x)^3}{3!}g''(x_j) + \frac{(x - x_j)^3}{3!}g''(x_{j+1})\right] + a_j(x_{j+1} - x) + b_j(x - x_j) \tag{1}$$

The constants of integration a_j, b_j are determined by substituting the values of $y(x) = f(x) = g(x)$ at x_j and x_{j+1}.

Thus $a_j = \frac{1}{h}\left[y_j - \frac{h^2}{3!}g''(x_j)\right]$ and $b_j = \frac{1}{h}\left[y_{j+1} - \frac{h^2}{3!}g''(x_{j+1})\right]$

Substituting the values of a_j and b_j and writing $g''(x_j) = W_j$, (1) takes the form

$$g(x) = \frac{(x_{j+1} - x)^3}{6h}W_j + \frac{(x - x_j)^3}{6h}W_{j+1}$$

$$+ \frac{x_{j+1} - x}{h}\left(y_j - \frac{h^2}{6}W_j\right) + \frac{x - x_j}{h}\left(y_{j+1} - \frac{h^2}{6}W_{j+1}\right) \tag{2}$$

$$\therefore \ g'(x) = \frac{(x_{j+1} - x)^2}{2h}W_j + \frac{(x - x_j)^2}{2h}W_{j+1} - \frac{h}{6}(W_{j+1} - W_j) + \frac{1}{h}(y_{j+1} - y_j)$$

To impose the continuity of $g'(x)$, we set

$g'(x - \epsilon) = g'(x + \epsilon)$ as $\epsilon \to 0$

$$\therefore \ \frac{h}{6}(2W_j + W_{j-1}) + \frac{1}{h}(y_j - y_{j-1}) = -\frac{h}{6}(2W_j + W_{j+1}) + \frac{1}{h}(y_{j+1} - y_j)$$

or $W_{j-1} + 4W_j + W_{j+1} = \dfrac{6}{h^2}(y_{j-1} - 2y_j + y_{j+1})$, $j = 1,2,\cdots,n-1$ (3)

Since the graph is linear for $x < x_0$ and $x > x_n$, we have $W_0 = 0$, $W_n = 0$ (4)

(3) and (4) give $(n+1)$ equations in $n+1$ unknowns $W_j (j = 0,1,2,\cdots,n)$ which can be solved.

Substituting the value of W_j in (2), gives the required cubic spline in each of the n subintervals, (x_j, x_{j+1}), $j = 0,1,2,\cdots n-1$

EXAM. 1: Find the cubic splines for the following table of values

x: 1 2 3 4
y: 1 5 4 8   here $x = 1$ $x = 3$

Using the relation

$W_{j-1} + 4W_j + W_{j+1} = 6(y_{j-1} - 2y_j + y_{j+1})$, $j = 1,2$

$j = 1$ $W_0 + 4W_1 + W_2 = 6(y_0 - 2y_1 + y_2) = 6X_2 = 12$

$j = 2$ $W_1 + 4W_2 + W_3 = 6(y_1 - 2y_2 + y_3) = 6x - 9 = -54$

$$4W_1 + W_2 = 12$$
$$W_1 + 4W_2 = -54$$

$W_0 = 0$, $W_3 = 0$ $\therefore$ $\dfrac{16W_1 + 4W_2 = 48}{-15W_1 = -102}$ $W_1 = \dfrac{102}{15} = \dfrac{34}{5}$

$W_2 = 12 - \dfrac{136}{5} = \dfrac{60 - 136}{5}$

$W_1 = \dfrac{34}{5}$; $W_2 = -\dfrac{76}{5}$

The cubic spline in the interval $(1,2)$: (x_0, x_1) is given by

$$g(x) = \dfrac{(x_{j+1} - x)^3}{6h} \cdot W_j + \dfrac{(x - x_j)^3}{6h} W_{j+1} + \dfrac{(x_{j+1} - x)}{h}\left(y_j - \dfrac{h^2}{6}W_j\right)$$

$$+ \dfrac{(x - x_j)}{h}\left(y_{0+1} - \dfrac{h^2}{6}W_{j+1}\right) \qquad j = 0,1,2,\cdots$$

$j = 0$

$$g(x) = \frac{(x_1-x)^3}{6}W_0 + \frac{(x-x_0)^3}{6}W_1 + (x_1-x)(y_0 - \frac{1}{6}W_0) + (x-x_0)(y_1 - \frac{1}{6}W_1)$$

$$= \frac{(x-1)^3}{6} \cdot \frac{34}{5} + (2-x)(1) + (x-7)(5 - \frac{1}{6} \cdot \frac{34}{5})$$

$$= \frac{34}{30}(x-1)^3 + 2 - x + \frac{116}{30}x - \frac{116}{30}$$

$$= \frac{17}{15}(x-1)^3 + \frac{86}{30}x - \frac{56}{30}$$

$$= \frac{17}{15}(x-1)^3 + \frac{43}{15}x - \frac{28}{15}$$

$$= \frac{17}{15}\left[x^3 - 3x^2 + 3x - 1 + \frac{43}{17}x - \frac{28}{17}\right]$$

$$g(x) = \frac{17}{15}[x^3 - 3x^2 + \frac{94}{17}x - \frac{45}{17}]$$

When $x=1$ $\quad g(x) = \frac{17}{15}\left[1 - 3 + \frac{94}{17} - \frac{45}{17}\right]$

When $x=2$ $\qquad = \frac{17}{15} \times \frac{15}{17} = 1$ or $(y=1)$

Hence the cubic spline $g(x) = \frac{17}{15}(x^3 - 3x^2 + \frac{94}{17}x - \frac{45}{17})$ in $(1 \le x \le 2)$

Similarly the cubic splines can be obtained for the other intervals.

Exercises (Splines)

1. Find the cubic spline for the data: $f(0) = 1$; $f(2) = 5$, $k_0 = -1$, $k_2 = 5$

2. Find the cubic spline for the data: $f(-1) = 2$; $f(1) = 0$; $f(3) = 22$; $f(4) = 44$, with $k_0 = k_3 = 0$

3. Find the cubic spline $g(x)$ to the given data, with k_0 and k_n as indicated.

   i. $f_0 = f(-2) = 1$; $f_1 = f(0) = 5$; $f(2) = 17$, $k_0 = -2$, $k_2 = -14$

   ii. $f_0 = f(0) = 1$; $f_1 = f(1) = 2$; $f_2 = f(2) = 7$; $f_3 = f(3) = 11$, $k_0 = 0$, $k_3 = 0$

   iii. $f(-2) = f(-1) = f(1) = f(2) = 0$, $f(0) = 1$,

   $k_0 = k_4 = 0$. Is $g(x)$ the approximation by the spline to $f(x)$ is even? Why? Also find the interpolation polynomial corresponding to the give f – values

4. Obtain the cubic spline approximation for the function given in the following table:

x: 1 2 3 4

$f(x)$: 1 5 11 8

Use the end conditions $f''(1) = 0 = f''(4)$

5. Obtain the cubic spline approximation for the function given in the tabular form:

x: 0 1 2 3

$f(x)$: 1 2 33 244 and $k_0 = f''(0) = 0$, $k_3 = f''(3) = 0$

[Ans: Interval Corresponding cubic spline

[0,1] $1 + 5x - 4x^2$

[1,2] $-53 + 167x - 162x^2 + 50x^3$

[2,3] $715 - 985x + 414x^2 - 46x^3$]

6. Fit the following four points by the cubic splines:

i: 0 1 2 3

x_i: 1 2 3 4

y_i: 1 5 11 8

Use the end conditions $y''_0 = y''_3 = 0$

Hence compute (i) $y(1.5)$ and (ii) $y'(2)$

7. Fit for the following four points the cubic splines with end conditions $f''(0.2) = 0$ and

$f''(0.5) = 0$

x: 0.2   0.3   0.4   0.5

$f(x)$: 0.185 0.106 0.093 0.240

{Hint: The equations for finding k_1 and k_2 are given by $k_{j-1} + 4k_j + k_{j+1} = \dfrac{3}{h}[f_{j+1} - f_{j-1}]$],

$j = 1, 2$. We get $4k_1 + k_2 = 2.76$, $k_1 + 4k_2 = 4.02$, $k_1 = 0.463$, $k_2 = 0.888$

$p_0(x) = 0.2 + 1.94(x - 0.2)^2 + 111.2(x - 0.2)^3$ in $[2,3]$

$p_1(x) = 0.106 + 0.466(x - 0.3) + 22.24(x - 3)^2 + 30.20(x - 0.3)^3 \cdot [0.3, 0.4]$ in $[3,5]$}

8. Determine the parameters a, b, c, d so that $g(x)$ is a natural cubic spline:

$g(x) = a + b(x - 1) + c(n - 1)^2 + d(x - 1)^3$, $x \in [0,1]$

[Ans: $a = -4$, $b = -6$, $c = -3$, $d = -1$]

9. Determine the coefficients in the function

$$g(x) = \begin{cases} x^3 - 1, & -a \le x \le 0 \\ ax^3 + bx^2 + cx + d, & 0 \le x \le 5 \end{cases}$$

so that it is a cubic spline that takes the value 2 when

13.7 NUMERICAL DIFFERENTIATION

The following are the three techniques used to obtain numerical differentiation.

(a) Methods based on interpolation.

(b) Methods based on finite difference operators.

(c) Methods based on undetermined co-efficients.

13.7.1 Methods based on interpolation

The values of the function $f(x)$ are given at a set of points $x_0, x_1, \ldots, x_n$. First the interpolating polynomials $P_n(x)$ are obtained and these polynomials are differentiated r times ($n \geq r$) to obtain $p_n^{(r)}(x)$. The value of $P_n^{(r)}(x_k)$ gives the approximante value of $f^{(r)}(x)$ at the nodal point x_k. Though $P_n(x)$ and $f(x)$ have the same values at the nodal points, yet the derivatives may differ considerably not only at these points, but also at any non-nodal point. The error of approximation of the *r-th* order at any point x is

$$E^{(r)}(x) = f^{(r)}(x) - P_n^{(r)}(x) \tag{1}$$

In the case of **non-uniform nodal points,** if (x_i, f_i), i=0,1, n are n+1 distinct tabular points, then the Lagrange interpolating polynomial fitting this data is given by

$$P_n(x) = \sum_{k=0}^{n} l_k(x) f_k, \tag{2}$$

where $lk(x)$ is the Lagrange fundamental polynomial, given by

$$l_k(x) = \frac{\pi(x)}{(x - x_k) \pi'(x_k)}, \quad with$$

$fk = f(x_k) =$ and $(x) = (x = x_0)(x - x_1) \ldots (x - x_n)$

The error of approximation will be given by

$$E_n(x) = f(x) - P_n(x) = \frac{\pi(x)}{(n+1)!} f^{(n+1)}(\xi), \tag{3}$$

$x_0 < \zeta < x_n$, for any point x.

Differentiating (2) and (3) with respect f_0 x, we obtain

$$P_n'(x) = \sum_{k \geq 0}^{n} l_k'(x) f_k \quad and \tag{4}$$

$$E_n'(x) = \frac{\pi'(x)}{(n+1)!} f^{(n+1)}(\xi) + \frac{\pi(x)}{(n+1)!} \frac{d}{dx} \left\{ f^{(n+1)}(\xi) \right\} \tag{5}$$

The right hand side of (5) can not be directly evaluated, due to the function $\zeta(x)$ being unknown. However at a nodal point x_k, $\pi(x_k) = 0$ and we get

$$E_n'(x_k) = \frac{\pi'(x_k)}{(n+1)!} f^{(n+1)}(\xi), \quad x_0 < \xi < x_n, \tag{6}$$

provided $\dfrac{d}{dx} \left\{ f^{(n+1)}(\xi) \right\}$ remains bounded.

Hence for any r, $1 \le r < n$, we obtain

$$f^{(r)}(x) \approx P_n^{(r)}(x) = \sum_{k=0}^{n} l_k^{(r)}(x) f_k \quad \text{at any point x.} \tag{7}$$

If we use **linear interpolation,** we have

$$l_0(x) = \frac{x - x_1}{x_0 - x_1}; \ l_1(x) = \frac{x - x_0}{x_1 - x_0} \quad and$$

$$P_1(x) = \frac{(x - x_1)}{(x_0 - x_1)} f_0 + \frac{(x - x_0)}{(x_1 - x_0)} f_1 \tag{8}$$

$$and \ P_1'(x) = \frac{f_1 - f_0}{x_1 - x_0}, \text{ a constant for all } x \in [x_0, x_1] \tag{9}$$

The respective errors are $E_1'(x_0) = \dfrac{x_0 - x_1}{2} f''(\xi)$ and

$$E_1'(x) = \frac{x_1 - x_0}{2} f''(\xi), \quad x_0 < \xi < x_1$$

In the case of a quadratic interpolation, we will get

$$P_2(x) = l_0(x) f_0 + l_1(x) f_1 + l_2(x) f_2$$

and $P_2'(x) = l_0'(x) f_0 + l_1'(x) f_1 + l_2'(x) f_2$ and on simplification, we get

$$P_2'(x_0) = \frac{2x_0 - x_1 - x_2}{(x_0 - x_1)(x_0 - x_2)} f_0 + \frac{x_0 - x_2}{(x_1 - x_0)(x_1 - x_2)} f_1 + \frac{x_0 - x_1}{(x_2 - x_0)(x_2 - x_1)} f_2 \tag{10}$$

The error term here is $E_2^1(x_0) = \dfrac{1}{6} (x_0 - x_1)(x_0 - x_2) f''(\xi), x_0 < \xi < x_2$

Similarly, we can obtain

$$P''(x) = 2 \left[\frac{f_0}{(x_0 - x_1)(x_0 - x_2)} + \frac{f_1}{(x_1 - x_0)(x_1 - x_2)} + \frac{f_2}{(x_2 - x_0)(x_2 - x_1)} \right]$$

which is constant for all $x \in [x_0, x_2]$. The error h at the tabular points x_0, x_1, and x_2 can be obtained.

In the case of uniform nodal points, if h is the step length between any two consecutive nodal points, We have $x_i = x_0 + ih$, and $f_i = f(x_i)$ for i = 1,2.....n

Using **Linear interpolation,** we have

$$f'(x_0) \approx P_1'(x_0) = \frac{f_1 - f_0}{h} \quad and \quad E_1'(x_0) = f'(x_0) - \frac{f_1 - f_0}{h}$$

$$= f'(x_0) - \frac{1}{h} \left[f(x_0) + \frac{h}{1!} f'(x_0) + \frac{h^2}{2!} f''(\xi) - f(x_0) \right]$$

$$= -\frac{h}{2} \; f'' \; l(\xi), \; x_0 \; < \; \xi \; < \; x_1$$

Similarly, using quadratic interpolation, we have

$$f'(x_0) \; = \; p_2' \left(x_0\right) = \frac{-3f_0 + 4f_1 - f_2}{2h}, \quad \text{with (local) truncation and as}$$

$$E_2{}'(x_0) = f'(x_0) \; -\frac{-3f_0 + 4f_1 - f_2}{2h^2} = -\frac{h^2}{3} f'''(\xi), x_0 < \xi < x_2$$

and $\; f''(x_0) = P_2''(x_0) \; = \frac{f_0 - 2f_1 + f_2}{h^2},$

with local truncation error $\; E_2'\left(x_0\right) = hf''(\xi) \; + \mathrm{O}\left(h^2\right), x_0 < \xi < x_2$

Definition

A Numerical differentiation method is said to be of order p if

$\left| f^{(r)}(x) - p^{(r)}(x) \right| < c.h^p$ where c is a constant independent of x.

13.7.2 *Methods based on finite differences*

Consider the relation

$$Ef\left(x\right) = f\left(x + h\right) = f\left(x\right) + hf'\left(x\right) + \frac{h^2}{21} f''(x) \rightarrow .....$$

$$= \left(1 + hD + \frac{1}{21}h^2D^2 + ... \right) f(x) = e^{hD} f(x),$$

where $D = \dfrac{d}{dx}$ is called the differential operator. Symbolically, we write $e^{hD} = E$ or

$hd = \log^E$. Consequently, we have

$$hD = \log E = \begin{cases} \log(1 + \Delta) = \Delta - \dfrac{1}{2}\Delta^2 + \dfrac{1}{3}\Delta^3 - ...... \\[2mm] -\log(1 - \nabla) = \nabla + \dfrac{1}{2}\nabla^2 + \dfrac{1}{3}\nabla^3 + ... \\[2mm] 2shi \;\; h^{-1}\left(\dfrac{\delta}{2}\right) = \delta - \dfrac{1}{2^2}, \dfrac{1}{3!}\delta^3 + ... \end{cases} \tag{1}$$

Also we can write.

$$hf'(x_k) = hDf(x_k) = \begin{cases} \Delta f k - \dfrac{1}{2}\Delta^2 f_k + \dfrac{1}{3}\Delta^3 f k - \ldots \\[2ex] \nabla f k + \dfrac{1}{2}\nabla^2 f_k + \dfrac{1}{3}\nabla^3 f_k + \ldots \\[2ex] \delta f_k - \dfrac{1}{2^2}\cdot\dfrac{1}{3!}\delta^3 f_k + \ldots \end{cases}$$

(2)

In addition, since at write $\mu = \sqrt{1 + \dfrac{\delta^2}{4}}$, we can also write

$$hD = \dfrac{\mu}{\sqrt{1 + \dfrac{\delta^2}{4}}}\left(2shi \quad h^{-1}\dfrac{\delta}{2}\right)$$

$$= \mu\left(\delta - \dfrac{1^2}{3!}\delta^3 + \dfrac{1^2.2^2}{5!}\delta^5 - \ldots\right)$$

(3)

Thus we get $hf'(x_k) = \mu\delta f_k - \dfrac{1^2}{3!}\mu\delta^3 f_k + \dfrac{1^2.2^2}{5!}\mu\delta^5 f_k \ldots$

Similarly, we obtain

$$h^r D^r = \begin{cases} \Delta^r - \dfrac{1}{2}r\Delta^{r+1} + \dfrac{r(3r+5)}{24}\Delta^{r+2} - \ldots \\[2ex] \nabla^r + \dfrac{1}{2}r\nabla^{r+1} + \dfrac{r(3r+5)}{24}\nabla^{r+2} + \ldots \end{cases}$$

(4)

$$\begin{cases} \mu\delta^r - \dfrac{r+3}{24}\mu\delta^{r+2} + \dfrac{5r^2+(52r+135)}{5760}\mu\delta^{r+4} - \ldots, \\[2ex] \delta^r - \dfrac{r}{24}\delta^{r+2} + \dfrac{r(52r+22)}{5760}\delta^{r+4} - \ldots, \end{cases}$$

In particular, the finite difference methods yield for the first two derivatives (i.e, , $r = 1$ and $r = 2$) the following results.

$$hf'(x_k) = \begin{cases} \Delta f_k - \dfrac{1}{2}\Delta^2 f_k + \dfrac{1}{3}\Delta^3 f_k - \ldots\ldots \\[2mm] \nabla f_k + \dfrac{1}{2}\nabla^2 f_k + \dfrac{1}{3}\nabla^3 f_k + \ldots\ldots, \\[2mm] \delta f_k - \dfrac{1}{6}.\mu\delta^3 f_k + \dfrac{1}{30}\mu\delta^5 f_k + \ldots\ldots \end{cases}$$

$$(5)$$

$$h^2 f''(x_k) = \begin{cases} \Delta^2 f_k - dt\ \ \Delta^3 f_k + \dfrac{11}{12}\Delta^4 fk - \ldots\ldots \\[2mm] \nabla^2 f_k + \nabla^3 f_k + \dfrac{11}{12}\nabla^4 f_k + \ldots\ldots, \\[2mm] \delta^2 f_k - \dfrac{1}{12}\delta^4 f_k + \dfrac{1}{90}\delta^6 f_k + \ldots\ldots \end{cases}$$

For practical purposes maintaining only the first term in the above formulas, we set

$$f'(x_k) = \begin{cases} \dfrac{1}{h}\ \ (f_{k+1} - f_k) \\[2mm] \dfrac{1}{h}\ \ (f_k - f_{k-1}) \\[2mm] \dfrac{1}{2h}\ (f_{k+1} - f_{k-1}) \end{cases}$$

$$(7)$$

$$f''(x_k) = \begin{cases} \dfrac{1}{h^2}\ (f_{k+2} - 2f_{k+1} + f_k) \\[2mm] \dfrac{1}{h^2}\ (f_k - 2f_{k-1} + f_{k-2}) \\[2mm] \dfrac{1}{2h}\ (f_{k-1} - 2f_k + f_{k+1}) \end{cases}$$

$$(8)$$

13.7.3 Methods based on undetermined coefficients.

Here $f^{(r)}(x)$ is expressed as a linear combination of the values of $f(x)$ at an arbitrarily chosen set of tabular points. Assuming that the tabular points are equispaced with step length h, we write as

$$h^r f^{(r)}(x_k) = \sum_{j=-p}^{p} a_j f_{k+j} \quad \text{(1) for symmetric arrangement of tabular points.} \quad (1)$$

or $h^r f^{(r)}(x_k) = \sum_{j=\pm\lambda}^{p} a_j f_{k+j}$ (ii) for non- symmetric arrangement of tabular points . The local truncation error is defined by (2)

$$E^{(r)}(x_k) = \frac{1}{h^r}\left\{ h^r f^{(r)}(x_k) - \sum_{j=-p}^{p} a_j f_{k+j} \right\} \quad (3)$$

or

$$E^{(r)}(x_k) = \frac{1}{h^r}\left\{ h^r f^{(r)}(x_k) - \sum_{j=\pm\lambda}^{p} a_j f_{k+j} \right\} \tag{4}$$

The coefficients a_j's are to be determinded by using Taylor's series about the point xk and equating the coefficients of various order derivatives on both sides we obtain the required number of equations to determine them. This method yields,

$$f''(x_k) = \frac{1}{12h^2}\left\{ -f_{k-2} + 16f_{k-1} - 30f_k + 16f_{k+1} - f_{k+2} \right\} \tag{5}$$

and $Error = -\dfrac{h^4}{90} f^{(6)}(\xi), x_{k-2} < \xi < x_{k+2}$ (6)

13.7.4 *Numerical partial differential methods*

We consider only one variable at a time and treat the other variables. Constant, consider a function $f(x,y)$ of two variables x and y. In the (x,y) plane the values of the function $f(x_i, y_i)$ with spacing h and k in the x and y directions respectively. We have

$x_i = x_0 + i\,h,\ y_i = y_0 + y\,k,\ i,j = 1,2\ldots.$

We write the various partial derivatives in terms of the $f(x_i, y_j) = f_{i,j}$ for $i, j = 1,2,\ldots$ as follows:

$$\left(\frac{\partial f}{\partial x}\right)_{(xi,yj)} = \begin{cases} \dfrac{1}{h}\ \left(f_{i+1,j} - f_{i,j} \right) + O(h) \\[2mm] \dfrac{1}{h}\ \left(f_{i,j} - f_{i-1,j} \right) + O(h) \\[2mm] \dfrac{1}{2h}\ \left(f_{i+1,j} - f_{i-1,j} \right) + O\!\left(h^2 \right) \end{cases} \tag{1}$$

$$\left(\frac{\partial f}{\partial y}\right)_{(xi,yj)} = \begin{cases} \dfrac{1}{k}\ \left(f_{i,j+1} - f_{i,j} \right) + O(k) \\[2mm] \dfrac{1}{k}\ \left(f_{i,j} - f_{i,j-1} \right) + O(k) \\[2mm] \dfrac{1}{2k}\ \left(f_{i,j+1} - f_{i,j-1} \right) + O\!\left(k^2 \right) \end{cases} \tag{2}$$

$$\left(\frac{\partial^2 f}{\partial x^2}\right)_{(xi,yj)} = \left(\frac{1}{h^2}\right)\left[f_{i-1,j} - 2f_{i,j} + f_{i+1,j} \right] + O\!\left(h^2 \right) \tag{3}$$

$$\left(\frac{\partial^2 f}{\partial y^2}\right)_{(xi,yj)} = \left(\frac{1}{k^2}\right)\left[f_{i,j-1} - 2f_{i,j} + f_{i,j+1} \right] + O\!\left(k^2 \right) \tag{4}$$

Since $\dfrac{\partial^2 f}{\partial x \partial y} = \dfrac{\partial}{\partial x}\left(\dfrac{\partial f}{\partial y}\right) = \dfrac{\partial}{\partial y}\left(\dfrac{\partial f}{\partial x}\right)$

We can write

$$\left(\dfrac{\partial^2 f}{\partial x \partial y}\right)_{(xi,yj)} = \dfrac{\partial}{\partial x}\left(\dfrac{\partial f}{\partial y}\right)_{(xi,yj)}$$

$$\approx \dfrac{\partial}{\partial x}\left[\left(\dfrac{1}{2k} f_{i,j+1} - f_{i,j-1}\right)\right]$$

$$\approx \dfrac{1}{2k}\left[\dfrac{1}{2h}\left\{f_{i+1,j+1} - f_{i-1,j+1}\right\} - \dfrac{1}{2h}\left\{f_{i+1,j-1} - f_{i-1,j-1}\right\}\right]$$

$$\approx \dfrac{1}{4hk}\left\{f_{i+1,j+1} - f_{i-1,j+1} - f_{i+1,j-1} + f_{i-1,j-1}\right\} \tag{5}$$

and $\left(\dfrac{\partial^2 f}{\partial x \partial y}\right)_{(xi,yj)} = \dfrac{\partial}{\partial y}\left(\dfrac{\partial f}{\partial x}\right)_{(xi,yj)} = \dfrac{\partial}{\partial y}\left\{\dfrac{1}{2h}\left(f_{i+1,j} - f_{i-1,j}\right)\right\}$

$$\approx \dfrac{1}{4hk}\left\{f_{i+1,j+1} - f_{i+1,j-1} - f_{i-1,j+1} + f_{i-1,j-1}\right\} \tag{6}$$

Which is same as (5)

(5) and (6) are of order $O\left(h^2 + k^2\right)$

Jacobian matrix:

If $f_1(x,y)$ and $f_2(x,y)$ are two given functions, then the Jacobian matrix at the point (x,y) is given as

$$J = \begin{bmatrix} \dfrac{\partial f_1}{\partial x} & \dfrac{\partial f_1}{\partial y} \\[2mm] \dfrac{\partial f_2}{\partial x} & \dfrac{\partial f_2}{\partial y} \end{bmatrix}$$

13.7.5 *Examples*

EXAM. 1: Find the first and second derivatives of the function tabulated below, at the point x=1.5

x	f(x)	Δ	Δ	Δ	
1.5	3.375				
		3.625			
2.0	7.000		3.000		
		6.625		0.750	
2.5	13.625		3.750		0
		10.375		0.750	
3.0	24.000		4.500		0
		14.875		0.750	
3.5	38.875		5.250		
		20.125			
4.0	59.000				

The Newton-Gregory forward difference formula is

$$f(a+xh) = f(a) + x\ \Delta f(a) + \frac{x(x-1)}{2!}\Delta^2 f(a) + \frac{x(x-1)(x-2)}{3!}\Delta^3 f(a) + \ldots \ldots \ldots (1)$$

Differentiating with respect to x thrice and then putting $x = 0$ in the equations obtained, we get.

$$h\ f'(a) = \Delta f(a) - \frac{1}{2}\Delta^2 f(a) + \frac{1}{3}\Delta^3 f(a)$$

$$h^2 f''(a) = \Delta^2 f(a) - \Delta^3 f(a)$$

$$h^2 f'''(a) = \Delta^3 f(a)$$

Here $a = 1.5$ and $h = 0.5$.

$$\therefore f'(1.5) = \frac{1}{0.5}\left[3.625 - \frac{1}{2}(3.000) + \frac{1}{3}(0.750)\right] = 4.750$$

$$f''(1.5) = \frac{1}{0.25}[3.000 - 0.750] = 9.000$$

$$f'''(1.5) = \frac{1}{0.125}[0.750] = 6.000$$

The function $f(x)$ which is tabulated above is

$$f(x) = f_0 + u.\Delta f_0 + \frac{u(u-1)}{2!}\Delta^2 f_0 + \frac{u(u-1)(u-2)}{3!}\Delta^3 f_0$$

where $h = \dfrac{x-x_o}{h} = \dfrac{x-1.5}{0.5} = 2(x-1.5)$

$\therefore f(x) = 3.375 + 2(x-1.5)3.625 + \dfrac{2(x-1.5)(2x-8.5)}{2} \times 3 + \dfrac{2(x-1.5)(2x-8.5)(2x-9.5)}{6}$

$= x^3 - 2x + 3$ on simplification

$\therefore f'|_{1.5} = (3x^2 - 2)$ at $x = 1.5$ is $= 4.750$

$f''(x)|_{1.5} = 6x$ at $x = 1.5$ is $= 9.000$

$f'''|_{1.5} = 0$

These values are exactly the same as obtained by numerical differentiation.

EXAM. 2: Find the first and second derivatives of the function tabulated below at the point $x = 1.1$.

x	f(x)	Δ	Δ	Δ	Δ
1.0	0				
		0.1280			
1.2	.1280		0.2880		
		0.3360		0.480	
1.4	.6440		0.3360		0
		0.7520		0.480	
1.6	1.2960		0.3840		0
		1.1360		0.480	
1.8	2.4320		0.4320		
		1.5680			
2.0	4.0000				

SOLUTION

Newton's-Gregory for formula is

$$f(a+xh) = f(a) + x\Delta f(a) + \frac{x(x-1)}{2}\Delta^2 f(a) + \frac{x(x-1)(x-2)}{6}\Delta^3 f(a) + .$$

$$= f(a) + x\Delta f(a) + \frac{(x^2-x)}{2}\Delta^2 f(a) + \frac{x^3-3x^2+2}{6}\Delta^3 f(a) + \tag{1}$$

Differentiating (1) twice with respect to , we get

$$hf'(a+xh) = \Delta f(a) + \frac{(2x-1)}{2}\Delta^2 f(a) + \frac{(3x^2-6x+2)}{6}\Delta^3 f(a) + \tag{2}$$

and

$$h^2 f''(a+xh) = \Delta^2 f(a) + (x-1)\Delta^3 f(a) +$$

Here $h = 0.2, a = 1.0$; set $x = \dfrac{1}{2}$ in the above equations (2) and (3), we have

$$(2.0)f'(1.1) = 0.1280 + \frac{\left(2x\dfrac{1}{2}-1\right)}{2}(0.2880) + \frac{\left(3.\dfrac{1}{4}-6.\dfrac{1}{2}+2\right)}{6}(0.0480) + 0$$

$= 0.1280 + 0 - 0.002 + 0 \ (0.2)f\ (1.1) = 0.126$

$\therefore (0.2)f\ (1.1) = 0.630$

$$(.04)\ f''(1.1) = 0.2880 + \left(\frac{1}{2}-1\right)(0.0480) + 0$$

$$= 0.2880 - 0.024 = 0.264$$

$\therefore f(1.1) = 6.60$

The interpolating polynomial is given by

$$P(x) = f(x_0) + \frac{(x-x_0)}{1.h}\Delta f(x_0) + \frac{(x-x_0)(x-x_1)}{2..h^2}\Delta^2 f(x) + \frac{(x-x_0)(x-x_1)(x-x_2)}{6}\frac{1}{h^3}\Delta^3 f(x_0) + 0$$

$$= 0 + \frac{x-1.0}{1}\times\frac{0.128}{0.2} + \frac{(x-1.0)(x-1.2)}{2(.04)}\times0.288 + \frac{(x-1)(x-1.2)(x-1.4)}{6}\frac{1}{(1.008)}\times0.048$$

$$= (x-1)\left(\frac{0.128}{.2}\right)\times\frac{(x-1)(x-1.2)}{2} + \frac{(x-1)(x-1.2)(x-1.4)}{6}$$

$= (x-1)\ (0.64) + (x^2-2.2x+1.2)(3.6) + (x-1)\ (x^2-2.6x+1.68)$

$= -0.64 + 0.64x + 3.6x2 + 4.32 - 7.92$

$= x^3 + 1.68x - 2.6x^2$

$\quad -1.68 + 2.6x2 - x2$

$= 2 - 3x + x3$

$P(x) = x^3-3x + 2$

$P'(x) = 3 x 2-3$ $\quad$ P" $(x) = 6x$

$P(1.1)=3$ $1.21-2=3.63-3=0.63$ $P"(1.1)=6x1.1=6.6$

These are the same as obtained by numerical differentiation.

EXAM. 3: The Newton- Gregory formula for backward interpolation is given by

$$P_n(x) = f(a + nh) + \frac{1}{h}\nabla f(a + nh)(x - \overline{a + nh})$$

$$+ \frac{1}{2!} \cdot \frac{1}{h^2} \cdot \nabla^2 f(a + nh).(x - \overline{a + nh})(x - \overline{a + nh - h})$$

$$+ \ldots\ldots\ldots\ldots\ldots$$

$$+ \frac{1}{n!} \cdot \frac{1}{h^n} \cdot \nabla^n f(a + nh).(x - \overline{a + nh})(x - \overline{a + nh - h}) \ldots\ldots\ldots\ldots\ldots (x - a - h)$$

Putting $u = \dfrac{x - (a - nh)}{h}$ or.$x = a + nh + hu$ in the above formula, we set.

$$P_n(x) = P_n(a + nh + hu)$$

$$f(a + nh) + u\nabla f(a + nh) + \frac{u(u+1)}{2!}\nabla^2 f(a + nh)$$

$$+ \frac{u(u+1)(h+2)}{3!}\nabla^3 f(a + nh) + \ldots\ldots$$

$$+ \frac{u(u+1)(u+2)\ldots\ldots(u+n-1)}{n!}\nabla^n f(a + nh)$$

Since In$^{(x)}$ is the backward interpolation formula for f(x), we can write the Newton-Gregory backward formula as

$$f(a + nh + xh) = f(a + nh) + x\nabla f(a + nh) + \frac{(x^2 + x)}{2!}\nabla^2 f(a + nh)$$

$$+ \frac{x^3 + 3x^2 + 2}{3!}\nabla^3 f(a + nh) + \qquad\qquad\qquad (1)$$

Here it is to be noted that a+nh in the last term and h is the differencing interval in the tabular representation.

Differentiating (1) with respect fox, we get

$$hf'(a + nh + xh) = \nabla f\left(a + nh + \frac{2x+1}{2}\right)\nabla^2 f(a + nh) + (a + nh) \qquad (2)$$

Differentiating (2) with respect to x, we get

$$h^2 f"(a + nh + xh) = \nabla^2 f(a + nh) + (x + 1)\nabla^3 f(a + nh) + \qquad (3)$$

EXA. 4: Find the first and second derivatives of f(x) at =0.4 from the following table.

x :	0.1	0.2	0.3	0.4
$y_2=(f)$:	1.10517	1.22140	1.34986	1.49182

SOLUTION

Since =0.4 lies near the end of the table, we will use Newton's backward formula. The difference table is as below

x	y	Δy	Δ	Δ
0.1	1.10517			
		0.11623		
0.2	1.22140		.01223	
		0.12846		.00127
0.3	1.34986		.01350	
		0.14196		
0.4	1.49182			

Newton's back ward formula us

$$f(a+nh+xh)=f(a+nh)+\nabla f(a+nh)+\frac{x^2+x}{2!}\nabla^2 f(a+nh)+\frac{x^3+3x^2+2x}{3!}\nabla^3 f(a+nh)+ \quad (1)$$

Differentiating (1) with respect to x, we get

$$hf'(a+nh+xh)=\nabla f(a+nh)+\frac{2x+1}{2!}\nabla^2 f(a+nh)+\frac{3x^3+6x^2+2}{6!}\nabla^3 f(a+nh)+ \quad (2)$$

Differentiating (2) with respect to x, we get

$$h^2 f''(a+nh+xh)=\nabla^2 f(a+nh)+(x+1)\nabla^3 f(a+nh)+$$

Put = 0 in (2), we have $\qquad\qquad\qquad\qquad\qquad\qquad (3)$

$$(0.1)f'(0.4)=0.14196+\frac{1}{2}(0.01350)+\frac{1}{3}(0.00127) =0.14913$$

$$\therefore f'(0.4)=\frac{0.14913}{0.1}=1.4913$$

Put x = 0 in (3)

$(-1)^2 f''(0.4) = 0.1350 + 0.00127 = 0.01477$

$$\therefore f''(0.4) = \frac{0.01477}{0.01} = 1.477$$

EXAM. 5: Given the following pairs of values of x and y = f(x)

x :	1	2	4	8	10
y=f(x) :	0	1	5	21	27

Determine numerically the first derivative at = 4.

SOLUTION:

In this case the values of the argument are not-equally spaced, so the Newton's divided difference formula is used. The divided difference table

x	y	Δ	Δ	Δ	Δ
1	0				
		1			
2	1		$\dfrac{1}{3}$		
		2		0	
4	5		$\dfrac{1}{3}$		$-\dfrac{1}{144}$
		4		$-\dfrac{1}{6}$	
8	21		$-\dfrac{1}{6}$		
		3		0.480	
10	27				

Newton's divided difference formula is

f(x)= f(a)+f(x-a)Δf(a)+(x-a) (x-b) Δ²f(a)+(x-a) (x-b) (x-c) Δ³f(a)+(x-a) (x-b) (x-c) (x-d)Δ⁴f(a)+..........................

Differentiating the above with respect to and putting x=c. we have

f (c)= Δf(a)+(2c-a-b)Δ²f(a)+(c-a)(c-b) Δ³f(a)+(c-a) (c-b) (c-d) Δ⁴f(a)..............+

Putting a=1, b=2, c=4, d=8, e=10, and using the values of the differences, we get

$$f'(4)=1+5\left(\frac{1}{3}\right)+3(2)(0)+3(0)+3(2)(-4)\left(-\frac{1}{144}\right)$$

$$=1+1.666+0.1666$$
$$=2.833$$

13.7.6. *Exercises*

1. Write down the various formulas for numerical differentiation.
   A slider in a machine moves along a fixed rod. It is distance along the rod are given in of the following table for various values of the time t seconds.

t sec	:	0.0	0.1	0.2	0.3	0.4	0.5
x cm	:	30.1	31.6	32.9	33.6	40.0	33.8

   Find the velocity of the slide and also its acceleration at time t=0.3 second.
   [Ans: 47.5cm/sec;857cm/sce²]

2. Establish a formula for finding the derivation of a tabulated function.

3. Using Lagrange polynomial $p_2(x)$ which approximates f(x) in the interval $x_0 \le x \le x_2 = x_0 + 2h$, through (x_0, t_0), (x_1, t_1), (x_2, t_2), where $f_j(x)$ f_j, establish

$$f'(x) \approx \frac{1}{2h}P_2^1(n) = \frac{1}{2h^2}(2x - x_1 - x_2)f_0 - \frac{1}{h^2}(2x - x_0 - x_2)f_1$$

$$+\frac{1}{2h^2}(2x - x_0 - x_2 - x_1)f_2$$

By evaluating f'(x) at x_0, x_1, x_2 obtain the "three-point formulas" (a)

(b) $f_0' \approx \frac{1}{2h}(-3f_0 + 4f_1 - f_2)$;

(c) $f_1' \approx \frac{1}{2h}(-f_0 + f_2)$;

4. Obtain f'(x) and f''(x) at x=1 using the data given in the following table.

x	:	0	1	2	3	4
f(x)	:	1	1.5	2.2	3.1	4.3

5. Find the derivative of y at x = 300 from the following table

x:	300	301	302	303	304	305	306	307
y:	5.704	5.707	5.711	5.714	5.717	5.720	5.724	5.727

6. Show that a "four-point formula" for the derivative is7. Consider $f(x)=x4$ for $x_0 = 0$, $x_1 = 0.2$, $x_2 = 0.4$, $x_3 = 0.6$, $x_4 = 0.8$. Calculate f_2^1 and determine the error.

$$f_2' \approx \frac{1}{6h}\left(-2f_1 - 3f_2 + 6f_3 - f_4\right)$$

7. Confider $f(x) = x^4$ for $x_0 = 0$, $x_1 = 0.2$, $x_2 = 0.4$, $x_3 = 0.6$, $x_4 = 0.8$. Calculate f_2^1 and determine the error.

8. A differentiation in rule of the from $f'(x_0) = a_0 f_0 + a_1 f_1 + a_2 f_2$, $(x_k = x_0 + kh)$ is given. Find the values of a_0, a_1, and a_2 so that rule is exact for $f \in p_3$ Find the error term.

9. Find the maximum and minimum values of y tabulated below:

x	:	0	1	2	3	4	5
y	:	0	0.25	0	2.25	16.00	56.25

10. Find the first and second derivatives of the following tabulated function at the point x = 1.5.

x	:	1.5	2	2.5	3.0	3.5	4.0
f(x)	:	3.6	7.225	13.85	24.225	39.1	59.225

[Ans: f' (1.5)=4.75: f'' (1.5)=9.0]

11. Find the first and second derivatives of the following tabulated function f(x) at the point x=1.1

x	:	1.0	1.2	1.4	1.6	1.8	2.0
f(x)	:	0.00	0.128	0.544	1.296	1.432	4.000

[Ans: f' (1.1)=0.2520; f'' (1.1)=1.506]

12 A rod is rotating in a plane about one of its ends. If the following table gives the angle radians though which the roe is turned for different values of time t seconds, find its angular velocity when t=0.7.

t seconds:	0.0	0.2	0.4	0.6	0.8	1.0
θ radius	0.00	0.12	0.48	1.10	2.00	3.20

[Ans 4.495 radians per second]

14. starting from Newton's backward interpolation formula, obtain a formula for finding the derivative of a tabulated function. Find y' (2.0) from the following table.

x	:	1.0	1.2	1.4	1.6	1.8	2.0
y	:	-3.000	0.156	5.392	13.831	27.019	47.000

15. Using the following data find f' (6.0) error = O(h) and f'' (6.3), error = O(h)².

x	:	6.0	6.1	6.2	6.3	6.4
f(x)	:	0.3500	-0.3996	-0.44469	-0.4844	-0.5192

16. Find the numerical value of the first derivative at x = 0.4 of the function f(x) defined as under

x	:	0.1	0.2	0.3	0.4
f(x)	:	1.10545	1.22168	1.35014	1.49210

[Ans: f' (0.4)=1.49133]

17. The following table gives the velocities of a particular at time t:

t seconds:	0	2	4	6	8	10	12
r(meters/sec)	4	6	16	34	60	94	136

18. Find the distance moved by the particle in 12 seconds and also the acceleration at t=2 seconds from the data given in question 17.

19. Using the following data, find f' (5):

x	:	0	2	3	4	7	9
f'(x)	:	4	26	58	112	466	922

[Ans: f (5)=98]

20. Using divided differences find the values of f (8), given that
f(6)=1.556;   f(7)=1.690;   f(9)=1.908
f(12)=2.158;   [Ans: f (8)=0.109]

20. Consider the following table of values:

x	:	0	1	2	4
y	:	1	1	2	5

and estimate the value of y and y' at x=3, using Lagrangian interpolation formula.

21. The following data gives corresponding values of pressure and specific volume of a super heated steam.

V	:	2	4	6	8	10
P	:	105	42.7	25.3	16.7	13.0

(a) Find the rate of change of pressure with respect to volume when V=2.
(b) Find the rate of change of volume with respect to pressure when P=105.

$$\left[\frac{dV}{dP} = \frac{1}{\frac{dP}{dV}} \right]$$

22. From the following data, calculate $\dfrac{d^2y}{dx^2}$ at x=3.5

x	:	3	3.5	4	4.5
y	:	1.4843	1.55023	1.60746	1.65801

13.8 NUMERICAL INTEGRATION

Let $I = \displaystyle\int_a^b w(x)f(x)dx$, $\qquad\qquad$ (1)

where in [a,b] is called the **weight function.** It is assumed that w(x) and w(x) f(x) are integrable, in the Riemann sense on [a,b]. The limits of integration may be finite, semi-infinite or infinite. It is desirable to approximate the above definite integral by a finite linear combination of the values of f(x) in the form.

$$I = \int_a^b w(x)f(x)dx \approx \sum_{k=0}^{n} \lambda_k f_k \text{ , where xk,} \qquad\qquad (2)$$

h = 0(1)n are called **nodes** distributed within the limits of integration [a,b] and λk, k=0(1)n are called the weights in the **quadrature formula** (2). The **error** of approximation is given as

$$R_n = \int_a^b w(x)f(x)dx - \sum_{k=0}^{n} \lambda_k f_k \qquad\qquad (3)$$

When the weight function w(x)=1, the integral (1) represents geometrically the area under the curve f between a and b. An integration method of the from (i) is said to be of **order** p, if it produces exact results (R_n=0) for all polynomials of degree less than or equal to p. In (1), there are (2n+2) unknowns [(n+1) nodes x_k's and (n+1) weights λ_k's] and the method can be made exact for polynomials of degree ≤2n+1. Hence the method of the form (1) can be of maximum order (2n+1). If all the (n+1) nodes or abscissae are prescribed in advance, then only the (n+1) weights are to be determined and the corresponding method will be of maximum order n.

13.8.1 Methods Based on Interpolation

The (n+1) abscissa xk's and the corresponding values fk's are given and hence the Lagrange interpolating formula fitting the data (x_k, f_k), k = 0(1) n is given by

$$f(x) = P_n(x) = \sum_{k=0}^{n} l_k(x)f_k + \frac{\pi(x)}{(n+1)!}f^{(n+1)}(\xi), x_0 < \xi < x_n, \qquad (2)$$

where $l_k(x)$ is the Lagrange fundamental polynomial

$$l_k(x) = \frac{\pi(x)}{(x - x_k)\pi'(x)},$$

$\pi(x) = (x - x_0)(x - x1) \ldots (x - x_n)$

Replacing the function $f(x)$ in (i) by the interpolating polynomial (2) and integrating within the given limits, we obtain

$$I = \int_a^b w(x)f(x)dx = \sum_{k=0}^n \left[\int_a^b w(x)l_k(x)dx + \int_a^b w(x)\frac{\pi(x)}{(n+1)!}f^{(n+1)}(\xi)dx \right]$$

$$= \sum_{k=0}^n \lambda_k f_k + Rn \,, \tag{3}$$

where $\lambda_k = \int_a^b w(x)l_k dx$ and $R_n = \frac{1}{(n+1)!}\int_a^b w(x)\pi(x)f^{(n+1)}(\xi)dx$

$$= \frac{f^{(n+1)}(\eta)}{(n+1)!}\int_a^b w(x)\pi(x)dx, \eta \in [a,b],$$

using the Mean Value Theorem of integral calculus.

or $|R_n| \le \frac{1}{(n+1)!}\int_a^b w(x)|\pi(x)||f^{(n+1)}(\xi)|dx$

if $\pi(x)$ changes sign in [a,b]

$$\le \frac{M_{n+1}}{(n+1)!}\int_a^b w(x)|\pi(x)|dx, \quad |f^{n+1}(x)| \le M_{n+1}, x \in [a,b]$$

Neglecting the error term in (3), we write the integration method as

$$\int_a^b w(x)f(x)dx = \sum_{k=0}^n \lambda_k f_k \tag{4}$$

13.8.2 *Newton-Cotes Method*

When $w(x) = 1$ and x_k's are equispaced with $x_0 = a$, $x_n = b$, with spacing $h = (b-a)/n$, the methods (4) are called **Newton-cotes Methods** and the weights λk's are called **Cotes numbers**. Setting $x_{s+1} = x_0 + sh$, we get

$$\pi(x) = h^{n+1}s(s-1)\ldots(s-n) \quad \text{and}$$

$$l_k(x) = \frac{(-1)^{n-k}}{k!(n-k)!}h[s(s-s)\ldots(s-k+1)(s-k-1)\ldots(s-n)]$$

and $\lambda_k = \dfrac{(-1)^{n-k}}{k!(n-k)!} h \displaystyle\int_0^n s(s-1)....(s-k+1)(s-k-1)....(s-n)ds$

AlsoRn = $\dfrac{h^{n+2}}{(n+1)!} \displaystyle\int_0^n s(s-1)...(s-n)f^{(n+1)}(\xi)ds$

For n = 1, we have x_0 = a, x_1 = b, h = b-a, we have λ_0 = -h

$\displaystyle\int_0^L (s-1)ds = \dfrac{h}{2}$ and $\lambda_1 = h\displaystyle\int_0^1 sds = \dfrac{h}{2}$

The method yields $\displaystyle\int_a^b f(x)dx = \dfrac{b-a}{2}[f(a)+f(b)]$ and $\qquad\qquad$ (5)

It is called the **trapezoidal rule**

Geometrically, it is the area of the trapezoid with width (b-a) ordinates f(a), f(b),

This is an approximation to the area under the curve y = f(x), above the x-axis and the ordinates x = a and x = b. The error in the trapezoidal rule becomes.

$$R_1 = \dfrac{h^3}{2}\int_0^1 s(s-1)f''(\xi)ds$$

$$= \dfrac{h^3}{2}f''(\eta)\int_0^1 s(s-1)ds, \eta \in (a,b)$$

$$= \dfrac{h^3}{12}f''(\eta) = -\dfrac{(b-a)^3}{12}f''(\eta)$$

Thus the trapezoidal rule is exact for polynomials of degree ≤ 1 and hence it is of order 1.

For n = 2, We have $h = \dfrac{(b-a)}{2}, x_0 = a, x_1 = \dfrac{a+b}{2}, x_2 = b$

$\therefore \quad \lambda_0 = \dfrac{h}{2}\displaystyle\int_0^2 (s-1),(s-2) = ds = \dfrac{h}{3}, \lambda_1 = \dfrac{h}{2}\displaystyle\int_0^2 s(s-2)ds = 4\dfrac{h}{3}$

and $\lambda_2 = \dfrac{h}{2}\displaystyle\int_0^2 s(s-1)ds = \dfrac{h}{3}$

$\therefore$ The method yields, $I = \displaystyle\int_a^b f(x)dx = \dfrac{b-a}{6}\left[f(a)+4f\left(\dfrac{a+b}{2}\right)+f(b)\right] \qquad$ (6)

and it is called the **Simpson's rule** The error associated with this method is given by

$$R_2 = \frac{h^4}{3!} \int_0^2 s(s-1)(s-2)f'''(\xi)ds = \frac{h^4}{3!}f'''(\eta)\left[\frac{s^4}{4} - s^3 + s^2\right]_0^2 = 0$$

∴ This method is exact for polynomials of degree 2.
For Polynominals of degree 3, by taking $f(x) = x^4$. The error term becomes

$$R_2 = \frac{c}{4!}f^{(IV)}(\eta), \eta \in [0.2] \text{ where}$$

$$c = \int_a^b x^4 dx - \left[\frac{(b-a)}{6}\right]\left[a^4 + 4\left(\frac{a+b}{2}\right)^4 + b^4\right]$$

$$= -\frac{(b-a)^5}{120}$$

There fore the error of approximation in the Simpson's rule becomes

$$R_2 = \frac{(b-a)^5}{2880}f^{(IV)}(\eta)$$

When n = 3, the corresponding integration method is called $\frac{3}{8}$ **the Simpson's rule.**

Here $h = \frac{b-a}{3}, x_0 = a, x_1 = x_0 + h = \frac{2a+b}{3}, x_2 = \frac{a+2b}{3}, x_3 = b$

For n = 3, k = 0, $\lambda_0 = \frac{(-1)^3}{0!3!}h\int_0^3 (s-1)(s-2)(s-3)ds = \frac{3}{8}h$

For n = 3, k = 1, $\lambda_1 = \frac{(-1)^{3-1}}{1!(3-1)!}h\int_0^3 s(s-1)(s-3)ds = \frac{9}{8}h$

For n = 3, k = 2, $\lambda_2 = \frac{(-1)^{3-2}}{2!(3-2)!}h\int_0^3 s(s-1)(s-3)ds = \frac{9}{8}h$

For n = 3, k = 3, $\lambda_3 \frac{(-1)^0}{3!0!}h\int_0^3 (s-1)(s-2)ds = \frac{3}{8}h$

$$\int_{x_0}^{x_3} f(x)dx = \lambda_0 f_0 + \lambda_1 f_1 + \lambda_2 f_2 + \lambda_3 f_3 = \frac{3h}{8}[f_0 + 3f_1 + 3f_2 + f_3]$$

The error term is given by

$$\text{Error} = \int_0^3 \frac{1}{24} s(s-1)(s-2(s-3))h^4 f^{(IV)}(\xi)ds$$

$$= -\frac{3}{80} h^5 f^{(IV)}(\xi), x_0 < \xi < x_3.$$

In general, the Newton-Cotes integration formulas are

$$\int_a^b f(x)dx = \int_a^b P_n(x)dx.$$

and $E_n(f; x) = \int_a^b \binom{s}{n+1} h^{n+1} f^{n+1}(\xi)ds, x_0 < \xi < x_n,$

and $s = \dfrac{x - n_0}{h}.$

Also, the Trapezoidal rule for

$$\int_{x_i}^{x_{i+1}} f(x)dx = \frac{f(x_i) + f(x_{i+1})}{2} h = \frac{h}{2}[f_i + f_{i+1}], i = 0,1,2........n-1.,$$

or $\displaystyle\int_{x_{i-1}}^{x_i} f(x)dx = \frac{h}{2}[f_{i-1} + f_i]; i = 1,2......., n$

If [a,b] is divided into n sub-intervals of size h, then we have the Trapezoidal formula

for numerical integration is given by $\displaystyle\int_a^b f(x)dx = \sum_{i=1}^n \frac{h}{2}[f_{i-1} + f_i]$

$$= \frac{h}{2}[f_0 + f_1.... + f_n] \tag{7}$$

The global Truncation error is $= -\dfrac{h^3}{12}[f''(\xi_1) + ... + f''(\xi_n)]$

$$= -\frac{h^3}{12} nf''(\xi) = O(h^3)$$

Simpson's $\frac{1}{3}$rd reduce for numerical integration

$$\int_{x_1}^{x_2} f(x)dx = \frac{h}{3}[f_0 + 4f_1 + f_2] - \frac{h^5}{90}f^{(IV)}(\xi), x_0 < \xi < x_2$$

$$\therefore \int_a^b f(x)dx \int_{x_0=a}^{x_n=b} f(x)dx$$

$$= \frac{h}{3}[f_0 + 4f_1 + f_2] + \frac{h}{3}[f_2 + 4f_3 + f_4] + \frac{h}{3}[f_4 + 4f_5 + f_6] + \ldots + \frac{h}{3}[f_{2n-2} + 4f_{2h-1} + f_{2n}]$$

$$= \frac{h}{3}[f_0 + 4(f_1 + f_3 + f_5 + \ldots + f_{2n-1}) + 2(f_2 + f_4 + \ldots + f_{2n-2}) + f_{2n}]$$

$$= \frac{h}{3} \text{ [first ordinate + 4 (odd ordinates) + 2 (even ordinates) + last ordinate]}$$

$$-\frac{(b-a)}{180}h^4 f^{(IV)}(\xi), a < \xi < b. \tag{8}$$

Simpson's $\frac{3}{8}$th rule

$$\int_{x_0}^{x_3} f(x)dx = \frac{3h}{8}[f_0 + 3f_1 + 3f_2 + f_3] - \frac{3h}{8}h^5 f^{(IV)}(\xi), x_0 < \xi < x_3$$

$$\therefore \int_a^b f(x)dx \int_{x_0}^{x_n} f(x)dx$$

$$= \int_{x_0}^{x_3} + \int_{x_3}^{x_6} + \ldots$$

$$= \frac{3h}{8}[(f_0 + 3f_1 + 3f_2 + f_3) + (f_3 + 3f_4 + 3f_5 + f_6) + (f_6 + 3f_7 + 3f_8 + f_9) + \ldots]$$

$$= \frac{3h}{8}[f_0 + 3(f_1 + f_2 + f_4 + f_5 + f_7 + f_8 + \ldots) + 2(f_3 + f_6 + f_9 + \ldots)]$$

$$-\frac{(b-a)}{80}h^4 f^{(IV)}(\xi), x0 < \xi < x_n.$$

Examples

EXAM. 1: Evaluate $\int\limits_0^1 \dfrac{dx}{1+x^2}$, by using (i)Trapezoidal rule ;(ii) Simposon's " $\dfrac{1}{3}$ "

rule, to obtain the approximate value of π in each case.

SOLUTION:

Divide the range into six equal parts with h = $\dfrac{1}{6}$ and compute the values of y =

$\dfrac{1}{1+x^2}$ at each point of subdivision. The following table gives these values:

x :	0	$\dfrac{1}{6}$	$\dfrac{2}{6}$	$\dfrac{3}{6}$	$\dfrac{4}{6}$	$\dfrac{5}{6}$	1

$Y_x = \dfrac{1}{1+x^2}$: 1.0000 0.97297 0.90 0.80 0.69231 0.59016 0.5000

(i) Trapezoidal rule.

$$\int\limits_a^b f(x)\, dx = \sum_{i=1}^{n=6} \frac{h}{2} \left[f_{i-1} + f_i \right]$$

$$= \frac{h}{2} \left[f_0 + 2\left(f_1 + f_2 + f_3 + f_4 + f_5 \right) + f_6 \right]$$

$$= \frac{1}{12} \left[\begin{array}{l} 1.00000 \\ 0.50000 \end{array} + 2\left(\begin{array}{l} 0.97297 + 0.90000 \\ 0.80000 + 0.69231 \end{array} + 0.59016 \right) \right]$$

$$\int\limits_6^1 \frac{\partial x}{1+x^2} = \frac{1}{12} \, (9.41088)$$

$$\frac{\pi}{4} = \frac{1}{12} \, (9.41088) \quad \text{or} \quad \pi = 3.13696$$

(ii) By Simpson's " $\dfrac{1}{3}$ " rule, we have

$$\int\limits_{x_0}^{x_0+6h} y\, dx = \frac{1}{3} \left[y_0 + y_6 + 4\left(y_1 + y_3 + y_5 \right) + 2\left(y_2 + y_4 \right) \right]$$

$$\int_0^1 \frac{dy}{1+x^2} = \frac{1}{18}[1.50000 + 4(.97297 + 0.80000 + 1 + 0.59016) + 2(0.90000 + 0.69231)]$$

$$\frac{\pi}{4} = \frac{1}{18}(14.13714) \text{ or } \pi = 3.14165$$

and by Simpson's "$\frac{3}{8}$" rule

$$\int_{x_0}^{x_0+6h} ydx = \frac{3h}{8}[y_0 + y_6 + 3(y_1 + y_2 + y_4 + y_5) + 2(y_3)]$$

$$\int_0^1 \frac{dx}{1+1x^2} = \frac{1}{16}[1.50000 + 3(.97297 + 0.90000 + 0.69231 + 0.59016) + 2(0.80000)]$$

$$\frac{\pi}{4} = \frac{1}{16}[12.56632] \text{ or } \pi = 3.14165$$

13.8.3 Gaussian Quadrature

The formulas for numerical integration were all predicated on evenly spaced x- values; this means the x-values were predetermined. With a formula of three terms, then, there were three parameters, the coefficients (weighting factors) applied to each of the functional values. A formula with three parameters corresponds to a polynomial of the second degree, one less than the number of parameters.

Gauss observed that if we remove the requirement that the function be evaluated at predetermined x- values, a three term formula will contain six parameters (the three x-values, now unknows, plus the three weights) and should correspond to an interpolating polynomial of degree five.

Formulas based on this principle are called **Gaussian Quadrature Formulas.** They can be applied only when f(x)is explicitly known, so that it can be evaluated at any desired value of x.

We shall determine the parameters in the simple case of a **Two-term formula,** **containing four unknown parameters:**

$$\int_{-1}^{+1} f(t)dt = a\ f(t_1) + bf(t_2).$$

We use a symmetrical interval of integration to simplify the arithmetic and call our variable as t. Our formula is to be valid for any polynomial of degree three. Hence it will hold if

$f(t) = t^3$, $f(t) = t^2$, $f(t) = t$ and $f(t) = 1$.

For $f(t) = t^3$ $\displaystyle\int_{-1}^{+1} t^3 dt = 0 = at_1^3 + bt_2^3$

$$f(t) = t^2 \quad \int_{-1}^{+1} t^2 dt = \frac{2}{3} = at_1^2 + bt_2^3,$$

$$f(t) = t \quad \int_{-1}^{+1} dt = 0 = at_1 + bt_2$$

$$f(t) = 1 \quad \int_{-1}^{+1} dt = 2 = a + b$$

There are 4 unknowns and 4 conditions.

Solving the 4 equations we get $a = b = 1$, $t_2 = -t_1 = \dfrac{1}{\sqrt{3}} = 0.5773$

$$\therefore \int_{-1}^{+1} f(t) dt = f(-0.5773) + f(0.5773)$$

Suppose our limits of integration are from a to b, we try to change the interval [a.b] of integration to [-1,+1] by using a change of variable.

Let $x = \dfrac{(b-a)t + (b+a)}{2}$ $\therefore dx = \dfrac{(b-a)}{2} dt$

$$\therefore \int_a^b f(x) dx = \frac{b-a}{2} \int_{-1}^{+1} f\left[\frac{(b-a)t + (b+a)}{2}\right] dt$$

Consider the example

$$I = \int_0^{\pi/2} \sin x \, dx = (-\cos x)^{\pi/2}{}_0 = 1.0$$

Let $x = \dfrac{\dfrac{\pi}{2}t + \pi/2}{2} = \dfrac{\pi}{4}(t+1); dx = \dfrac{\pi}{4}$

when $t = -1$, $x d = 0$ and where $t = 1$, $x = \dfrac{\pi}{2}$

$$\therefore\ I = \frac{\pi}{4}\int_{-1}^{+1} \mathrm{Sin}\left(\frac{\pi t + \pi}{4}\right) dt$$

The Gaussian formula calculates the value of the new integral as a weighted sum of two values of the integrated at $t = -0.5773$ and $t = 0.5773$.

Hence $I = \dfrac{\pi}{4}\left[(1.0)\sin(0.105660.\pi) + (1.0)\sin(0.39434\pi)\right] = 0.99847$

$\therefore$ Error $= 1 - 0.99847 = 1.53 \times 10^{-3}$.

In the integration methods based on interpolation given by

$$= \int_{a}^{b} w(x)\, f(x)\, dx = \sum_{k=0}^{n} \lambda_k f_k \text{ the nodes } x_k \text{ and the weights } \lambda_k, k = 0(1)\,n \text{ can be both}$$

determined by making the formula exact for polynomials of degree almost equal to $(n + 1 + n + 1 - 1) = 2n + 1$.

When the nodes are known, the corresponding methods are called **Newton-cotes methods.** When the nodes are also to be determined, the methods are called **Gaussian Integration Methods** (or **Gauss Quachature methods**)

13.8.4 *Gauss-Legendre Integration Methods*

Any finite interval [a,b] can be transformed into [-1,1], using

$$x = \frac{(b - a)t + (b + a)}{2}$$

Consider the integral in the form,

$$\int_{-1}^{+1} w(x) f(x)\, dx = \sum_{k=0}^{n} \lambda_k f_k \ . \text{ For } w(x) = 1,$$

the integral reduces to $\displaystyle\int_{-1}^{+1} f(x)\, dx = \sum_{k=0}^{n} \lambda_k f_k$

In this case all the nodes x_k and weights λ_k, $k = 0(1)n$ are unknown.

For $n = 2$, the method becomes,

$$\int_{-1}^{+1} f(x)\, dx = \lambda_0 f(x_0) + \lambda_1 f(x_1) + \lambda_2 f(x_2)$$

There are six unknowns and this can be made exact for polynomials up to degree 5. Therefore for $f(x) = x^i$, $i = 0(1)\ 5$ we get the following system of equations

For i= 0, we have $\lambda_0 + \lambda_1 + \lambda_2 = 2$

For i= 1, we have $\lambda_0 \times \lambda_0 + \lambda_1 x_1 + \lambda_2 x_2 = 0$

For i=2, we have $\lambda_0 x_0^2 + \lambda_1 x_1^2 + \lambda_2 x_2^2 = \dfrac{2}{3}$

For i=3 we have $\lambda_0 x_0^3 + \lambda_1 x_1^3 + \lambda_2 x_2^3 = 0$

For i=4 we have $\lambda_0 x_0^4 + \lambda_1 x_1^4 + \lambda_2 x_2^4 = \dfrac{2}{5}$

For i=5, we have $\lambda_0 x_0^5 + \lambda_1 x_1^5 + \lambda_2 x_2^5 = 0$

Solving the above system of equations,

We get $x_0 = -\sqrt{\dfrac{3}{5}}, x_1 = 0, \ x_2 = \sqrt{\dfrac{3}{5}} = 0.77459$

$\lambda_0 = \dfrac{5}{9}, \lambda_1 = \dfrac{8}{9} = 0.8888, \lambda_2 = \dfrac{5}{9} = 0.5555$

$\therefore$ We have

$$\int_{-1}^{+1} f(x)dx = \dfrac{1}{9}\left[5f\left(-\sqrt{\dfrac{3}{5}}\right) + 8f(0) + 5f\left(\sqrt{\dfrac{3}{5}}\right) \right]$$

The Error term is given by

$R_5 = \dfrac{c}{6!} f^{(6)}(\xi), -1 < \xi < 1,$ where

$$c = \int_{-1}^{+1} x^6 dx - \left(\lambda_0 x_0^6 + \lambda_1 x_1^6 + \lambda_2 x_2^6 \right) = \dfrac{2}{7} - \dfrac{6}{25} = \dfrac{8}{175}$$

It is found that the nodes xk's are the roots of the Legendre polynomial $P_{n+1}(x)=0,$

where $P_{n+1}(x)= \dfrac{1}{2^{n+1}(n+1)!} \dfrac{d^{n+1}}{dx^{n+1}} \left[(x^2 - 1)^{n+1} \right], n = 0,1,\dots$. Using the recurrence formula

for Legendre Polynomials

$(n+1) L_{n+1}(x) - (2_{n+1}) \times L_n(x) + nL_{n-1}(x) = 0,$ we get $L_0(x)=1, L_1(x)= \dfrac{3}{2}x^2 - \dfrac{1}{2}.$

For $L_2(x) = \dfrac{1}{2} = 0$, the roots are $x = \pm\sqrt{\dfrac{1}{3}} = \pm0.5773$

Also $L_3(x) = \dfrac{5x^3 - 3x}{2}, L_4(x) = \dfrac{35x^4 - 30x^2 + 3}{8},$ etc

13.8.5 Gauss - ChebySher Integration Methods

Here the weight is $w(x) = \dfrac{1}{\sqrt{1-x^2}}$ and the methods are of the form

$$\int_{-1}^{+1} \frac{1}{\sqrt{1-x^2}} f(x)dx = \sum_{k=0}^{n} \lambda_k f_k,$$

(1)

called Gauss-Chebyshev integration methods. The methods are exact for polynomials of degree upto 2n+1. The nodes x_k's are found to be the roots of the Chebyshev polynomials

$$T_{n+1}(x) = Cos[(n+1)\, Cos^{-1}x] = 0 \text{ and are given by } x_k = Cos\left[\frac{(2k+1)\pi}{2n+2}\right], k = 0,1....n \text{ (2)}$$

Taking $n = 2$, we get from (1)

$$\int_{-1}^{+1} \frac{f(x)}{\sqrt{1-x^2}} = dn = \lambda_0 f(x) + \lambda_1 f(x_1) + \lambda_2 f(x_2)$$

(3)

Since the methods are exact for $f(x) = x^i$, i=0(1)5,
We get the following system of equations.

$$\lambda_0 + \lambda_1 + \lambda_2 = \pi$$

$$\lambda_0 x_0 + \lambda_1 x_1 + \lambda_2 x_2 = 0$$

$$\lambda_0 x_0^2 + \lambda_1 x_1^2 + \lambda_2 x_2^2 = \frac{\pi}{2}$$

$$\lambda_0 x_0^3 + \lambda_1 x_1^3 + \lambda_2 x_2^3 = 0$$

$$\lambda_0 x_0^4 + \lambda_1 x_1^4 + \lambda_2 x_2^4 = \frac{3\pi}{8}$$

$$\lambda_0 x_0^5 + \lambda_1 x_1^5 + \lambda_2 x_2^5 = 0$$

The nodes are given by

$$x_k = Cos(2k+1)\frac{\pi}{6}, k = 0,1,2... \text{ or}$$

$$x_0 = Cos\frac{\pi}{6} = Cos\,30° = \frac{\sqrt{3}}{2}\ ; x_1 = 0, x_2 = Cos\frac{5\pi}{6} = -Cos\frac{\pi}{6} = -\frac{\sqrt{3}}{2}.$$

Solving for $\lambda_0, \lambda_1, \lambda_2$ from (4), we get $\lambda_0 = \lambda1 = \lambda_2 = \dfrac{\pi}{3}$.

Hence the method gives

$$\int_{-1}^{+1} \frac{x^6 dx}{\sqrt{1-x^2}} = \frac{\pi}{3}\left[f\left(\frac{\sqrt{3}}{2}\right) + f(0) + f\left(\frac{\sqrt{3}}{2}\right)\right] \tag{5}$$

With error term $R_3 = \frac{C}{6!} f^{(VI)}(\xi), -1 < \xi < 1,$

Where $C = \int_{-1}^{+1} \frac{x^6}{\sqrt{1-x^2}}\, dx - \left(\lambda_0 x_0^{\ 6} + \lambda_1 x_1^{\ 6} + \lambda_2 x_2^{\ 6}\right)$

$$= \frac{\pi}{32}.$$

[Note : all the weighte λk's are given by $\lambda k = \dfrac{\pi}{n+1}, k = 0,1,2...., n$ \hfill (6)

13.8.6 Gauss-Hermite Integration Methods

Method 1: Gauss-Her mite integration methods consist of

$$\int_{-\infty}^{\infty} \exp(-x^2) f(x) dx = \sum_{k=0}^{n} \lambda_k f_k \text{ , where } x_k\text{'s are nodes which are roots of the Hermite}$$

polynomial $H_{(n+1)}(x) = (-1)^{n+1} \dfrac{d^{n+1}}{dx^{n+1}}\left[\exp(-x^2)\right]$ and the weights are obtained from

$$\lambda_k = \int_{-\infty}^{\infty} \frac{\exp(-x^2) H_{k+1}(x)}{(x-x_k) H'_{k+1}(x)} dx$$

We get exact results for polynomials of degree upto $(2n+i)$

The Hermite polynomials are orthogonal with respect to the weight function $\exp(-x^2)$

on $(-\infty, \infty)$, i.e., $\int_{-\infty}^{\infty} \exp(-x^2) H_m(x) Hn(x) dx = 0$ for

The following table furnishes the values of the nodes and weights for $n = 0,1,2,$

n :	0	1	2
nodes x_k:	0	± 0.7071067812	± 1.2247448714
weights λ_k:	1.772453850	0.8862269255	1.1816359006
			0.2954089752

Method 2
Multiple Integration by using mapping function
Using a double integral notation we write

$$\iint_A f(x,y)dA = \int_a^b \left[\int_c^d f(x,y)dy \right] dx = \int_c^d \left[\int_a^b f(x,y))dx \right] dy,$$

where A is the rectangular region bounded by the lines x=a, x=b, y=c, y=d. The double integration by numerical means reduces to a double summation of weighted function values. Hence we can write

$$\iint_A f(x,y)dxdy = \sum_{j=1}^m v_j \sum_{i=1}^n w_i f_{ij}$$

Taking a rectangular region bounded by x=1.5, x=3.0, y=0.2, y=0.6 with x=0.5 and y=o.1, i=1(1)4 and j=1(1)5, the value of the integral by weighted summation can be written in the form

$$\iint_A \sum_{j=1}^4 v_j \sum_{i=1}^4 w_i f_i = \frac{\Delta y}{3} \frac{\Delta x}{2} \begin{bmatrix} (f_{1,1} + 2f_{2,1} + 2f_{3,1} + f_{4,1}) \\ + 4(f_{1,2} + 2f_{2,2} + 2f_{3,2} + f_{4,2}) \\ + 2(f_{1,3} + 2f_{2,3} + 2f_{3,3} + f_{4,3}) \\ + 4(f_{1,4} + 2f_{2,4} + 2f_{3,4} + f_{4,4}) \\ (f_{1,5} + 2f_{2,5} + 2f_{3,5} + f_{4,5}) \end{bmatrix}$$

- As it is convenient to write this in pictorial operator form, in which the weighting factors are displayed in an array that is a map to the location of the functional values to which they are applied.

- Since the number of panels in the x-diction is not even, Simpson's $\frac{1}{3}$ rule does not apply readily. As it is immaterial which integral we evaluate first, here start with y at 0.2, 0.3,0.4,0.5,and 0.6, Δy=0.1 and Δx=0.5

$$\int_{1.5}^{3.0} f(x,y)dx = \int_{1.5}^{3.0} f(x,0.2)dx = \frac{h}{2}[f, +2f_2 + 2f_3 + f4] \text{ (Trapezoidal rule)}$$

Then we now sum these in the y-direction according to the Simpson's $\frac{1}{3}$ rule, the weights being 1,2,3,4,1.

- Except for the difficulty of representation beyond two dimensions, this operator technique also applies to triple integrals.

$$\therefore \int_{D.2}^{.6} \int_{1.5}^{3.0} f(x,g)dxdy = \frac{\Delta y}{3} \frac{\Delta x}{2} \begin{Bmatrix} 1 & 4 & 2 & 4 & 1 \\ 2 & 8 & 4 & 8 & 2 \\ 2 & 8 & 4 & 8 & 2 \\ 1 & 4 & 2 & 4 & 1 \end{Bmatrix} f_{ij}$$

Examples:

EXAM.1: Evaluate the integral $I = \int_1^2 \int_1^2 \dfrac{dxdy}{x+y}$, using the trapezoidal rule with h = k=0.5, and h = k = 0.25.

SOLUTION:

With h = k = 0.5 , we find

$$I = \frac{1}{16}\left[f(1,1) + f(2,1) + f(1,2) + f(2,2) + 2\left\{ f\left(\frac{3}{2},1\right) + f\left(1,\frac{3}{2}\right) + f\left(2,\frac{3}{2}\right) + f\left(\frac{3}{2},2\right) \right\} + 4f\left(\frac{3}{2},\frac{3}{2}\right) \right]$$

$$= \frac{1}{16}\left[0.5 + \frac{1}{3} + \frac{1}{3} + 0.25 + 2\left(0.4 + 0.4 + \frac{2}{7} + \frac{2}{7}\right) + \frac{4}{3} \right]$$

$$= 0.343304$$

With h = k = $\dfrac{1}{4}$, we find

$$I = \frac{1}{64}\left[f(1,1) + f(2,1) + f(1,2) + f(2,2) + 2\left\{ f\left(\frac{5}{4},1\right) + f\left(\frac{3}{2},1\right) + f\left(\frac{7}{4},1\right) \right\} + f\left(1,\frac{5}{4}\right) \right]$$

$$= +f\left(1,\frac{3}{2}\right) + \left[f\left(1,\frac{7}{4}\right) + f\left(\frac{5}{4},2\right) + f\left(\frac{3}{2},2\right) + f\left(\frac{7}{4},2\right) \right] = 0.340668.$$

Method-3

There is an alternative representation to such mapping (pictorial) operators that is easier to translate into a computer program. This approach is explained below.

Consider the numerical in tegration formula for are variable

$$\int_{-1}^{+1} f(x)dx = \sum_{i=1}^{n} a_i f(x_i) \tag{1}$$

This formula can be made exact if f(x) is any polynomial of certain degree, say s.

Consider the multiple integral formula $\displaystyle\int_{-1}^{+1}\int_{-1}^{+1} f(x,y)dxdy = \sum_{i=1}^{n}\sum_{j=1}^{n} a_i a_j f(x_i, y_j)$ (2)

We can show that the equitation (2) is exact for all polynomials in x and y upto degrees s. Such a polynomial is a linear combination of terms of the form $f(x,y) = x^\alpha y^\beta$, where α and α are non-negative integers whose sum is equal to s or less. To establish, we assume that $f(x,y) = x^\alpha y^\beta$. Since the limits are constants and the integrand is factorable, let

$$I = \left(\int_{-1}^{+1}\int_{-1}^{+1} x^\alpha y^\beta \, dxdy \right) = \left(\int_{-1}^{+1} x^\alpha dx \right)\left(\int_{-1}^{+1} y^\beta dx \right)$$

$$= \left(\sum_{i=1}^{n} a_i x_i^{\alpha} \right) \left(\sum_{j=1}^{n} a_i y_i^{\beta} \right) \tag{3}$$

In order to get the product of summations we illustrate it for a simple case

$$= \left(\sum_{i=1}^{3} u_i \right) \left(\sum_{j=1}^{2} v_j \right) = \sum_{i=1}^{3} \left(\sum_{j=1}^{3} u_i v_i \right) \sum_{i=1}^{3} \sum_{j=1}^{2} u_i v_j$$

This is established as follows:

$$\left(\sum_{i=1}^{3} u_i \right) \left(\sum_{j=1}^{2} v_j \right) = \sum_{i=1}^{3} u_i \sum_{j=1}^{2} v_j$$

$$= (u_1 + u_2 + u_3)(v_1 + v_2)$$
$$= u_1 v_1 + u_1 v_2 + u_2 v_1 + u_2 v_2 + u_3 v_2 + u_3 v_1 + u_3 + v_2$$

And $\displaystyle\sum_{i=1}^{3} \sum_{j=1}^{2} u_i v_j = (u_1 v_1 + u_1 v_2) + (u_2 v_1 + u_2 v_2) + (u_3 v_1 + u_3 v_2)$

In removing the parentheses, we set that the sides are the same:

Using this principle, we can write the equation (3) in the form I$= \displaystyle\sum_{i=1}^{n} \sum_{j=1}^{n} a_i a_j x_i^{\alpha} y_j^{\beta}$

$$\tag{4}$$

Example

$I = \displaystyle\int_{-1}^{1} \int_{-1}^{0} e^x . y \, dx \, dy$ by Gaussian quadtrature using three-term formula for x and two term formula for y. We first make the change variable to adjust the limits for y to (-1,1)

Let $\qquad y = \dfrac{1}{2}(u-1) : dy = \dfrac{1}{2} du$

Our integral becomes

$$I = \frac{1}{8} \int_{-1}^{+1} \int_{-1}^{+1} (u-1) e^x dx dy$$

The two and three -point Gaussian formulas are,

$$\int_{-1}^{+1} f(x) dx = (1) \ f(-0.5774) + (1) f(0.5774)$$

and $\displaystyle\int_{-1}^{+1} f(x) dx = \frac{5}{9}(-0.7746) + \frac{8}{9} f(0) + \frac{5}{9}(0.7746).$

The integral is then

$$I = \frac{1}{8} \sum_{i=1}^{2} \sum_{j=1}^{3} a_i b_j (u_i + 1) e^{x_j}$$

$$a_1 = 1, a_2 = 1, b_1 = \frac{5}{9}, b_2 = \frac{8}{9}, b_3 = \frac{5}{9}$$

$$\therefore I = \frac{1}{8} \left[(1)\left(\frac{5}{9}\right)(-0.5774 + 1)e^{-0.7446} + (1)\left(\frac{8}{9}\right)(-0.5774 + 1)e^{0} + (1)\frac{5}{9}(0.5774 + 1)e^{-0.7446} \right]$$

$$= -0.58758$$

13.8.7 Exercise

1. A river is 80 meters wide. The depth "h" in meters at a distance of "d" meters from are bank is given in the following table . Find approximately the area of a cross-section of the river by

   (i) Simpson's $\frac{1}{3}$ rule and (ii) Simpson's "$\frac{3}{8}$" rule

d:	0.	10.	20.	30.	40.	50.	60.	70.	80
h:	0.	4.	7.	9.	12.	15.	14.	8.	3.

2. Compute the value of $\int_0^{\pi/2} \sqrt{1 - 0.162 \sin^2 \phi} \, d\phi$ by Simpson's $\frac{1}{3}$ rule taking

   $\phi = 0°, 15°, 30°, 45°, 60°, 75°, 90°$

3. Using composite trapezoidal rule for the function tabulated over the interval from x=1.8 to x=3.4 find the value of $\int_{1.8}^{3.4} f(x) \, dx$ and also find the global error, where $f(x) = e^x$

x:	1.8	2.0	2.2	2.4	2.6
f(x):	6.050	7.389	9.025	11.023	13.464^S

x:	2.8	3.0	3.2	3.4
f(x)	16.445	30.286	24.533	9.464

   $$\left[\text{Ans:} 23.9944, E = -\frac{1}{12} h^3 n . f(\xi) = -\frac{1}{12}(b-a)h^2 f''(\xi), 1.8 < \xi < 3.4, = \begin{cases} -0.03 & 23 \\ -0.15 & 98 \end{cases} (\text{Min} - \text{max}) \right]$$

   (Actual error = -0.080)

4. Use the method of undetermined coefficients to derive the trapezoidal rule

   [Ans: Assume the value of the integral $\int_a^b f(x) dx = af_0 + bf_1$; change the limits x = 0 to x =

n and take f(x) = 1 and f(x)=x; a=$\dfrac{h}{2}$, b = $\dfrac{h}{2}$]

The following table gives the velocity v of a particle at time t:

t(sec onds): 0 2 4 6 8 10 12

v(metres/Sin): 4 6 16 34 60 94 136

Find the distance moved by the particle in 12 seconds and also the acceleration at t=2 seconds.

5. Find the first and second derivative of the function tabulated below at the point X=0.7

x: 0.4   0.5   0.6   0.7   0.8

y: 1.5836 1.7974 2.0442 2.3275 2.6511

6. Use Simpson's one third rule, to evaluate $\int_{7.47}^{7.52} f(x)dx$ from the following table.

x: 7.47 7.48 7.49 7.50 7.51 7.52

y: 1.93 1.95 1.98 2.01 2.03 2.06

7. Determine abscissas and weights with gauss Guadrature formula

$$\int_0^1 \frac{f(x)}{\sqrt{x}} dx = A_0\, f(x_0) + A_1\, f(x_1)$$

Use two point Gauss quadrature formula to approximate the integral $\int_{-1}^{2} \exp(-x^2)\, Cosx dx$.

13.9 Eigen value problems

Let A = [a_{ij}] be a given n x n matrix and consider the vector equation.

A x = λ x, Where λ is a number (1). It is clear that zero vector **x = 0** is a solution of (1) for every value of λ. A value of λ for which (1) has a solution **x ≠ 0** is called an **Eigen value** on characteristic value (or latent root) of the matrix **A**. The corresponding solutions **x ≠ 0** of (1) are called **eigenvectors or characteristic vectors** of A corresponding to that eigen value λ. The problem of determining the eigen values and eigenvectors of a matrix is called an **algebraic eigen value problem.**

Consider a system of equations **A x = λ x** or $\sum a_{ij} x = \lambda x$ or **(A - λI) x = 0**, (2), where I is a unit matrix in matrix notation. This homogeneous system of equations has a non-trivial solution if and only if the corresponding determinant of the coefficients is Zero :

D (λ) = det (A - λI) = 0   (3)

D (λ) is called the characteristic determinant and the equation (3) is called the characteristic equation corresponding to the matrix A. D(λ) is a polynomial of n the degree in λ and is called the **characteristic polynomial.** The eigen values of a square matrix A are the roots of the corresponding Characteristic equation (3). The n roots $\lambda_1, \lambda_2, \ldots, \lambda_n$ are called the eigen values of A and may be district or repeated, real or complex. The modulus of the largest eigen value is called the **spectral radius** of A. Corresponding to each eigen value λ_i there exists an eigen vector x_i which is a non-trivial solution of $(A - \lambda I) x_i = 0$ (4). x_i are the corresponding eigenvectors then any vector y, in the eigen space can be expressed as a linear combination of the x_i.

$$Y = C_1 X_1 + C_2 X_2 + \ldots\ldots + C_n X_n \tag{5}$$

The eigen values of a matrix a are given by the roots of the characteristic equation det $(A - \lambda I) = 0$ which when simplified gives the polynomial equation.

$$P(\lambda) = (-1)^n \lambda^n + q_1 \lambda^{n-1} + q_2 \lambda^{n-2} + \ldots\ldots + q_n = 0 \tag{6},$$

where the sign $(-1)^n$ is used to give the terms of the polynomial the same sign that they would have if the polynomial were generated by expanding the determinant. The coefficients of the polynomial (6) can be determined with the help of the Faddier Leverrier method.

13.9.1 Faddeev-Leverrier method

Let $B_1 = A$ and we have $a_1 = t_r B_1$

$B_2 = A (B_1 - a_1 I)$ and $a_2 = \dfrac{1}{2} t_r B_2$

$B_3 = A (B_2 - a_2 I)$ and $a_3 = \dfrac{1}{3} t_r B_3$ $\qquad(7)$

$B_k = A (B_{k-1} - a_{k-1} I)$ and $a_k = \dfrac{1}{k} t_r B_k$

$B_n = A (B_{n-1} - a_{n-1} I)$ and $an = \dfrac{1}{n} t_r B_n$,

where tr $A = a_{11} + a_{22} + \ldots\ldots + a_{nn}$

The determination of the eigen values and the corresponding eigenvectors of the system $(A - \lambda I) x = 0$ may be classified in to two types: (1) Direct methods and (ii) iterative methods.

Hence the roots of the polynomial equation (6) can be determined by any one of the above two methods.

A non-zero victor x_i such that $A x_i = \lambda_i x_i$ $\qquad\qquad(8)$

is called the eigen vector the characteristic vector corresponding to λ_i. Multiplying (8) by an arbitrary constant c and putting $y_i = cx_i$,

we get $A y_i = \lambda_i y_i$ $\qquad\qquad(9)$

which shows that an eigen vector is determined only to within an arbitrary multiplicative constant.

Also we obtain $A^m x = \lambda^m x$ $\qquad\qquad(10)$

which shows that λ^m is an eigen value of A^m and x is the corresponding eigen vector.

Substituting (10) into (6), we get $p(A) = 0$ $\qquad\qquad(11)$

Which gives the result that a square matrix A statistics its own characteristic equation. This results is known as the **Cayley - Hamilton** theorem. Replacing the matrix A in (2) by the transpose matrix A^T we find

$$\det (A^T - \lambda I) = \det (A - I) = 0 \tag{12}$$

Thus **A** and A^T have the same eigen values. For distinct eigen values, if $u_1, u_2, \ldots, u_n$ are the eigen vectors of A and $v_1, v_2, \ldots, v_n$ are the eigen vectors A^T, we find

$$\det (A^T - \lambda I) = \det((A - \lambda I)) = 0, \tag{13}$$

then we have $Au_i = \lambda_i u_i$ and $A^T v_j = \lambda_j v_j$ and $v_j^T A u_i = \lambda_i A_j^T u_i$ and $v_j^T A = \lambda_j v_j^T$ (14)

or $v_j^T A u_i = \lambda_i v_j^T u_i$ (15)

From (14) and (15) we get

$$(\lambda_i - \lambda_j) v_j^T u_i = 0 \tag{16}$$

If $i \neq j$, $\lambda_i \neq \lambda_j$ and so we have $v_j^T u_i = 0$ (17)

¶If $i = j$, $v_j^T u_i \neq 0$ and since the length of eigen vectors is arbitrary, we normalize them such that $v_i^T u_i = 1$ (18)

Thus we have $v_j^T u_i = \begin{cases} 0, & i \neq j \\ 1, & i = j \end{cases}$ (19)

We conclude therefore that the eigenvectors corresponding to different eigen values of a matrix and of its transpose are orthogonal. Such sets of mutually orthogonal vectors are called **bi-orthogonal sets.**

When A is a symmetric matrix, $A = A^T$, $u_i = v_i$ and the eigen vectors corresponding to different eigenvalues are orthogonal.

Also from $Au_i = \lambda_i u_i$ $u_j^T A u_i = \lambda_i u_i^T u_i$ or $\lambda_i = \dfrac{u_j^T A u_i}{u_j^T u_i}$ (20)

which gives an expression for the eigen values in terms of the eigen vectors. For arbitrary u, (20) is called the **Rayleigh quotient.**

13.9.2 *Similarity Transformation*

Consider two matrices A and B of the same order. If a nonsingular matrix S can be determined such that B = S AS, (1) then the matrices A and B are said to be similar and (1) is called a similarity transformation. The matrix S is called the similarity matrix.

From (1), we Can write

$$SBS^{-1} = S (S^{-1} AS) S^{-1} = (SS^{-1}) A (SS^{-1}) = A$$

$\therefore$ $A = SBS^{-1}$

If λ_i is an eigen value of A and u_i. is the corresponding eigenvector, then $Au_i = \lambda_i u_i$

or $S^{-1} Au_i = \lambda_i S^{-1} u_i$ . . . . (3)

Substituting $u_i = Sv_i$ in (3) and using (1)

we get $\qquad$ $Bv_i = S^{-1} ASVi = S-1\ Aui = i\ S-1\ ui$

$\qquad\qquad\qquad = \lambda_i S^{-1} Sv_i = \lambda_i v_i$

$\therefore \qquad\qquad Bv_i = \lambda_i v_i$ $\qquad\qquad\qquad\qquad\qquad\qquad\qquad$ (4)

Thus S^{-1} as has the same eigen values as A and its eigen vectors v_i are obtained from the relation $v_i = S^{-1}u_i$ $\qquad\qquad\qquad\qquad\qquad\qquad\qquad\qquad\qquad\qquad\qquad$ (5)

Suppose that the matrix A has eigen values λ_i with eigen vectors u_i and that A has an

inverse A^{-1}. Then $Au_i = \lambda_i u_i$ which may be written $A^{-1} u_i = \dfrac{1}{\lambda_i} u_i$ $\qquad\qquad$ (6)

The inverse matrix A^{-1} has the same eigen vectors as A but has eigen values $\dfrac{1}{\lambda_i}$

A similarity transformation, where S is the matrix of eigenvectors, reduces a matrix A to its diagonal form. The eigen values of A are located on the leading diagonal of this diagonal

Define similarity transformation. Show that the eigen values of a matrix are invariant under similarity transformation.

Construct a similar matrix corresponding to the matrix $A = \begin{bmatrix} 6 & 3 & 4 \\ 3 & 6 & 4 \\ 4 & 4 & 5 \end{bmatrix}$

Power Method

Power method is used to determine the largest eigen value (in magnitude) and the corresponding eigen vector of the system $A X = \lambda X$ $\qquad\qquad\qquad\qquad\qquad\qquad$ (1)

Let $\lambda_1, \lambda_2, \ldots, \lambda_n$ be the district eigen values such that $1\ \lambda_1 |> | \lambda_2 |> \ldots > | \lambda_n$ and $v_1, v_2, \ldots, v_n$ be the corresponding eigen vectors. The procedure explained here is applicable if a complete system of n independent eigen vectors exists, even though some of the eigen values $\lambda_2, \ldots, \lambda_n$ may not be district. Then any vector v in the space of eigen vectors $v_1, v_2, \ldots, v_n$ can be written as

$$\underline{V} = C_1 V_1 + C_2 V_2 + \ldots + C_n V_n \qquad\qquad\qquad\qquad (2)$$

Now $A\underline{V} = C_1 \lambda_1 V_1 + C_2 V_2 + \ldots + C_n \lambda_n V_n$

$$= \lambda_1 \left[C_1 v_1 + C_2 \left(\frac{\lambda_2}{\lambda_1} \right) V_2 + C_n \left(\frac{\lambda_n}{\lambda_1} \right) V_n \right]$$

$$A^2\underline{V} = \lambda_1^2 \left[C_1 v_1 + C_2 \left(\frac{\lambda_2}{\lambda_i} \right)^2 V_2 + C_n \left(\frac{\lambda_n}{\lambda_1} \right)^2 V_n \right] \tag{3}$$

$$A^k\underline{V} = \lambda_1^k \left[C_1 v_1 + C_2 \left(\frac{\lambda_2}{\lambda_i} \right)^k V_2 + C_n \left(\frac{\lambda_n}{\lambda_1} \right)^k V_n \right]$$

$$A^{k-1}\underline{V} = \lambda_1^{k+1} \left[C_1 v_1 + C_2 \left(\frac{\lambda_2}{\lambda_i} \right)^{k+1} V_2 + C_n \left(\frac{\lambda_n}{\lambda_1} \right)^{k+1} V_n \right] \tag{4}$$

As $k \to \infty$, the righthand sides of (3) and (4) tend to

$\lambda_1^k C \underline{V}$ and $\lambda_1^{k+1} C_1 V_1$, since $\left| \frac{\lambda_i}{\lambda} \right| < 1, i = 2,3...n$

Consequently the victor $C_1, V_1 + C_2 \left| \frac{\lambda_2}{\lambda_1} \right|^k V_2 + ... + C_n \left(\frac{\lambda_2}{\lambda_1} \right)^k V_n$

tends to $C_1 V_1$ which is the eigen victor Corresponding to λ_1. The eigen value λ_1 is obtained as the natro of the corresponding components of $A^k V$ and $A^{k+1} V$. That is

$$\lambda_1 = \lim_{k \to \infty} \frac{\left(A^{k+1}V \right)_r}{\left(A^k V \right)_r} r = 1,2...n \tag{5}$$

where the suffix r denotes the r th Component of the vector.

If $|\lambda_2| << |\lambda_1|$, then faster convergence is obtained. The round off error will be in control. If we normalize (such that the largest element is unity) the victor before pre multiplying by A We can write the method as follows: Let $y_{k+1} = A_{vk}$, $V_{k+1} = \frac{y_{k+1}}{m_{k+1}}$, where $mk+1$ is the largest element in magnitude of y_{k+1}.

Then $?_1 = \lim_{k \to \infty} \frac{(y^{k+1})_r}{(\underline{V}^k)_r} r = 1, 2....n$

And v_{k+1} is the required eigenvector. The initial viector is usually chosen as a vector with all components equal to unity.

Examples

EXAM.1: Obtain the largest eigenvalue (in maguetude) and the corresponding eigenvector of the matrix given below by the Power method.

$A = \begin{bmatrix} 1 & 2 \\ 5 & 4 \end{bmatrix}$ The Characteristic equation is

SOLUTION

$\begin{bmatrix} 1-\lambda & 2 \\ 5 & 4-\lambda \end{bmatrix} = 0 \Rightarrow \lambda^2 - 5\lambda - 6 = 0 \therefore \lambda = 6 \text{ and } \lambda = -1:$

The eigen victor corresponding to $\lambda = 6$ is given by $\begin{bmatrix} -5 & 2 \\ 5 & -2 \end{bmatrix}$

$\begin{bmatrix} x_1 \\ x_2 \end{bmatrix} = 0 \Rightarrow -5x_1 + 2x_2 = 0 \Rightarrow x_2 = 1, x_1 = \dfrac{2}{5}$ or Eigen vector is $[2/5,1]^T$ Consider the power

method with starting vector Vo = $[1,1]^T$

$\therefore y_1 = AV_0 = \begin{bmatrix} 1 & 2 \\ 5 & 4 \end{bmatrix} \begin{bmatrix} 1 \\ 1 \end{bmatrix} = \begin{bmatrix} 3 \\ 9 \end{bmatrix}$ or $V_1 = \begin{bmatrix} \frac{1}{3} \\ 1 \end{bmatrix}$

$\therefore y_2 = AV_1 = \begin{bmatrix} 1 & 2 \\ 5 & 4 \end{bmatrix} \begin{bmatrix} \frac{1}{3} \\ 1 \end{bmatrix} = \begin{bmatrix} \frac{7}{3} \\ \frac{17}{3} \end{bmatrix}$ or $V_2 = \begin{bmatrix} \frac{7}{17} \\ 1 \end{bmatrix}$

$\therefore y_3 = AV_2 = \begin{bmatrix} 1 & 2 \\ 5 & 4 \end{bmatrix} \begin{bmatrix} \frac{7}{17} \\ 1 \end{bmatrix} = \begin{bmatrix} \frac{41}{17} \\ \frac{103}{17} \end{bmatrix}$ or $V_3 = \begin{bmatrix} \frac{41}{103} \\ 1 \end{bmatrix}$

$\therefore y_4 = AV_3 = \begin{bmatrix} 1 & 2 \\ 5 & 4 \end{bmatrix} \begin{bmatrix} \frac{41}{103} \\ 1 \end{bmatrix} = \begin{bmatrix} \frac{247}{103} \\ \frac{617}{103} \end{bmatrix}$ or $V_4 = \begin{bmatrix} \frac{247}{617} \\ 1 \end{bmatrix}$

$\therefore y_5 = AV_4 = \begin{bmatrix} 1 & 2 \\ 5 & 4 \end{bmatrix} \begin{bmatrix} \frac{247}{617} \\ 1 \end{bmatrix} = \begin{bmatrix} \frac{1481}{617} \\ \frac{3703}{617} \end{bmatrix}$

Now $V_{k+1} = \dfrac{y_{k+1}}{m_{k+1}}$ becomes for k = 4, $V_5 = \begin{bmatrix} \dfrac{1481}{617} \\ \dfrac{3703}{617} \end{bmatrix} \div \begin{bmatrix} \dfrac{3703}{617} \end{bmatrix} = \begin{bmatrix} \dfrac{1481}{3703} \\ 1 \end{bmatrix} = \begin{bmatrix} \dfrac{2}{5} \\ 1 \end{bmatrix}$

$\lambda_1 = \displaystyle\lim_{k \to \infty} \dfrac{(y_{k+1})_r}{(V_k)_r}$, r = 1,2....and V_{k+1} in the required eigen victor Here $y_5 = \begin{bmatrix} \dfrac{1481}{617} \\ \dfrac{3703}{617} \end{bmatrix}$

and $V_4 = \begin{bmatrix} \dfrac{247}{617} \\ 1 \end{bmatrix}$ $\therefore \lambda_1 = \begin{bmatrix} \dfrac{1481}{617} \\ \dfrac{3703}{617} \end{bmatrix} \div \begin{bmatrix} \dfrac{247}{617} \\ 1 \end{bmatrix} = \begin{bmatrix} \dfrac{1481}{247} \\ \dfrac{3703}{617} \end{bmatrix} \div \begin{bmatrix} 6 \\ 6 \end{bmatrix}$

Hence by power method the largest eigen value is 6 and the corresponding eigen victor is $[2/5,1]^T$

They are the same as the exact eigen value and the eigen victor.

EXAM 2 : power method for the matrix A= $\begin{bmatrix} 3 & 2 \\ 1 & 2 \end{bmatrix}$

The characteristic equation is $(3-\lambda)(2-\lambda) -2 = 0$ or $\lambda^2-5\lambda+4 = 0$ $\lambda = 4$ and $\lambda = 1$.

$\therefore$ The largest eigenvalue is 4. and the corresponding eigen victor is $[2\ 1]^T$ or $\left[1, \dfrac{1}{2}\right]^T$

we now consider the power method $V_0 = [1,1]^T$

$$y_1 = AV_0 = \begin{bmatrix} 3 & 2 \\ 1 & 2 \end{bmatrix}\begin{bmatrix} 1 \\ 1 \end{bmatrix} = \begin{bmatrix} 5 \\ 3 \end{bmatrix} \therefore V_1 = \begin{bmatrix} 1 \\ \dfrac{3}{5} \end{bmatrix}$$

$$y_2 = \begin{bmatrix} 3 & 2 \\ 1 & 2 \end{bmatrix} = \begin{bmatrix} 1 \\ \dfrac{3}{5} \end{bmatrix} = \begin{bmatrix} 21/5 \\ 11/5 \end{bmatrix} \therefore V_2 = \begin{bmatrix} 1 \\ \dfrac{11}{21} \end{bmatrix}$$

$$y_3 = \begin{bmatrix} 3 & 2 \\ 1 & 2 \end{bmatrix} = \begin{bmatrix} 1 \\ \dfrac{11}{21} \end{bmatrix} = \begin{bmatrix} 85/21 \\ 43/21 \end{bmatrix} \therefore V_3 = \begin{bmatrix} 1 \\ \dfrac{43}{85} \end{bmatrix}$$

$$y_4 = \begin{bmatrix} 3 & 2 \\ 1 & 2 \end{bmatrix} = \begin{bmatrix} 1 \\ \dfrac{43}{85} \end{bmatrix} = \begin{bmatrix} \dfrac{341}{85} \\ \dfrac{171}{85} \end{bmatrix} \therefore V_4 = \begin{bmatrix} 1 \\ \dfrac{171}{341} \end{bmatrix}$$

$$y_5 = \begin{bmatrix} 3 & 2 \\ 1 & 2 \end{bmatrix} = \begin{bmatrix} 1 \\ \dfrac{171}{341} \end{bmatrix} = \begin{bmatrix} \dfrac{1365}{341} \\ \dfrac{683}{341} \end{bmatrix} \quad \therefore V_5 = \begin{bmatrix} 1 \\ 683 \\ 1365 \end{bmatrix}$$

$$\therefore \quad \lambda_1 = \lim_{k \to \infty} \frac{(y_{k+1})_r}{(V_k)_r}, r = 1,2\ldots$$

Here $\lambda_1 = \dfrac{y_5}{V_4} = \begin{bmatrix} \dfrac{1365}{341} \\ \dfrac{683}{341} \end{bmatrix} \div \begin{bmatrix} 1 \\ \dfrac{171}{341} \end{bmatrix}$

$$= \begin{bmatrix} \dfrac{1361}{341} \\ \dfrac{683}{341} \end{bmatrix} \approx \begin{bmatrix} 4 \\ 2 \end{bmatrix}$$

And the corresponding eigenvector

is $\begin{bmatrix} 1 \\ \dfrac{683}{1364} \end{bmatrix} \approx \begin{bmatrix} 1 \\ \dfrac{1}{2} \end{bmatrix}$

$\therefore$ They are the same as the exact eigen value and eigen vector.

EXAM. 3: Use the power method to find the largest eigen value (in magnitude) and thecorresponding eigen vector for the matrix

$A = \begin{bmatrix} 3 & -4 & 4 \\ 1 & -2 & 4 \\ 1 & -1 & 3 \end{bmatrix}$ the characteristic equation is

$\begin{vmatrix} 3-\lambda & -4 & 4 \\ 1 & -2-\lambda & 4 \\ 1 & -1 & 3-\lambda \end{vmatrix} = 0$ or $[(3-?)[-(2+-?)(3-?)+4] = 0$

$$(3-\lambda)-(\lambda^2 +-\lambda -2) = 0$$

or $(3-\lambda)(\lambda+1)(\lambda-2) = 0$

$\therefore \lambda = -1, 2$ q 3,

$\therefore$ The largest eigen value is $\lambda = 3$ and the corresponding eigen vector is

$$\begin{bmatrix} 0 & -4 & 4 \\ 1 & -5 & 4 \\ 1 & -1 & 0 \end{bmatrix} \begin{bmatrix} x_1 \\ x_2 \\ x_3 \end{bmatrix} = 0 \qquad \begin{aligned} -4x_2 + 4x_3 &= 0 \Rightarrow x_2 = x_3 \\ x_1 - 5x_2 + 4x_3 &= 0 \Rightarrow -4x_2 + 4x_3 x_2 = x_3 \\ x_1 - x_2 &= 0 \end{aligned}$$

$\therefore x_1 = x_2 = x_3 = 1$ say

$\therefore$ the eigen vector is $[1,1,1]^T$

We now use the power method.

Let $v_0 = [1,1,1]^T$

$$\therefore y_1 = Av_0 = \begin{bmatrix} 3 & -4 & 4 \\ 1 & -2 & 4 \\ 1 & -1 & 3 \end{bmatrix} \begin{bmatrix} 1 \\ 1 \\ 1 \end{bmatrix} = \begin{bmatrix} 3 \\ 3 \\ 3 \end{bmatrix} \therefore V_1 = \begin{bmatrix} 1 \\ 1 \\ 1 \end{bmatrix}$$

$$\therefore y_2 = Av_1 = \begin{bmatrix} 3 & -4 & 4 \\ 1 & -2 & 4 \\ 1 & -1 & 3 \end{bmatrix} \begin{bmatrix} 1 \\ 1 \\ 1 \end{bmatrix} = \begin{bmatrix} 3 \\ 3 \\ 3 \end{bmatrix} \therefore V_2 = \begin{bmatrix} 1 \\ 1 \\ 1 \end{bmatrix}$$

All the vectors v_0, v_1, v_2 .... Are the same $[1,1,1]^T$. Hence the iteration can be stepprd.

$$\therefore \lambda = \frac{y_2}{v_1} = \begin{bmatrix} 3 \\ 3 \\ 3 \end{bmatrix} \div \begin{bmatrix} 1 \\ 1 \\ 1 \end{bmatrix} = \begin{bmatrix} 3 \\ 3 \\ 3 \end{bmatrix}$$

$$V_2 = \begin{bmatrix} 1 \\ 1 \\ 1 \end{bmatrix}$$

$\therefore$ The eigen value is 3 and the corresponding eigen vector is $[1,t,T]^T$

They are the same as the exact eigen value and eigen vector.

EXAM. 4: Use the power method to find the largest eigen value [ni magnitude] and the corresponding eigenvector for the matrix

The characteristic equation is

$$A = \begin{vmatrix} 7-\lambda & -2 & 0 \\ -2 & 6-\lambda & -2 \\ 0 & -2 & 5-\lambda \end{vmatrix} = 0 \Rightarrow \lambda^3 - 18\lambda^2 + 99\lambda - 162 = 0f$$

Set $\lambda = 3 \Rightarrow f(\lambda)=0$ the remaining rests are given by $(\lambda - 6)(\lambda - 9) = 0$

$\therefore$ The largest reset is $\lambda = 9$. $\therefore$ The largest eigen value of the matrix is $\lambda = 9$. The corresponding eigen vector is $[2,-2,1]^T$

Now we consider the power method, $v_0\,[1,1,T]^T$

$$y_1 = Av_0 = \begin{bmatrix} 7 & -2 & 0 \\ -2 & 6 & -2 \\ 0 & -2 & 5 \end{bmatrix}\begin{bmatrix} 1 \\ 1 \\ 1 \end{bmatrix} = \begin{matrix} 5 \\ 2 \\ 3 \end{matrix} = 2v_1 = \begin{bmatrix} 1 \\ \dfrac{2}{5} \\ \dfrac{3}{5} \end{bmatrix}$$

$$y_2 = Av_1 = \begin{bmatrix} \dfrac{31}{5} \\ \dfrac{-4}{5} \\ \dfrac{11}{5} \end{bmatrix}; v_2 = \begin{bmatrix} 1 \\ \dfrac{-4}{31} \\ \dfrac{11}{31} \end{bmatrix}$$

$$y_3 = Av_2 = \begin{bmatrix} \dfrac{225}{31} \\ \dfrac{-108}{31} \\ \dfrac{63}{31} \end{bmatrix}; v_3 = \begin{bmatrix} 1 \\ \dfrac{-108}{225} \\ \dfrac{63}{225} \end{bmatrix} = \begin{bmatrix} 1 \\ \dfrac{-36}{75} \\ \dfrac{+21}{75} \end{bmatrix} = \begin{bmatrix} 1 \\ \dfrac{-12}{25} \\ \dfrac{7}{25-} \end{bmatrix}$$

$$y_4 = Av_3 = \begin{bmatrix} 7 & -2 & 0 \\ -2 & 6 & -2 \\ 0 & -2 & 5 \end{bmatrix}\begin{bmatrix} 1 \\ \dfrac{-12}{25} \\ \dfrac{7}{25} \end{bmatrix} = \begin{bmatrix} \dfrac{199}{25} \\ \dfrac{-136}{25} \\ \dfrac{59}{25} \end{bmatrix} =$$

$$v_4 = \begin{bmatrix} 1 \\ \dfrac{-136}{199} \\ \dfrac{59}{199} \end{bmatrix}$$

$$y_5 = Av_4 = \begin{bmatrix} \dfrac{1665}{199} \\ \dfrac{-1332}{199} \\ \dfrac{567}{199} \end{bmatrix}$$

$$\lambda = \frac{y_5}{v_y} = \begin{bmatrix} \dfrac{1665}{199} \\ \dfrac{-1332}{136} \\ \dfrac{567}{59} \end{bmatrix} \approx \begin{bmatrix} 8.37 \\ 9.8 \\ 9.6 \end{bmatrix}$$

$$\lambda_1 \approx 9$$

The corresponding eigen vector using the power method is given by $v_5 = \dfrac{y_5}{m_5}$

$$\approx [1,-08, 0.34]^T$$
$$\approx [2,-1.6, 0.68]^T$$

EXAM. 1: Explain the power method for producing the dominant eigen value and eigen victor of a matrix.

EXAM. 2: Using the power method, obtain the largest eigen value for the matrix

$\begin{bmatrix} 4 & 1 \\ 1 & 3 \end{bmatrix}$ and compare your result with the explicit solution of the characteristic equation.

Applying the power method, find the dominant eigen value and the corresponding eigen victor of each me of the following matrices:

$$\begin{bmatrix} 3 & 2 & 6 \\ -1 & 12 & 1 \\ 4 & 2 & 1 \end{bmatrix} \begin{bmatrix} 11 & 6 & -2 \\ -2 & 18 & 1 \\ -12 & 24 & 13 \end{bmatrix} \begin{bmatrix} 2 & -1 & 0 \\ -1 & 2 & -1 \\ 0 & -1 & 2 \end{bmatrix} \begin{bmatrix} 15 & 4 & 3 \\ 10 & -12 & 6 \\ 20 & -4 & 2 \end{bmatrix}$$

$$\begin{bmatrix} 3 & 2 & 5 \\ 6 & -5 & 3 \\ -24 & 38 & 2 \end{bmatrix} \begin{bmatrix} 1 & -3 & 2 \\ 4 & 4 & -1 \\ 6 & 3 & 5 \end{bmatrix} \begin{bmatrix} 4 & 0 & 5 \\ 6 & 4 & -4 \\ 5 & -4 & 6 \end{bmatrix} \begin{bmatrix} 5 & 2 & 2 \\ 2 & 2 & 1 \\ 2 & 1 & 2 \end{bmatrix}$$

$$\begin{bmatrix} 25 & -40 & 5 & -6 \\ -40 & 60 & -15 & 10 \\ 5 & -15 & 6 & 3 \\ -6 & 10 & -3 & 2 \end{bmatrix}$$

(Four iterations will do)

$$\begin{bmatrix} 4 & 1 & 0 \\ 1 & 20 & 1 \\ 0 & 1 & 4 \end{bmatrix}, \begin{bmatrix} 4 & 1 & 0 \\ 1 & 2 & 1 \\ 0 & 1 & 1 \end{bmatrix}, \begin{bmatrix} 1 & 2 \\ 5 & 4 \end{bmatrix}$$

Corrrect to 3 decimal places   correct one decimal place

$$\begin{bmatrix} 1 & 3 & -1 \\ 3 & 2 & 4-1 \\ -1 & 4 & 10 \end{bmatrix}, \begin{bmatrix} 2.5 & 3.1 & 1.6 \\ 3.1 & 8.9 & 7.2 \\ 1.6 & 7.2 & 23.8 \end{bmatrix}, \begin{bmatrix} 2 & 3 & 5 \\ 3 & 4 & 6 \\ 5 & 6 & 7 \end{bmatrix}, \begin{bmatrix} 3 & 1 & 1 \\ 2 & 1 & 1 \\ -1 & -1 & 3 \end{bmatrix}$$

Ans: 11.66 Ans: 26.99797

(B.025,0.425,1.00)^T (0.11819, 0.418021, 1)^T

EXAM. 3: Show that the eigen values of a matrix are invariant under similarity transformation.

EXAM. 4: Construct a similar matrix to $\begin{bmatrix} 1 & 0 & 2 \\ 0 & 2 & 1 \\ 2 & 1 & 1 \end{bmatrix}$ and verify that the eigen values

are preserved under similarity transformation.

EXAM. 5: Determine the least eigen value of the matrix $\begin{bmatrix} 4 & 0 & 5 \\ 0 & 5 & -4 \\ 5 & -4 & 6 \end{bmatrix}$ by the power

method, correct to two decimal places. Find also the corresponding eigen vector.

EXAM. 6: Use the power method to estimate the eigen value of maximum modulus and the corresponding eigenvector for the tridiagenal matrix A of order 3 with

$$a_{ii} = 4; \ a_i + 1_i = a_{i,i+1} = -1 \ \text{for} \ i = 1,2 \ A = \begin{bmatrix} 4 & -1 & -1 \\ -1 & 4 & -1 \\ -1 & -1 & 4 \end{bmatrix}$$

13.10 NUMERICAL SOLUTION OF ORDINARY DIFFERENTIAL EQUATIONS

Although many methods exist for obtaining analytical solution of differential equations, they are primary limited to special differential equations. In practical problems, frequently, a differential equations is not amenable to solution by special methods and a numerical solution must be sought, as numerical methods have no such limitations.

We consider some basic methods for the numerical solution of ordinary differential equations. The important methods of solving ordinary differential equations of first order are as follows.

(i) Picard's methods of successive approximations

Integrating the differential equation $\dfrac{dy}{dx} = f(x, y)$, with initial condition $y = y_o$ at $x =$

x_o, between the limits xo and x1 we get $\displaystyle\int_{yo}^{y} dy = \int_{xo}^{x} f(x, y)dx$ or $y - y_o = \displaystyle\int_{xo}^{x} f(x, y)dx$ or y

$$= y_o + \int_{xo}^{x} f(x, y)dx$$

As a first approximation, we replace y by y_o in f(x, y) and $y^{(1)} = y_o + \displaystyle\int_{xo}^{x} f(x, y_o)dx$; a

second approximation, as $y^{(2)} = y_o + \displaystyle\int_{xo}^{x} f(x, y^{(1)})dx$ and in general the nth approximation as

$$y^{(x)} = y_o + \int_{xo}^{x} f(x, y^{(n-1)})dx$$

The process is stopped when the two values of y, viz $y^{(n-1)}$ and $y^{(n)}$ are the same to the desired degree of accuracy.

[Note: Each approximation gives a better approximation of the required solution than the preceding one]

(ii) Euler's method

Consider the differential equations $\dfrac{dy}{dx} = f(x, y)$, with $y = y_o$ at $x = x_o$. If $y = F(x)$ is the

solution, divide the interval $[x_o, 1]$ into n equal sub intervals. Now in the interval (x_o, x_1) we have an approximation

$$y_1 - y_o \approx (x_1 - x_o)\left(\frac{dy}{dx}\right)_{x=xo} = (x_1 - x_o)f(x_o, y_o) \text{ or } y_1 - y_o \approx +hf(x_o, y_o), h = x_1 - x_o$$

$$\Rightarrow y_1 = y_o + h\left(\frac{\Delta y}{\Delta x}\right)$$

In general $y_n \approx y_{n-1} + hf(x_{n-1}, y_{n-1})$

[Note (i) A great disadvantage of the Euler's method lies in the fact that if $\dfrac{dy}{dx}$ changes rapidly over an interval, then its value over the interval may give a poor approximation as compared to the average value over the interval and the calculated value will be in much error than its true value.

(ii) The actual solution curve is approximated by a sequence of short straight lines and at times these may deviate from the solution curve significantly.]

Euler method is called a **first-order method,** because we have considered the constant term and the term containing the first power of h, in the expansion of $y(x+h) \approx y(x) + hy'(x) + h^2 y''(x) + \dots$. The omission causes an error called **the truncation error** of this method : $T.E = O(h^2)$.

(iii) Improved Euler's method

Here we consider a line passing through (x_o, y_o) whose slope is the average of the slopes at (x_o, y_o) and $(x_1, y_1^{(1)})$ such that

$$y_1^{(1)} = y_o + h + (x_o, y_o)$$

or $y_1 - y_o = (x_1 - x_0)\left[\dfrac{f(x_0, y_0) + f(x_1, y_1^{(1)})}{2}\right]$

In general

$$y_{m+1} = y_m + \frac{h}{2}\left\{f(x_m, y_m) + f(x_{m+1}, y^{(1)}{}_{m+1})\right\} \text{ Where } x_{m+1} - x_m = h$$

When f is a function of x alone, we can wile $y_{m+1} = y_m + \dfrac{h}{2}[f(x_m) + f(x_{m+1})]$.

This agrees with the term, obtained by integrating of $y = f(x)$ using trapezoidal rule.

(iv) Modified Euler's method

In the Euler's method, the curve of solution in the interval (x_0, x_1) =segment interval say LL_1 is approximated by the tangent at the point $P(x_0, y_0)$ such that at $P_1(x_1 = x_0 + h, y_1)$, we have $y_1 = y_0 + hf(x_0 y_0)$.

Then the slope of the curve of solution through $P_1 \left[\text{i.e.,} \left(\dfrac{dy}{dx} \right)_{P1} = f(x_0 + h, y_1) \right]$ is

computed and the tangents at P_1 to the curve is drawn meeting the ordinate through L_2 in P_2, $(x_0 + 2_h, y_2)$. Hence we find a better approximation $y_1(1)$ of $y(x_0 + h)$ by taking the slope of the curve as the mean of the slopes of the tangents at P and P_1, i.e.,

$$y_1^{(1)} = y_o + \frac{h}{2} [f(x_0, y_0) + f(x_0 + h, y_1)]$$

$$= y_o + \frac{h}{2} [f(x_0, y_0) + f(x_0 + h, y_1^{(1)})]$$

Or, in general $y_1^{(1)} = y_o + \dfrac{h}{2} \left[\left(\dfrac{dy}{dx} \right)_{x_0} + \left(\dfrac{dy}{dx} \right)_{x_1}^{(n-1)} \right]$

The truncation error in the modified Euler's method is found by comparing it with the

Taylor series $y_{n+1} = y_n + y'_n h + \dfrac{1}{2} y''_n h^2 + \dfrac{1}{6} y'''_n (\xi) h^3, x < \xi < x_{n+h}$

Replace the second derivative by the forward difference approximation for y'' us $(y'_{n+1} - y'_n)/h$ which has error for O(h) and write the error term as $O(h^3)$.

$$y_{n+1} = y_n + h \left[y'_n + \frac{1}{2} \left\{ \frac{y'_{n+1} - y'_n}{h} + O(h) \right\} h \right] + O(h^3)$$

$$= y_n + h \left(y'_n + \frac{1}{2} y'_{n+1} - \frac{1}{2} y'_n \right) + O(h^3)$$

$$= y_n + h \frac{(y'_n + y'_{n+1})}{2} + O(h^3)$$

(v) Taylor's series method

Consider the first order differential equation $\dfrac{dy}{dx} = f(x, y)$, with $y = y_0$ at $x = x_0$.

Differentiating the above, we have $\dfrac{d^2y}{dx^2} = \dfrac{\partial f}{\partial x} + \dfrac{\partial f}{\partial y} \dfrac{dy}{dx}$ or $y'' = f_x + f_y f'$. Continuing the differentiation, successively. We can set y''', $y^{(iv)}$,........ Putting $x = x_0$ and $y = y_0$, the values of $(y')_0$, $(y'')_0$, $(y''')_0$....can be obtained. Hence the Taylor's

series $y = y_0 + (x - x_0)(y')_0 + \dfrac{(x - x_0)^2}{2!}(y'')_0 + \dfrac{(x - x_0)^3}{3!}(y''')_0 + ...$ gives the values of y for

every value of x for which the Taylor expansion converses.

On finding the value of y1 for $x = x_1$ in the Taylor expansion, y'', y''' can be expanded about the point $x = x_1$. In this way the solution can be extended beyond the range of conversance of the Taylor series.

[Note: If f(x,y) is some what complicated and calculation higher order derivatives becomes tedious, the Taylor's series method cannot be used gainfully].

(vi) Runge-Kutta methods

The methods named after Carl Runge and Wilhelm Kutta are designed analogus to the Taylor series methods without requiring analytic differentiation of the original differential equation.

Consider the Taylor series expansion for two variables. The infinite, series is

$$f(x + h, y + k) = \sum_{i=0}^{\infty} \frac{1}{i!}\left(h\frac{\partial}{\partial x} + k\frac{\partial}{\partial y}\right)^i f(x, y)$$

$$= \sum_{i=0}^{n-1} \frac{1}{i!}\left(h\frac{\partial}{\partial x} + k\frac{\partial}{\partial y}\right)^i f(x, y)$$

$$+ \frac{1}{n!}\left(h\frac{\partial}{\partial x} + k\frac{\partial}{\partial y}\right)^n f(\xi, \eta)$$

Where the point (ξ, η) lies on the line segment that joins (x,y) to (x+h,y+k) in the plane. We use subscripts to denote partial derivatives, so

$$f(x + h, y + k) = f + (hf_x + k_y) + \frac{1}{2!}\left[h^2 f_{xx} + 2hkf_{xy} + k^2 f_{yy}\right].$$

$$+ \frac{1}{3!}\left(h^3 f_{xxx} + 3h^2 kf_{xxy} + 3hk^3 f_{xyy} + k^3 f_{yyy}\right) + ...$$

As a special case, we set

$$f(x + h, y) = f + hf_x + \frac{h^2}{2!}f_{xx} + \frac{h^3}{3!}f_{xxx} + ...$$

and $$f(x, y + k) = f + kf_y + \frac{k^2}{2!}f_{yy} + \frac{k^3}{3!}f_{yyy} + ...$$

● Runge-Kutta method of second order

To import some idea of how Runge-Kutta methods are developed, we will show the derivation of a **simple second-order method.** Here, the increment to the y is a weighted average of two estimates of the increment which we call k_1 and k_2. Thus for the

equations $\dfrac{dy}{dx} = f(x, y)$,

$$y_{n+1} = y_n + ak_1 + bk_2,$$

$$k_1 = hf(x_n, y_n) \tag{1}$$
$$k_2 = hf(x_n + \alpha h, y_n + \beta k)$$

Our problem is to devise a scheme of choosing the form parameters a, b, α, β. We do so by making equation (1) agree as well as possible with the Taylor series expansion, is

which the y-derivatives we written in terms of f, from. $\dfrac{dy}{dx} = f(x, y)$,

$$y_{n+1} = y_n + h(x_n, y_n) + \frac{h^2}{2!} f'(x_n, y_n) + \ldots \ldots$$

$$= y_n + hf_x + \frac{h^2}{2!}(f_x + f_y f)_n \tag{2}$$

$$= y_n + ahf(x_n, y_n) + bhf[x_n + \alpha h, y_n + \beta hf(x_n, y_n)]$$

$$= y_n + ahf_x + bh(f + f_x \alpha h + f_y \beta hf)_n$$

$$= y_n + (a + b)hf_n + h^2(\alpha b f_x + \beta b f_y f)_n \tag{3}$$

Equation (3) will be identical to equation (2), if $a + b = 1, \alpha b = \dfrac{1}{2}, \beta b = \dfrac{1}{2}$

Note that only three equations need to be satisfied by the four unknowns. We can choose one value arbitrarily; hence we have a set of second order methods. For example, taking $a = \dfrac{2}{3}$, we have $b = \dfrac{1}{3}$, $\alpha = \dfrac{3}{2}$, $\beta = \dfrac{3}{2}$. Other choices given other sets of parameters that agree with the Taylor series expansion. If we take $a = \dfrac{1}{2}$, the other variables we $b = \dfrac{1}{2}, \alpha = 1, \beta = 1$. This last set of parameters gives the modified Euler algorithm that we have previously discussed; the modified Eular method is a special case of a second order Runge-Kutta method.

● **Fourth-order Runge-Kutta methods.**

They are most widely used and are derived in similar fashion. Greater complexity results from having to compare terms through h^4 and gives a set of 11 equations in 13 unknowns. The set of 11 equations can be solved with 2 unknowns being chosen arbitrarily. The most commonly used set of values leads to the algorithm.

$$y_{n+1} = y_n + \frac{1}{6}(k_1 + 2k_2 + 2k_3 + k_4),$$

$$k_1 = hf(x_n, y_n)$$

$$k_2 = hf\left(x_n + \frac{h}{2}, y_n + \frac{k_1}{2}\right),$$

$$k_3 = hf + \left(k_n + \frac{1}{2}h, y_n + \frac{1}{2}k_2\right)$$

$$k_4 = hf + (x_n + h, y_n + k_3)$$

[Note: The local error term for the second order Runge-Kutta method is $O(h^3)$ and for the fourth order Runge-Kutta methods is $O(h^5)$ and the global error could be $O(h^4)$ for the fourth order.

The methods so for discussed are called single step methods. The use only the information at (x_n, y_n) to get to (x_{n+1}, y_{n+1})

[The third order Runge-Kutta will be $y(x+h) = y(x) + \frac{1}{9}(2k_1 + 3k_2 + 4k_3)$, where

$k_1 = hf(x,y)$

$$k_2 = hf\left(x + \frac{1}{2}h, y + \frac{1}{2}k_1\right) \quad k_3 = hf\left(x + \frac{3}{4}h, y + \frac{3}{4}k_2\right)]$$

● **Distinguishing properties of Runge-Kutta methods.**

- The formulae do not require prior calculations of higher order derivatives of $y(x)$ as the Taylor series method does calculation of derivations which may be difficult.
- The formulae involve the computation of $f(x,y)$ at various positions instead of derivatives and this function occcers in the given differential equation.
- To evaluate ym+1, we need information only at the point ym.. The previous values ym-1, ym-2......, are not required. Hence the Runge-Kutta methods are one-step methods.

These methods agree with the Taylor series solution up to the term in hr, where r differs from method to method and is known as the order of the Runge-Kutta method.

(vii) Multistep Methods

Runge-Kutta methods (which include Euler and modified Euler as special cases) are called single-step methods, because they use only the information from the last step computed. In this they have the ability to perform the next step with a different step size and are ideal for beginning the solution where only the initial conditions are available. After the solution has begun, however, there is additional information available about the function (and its derivative), if we are wise enough to retain it in the memory of the computer. A multi step method is one that takes advantage of this fact.

The principle behind a multi step method is to utilize the past values of y and /or y to construct a polynomial that approximates the derivative function and extrapolate this into the next interval. The number of past points that are used sets the degree of the polynomial and is there for responsible for the truncation error.

The general multi step method may be written as

$$y_{j+1} = a_1 y_j + a_2 y_{j-1} + \ldots\ldots a_k y_{jk+1}$$

$$+ h\left(b_0 y'_{j+1} + b_1 y'_j + \ldots\ldots\ldots b_k y'_{j-k+1}\right)$$

$$\text{or } y_{j+i} = \sum_{i=1}^{k} a_i y_{j-i+1} + h_{j+i} \sum_{i=0}^{k} b_i y'_{j-i+1} \tag{1}$$

Symbolically we can write (1) as $p(E)y_{j-k+1} - hq(E)y'_{j-k+1} = 0$,

where p and q are polynomials defined by

$$p(\xi) = \xi^k - a_1 \xi^{k-1} - a_2 \xi^{k21} - \ldots\ldots - a_k$$

$$q(\xi) = b_0 \xi^k + b_1 \xi^{k-1} + \ldots\ldots + b_k$$

The formula (1) can only be used is we know the values of the solution y(x) and y'(x) at k success given points. These values will be assumed to be given. Further, if $b_0 = 0$, the method. (1) is called an **explicit** or **predictor method**. When $b_0 \neq 0$, it is called an implicit or corrector method. We also assume that the polynomials $p(\zeta)$ and $q(\zeta)$ have no common factor, since, otherwise (1) can be reduced to a difference equation of lower order.

In order to determine the coefficients a's and b's, we write the local truncation error as

$$T_{j+i} = y(x_{j+1}) - \sum_{i=1}^{k} a_i y(x_{j-i+1}) - h\sum_{i=0}^{k} b_i y'(x_{j-i+1})$$

We assume that the function y(x) has continuous derivatives of sufficiently high order. Expanding $y(x_{j-i+1})$ and $y'(x_{j-i+1})$ in Taylor series and substituting them in (2), we get

$$T_{j+i} = c_0 y(x_j) + c_1 h y'(x_j) + c_2 h^2 y''(x_j) + \ldots c_p h^p y^{(p)}(x_j) + R_{p+1} \tag{3}$$

Where $c_0 = 1 - \sum_{i=1}^{k} a_i \tag{4}$

$$c_q = \frac{1}{q!}\left[1 - \sum_{i=1}^{k} a_i (1-i)^q\right] - \frac{1}{(q-1)!} - \sum_{i=0}^{k} b_i (1-i)^{q-1}, q = 1(1)b$$

$$R_{p+1} = \frac{1}{p!}\left[\int_{x_j}^{x_{j+i}} (x_{j+i} - s)^p y^{p+1}(s)ds - \sum_{i=1}^{k} a_i \int_{x_j}^{x_{j+1}} (x_{(j-i+1)} - s)^p y^{(p+1)}(s)ds\right.$$

$$-hp\int_{x_j}^{x_{j+i}} b_0(x_{j+i} - s)^{p-1} y^{p+1}(s)ds - hp\sum_{i=1}^{k} b_i \int_{x_j}^{x_{j+1}} (x_{j+i} - s)^{p-1} y^{(p+1)}(s)ds \quad (5)$$

We define the following:

1. The linear multi step method (1) is said to be of **order p** if, in (3) $c_0 = c_1 = c_2$ $=\ldots\ldots= c_f= 0$ and $c_0 \neq 0$ and the p the order difference scheme is given by (1). The truncation error becomes.

    $$T_{j+i} = c_{p+1}h^{p+1}y^{p+1}(x_j) + O(h^{p+2}), \text{ where } (c_{p+1})/q^{(1)} \text{ is called the error constant.}$$

2. The linear multi step method is said to be **consistent** if it has order p≥1. Form the definition 1, we get $c_0 = 0 = c_1$, which implies $p(1) = 0$ and $p'(1)=q(1)$.

3. The linear multi step method is said to satisfy the **root condition,** if the roots of the equation $p(\zeta)=0$ lie inside the unit circle in the complex plane, and are simple is they lie on the unit circle

Determination of a_i and b_i

Choosing $y(x) = e^x$ and substituting $y_j = e^{xj}$ and simplifying (1), we get

$$e^{kh} - a_1 e^{(k-1)h} - \ldots - a_k - h(b_0 e^{kh} + b_1 e^{(k-1)h} + \ldots\ldots b_k = 0)$$ Putting $e^h = \zeta$, we

obtain

$$p(\xi) - \log[\xi q(\xi)] = 0 \quad (6)$$

As $h \to 0, \xi \to 1$, we may use (6) for determining $p(\zeta)$ or $q(\zeta)$ of maximum order is $q(\zeta)$ or $p(\zeta)$ in given. The technique of refining an initially crude estimate of y_j, by means of a more accurate formula is known as Predictor-Corrector Method.

(viii) Adams-Bash forth Methods

$$p(\xi) = \xi^{k-1}(\xi - 1) \text{ and } q(\xi) \text{ of degree k-1, } q(\xi) = \xi^{k-1}\sum_{m=0}^{k-1}\gamma_m(1-\xi^{-1})^m$$

where $\gamma_m + \frac{1}{2}\gamma_{m-1} + \ldots\ldots + \frac{1}{m+1}\gamma_0 = 1, m = 0,1,2\ldots\ldots$

Given $\dfrac{dy}{dx} = f(x,y)$ and $y_0 = y(x_0)$, we compute y_{-1}, y_{-2} and y_{-3} by any first order method.

We use Newton's back word interpolation formula

$$f(x,y) = f_0 + n\nabla f_0 + \frac{n(n+1)}{2}\nabla^2 f_0 + \frac{n(n+1)(n+2)}{6}\nabla^3 f_0 + ...$$

In $y_1 = y_0 + \displaystyle\int_{x_0}^{x_0+h} f(x,y)dx \quad \therefore dx = hdn, x = x_0 + nh = 0 + nh$

$$= y_0 + \int_0^1 \left(f_0 + n\nabla f_0 + \frac{n(n+1)}{2}\nabla^2 f_0 + ... \right) dn$$

Neglecting the fun the and higher order difference and expressing $\nabla f_0, \nabla^2 f_0$ and $\nabla^3 f_0$

in terms of function values, we get $y_1 = y_0 + \dfrac{h}{24}\left(55f_0 - 59f_{-1} + 37f_{-2} - 9f_{-3}\right)$

This is called **Adam-Bash forth predictor formula.** Having found y_1, we find $f_1 = f(x_0+h, y_1)$. Then to find a better value of y_1, we derive a corrector formula by substituting Newton's back ward formula at $f1$, i.e.,

$$f(x,y) = y_1 + n\nabla f_1 + \frac{n(n+1)}{2}\nabla^2 f_1 + \frac{n(n+1)(n+2)}{6}\nabla^3 f_1 + ...$$

$$\therefore y_1 = y_0 + \int_{x_0}^{x_1} \left(f_1 + n\nabla f_1 + \frac{n(n+1)}{2}\nabla^2 f_1 + \frac{n(n+1)(n+2)}{6}\nabla^3 f_1 + ... \right) dx$$

Put $x = x_1 + nh$, $dx = hdn$.

$$\therefore y_1 = y_0 + \int_{-1}^{0} \left(f_1 + n\nabla f_1 + \frac{n(n+1)}{2}\nabla^2 f_1 + ... \right) dn$$

$$= y_0 + h\left(f_1 - \frac{1}{2}\nabla f_1 \frac{1}{12}\nabla^2 f_1 - \frac{1}{24}\nabla^3 f_1 - .... \right)$$

or $y_1 \approx +y_0 + \dfrac{h}{24}\left(9f_1 + 19f_0 - 5f_{-1} + f_{-2}\right)$,

which is called **Adams-Moulton Corrector formula**

(ix) Milne's Method

Given $\dfrac{dy}{dx} = f(x, y)$ and $y = y_0$, $x = x_0$, to find an approximate value for y for $x = x_0 + nh$

by Milne's method, we proceed as follows:

The value $y_0 = y(x_0)$ being given, we compute $y_1 = y(x_0 + h)$, $y_2 = y(x_0 + 2h)$, $y_3 = y(x_0 + 3h)$ by Taylor's series method. Next, we calculate $f_0 = f(x_0, y_0)$, $f_1 = f(x_0 + h, y_1)$, $f_2 = f(x_0 + 2h, y_2)$; $f_3 = f(x_0 + 3h, y_3)$. Then to find $y_4 = y(x_0 + 4h)$, we substitute Newtons forward interpolation formula.

$$f(x, y) = f_0 + n\Delta f_0 + \frac{n(n-1)}{2}\Delta^2 f_0 + \frac{n(n-1)(n-2)}{6}\Delta^3 f_0 + \ldots\ldots$$

In the relation

$$y_4 \quad = \quad y_0 + \int_{x_0}^{x_0+4h} f(x, y)dx$$

$$= \quad y_0 + \int_{x_0}^{x_0+4h}\left[f_0 + n\Delta f_0 + \frac{n(n-1)}{2}\Delta^2 f_0 + \frac{n(n-1)(n-2)}{6}\Delta^3 f_0 + \ldots\ldots\right]dx$$

Set $x = x_0 + nh$, $dx = hdn$

$$\therefore y_4 \quad = \quad y_0 + h\int_0^4\left[f_0 + n\Delta f_0 + \frac{n(n-1)}{2}\Delta^2 f_0 + \ldots\ldots\ldots\right]dn$$

$$= \quad y_0 + h\left[4f_0 + 8\Delta f_0 + \frac{20}{3}\Delta^2 f_0 + \frac{8}{3}\Delta^3 f_0 + \ldots\ldots\ldots\right]$$

or $\quad y_4 \quad = \quad y_0 + \dfrac{4h}{3}(2f_1 - f_2 + 2f_3)$

which is called a **predictor**. Having found y_4, we obtain a first approximation to $f_4 = f(x_0 + 4h, y_4)$. Then a better value of y_4 is found by Simpson's $\dfrac{1}{3}$ rule

as $y_4 = y_2 + \dfrac{h}{3}(f_2 + 4f_3 + f_4)$ which is called a **corrector**. We repeat this step until y_4 remains unchanged. Similarly, $y_4, y_5, \ldots$ are calculated as before. This is **Milne's Predictor - Corrector method** The general Milne's Predictor - Corrector formula pair is given by

$$y_{n+1,p} = y_{n-3} + \frac{4h}{3}[2f_n - f_{n-1} + 2f_{n-2}] \text{ with truncation error} = \frac{28}{90}h^5 y^{(5)}(\xi_1), \ x_{n-3} < \zeta < x_{n+1}$$

and $y_{n+1,c} = y_{n-1} + \dfrac{h}{3}[f_{n+1} + 4fn + f_{n-1}]$ with truncation error $= -\dfrac{h^5}{90}y^{(5)}(\xi_2), \ x_{n-1} < \xi_2 < x_{n+1}$.

(x) Differential equations of second order

Consider the second order differential equation $\dfrac{d^2y}{dx^2} = f\left(x, y, \dfrac{dy}{dx}\right)$

write $\dfrac{dy}{dx} = z$ and the second order equation can be reduced to two first order simultaneous

differential equations : $\dfrac{dy}{dx} = z, \dfrac{dz}{dx} = f(x, y, z)$ These equations can be solved using the

first order method.

The procedure is explained by considering the example

$\dfrac{d^2y}{dx^2} = y + x\dfrac{dy}{dx}, y(0) = 1, \dfrac{dy}{dx}(0) = 0$ to find $y(0.2)$ and $\dfrac{dy}{dx}(0,2)$

SOLUTION:

The given differential equation is $\dfrac{d^2y}{dx^2} = y + x\dfrac{dy}{dx}$. Let $\dfrac{dy}{dx} = z = f(x, y, z)$

Here $\dfrac{dz}{dx} = y + xz = \phi(x, y, z)$ say

The initial conditions are $x_0 = 0$, $y(0) = 1$, $y_0 = 1$, $z_0 = 0$ and $h = 0.2$ here.

The fourth order Rungekutta formulae become

$k_1 = hf(x_0, y_0, z_0) = \quad 0.2f(0,1,0) = 0.2(1) = 0.2$

$k_2 = hf(x_0 + \dfrac{h}{2}, y_0 + \dfrac{k_1}{2}, z_0 + \dfrac{\ell_1}{2})$ $= 0.2f(.1,1,.1) = 0.02$	$\ell_1 = h\phi(x_0, y_0, z_0)$ $= h\phi(0,1,0) = 0.2(1) = 0.2$
$k_3 = hf(x_0 + \dfrac{h}{2}, y_0 + \dfrac{k_2}{2}, z_0 + \ell_2)$ $= (.2)f(.1,1.01,0.101)$ $= (.2)(0.101)$ $= 0.0202$	$\ell_2 = h\phi(x_0 + \dfrac{h}{2}, y_0 + \dfrac{k_1}{2}, z_0 + \dfrac{\ell_1}{2})$ $= (.2)(.1,1.0,.1)$ $= (.2)(1 + 0.01) = 0.202$
$k_4 = h + (x_0 + h, y_0 + k_3, z_0 + \ell_3)$ $= (.2)f(.2,1.0202,.2040)$ $= 0.040804$	$\ell_3 = (.2)\phi(x_0 + \dfrac{h}{2}, y_0 + \dfrac{k_2}{2}, z_0 + \dfrac{\ell_2}{2})$ $= (.2)[.1,.01,0.101]$ $= (.2)[1.01 + .0101]$ $= .20402$ $\ell_4 = h\phi(x_0 + h, y_0 + k_3, z_0 + \ell_3)$ $= (.2)(1.0202 + 0.04808)$ $= .213656$

$\therefore k = \dfrac{1}{6}(k_1 + 2k_2 + 2k_3 + k_4)$	$\ell = \dfrac{1}{6}(\ell_1 + 2\ell_2 + 2\ell_3 + \ell_4)$
$= \dfrac{1}{6}(0 + 0.04 + 0.0404 + 0.0408)$	$= \dfrac{1}{6}(0.2000 + 0.4040 + 0.40804 + .213656)$
$= 0.0202$	$= 0.204282$
$x = 0.2 \qquad\qquad y = 1.0202$	$z = 0 + 0.204282$

$$\text{or } \frac{dy}{dx}(0.2) = 0.204282$$

13.10.1 Examples

EXAM. 1: Use Picards method to approximate y when x=0.2, given that y=1 when

x=0 and $\dfrac{dy}{dx} = x - y$

SOLUTION :

We have $f(x,y) = x-y$, $x_0 = 0$, $y_0 = 1$

The first approximation is $y^{(1)} = y_0 + \int\limits_0^x f(x,y_0)\,dx = 1 + \int\limits_0^x (x-1)dx = \dfrac{x^2}{2} - x + 1$

The Second approximation is $y^{(2)} = y_0 + \int\limits_0^x (x - \dfrac{x^2}{2} + x - 1)\,dx = 1 - \dfrac{x^3}{6} + x^2 - x$

$$= -\frac{x^3}{6} + x^2 - x + 1$$

The Third approximation is

$$y^{(3)} = 1 + \int\limits_0^x (x + \frac{x^3}{6} - x^2 + x - 1)dx = \frac{x^4}{24} - \frac{x^3}{3} + x^2 - x + 1$$

$$y^{(4)} = 1 + \int\limits_0^x (x - \frac{x^4}{24} + \frac{x^3}{3} - x^2 + x - 1)dx$$

The Fourth approximation is

$$= -\frac{x^5}{120} + \frac{x^4}{12} - \frac{x^3}{3} + x^2 - x + 1$$

∴ where x = 0.2; $y^{(1)} = 0.82$; $y^{(2)}=0.83867$, $y^{(3)}=0.83740$; $y^{(4)} = 0.83746$ and so on

EXAM: 2: Using Euler's method, find an approximate value of y corresponding to x =1.2, choosing the step length = 0.2, given that $\frac{dy}{dx} = x + y, y(0)=0$

SOLUTION:

$y_n = y_{n-1} + hf[x_0 + (n-1)h, y_{n-1}]n = 1,2,3,\ldots\ldots$

$\frac{dy}{dx} = x + y, y(0) = 0$ Set n=1 in

$y_n = y_{n-1} + hf(x_0 + (n-1)h, y_{n-1}) = y_0 + (.2) + (0,0), y_0 = 0.$

$y_2 = y_1 + .2f(.2,0) = 0 + .04 = 0.04$

$y_3 = y_2 + .2f(.4,.04) = 0.04 + 0.088 = 0.128$

$y_4 = y_3 + .2f(.6,0.128) = 0.2736$

$y_5 = y_4 + .2f(.8,0.2736) = .48832$

$y_6 = y_5 + .2f(1.0,.48832) = .785984$

$y_7 = y_6 + .2f(1.2,0.785984) = 0.785984 + 0.3971968 = 1.1831858$

$y_7 = 1.1831858$

EXAM. 3: Solve the equation $\frac{dy}{dx} = x + y^2$, x=0, y(0)=1 by Improved Euler method with step length h=0.1

SOLUTION

The algorithm for the improved Euler method is

$$y_{m+1} = y_m + \frac{h}{2}[f(x_m, y_m) + f(x_{m+1}, y_{m+1}^{(1)})]$$

For m=0, we have

$$y_1 = y_0 + \frac{h}{2}[f(x_0, y_0) + f(x_1, y_1^{(1)})],$$

where $y_1^{(1)} = y_0 + hf(x_0, y_0) = 1 + .1(1)1.1$

$\therefore y_1 = 1 + \frac{.1}{2}[1 + f(.1,1.1)] = 1.1155$

set m= 1

$$y_2 = y_1 + \frac{h}{2}[f(x_1, y_1) + f(x_2, y_2^{(1)})]$$

$$\text{where } y_2^{(1)} = y_1 + hf(x_1, y_1)$$

$$= 1.1155 + (.1)f(.1, 1.1155) = 1.24993$$

$$y_2 = 1.1155 + \frac{.1}{2}[f(.1, 1.1155) + f(.2, 1.24993)]$$

$$= 1.1155 + \frac{.1}{2}(3.1068) = 1.1155 + 0.15534 = 1.27084. \text{and so as}$$

EXAM. 4: Given that $\dfrac{dy}{dx} = x + y^2$, $y(0) = 1$ find an approximate value of y at x = 0.1 by Modified Euler's method.

SOLUTION

The algorithm for the modified Euler's method is

$$y_1^{(n)} = y_0 + \frac{h}{2}\left[\left(\frac{dy}{dx}\right)_{x_0} + \left(\frac{dy}{dx}\right)_{x_1}^{(n-1)}\right], n = 1, 2, \ldots \ldots$$

Taking h = 0.1, the variance calculated are arranged as follows

x x+y²= y	Mean Slope	Old y+0.1(Mean slope)=new y
0.0 0+1=1	-	1.00+0.1(1.00) =1.10
0.1 0.1+(1.1)²=1.31	$\frac{1}{2}(1+1.31) = 1.155$	1.00+0.1(1.155)=1.1155
.1 .1+(1.1155)² =1.3442	$\frac{1}{2}(1+1.3442) = 1.1721$	1.00+.11721=1.1172
.1 .1+(1.1172)² = 1.3481	$\frac{1}{2}(1+1.3481)$	1.00+ (.1) (1.174)
	$= \frac{2.3481}{2} = 1.174$	=1.0000+.1174
		= 1.1174

Hence y(0.1) = 1.1174 approximately. Using this value y(0.1), we can find successively the new values of y at 0.2 and so on.

EXAM. 5: Apply the fourth order Rungekutta method to the differential equation $\dfrac{dy}{dx} = -2x - y$, $y(0) = -1$, with h = 0.1 to obtain the values of y at x=0.1 and x=0.2.

SOLUTION:

$$\frac{dy}{dx} = -2x - y, \ y(0) = -1, h = 0.1$$

$$y_{n+1} = y_n + \frac{1}{6}(k_1 + 2k_2 + 2k_3 + k_4)$$

$$k_1 = hf(x_0, y_0) = (.1)f(0, -1) = (.1)(1) = .1$$

$$k_2 = hf(x_0 + \frac{h}{2}, y_0 + \frac{k_1}{2}) = (.1)f(.05, -0.95) = (.1)(-0.1 + 0.95) = 0.085$$

$$k_3 = hf(x_0 + \frac{h}{2}, y_0 + \frac{k_2}{2}) = (.1)f(.05, -0.952) = (.1)(-1 + .9575) = 0.8575$$

$$k_4 = hf(.1, -1 + 0.08575) = (.1)f(.1, -.91425) = 0.071475$$

$$k = \frac{1}{6}(k_1 + 2k_2 + 2k_3 + k_4) = \frac{0.512975}{6} = 0.08549$$

$$\therefore y_1 = y_0 + 0.08549 = -1 + 0.08549 = -0.91451 \ -0.9145$$

$$\therefore y(.1) = -0.9145$$

Similarly y(0.2) = -0.8562.

EXAM. 6: Solve the differential equation

$$\frac{dy}{dx} = x - y^2, \ y(0) = 1 \text{ to find } y(0.4)$$

By Adam's - Bashforth predictor and Adam's Mouton corrector formula, starting with the values

y(0.1) = 0.9117, y(0.2) = 0.8494, y(0.3) = 0.8061

SOLUTION:

$$y_{n+1, p} = y_n + \frac{h}{24}\left[55f_n - 59f_{n-1} + 37f_{n-2} - 9f_{n-3}\right]$$

$$y_{n+1, C} = y_n + \frac{h}{24}\left[9f_{n+1} + 19f_n + 5f_{n-1} + f_{n-2}\right]$$

$$n = 0, y_{1,p} = y_0 + \frac{h}{24}\left[55f_0 - 59f_{-1} + 37f_{-2} - 9f_{-3}\right]$$

$$y_{1,c} = y_0 + \frac{h}{24}\left[9f_1 + 19f_0 - 5f_{-1} + f_{-2}\right]$$

The table below furbishes the Adam's method with h = 0.1 and starting values.

x = 0.0	y-3 = 1.0000	f-3 = 0.0 - (1.0)2 =	- 1.0000
x = 0.1	y-2 = 0.9117	f-2 = 0.1 - (0.9117)2 =	- 0.7312
x = 0.2	y-1 = 0.8494	f-1 = 0.2 - (0.8494)2 =	- 0.5215
x = 0.3	y0 = 0.8061	f-0 = 0.3 - (0.8061)2 =	- 0.3498

$\therefore$ The predictor is

$$x = 0.4 \Longrightarrow y1, P = yo + \frac{h}{24}(55f_0 - 59f_{-1} + 37f_{-2} - 9f_{-3}) = 0.7789$$

$$x = 0.4, \Longrightarrow y1, c = y0 + \frac{h}{24}[9f_1 + 19f_0 - 5f_{-1} + f_{-2}] = 0.7785$$

Hence y(0.4) = 0.7785.

EXAM. 7: Use Milne's (predictor - corrector) method to find the solution at x = 0.4, for the differential equation

$$\frac{dy}{dx} = xy + y^2, \ y(0) = 1, \ y(0.1) = 1.1169 \ ; \ y(0.2) = 1.2773, \ y(0.3) = 1.504$$

Now the starting values for the Milne's Method are:

$$
\begin{array}{lll}
x = 0.0 & y_0 = 1.0000 & f_0 = 1.0000 \\
x = 0.1 & y_1 = 1.1169 & f_1 = 1.3591 \\
x = 0.2 & y_2 = 1.2773 & f_2 = 1.8869 \\
x = 0.3 & y_3 - 1.5049 & f_3 = 2.7132
\end{array}
$$

Predictor : $y_4, p = y_0 + \frac{4h}{3}(2f_1 + f_2 + 2f_3)$

$\therefore x_4, y_{4,p} = 1.8344 \ f_4 = 4.0988$

Corrector : $y_{4,c} = y_2 + \frac{h}{3}(f_2 + 4f_3 + f_4) = 1.8386, \ f_4 = 4.1159$

Again using the corrector

$$y_4 = 1.2773 + \frac{0.1}{3}(1.8869 + 4(2.7132) + 4.1159)) = 1.8391,$$

Again using the corrector

$$y4 = 1.2773 + \frac{0.1}{3}[1.8869 + 4(2.7132) + 4.1182] = 1.8392$$

Hence y4 = 1.8392

13.10.2. *Exercises*

(A) *Short answer questions*

1　Explain briefly the following terms (i) Local truncation error ; (ii) Global truncation error ; (iii) Round - off error; (iv) Single step methods ; (v) multistep methods (vi) Predictor - corrector methods; (vii) Stable method - (a method is said to be stable if it produces a bounded solution which imitates the exact solution) (viii) convergence criteria -(if yp, yc are the value of yn+1 from the predictor formula, and corrector formula, and if successive re-corrections are made, the limit to which the successive re-corrections converge is termed the convergence criterion of the procedure); (ix) convergence of a method-(any numerical method for solving a differential equation is said to convergent if the approximate solution yn approaches the exact solution y(xn) as h (the step length) tends to zero, provided the rounding errors arising from the initial conditions approach zero).

2　Write down the most widely used the Fourth - Order Runge-Kutta method algorithm.

- Tell why the global error is different from the local error.
- State the condition a differential equation must meet to have a unique solution.

[Ans: Let f(x,y) in the differential equation $\frac{dy}{dx} = f(x, y)$ with y(x0)= y0 be defined and continuous or a region R that contains the point (x0,y0). We assume that the region is a closed and bounded rectangle. Then f(x,y) is said to satisfy the Lipschitz condition.

There is an L>0 so that for all x,y1,y2 in R we have |f(x,y1)-f(x,y2)| < L|y1-y2|. If this condition can be established, then the solution to the given differential equation is unique.]

3.　Write the following differential equation into a system of first order equations suitable applying the Runge-Kutta method

$$x''' = 2x' + \log(x'') + \cos(x)$$

$$x(0) = 1,\ x'(0) = -3,\ x''(0) = 5$$

$$[Ans:\quad x' = \begin{bmatrix} x_2 \\ x_3 \\ 2x_2 + \log x_3 + \cos(x_1) \end{bmatrix},\ x(0) = (1,-3,5)T$$

4.　Consider x = x -x, x(0) = 0, x (0) = 1. Determine the associated first order system and its auxiliary conditions

$$[Ans: X = \begin{bmatrix} x_2 \\ x_2 - x_1 \end{bmatrix};\ X'(0) = [0,1]^T$$

Consider $\begin{cases} x'=y \\ y'=x \end{cases}$ with $\begin{cases} x(0)=-1 \\ y(0)=0 \end{cases}$

Write down the equations without derivatives to be used in the Taylor Series

[Ans: $x_0=t,\ x_1=x,\ x_2=y\quad X'=\begin{bmatrix} 1 \\ x_2 \\ x_1 \end{bmatrix} X(0)=[-1,0]^T$

(B) Long Answer Questions

1. Write the algorithm of Euler and its modified version in the numerical solution of

   the problem $\dfrac{dy}{dx}=f(x,y),\ y(x_0)=y_0$

2. Find y at x = 0.2, if $\dfrac{dy}{dx}=x+y^2$, y(0) = 0. Use step length h = 0.1

3. Using Euler's predictor-corrector method find y(0.1) and y(0.2) given that

   $2\dfrac{dy}{dx}=xy,\ y(0)=1.$

4. Using Euler method, obtain the approximate solution of $\dfrac{dy}{dx}+\dfrac{1}{x}y=e^x$, y(1)=1

5. Given a geometrical interpretation of Euler's method and its modified form for solving first order differential equations. Using the above method,
   (a) Obtain y(0.2), y(0.4) and y(0.6), correct to there decimal places, if

   $y\dfrac{dy}{dx}=y-x^2$, y(0)=1 satisfies

   (b) $\dfrac{dy}{dx}=x^2+y$, y(0)=1, obtain y(0.02) and y(0.04)

   (c) $\dfrac{dy}{dx}=-2\,x\,y^2$, y(0)=1, h = 0.1, find y(0.1) and y(0.2)

   (d) $\dfrac{dy}{dy}=x+y+xy$, y(0)=1 with h = 0.1 to get y (0.1) [Ans : y(0.1) = 1.1142].

6. Show that Euler method is stable if $\left|1+h\dfrac{\partial f}{\partial y}\right|<1$, with the usual notation

7. Use Picard's method of successive approximation to solve the following differential equation

   (1) $\dfrac{dy}{dx} = x + x^4 y, y(0) = 3.$ for x = 0.5

   (2) $\dfrac{dy}{dx} = \dfrac{x^2}{y^2 + 1}, y(0) = 0$ to obtain y at x = 0.25

8. Solve the following differential equation by Taylor Series method

   a. $\dfrac{dy}{dx} = x + y, y(0) = 1,$ by taking h = 0.1 and find the values of y

      corresponding to x = 0.1 and x = 0.2 up to 4 significant figures.

   b. $5x\dfrac{dy}{dx} + y^2 - 2 = 0, y(4) = -1,$

      y(4.1) = 1.049, find y(4.2) by (i) by Taylor series method ; (ii) by the Range-Kutta method of fourth order; (iii) y(4 -4) by Milne's method.

9. Apply Taylor series method obtain the value of y correct to four decimal places for

   (i) x = 1.5 and (ii) x = 3.5 from the differential equation $\dfrac{dy}{dx} = x.y^{1/3}$ with y(1) = 1.

10. $\dfrac{d^2 x}{dt^2} = K \sin x = 0; \; x(0) = 2.1, \left.\dfrac{dx}{df}\right|_{t=0} = 0,$ taking the first three terms in Taylor

    series.

11. Using Taylor series method, obtain correct to three significant figures the value

    of y at x = 1.1 and 1.2 if y satisfies the equation $\dfrac{dy}{dx} + \dfrac{y}{x} = \dfrac{1}{x^2}, y(0) = 1,$ Hence

    by Adam's method obtain the value of y at x = 1.3

12. Using Taylor series method solve the following differential equations

    i. $5x\dfrac{dy}{dx} + y^2 - 2 = 0, y(4) = 1,$ to find y(4.3) correct to 4 places of decimals.

    ii. $\dfrac{dy}{dx} = x + y + xy, y(0) = 1$ to find y(0.1) upto 4 places of decimals.

    [Ans : $y = 1 + x + \left(\dfrac{3}{2}\right)x^2 + \left(\dfrac{5}{6}\right)x^3 + \left(\dfrac{7}{12}\right)x^4$; y(0.1) = 1.1159]

iii. $\dfrac{d^2y}{dx^2} = xy,\ y(0)=1, \dfrac{dy}{dx}$ at $x=0$ is equal to 1 to find $y(0.2)$, $h=0.2$,

[Ans : $y(0.2) = 1.2015$].

13 Write the fourth order Runge-Kutta algorithm for the numerical solution of the

problem $\dfrac{dy}{dx} = f(x,y), y(x_o) = y_0$

14 Use the R-K algorithm to find $y(0.1)$ and $y(0.2)$ correct to four decimal places,

giventhat $\dfrac{dy}{dx} = y - x, y(0) = 2$.

15. (i) $\dfrac{dy}{dx} = x + y^2, y(0) = 0;\ h = 0.1,$ Find y when $x = 0.2$

(ii) $\dfrac{dy}{dx} = x + y,\ y(0) = 1;\ h = 0.1,$ Find $y(0.1)$ correct to four decimal places.

16 Develop a second order Runge - Kutla algorithm for solving $\dfrac{d^2y}{dx^2} + a\dfrac{dy}{dx} + b\,y = f(x)$.

Solve the equation $\dfrac{d^2y}{dx^2} - x^2\dfrac{dy}{dx} - 2xy = 1$ for $x = 0.1$, given that $y(0) = 1$ and

$\left.\dfrac{dy}{dx}\right|_{x=0} = 0,$ by the fourth order R.K. Method.

17 Find the solution of the differential equation $\dfrac{d^2y}{dx^2} = x\left(\dfrac{dy}{dx}\right)^2 - y^2,\ y = 1, \dfrac{dy}{dx} = 0$

when
$x = 0$ at $x = 0.2$ and $x = 0.4$ by using the fourth order Runge - Kutla method correct to three decimal places.

(a). Using Runge - Kutta method with fourth order accuracy, obtain $y(0.1)$ and

$y(0.2)$ to three significant figure, when y satisfies $\dfrac{dy}{dx} = x^2 + y^2, y(0) = 0$

(b). Check the answer obtained for $y(0.1)$ in the above problem by Taylor series method:

18 Using Runge - Kutta method with fourth order accuracy, determined correct to threedecimal places, the approximate volumes of y at $x = 0.1$ and $x = 0.2$ if y

satisfies $\dfrac{dy}{dx} - x^2y = x$, $y(0) = 1$.

19 Solve by the Runge - Kutta method of fourth order accuracy of the following differential equations:

1. $\dfrac{dy}{dx} = 0.1(x+)e^{-x}$, $y(0) = 5$ in the interval (0.1) taking a step size of h = 0.2

2. $\dfrac{dy}{dx} = \dfrac{1}{x+y}$, $y(0) = 1$ for x = 0.5 (0.5)

3. $\dfrac{dy}{dx} = xy2$, $y(1) = 3$. h = 0.5, obtain y at x = 1.5 and 2.0

4. $\dfrac{dy}{dx} = 1 + y2$, $y(0) = 0$, find y(0.2) and y(0.4)

5. $y\dfrac{dy}{dx} = y^2 - 2x$, $y(0) = 1$, with h = 0.2 compute y for x = 0.2 and x = 0.4

6. $\dfrac{dy}{dx} = 0.25 \ y^2 + x^2$, $y(0) = 1$, h = 0.1 over the interval [0, 0.2]

7. $\dfrac{dy}{dx} = 3x + 0.5 y$, $y(0) = 1$, compute y for x = 0.2

8. $\dfrac{dy}{dx} = 2x^2 - y$, $y(0) = -1$; compute y for x = 0.1 and x = 0.2, with h = 0.1

[Ans : y(1) = -0.9042 ; Exact solution is y(x) = $5e^{-x} + 2x^2$ - 4x + 4]

20. Write down the equation to represent the linear multistep method for solving first order differential equation numerically indicating the concept of explicit and implicit representation.

21. Write the Adam's [Predictor] algorithm for the numerical integration of the differential equation $\dfrac{dy}{dx} = f(x, y)$ and indicate the local truncation error.

22. Use Adam's predicator to calculate y(1.4) and y(1.5), given that

$\dfrac{dy}{dx} = x + x^2y$, $y(1) = 1$, $y(1.1) = 1.24$, $y(1.2) = 1.58$ and $y(1.3) = 2.10$

23. Using the fourth order Runge - Kutla method, find y(0.1), y(0.2) and y(0.3) correct

to four decimal place given $\dfrac{dy}{dx} = x + y$, $y(0) = 1$ and step size $h = 0.1$

24. Use the following predictor - corrector formulas due to Adam - Moulton, to find $y(0.4)$

$$y_{n+1p} = y_n + \frac{h}{24}\left(55f_n - 59f_{n-1} + 37f_{n-2} - 9f_{n-3}\right)$$

$$y_{n+1c} = y_n + \frac{h}{24}\left(9f_{n+1} + 19f_n - 5f_{n-1} - f_{n-2}\right)$$

25. Solve by Adam's method correct to three decimal places the equation $\dfrac{dy}{dx} = y - x^2$, $y(0) = 1$ for $x = 0.0\,(0.2)\,(0.8)$, calculating the two starting values by the Taylor's series method.

26. Obtain the solution to the initial value problem $\dfrac{dy}{dx} = x^2 + yx^2$, $y(1) = 1$ at $x = 1.0\,(0.1)(1.3)$ by Taylor series method and at $x = 1.4$ by Adam's predictor corrector method.

27. Using Adam's-Bashforth predictor - corrector method, solve the following differential equations

(a) $\dfrac{dy}{dx} = y + x^2$, $y(0) = 1$

from $x = 0$ to 0.4 with $h = 0.1$ The starting values are
$y(0) = 1.0000$, $y(0.1) = 1.1055$,
$y(0.2) = 1.2242$, $y(0.3) = 1.3596$.

(b) $\dfrac{dy}{dx} = y - x^2$, $y(0) = 1$, $y(0.2) = 1.222$

$y(0.4) = 1.468$, obtain $y(0.6)$ and $y(0.8)$

(c) $\dfrac{dy}{dx} = x + y$, $y(0) = 1$, $y(0.1) = 1.21034$

$y(0.2) = 1.44280$, $y(.3) = 1.6971$,
$y(.4) = 1.9836$, $y(.5) = 2.2970$.
(Ans: Adam's prediction for $y(0.6) = 2.0442$.)

(d) $\dfrac{dy}{dx} = -2xy^2$, $y(0) = 1$, $h = 0.2$ on the interval $[0.04]$ with starting values,

$y(0) = 1$, $y\,90.2) = 0.9615$, $\dfrac{dy}{dx}$ at $x = 0.26 = -0.3698$

(e) $\dfrac{dy}{dx} = x^3 + x^2$, $y(0) = 0$, $h = 0.2$ with $y(0.2) = 0.0004$, $y(0.4) = 0.0064$ and

$y(0.6 = 0.0325$ for $x = 0.8$

(f) $\dfrac{dy}{dx} = \dfrac{1}{x^2} - \dfrac{y}{x}$, $y(1) = 1$, $y(1.1) = 0.996$,

$y(1.2) = 0.986$; $Y(1.3) = 0.972$
find yp (1.4) and yc (1.4)
[Ans : y4,p = 0.8457, y41 c = 0.949.

28. Using Milne's method, solve the differential equation $\dfrac{dy}{dx} = y - x^2$ for $x = 0.8$

given that $y(0) = 1.000$, $y(0.2) = 1.220$: $y(0.4) = 1.468$ and $y(0.6) = 1.738$, correct to three decimal {places of accuracy.

29. Given that

$2\dfrac{dy}{dx} = (1 + x^2) y^2$ and $y(0) = 1$, $y(0.1) = 1.06$, $y(0.2) = 1.12$, $y(0.3) = 1.21$,

evaluate y(0.4) by Milne's predictor - Corrector method.

30. Use Milne's method to solve the differential equation $\dfrac{dy}{dx} + 2xy = 2x^3$, $y(0) = 0$

for $x = 0.1, 0.2, 0.3$

31 Solve the following differential equations by Milne's predictor - corrector method.

(a) $\dfrac{dy}{dx} = \dfrac{1}{2}(1 + x^2) y^2$; $y(0) = 1$,

$y(0.1) = 1.06$, $y(0.2) = 1.12$, $y(0.3) = 1.21$ and evaluate y (0.4)
[Ans : $y_{4,p} = 1.27715$; $y_{4,e} = 1.2797$]

(b) $\dfrac{dy}{dx} + \dfrac{y}{x} = \dfrac{1}{x^2}$, $y(1) = 1$,

$y(1.1) = 0.996$ y (1.2) = 0.986,
$y(1.3) = 0.972$, find y(1.4)

(c) $\dfrac{dy}{dx} = x^2 + y$, $y(0) = 1$, by finding the starting values y(0.2), y(0.4), y(0.6) by

any one of the known first order methods. Compute y(0.8)

(d) $\dfrac{dy}{dx} = f(x, y) = x + y$, $y(0) = 1$,

with $y(-0.1) = 1.11034$, $y(.2) = 1$.
$y(.3) = 1.39971$, find for $x = 0.4$, the predictor and corrector values

[Ans : 1.58364 and 1.79743]
(ii) Find for x = 0.5
[Ans : 1.79742, 179743].

13.10.3 *Shooting Method (ODE - Boundary value problem)*

Consider the solution of the boundary value problem $u'' = f(x,u,u')$, $x \in [a,b]$ with the boundary conditions $u(a) = r_1$, $u(b) = r_2$. (1)

In the shooting method, we solve the initial value problem $u'' = f(x,u,u')$, $u(a) = r_1$, $u'(a) = \alpha$ where α is some approximation of the initial shape. Using any of the methods for solving the initial value problems, the approximation $u^{(1)}(b)$ to the either smaller or larger than the required solution $u(b) = r_2$. The differential equation under consideration may or may not be a linear differential equation.

If $z(x,\alpha)$ is defined to be a solution of $L(z'',z',z,x) = 0$, $z(a,\alpha) = r_1$, $z(a,\alpha) = \alpha$ (2)

Then u(x) will be equal to $z(x,\alpha)$ for one or more values for α. Of course if L()=0 were a linear equation, then there would be single value of α'. The parameter must be determined so that $z(b, \alpha) = r_2$.

Using any single step method, say Euler's method, we get for some arbitrary initial guess for α say α_0, then if $z(b,\alpha_0) = r_2$ then $u(x) = z(x,\alpha_0)$

If $z(b, \alpha_0) = r_2$, then a new value of α must be chosen, say $\alpha1$. Then the equation (2) is then integrated for this new value of α. The process of choosing new values for α is repeated until the value of $z(b,\alpha)$ is sufficiently close to r_2.

If the new α's are chosen well, then $z(b,\alpha)$ will converge to r2 and a numerical approximation to one will have been obtained. One way to choose the sequence of α's is by Newton's Method.

$$\alpha_{n+1} = \alpha_n - \frac{z(b,\alpha_n) - r_2}{\frac{\partial}{\partial \alpha} z(b,\alpha) \Big|_{\alpha = \alpha_n}}$$ (3)

A numerical way to implement (3) might be

$$\alpha_{n+1} = \alpha_n - \frac{z(b,\alpha_n) - r_2}{\frac{1}{\epsilon}[z(b,\alpha_{n+\epsilon}) - z(b,\alpha_n)]}$$ (4)

Where is a small number.

Examples

EXAM. 1: Suppose we have the non linear second order ordinary differential equation

$$y'' + 2(y')^2 = 0; \qquad\qquad y(0) = 1, \qquad y(1) = \frac{1}{2}$$

(1)

Since (1) has explicit "dependence on y, the dependent variable missing" method can be used to solve this equation exactly. By this technique, the solution of (1) is

found to be $y(x) = 1 + \log(1 + \dfrac{1-e}{e} x)$ (2)

Hence, $y'(0) = \dfrac{1-e}{2e} \approx -0.31606.$

By use of the shooting method, a computer program should "discover" that For this purpose the equation (1) is turned into the two first order ordinary differential equations

$$\frac{dy}{dx} = z, \quad \frac{dz}{dx} = -2y^2 \tag{3}$$

and then integrated by the use of Euler's method. An initial guess of y'(0)=0 is used in the program. The successive approximation of appear below.

[Note: The computer program requires a larger number of steps in the interval [0,1] in order to achieve the accuracy.]

EXAM. 2: Find a solution of the boundary value problem.:

$u'' = u + x, \quad x \in [0,1], \quad u(0) = 0 = u(1)$

using the shooting method.

[use the fourth order Taylor series method to solve the initial value problem with h=0.2]

SOLUTION :

Take $u'(0) = \dfrac{1}{2}$

The Taylor series method give:

$$u_{n+1} = u_n + hu'_n + \frac{h^2}{2}u''_n + \frac{h^3}{6}u'''_n + \frac{h^4}{24}u^{iv}_n$$

$$= \left(1 + \frac{h^2}{2} + \frac{h^4}{24}\right)u_n + hu'_n\left(1 + \frac{h^2}{6}\right) + \left(\frac{h^2}{2} + \frac{h^4}{24}\right)x_n + \frac{h^3}{6}$$

$$\therefore u'_{n+1} = u'_n + hu''_n + \frac{h^2}{2}u'''_n + \frac{h^3}{6}u^{iv}_n$$

$$= \left(1 + \frac{h^2}{2}\right)u'_n + \left(h + \frac{h^3}{6}\right)u_n + \left(h + \frac{h^2}{6}\right)x_n + \frac{h^2}{2}$$

with h = 0.2, we have

$$u_{n+1} = 0.00133 + 0.2007x_n + 1.02007u_n + 0.20133u_n'$$

$$u_{n+1}' = 0.20 + 0.20133x_n + 0.20133u_n + 1.02u_n'$$

We get

$$u(0.2) \approx u_2 = 0.10200; \qquad\qquad u'(0.2) \approx u_2' = 0.53000$$

$$u(0.4) \approx u_3 = 0.21610; \qquad\qquad u'(0.4) \approx u_3' = 0.62140$$

$$\ldots\ldots\ldots\ldots\ldots\ldots\ldots\ldots \qquad\qquad \ldots\ldots\ldots\ldots\ldots\ldots\ldots\ldots$$

$$u(1.0) \approx u_6 = 0.76254; \qquad\qquad u'(1.0) \approx u_6' = 1.31397$$

$$u_1(b) = 0.76254$$

Let the second choice of the initial slope be $u'(0) = -\dfrac{1}{2}$

Find $u_2(b) = -0.41250$

The solution of the differential equation can be written as $u(x) = c_1 u_1(x) + c_2 u_2(x)$

we have $u(a) = r_1 = c_1 r_1 + c_2 r_2$ $\qquad$ or $\qquad$ $c_1 + c_2 = 1$

and $u(b) = r_2 = c_1 u_1(b) + c_2 u_2(b)$

$$\therefore c_2 = \frac{r_2 - u_1(b)}{u_2(b) - u_1(b)}, \quad c_1 = 1 - c_2$$

Here we have $c_2 = \dfrac{0 - 0.76254}{-0.41250 - 0.76254}$

$$= 0.64895$$

and $\qquad\qquad c_1 = 1 - c_2 = 0.35105$

∴ The required solution is

u(x) = 0.35105 $u_1(x)$ + 0.64895 $u_2(x)$.

Using the above, find the approximate solution of u(0.2), u(0.4),........u(0.8)

EXAM. 3: Use shooting method to solve

$$x^2 u'' - 2u + x = 0, \qquad x \in [2,3], \qquad u(2) = u(3) = 0.$$

with $u'(2) = 0.5$ and $u'(2) = -0.5$, $\quad$ h = 0.25

[Ans: u(2.25)=0.037887,etc].

EXAM. 4: Use the shooting method to solve

$$u'' = 6u^2, u(0) = 1, u(0.5) = 4/9.$$

Assume the initial approximations

$u'(0) = \alpha_0 = -1.8$ and $u'(0) = \alpha_1 = -1.9$,

with h=0.1

[Ans : $\alpha_2 = -2.0035$, u(0,1) = 0.82614,..........u(0.4) = 0.50985.]

13.11 Numerical Solution of Partial-Different Equations

13.11.1 Introduction

When the function being studied involves two or more independent variables, the differential equation is usually called a **partial** differential equation. There are two numerical techniques-**finite difference methods** and the **finite element method**.

Some partial differential equations from applied problems pertaining to the physical phenomena in three spatial variables (x, y, z) and time t (with appropriate initial and boundary Conditions) are

- **The wave equations is**

$$\frac{\partial^2 u}{\partial x^2} + \frac{\partial^2 u}{\partial y^2} + \frac{\partial^2 u}{\partial z^2} + \frac{\partial^2 u}{\partial t^2} = 0$$

- **The heat equation is**

$$\frac{\partial^2 u}{\partial x^2} + \frac{\partial^2 u}{\partial y^2} + \frac{\partial^2 u}{\partial z^2} - \frac{\partial u}{\partial t} = 0$$

- **Laplace's equations is**

$$\frac{\partial^2 u}{\partial x^2} + \frac{\partial^2 u}{\partial y^2} + \frac{\partial^2 u}{\partial z^2} = 0$$ in Cartasian co-ordinate system and similar in the other co-

ordinates systems like cylindrical and spherical forms.
- **The bi harmonic equation y is**

$$\frac{\partial^4 u}{\partial x^2} + \frac{\partial^4 u}{\partial x^2 \partial y^2} + \frac{\partial^4 u}{\partial y^4} = 0$$

The Nervier-Stokes equations are:

$$\frac{\partial u}{\partial t} + u \frac{\partial u}{\partial x} + v \frac{\partial u}{\partial y} + \frac{\partial p}{\partial x} = \frac{\partial^2 u}{\partial x^2} + \frac{\partial^2 u}{\partial y^2}$$

$$\frac{\partial v}{\partial t} + u \frac{\partial v}{\partial x} + v \frac{\partial v}{\partial y} + \frac{\partial p}{\partial y} = \frac{\partial^2 v}{\partial x^2} + \frac{\partial^2 v}{\partial y^2}$$

and other types.

We consider the general second order partial differential equation of the form

$$A\frac{\partial^2 u}{\partial x^2} + 2B\frac{\partial^2 u}{\partial x \partial y} + C\frac{\partial^2 u}{\partial y^2} - F\left(x, y, u, \frac{\partial u}{\partial x}, \frac{\partial u}{\partial y}\right) = 0 \tag{1}$$

If A,B, and C are functions of $x, y, u, \frac{\partial u}{\partial x}, \frac{\partial u}{\partial y}$, then (1) is called a **quasi linear** partial differential equation. When A, B, and C are function of x, y, and F is a linear function of $u, \frac{\partial u}{\partial x}$ and $\frac{\partial u}{\partial y}$ then (1) is called **linear**. A linear partial differential equation may be written as

$$A\frac{\partial^2 u}{\partial x^2} + 2B\frac{\partial^2 u}{\partial x \partial y} + C\frac{\partial^2 u}{\partial y^2} + D\frac{\partial u}{\partial x} + E\frac{\partial u}{\partial y} + Fu + G = 0 , \tag{2}$$

where A, B, C,........,G are constants or functions of x and y only. They equation (2) is **homogeneous** if G=0, otherwise in **homogeneous**. A solution of (2) is of the from

$$\phi(x, y, u) = 0 \tag{3}.$$

This represents a surface called an **integral surface** in the (x, y, u) space. If on the integral surface, there exist curves a cross which the partial derivatives $\frac{\partial^2 u}{\partial x^2}, \frac{\partial^2 u}{\partial x \partial y}, \frac{\partial^2 u}{\partial y^2}$ are discontinuous or indeterminate, then these curves are called the **characteristics** of the partial differential equation

13.11.2 *Classification*

Using the transformation $\xi = \xi(x, y), \eta = \eta(x, y)$ where $\zeta?, \eta$ are twice differentiable functions, in (2) and simplifying we get $\left[A\left(\frac{\partial\xi}{\partial x}\right)^2 + 2B\frac{\partial\xi}{\partial x}\cdot\frac{\partial\xi}{\partial y} + C\left(\frac{\partial\xi}{\partial y}\right)^2\right]\frac{\partial^2 u}{\partial\xi^2}$

$$+ 2\left[A\frac{\partial\xi}{\partial x}\cdot\frac{\partial\eta}{\partial x} + B\frac{\partial\xi}{\partial x}\cdot\frac{\partial\eta}{\partial y} + B\frac{\partial\xi}{\partial y}\cdot\frac{\partial\eta}{\partial x} + C\frac{\partial\xi}{\partial y}\cdot\frac{\partial\eta}{\partial y}\right]\frac{\partial^2 u}{\partial\xi\partial\mu}$$

$$+ \left[A\left(\frac{\partial\eta}{\partial x}\right)^2 + 2B\frac{\partial\eta}{\partial x}\cdot\frac{\partial\eta}{\partial y} + C\left(\frac{\partial\eta}{\partial y}\right)^2\right]\frac{\partial^2 u}{\partial\eta^2} + \ldots\ldots\ldots = 0 \tag{4}$$

Setting the co-efficient of $\dfrac{\partial^2 u}{\partial \xi^2}$ and $\dfrac{\partial^2 u}{\partial \mu^2}$ to zero and simplifying, we get

$$A\left(\frac{\partial \xi}{\partial x}\bigg/\frac{\partial \xi}{\partial y}\right)^2 + 2B\left(\frac{\partial \xi}{\partial x}\cdot\bigg/\frac{\partial \xi}{\partial y}\right) + C = 0 \qquad (5)$$

$$A\left(\frac{\partial \eta}{\partial x}\bigg/\frac{\partial \eta}{\partial y}\right)^2 + 2B\left(\frac{\partial \eta}{\partial x}\cdot\bigg/\frac{\partial \eta}{\partial y}\right) + C = 0 \qquad (6)$$

Since $\left(\dfrac{\partial \xi}{\partial x}\bigg/\dfrac{\partial \xi}{\partial x}\right) = -\dfrac{dy}{dx}$ and $\left(\dfrac{\partial \eta}{\partial x}\cdot\bigg/\dfrac{\partial \eta}{\partial y}\right) = -\dfrac{dy}{dx}$

both equations (5) and (6) reduce to the same ordinary differential equation

$$A\left(\frac{dy}{dx}\right)^2 - 2B\frac{dy}{dx} + C = 0 \qquad (7)$$

Hence $\zeta = $ a constant and $\eta = $ a constant are respectively the solutions of

$$\frac{dy}{dx} = \frac{+B - \sqrt{(B^2 - AC)}}{A} \qquad (8)$$

$$\frac{dy}{dx} = \frac{+B + \sqrt{(B^2 - AC)}}{A} \qquad (9)$$

The curves $\zeta = $ a constant and $\eta = $ a constant are called the characteristics. If $B^2 - AC > 0$ at every point (x, y), then ζ and η are real and (2) has two real characteristics. The equation (2) is called a **hyperbolic** partial differential equation.

If B^2 then equations (8) and (9) coincide and the resulting solution $\zeta = $ a constant is the only characteristic of the equation. The equation (2) is called a **parabolic** partial differential equation. If $B^2 - AC < 0$, then the characteristics are imaginary. The equation (2) is called an **elliptic** partial differential equation. The transformation $\xi = \xi(x, y)$ and $\eta = \eta(x, y)$ reduces the equation (2) to a standard form. We obtain

Hyperbolic equation as $\dfrac{\partial^2 u}{\partial \xi \partial \eta} = F_1\left(\xi, \eta, \dfrac{\partial u}{\partial \xi}, \dfrac{\partial u}{\partial \eta}\right)$ or $\dfrac{\partial^2 u}{\partial \xi^2} - \dfrac{\partial^2 u}{\partial \eta^2} = F_2\left(\xi, \eta, u, \dfrac{\partial u}{\partial \xi}, \dfrac{\partial u}{\partial \eta}\right)$

$$(10)$$

Parabolic equation

$$\frac{\partial^2 u}{\partial \xi^2} = F_3\left(\xi, \eta, u, \frac{\partial u}{\partial \xi}, \frac{\partial u}{\partial \eta}\right)$$

Or

$$\frac{\partial u}{\partial \eta^2} = F_4\left(\xi, \eta, u, \frac{\partial u}{\partial \xi}, \frac{\partial u}{\partial \eta}\right)$$ (11)

Elliptic equation

$$\frac{\partial^2 u}{\partial \xi^2} + \frac{\partial^2 u}{\partial \eta^2} = H\left(\xi, \eta, u, \frac{\partial u}{\partial \xi}, \frac{\partial u}{\partial \eta}\right)$$ (12)

The well known examples of the three types are:

(i) **Heat flow equation** $\dfrac{\partial u}{\partial t} = \dfrac{\partial^2 u}{\partial x^2}$ which is of the **parabolic** type (13)

(ii) **Wave equation** $\dfrac{\partial^2 u}{\partial t^2} = \dfrac{\partial^2 u}{\partial x^2}$ which is of the **hyperbolic** type

(iii) **Laplace equation** $\dfrac{\partial^2 u}{\partial x^2} + \dfrac{\partial^2 u}{\partial y^2} = 0$ which is of the **elliptic** type

[Note: 1. We have we used x and t as the independent variables in the case of parabolic and hyperbolic equation and x and y as independent variables in the elliptic equation

2. The parabolic and the hyperbolic type are either initial value of equation or initial boundary value problems, whereas the elliptic type of equation is always a boundary value problem

13.11.3 *Finite different approach*

We superimpose on the region R enclosed by the boundary ∂R a network or mesh of points formed by lines $x_m = a + mh$, m=0, 1, 2,.......$y_m = b + nk$, n=0, 1, 2, 3....... (1)
where the quantities h and k are mesh sizes in x and y directions respectively. The points of intersection of the network are called nodes. The **node** (x_m, y_n) will be denoted by (m, n). The partial derivatives in the differential equation are replaced by suitable finite difference quotients, converting the differential equation to a difference equation at each nodal point. The given data issued to modify the difference equation at the nodes near or on the boundary. The solution of this system of equations gives the **numerical solution** of the partial differential equation.

13.11.4 *Finite difference approximations to partial derivatives.*

Consider a rectangular region R in the (x, y)-plane. Divide this region into a rectangular network of sides $\Delta x = h$ and $\Delta y = k$ in the x and y direction. The points of inter section of the dividing lines are called **mesh points, nodal points** or **grid points.**

The finite difference approximations for the partial derivatives in the x-direction are

$$\frac{\partial u}{\partial x} = \frac{u(x+h, y) - u(x, y)}{h} + O(h) \text{ -forward difference}$$

$$= \frac{u(x, y) - u(x-h, y)}{h} + O(h) \text{ -backward difference}$$

$$= \frac{u(x+h, y) - u(x-h, y)}{2h} + O(h^2) \text{ -central difference}$$

and $$\frac{\partial^2 u}{\partial x^2} = \frac{u(x-h, y) - 2u(x, y) + u(x+h, y)}{h^2} + O(h^2)$$

Writing u(x,y)=u(ih,jk) as simply $u_{i,j}$, the above approximations become

$$u_{i,j} = \frac{1}{h}(u_{i+1,j} - u_{i,j}) + O(h) \quad = \quad \frac{1}{h}(u_{i,j} - u_{i-1,j}) + O(h)$$

$$= \frac{1}{2h}(u_{i+1,j} - u_{i-1,j}) + O(h^2) \quad \text{and} \quad u_{xx} = \frac{1}{h^2}(u_{i-1,j} - 2u_{ij} + u_{i+1,j}) + O(h^2)$$

Similarly, we have the approximations for the partial derivatives with respect to y are:

$$u_y = \frac{1}{k}(u_{i,j+1} - u_{i,j}) + O(k) = \frac{1}{k}(u_{i,j} - u_{i,j-1}) + O(k)$$

$$= \frac{1}{2k}(u_{i,j+1} - u_{i,j-1}) + O(h^2)$$

and $u_{yy} = \frac{1}{k^2}(u_{i,j-1} - 2u_{i,j} + u_{i,j+1}) + O(k^2)$

13.11.5 *Numerical Methods for parabolic equations.*

We use subscripts to denote position and superscripts for time. Consider the heat flow equation in one spatial variable accompanied by boundary conditions appropriate to certain physical phenomenon.

Let u represent the temperature at any point in the rod, whose distance from the left and is x. Heat is flowing from left to right under the influence of the temperature gradient

$\dfrac{\partial u}{\partial x}$. Make a balance of the rate of heat flow into and out of the element. Let k represent the thermal conductivity $cal/sec/[(cm)^2(C/cm)]$, which we assume to be constant.

Rate of heat flow "in": - $-kA\dfrac{\partial u}{\partial x}$

Rate of flow of heart "out" : $-kA\left[\dfrac{\partial u}{\partial x}+\dfrac{\partial}{\partial x}\left(\dfrac{\partial u}{\partial x}\right)dx\right]$

The difference between the rate of flow "in" and "out" is the rate at which heat is being stored in the element. If s is the heat capacity, $cal/[(g)(°C)]$ and is the density, $g/cm2$, we have, with t for time,

$$-kA\dfrac{\partial u}{\partial x}-\left\{-KA\left[\dfrac{\partial u}{\partial x}+\dfrac{\partial}{\partial x}\left(\dfrac{\partial u}{\partial x}\right)dx\right]\right\}=s.\rho\,(Adx)\dfrac{\partial u}{\partial t}$$

Simplifying we have

$$k\dfrac{\partial^2 u}{\partial x^2}=s.\rho\,\dfrac{\partial u}{\partial t}\ \text{or}\ \dfrac{\partial u}{\partial t}=c^2\,\dfrac{\partial^2 u}{\partial x^2}$$

When $c^2=\dfrac{k}{s\rho}$ is the diffusivity of the substance (cm2/sec)

This is the basic mathematical model for unsteady - state flow and we have derived it for heat flow.

Solution of one - dimensional heat equation

$\dfrac{\partial u}{\partial t}=c^2\,\dfrac{\partial^2 u}{\partial x^2}$, with boundary conditions u(0,t) = u(l,t) = 0, and u(x,0) = f(x). $0\le x\le 1$ = R (region), $0\le t\le T=T$ (the time). We superimpose on the arbitrary region R x [0, T], a rectilinear grid with grid lines parallel to the coordinate axes with spacing h and k in space and time directions, respectively. The grid points on R x [0, T] are given by

t_j = jk, j= 0, 1, 2, . . . ,N

x_i = a + ih, i = 0,1,2, .. , M for the general region R = $a\le x\le b$, where x0 = a, x_m = b, M = /h. The space nodes on the n-th time grid usually constitute the jth **layer** or **level**. Let the solution value u(x_i,t_j) be denoted by U_i^j and is approximate value by u_i^j .

In the rectangular mesh in the x-t plane with spacing along x-direction and along the time t direction, and denoting the mesh point (x,t) = (ih, jk) as simply i,j, we have

$$\frac{\partial u}{\partial t} = \frac{1}{k}\left(u_i^{j+1} - h_i^{j}\right) \text{ and } \frac{\partial^2 u}{\partial x^2} = \frac{1}{h^2}\left[u_{i-1}^{j} - 2u_i^{j} + u_{i+1}^{j}\right]$$

$$\therefore \frac{\partial u}{\partial t} = c^2\frac{\partial^2 u}{\partial x^2} \text{ becomes } u_i^{j+1} - u_i^{j} = \frac{kc^2}{h^2}\left[u_{i-1}^{j} - 2u_i^{j} + u_{i+1}^{j}\right] = \alpha\left[u_{i-1}^{j} - 2u_i^{j} + u_{i+1}^{j}\right]$$

or $u_i^{j+1} = \alpha u_{i-1}^{j} + (1 - 2\alpha)u_i^{j} + \alpha u_{i+1}^{j}$ \hfill (1)

This formula enables us to determine the value of u at the (i, j+1)th mesh point in terms of the known function values at the points x_i-1, x_i and xi+1 at the instant t_j. It is a relation between the function values at the two time levels j+1 and j and is therefore called a **2-level** formula the schematic form is shown in the diagram below

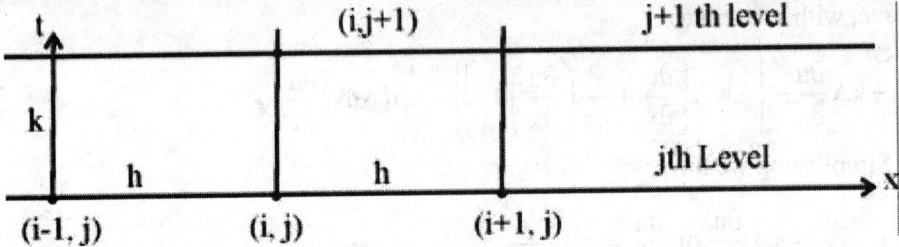

The formula (1) is called the **Schmidt explicit** formula which is valid only for

$$0 < \alpha \le \frac{1}{2}. \text{ For } \alpha = \frac{1}{2}, (1) \text{ reduces to}$$

$$u_i^{j+1} = \frac{1}{2}\left[u_{i-1}^{j} + u_{i+1}^{j}\right] \text{ which shows that the value of u at } x_i \text{ at time } t_{j+1} \text{ is the mean of the}$$

u-values at x_{i-j} and x_{i+1} at time tj. This relation known as **Bender Schmidt Recurrence Relation,** gives the values of u at the internal mesh points with the help of boundary conditions.

[h should be small, and k is necessarily very small, resulting the computations exceptionally lengthy, as more and more time - levels would be required to cover the region].

(b)Crank - Nicolson method

In this method, $\frac{\partial^2 u}{\partial x^2}$ is replaced by the average of the central - difference approximations

on the j th and (j + 1) th time rows. Thus $\frac{\partial u}{\partial t} = c^2\frac{\partial^2 u}{\partial x^2}, c^2 = k/s\rho$ is reduced to

$$\frac{1}{k}\left[u_i^{j+1} - u_i^{j}\right] = c^2\frac{1}{2}\left[\frac{1}{h^2}\left(u_{i-1}^{j} - 2u_i^{j} + u_{i+1}^{j}\right) + \frac{1}{h^2}\left(u_{i-1}^{j+1} - 2u_i^{j+1} + u_{i+1}^{j+1}\right)\right]$$

Or

$$-\alpha u_{i-1}^{j+1} + (2+2\alpha)u_i^{j+1} - \alpha u_{i+1}^{j+1} = \alpha u_{i-1}^j + (2-2\alpha)u_i^j + \alpha u_{i+1}^j, \text{ where } \alpha = \frac{kc^2}{h^2} \qquad (2)$$

Clearly the left side of (2) contains three unknown values of u at the (j+1)th level while all the three values as the right side are known values of u at the j th level. Thus (2) is a 2-level implicit relation and is known as Crank Nicolson formula. It is convergent for all finite values of . The computational formula of this method is given in the diagram below:

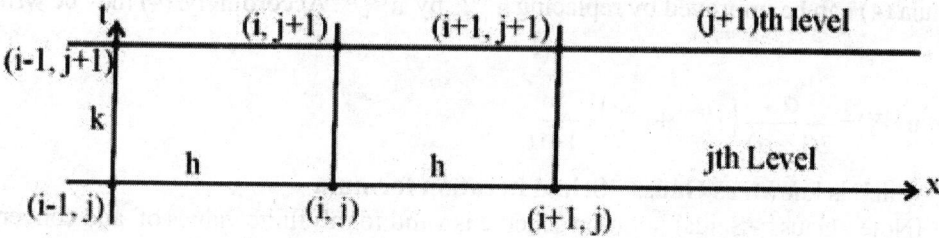

If there are n internal mesh points on each row, then the relation (2) gives n simultaneous equations for the n unknown values in terms of the known boundary values.

[Note1. The new values of u_i^{j+1} is not given directly in terms of known values one time step curlier, but is a function of unknown values of u's at adjacent positions as well. It is therefore termed an **implicit method** Hence interactive methods of solution have to be adopted.

2. For $\alpha = 1$, we get the simplification scheme

$$-u_{i-1}^{j+1} + 4u_i^{j+1} - u_{i+1}^{j+1} = u_{i-1}^j + u_{i+1}^j$$

3. One advantage of the Crank - Nicolson method is that it is stable for any values of α, although small values are more accurate]

(c) Iterative methods of solution for an implicit scheme

Crank Nicolson's formula (2) is written as

$$(1+\alpha)\, u_i^{j+1} = \frac{1}{2}\alpha\left(u_{i-1}^{j+1} + u_{i+1}^{j+1}\right) + u_i^j + \frac{1}{2}\alpha\left(u_{i-1}^j - 2u_i^j + u_{i+1}^j\right) \qquad (3)$$

Here only u_i^{j+1}, u_{i-1}^{j+1} and u_{i+1}^{j+1} are unknown while all others are known, since these were already competed in the j the step.

Writing $b_i = u_i^j + \frac{\alpha}{2}\left(u_{i-1}^j - 2u_i^j + u_{i+1}^j\right)$ and dropping j's, (3) becomes

$$u_i = \frac{\alpha}{2(1+\alpha)}(u_{i-1} + u_{i+1}) + \frac{b_i}{1+\alpha}$$

This gives the iteration formula

$$u_i^{(u+1)} = \frac{\alpha}{2(1+\alpha)}\left\{u_{i-1}^{(n)} + u_{i+1}^{(n)}\right\} + \frac{b_i}{1+\alpha}, \tag{4}$$

which expresses the (n+1)th iterates in terms of the n th iterates only.
This is known as the **Jacobi's iteration formula.**

As the latest value of u_i-1 i.e. $u_{i-1}^{(n+1)}$ is already available, the convergence of the iteration

formula (4) can be improved by replacing $u_{i-1}^{(n)}$ by $u_{i-1}^{(n+1)}$. Accordingly, (4) may be written as.

$$u_i^{(n+1)} = \frac{\alpha}{2(1+\alpha)}\left\{u_{i-1}^{(n+1)} + u_{i+1}^{(n)}\right\} + \frac{b_i}{1+\alpha} \tag{7}$$

which is known as **Gauss - Seidal iteration formula**
[Note : Gauss - Seidal iteration scheme is valid for all finite values of and converges twice as fast - as jacobi's Scheme].

Example

Consider the Molecular diffusion (alcohol vapour)

Equation $\dfrac{\partial c}{\partial t} = D\dfrac{\partial^2 c}{\partial x^2}$, where D is the diffusion coefficient equal to 0.119 cm²/ sec

with initial condition c (x,0) = 0 and boundary conditions are c (0,t) = 0 and c (20,t) = 10.0, for

$0 \le x \le 20.$

Explicit Solution: Choose Δx = 4cm. Using the maximum value permitted for Δt yields

D. $\dfrac{\Delta t}{(\Delta x)^2} = \dfrac{1}{2} = 0.119. \dfrac{\Delta t}{(4)^2}$, $\Delta t = \dfrac{8}{0.119} = 67.2$

The solution of the equation is

$$u_i^{j+1} = \frac{1}{2}\left\{u_{i-1}^j + u_{i+1}^j\right\}, j = 0,1,2,\ldots..$$

For t = 0 u (0,0) = 0, u(4,0) = u(8,0) = u(12,0) = u(16,0) = 2 and u (20,0) = 10

For Δ t = 67.2 sec, j = 0 $u_1^1 = \dfrac{1}{2}\left(u_0^0 + u_2^0\right) = \dfrac{1}{2}(0+2) = 1$

$u_2^1 = \dfrac{1}{2}\left(u_1^0 + u_3^0\right) = \dfrac{1}{2}(2+2) = 2$; $u_3^1 = \dfrac{1}{2}\left(u_2^0 + u_5^0\right) = \dfrac{1}{2}(2+2) = 2$

$u_4^1 = \dfrac{1}{2}\left(u_3^0 + u_5^0\right) = \dfrac{1}{2}(2+10) = 6$, $u_5^1 = 10$

$\therefore$ For t = 67.2 sec, the row the values of u are 0, 1, 2, 2, 4, 10

For t =134.4, the third row values of u are
0, 1, 1.5, 4, 6, 10 and so on

2. Solve the diffusion of alcohol vapours equation $\dfrac{\partial^2 u}{\partial x^2} = \dfrac{1}{D}\dfrac{\partial u}{\partial t}$, $u[x,0] = 2.0$,

and $u(0,t) = 0$, $u(20, t) = 10$, $D = 0.119$ cm^2 / sec

SOLUTION:

If $D\dfrac{\Delta t}{(\Delta x)^2} = \alpha = 1$, in the

Crank - Nicolson formula, it becomes

$$-u_{i-1}^{j+1} + 4u_i^{j+1} - u_{i+1}^{j+1} = u_{i-1}^{j} + u_{i+1}^{j}$$

$\therefore D\dfrac{\Delta x}{(\Delta t)^2} = \alpha = 1$ and $\Delta x = 4$cm, $\Delta t = 134.2$ sec at the end of the first Step.

For $j = 0$, and $i = 1$ we have $-u_0^1 + 4u_1^1 - u_2^1 = u_0^0 + u_2^0$

$-0.0 + 4u_1 - u_2$	$=$	$0.0 + 2.0 = 2$
$-u_1 + 4u_2 - u_3$	$=$	$2.0 + 2.0 = 4$
$-u_2 + 4u_3 - u_4$	$=$	$2.0 + 2.0 = 4$
$-u_3 + 4u_4 - 10$	$=$	$2.0 + 10.0 = 12$

The set of equations after simplification is

$4u_1 - u_2 = 2$ (1)
$-u_1 + 4u_2 - u_3 = 4$ (2)
$-u_2 + 4u_3 - u_4 = 4$ (3)
$4u_3 + 4u_4 = 22$ (4)

Solving the above system, we get
$u_1 = 1.005$; $u_2 = 2.02$; $u_3 = 3.075$; $u\backslash = 6.268$
$\therefore$ For the time step $t = 134.4$,
the calculated concentration values at $x = 0, 4, 8, 12, 16$ and 20 are
0, 1.005, 2.02, 3.075, 6.268, 10 respectively. We can find for the successive time steps
268.8, 403.2 . . . the values of u can be found.

Exercises in prapolic equations

1 Find the solution of the parabolic equation $u_{xx} = 2u_t$, when $u(0,t) = 4(4,t) = 0$ and
$u(x,0) = x(4-x)$, taking $h=1$, using explicit method to find the values of u upto $t=5$.

[Hint : $\alpha = \dfrac{1}{2} = \dfrac{k}{c^2 h^2} = \dfrac{k}{2}; k = 1$]

2 Find the values of u(x,t) satisfying the parabolic equation $u_t = 4u_{xx}$, with boundary conditions $u(0,t)$ = $u(8,t)$ = 0 and $u(x,0)$ = 4x-

$\dfrac{1}{2} x^2$, at the po int s $x = i$, $i = 0,1,2,...,7$, $t = \dfrac{1}{8} j, j = 0,1,2,..$ 5 [Hint $c^2 = 4, h = 1, k = \dfrac{1}{8}$ $\therefore \alpha$

$= \dfrac{1}{2}$]

3 Given uxx - u_t = 0, u(0,t) = 0 = u (5,t), u(x,0) = x^2 (25 - x^2), with h = 1, k = $\dfrac{1}{2}$,

using the explicit method.[Hint : $\therefore \alpha = \dfrac{1}{2}$]

4 Given $u_t = \dfrac{5}{33} u_{xx}$ with h = Δx = 0.25 and $\alpha = \dfrac{1}{2}$ u(x,0) = 100x for $0 \le x \le 1$ and
u(x,0) = 100 (2 -x) for $1 \le x \le 2$ and boundary conditions u(0,t) = 0 = u(2,t) using the explicit method given by

$u_i^{j+1} = \dfrac{1}{2}\left(u_{i-1}^j + u_{i+1}^j\right)$. Find the solution upto two time steps. [Hint k = Δt = 0.206 sec]

13.11.6 Numerical Methods for Hyperbolic equations

1. Solution by finite difference method

We consider the one dimensional **wave equation** as a model problem, given by

$$\dfrac{\partial^2 u}{\partial t^2} = \left(\dfrac{Tg}{w}\right)\dfrac{\partial^2 u}{\partial x^2} = c^2 \dfrac{\partial^2 u}{\partial x^2}, \text{ where } c = \sqrt{\dfrac{Tg}{w}}, \text{ subject to the initial conditions : } u = f(x),$$

$\dfrac{\partial u}{\partial t} = g(x), 0 \le x \le 1, \text{at } t = 0$ and the boundary conditions u(0,t) = u(1,t) = 0.

Replacing the derivations by difference quotients, using superscripts to denote time and subscript for position, we get.

$$\dfrac{u_i^{j+1} - 2u_i^j + u_i^{j-1}}{(\Delta t)^2} = \dfrac{Tg}{w}\left(\dfrac{u_{i+1}^j - 2u_i^j + u_{i-1}^j}{(\Delta x)^2}\right)$$

Solving for the displacement at the end of the current interval, at t = t_{j+1}, we get

$$u_i^{j+1} = \frac{Tg(\Delta t)^2}{w(\Delta x)^2}\left(u_{i+1}^j + u_{i-1}^j\right) - u_i^{j-1} + 2\left(1 - \frac{Tg(\Delta t)^2}{w(\Delta x)^2}\right)u_i^j \qquad (2)$$

Let $c^2 = \dfrac{Tg}{w},\; \alpha = \dfrac{k}{h}$

Then (2) becomes

$$u_i^{j+1} = 2(1 - \alpha^2 c^2)u_i^j + \alpha^2 c^2\left(u_i^j + u_{i+1}^j\right) - u_i^{j-1} \qquad (3)$$

Replacing the derivative $\dfrac{\partial u}{\partial t} = g(x)$ by its central difference approximation, we get

$$\frac{u_i^{j+1} - u_i^{j-1}}{2k} = \frac{\partial u}{\partial t} = g(x)$$

$$u_i^{j+1} = u_i^{j-1} + 2kg(x) \text{ at } t = 0 \qquad (4)$$

and (4) becomes for j = 0, we get $u_i^1 = u_i^{-1} + 2kg(x)$ \qquad (5)

also initial condition u = $f(x)$ at t = 0 becomes $u_i^{-1} = f(x)$ \qquad (6)

combining (5) and (6), we have $u_i^1 = f(x) + 2kg(x)$ \qquad (7)

The boundary conditions become $u_0^j = 0$ and $u_1^j = 0$ \qquad (8)

Hence the explicit form (3) gives the values of u_i^{j+1} at the (j+1) th level when the modal values at (j-1) th and jth levels are known from (6) and (7). The figure given below shows the relationships among the three levels (j-1, j and j+1)

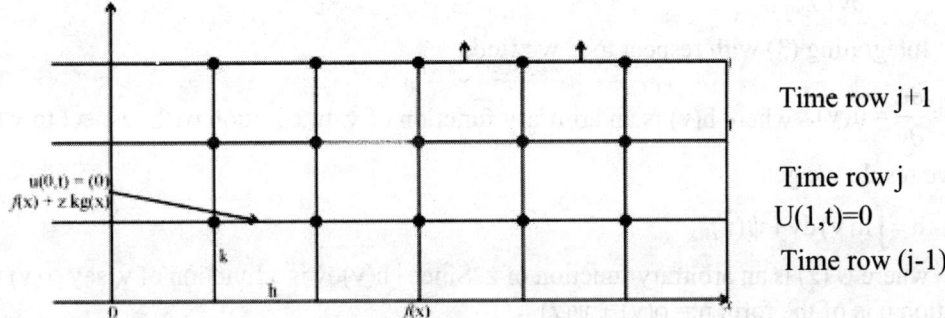

Time row j+1

Time row j

U(1,t)=0

Time row (j-1)

u(0,t) = (0)
f(x) + z kg(x)

[Note = 1. The coefficients in (3) will vanish if c = 1 or $k = \dfrac{h}{c}$. Then (3) reduces to the simple form

$$u_i^{j+1} = u_{i-1}^j + u_{i+1}^j - u_i^{j-1}$$

2. For $\alpha = \dfrac{1}{c}$, the solution (3) is stable and coincides with the solution of (1)

For $\alpha < \dfrac{1}{c}$, the solution is stable but in accurate

For $\alpha > \dfrac{1}{c}$, the solution is unstable.

3. The formula (3) converges for

$\alpha \le 1$ i,e $k \le h$.]

2. d' Alembert's solution of the wave equation.

This is an analytical solution. The one - dimensional wave equation, with

$$\sqrt{\dfrac{Tg}{w}} = c \text{ is } \dfrac{\partial^2 u}{\partial t^2} = c^2 \dfrac{\partial^2 u}{\partial x^2} \tag{1}$$

By direct substantiation, it is readily seen that for any arbitrary functions F and G equation (1) is solved by

$u(x,t)= \phi (x + ct) + \psi (x - ct)$ \hfill (2)

Let $v = x + ct$ and $z = x - ct$. Then u becomes a function of v and z.

Now $u_x = u_v v_x + u_z z_x = u_v + u_z$

and $u_{xx} = (u_v + u_z)_x = (u_v + u_z)_v v_x + (u_v + u_z)_z z_x$

$= u_{vv} + 2u_{vz} + u_{zz}$

we transform u_{tt} by the same procedure, getting $u_{tt} = c^2 (u_{vv} - 2u_{vz} + u_{zz})$

Inserting these two results in (1), we obtain

$$u_{vz} = \dfrac{\partial^2 u}{\partial v \partial z} = 0 \tag{3}$$

Integrating (3) with respect to z, we find

$\dfrac{\partial u}{\partial v} = h(v)$, where h(v) is an arbitrary function of v. Integration with respect to v the above result, we get.

$h = \displaystyle\int h(v)dv + \psi(z)$,

where ψ (z) is an arbitrary function of z. Since $\int h(v)dv$ is a function of v, say ϕ (v) the solution u is of the form $u = \phi(v) + \psi(z)$

or $u(x,t)= \phi(x + ct) + \psi (x - ct)$ \hfill (4)

This is known as **d'Alembert's solution** of the wave equation (1)

The functions ϕ and ψ can be determined from the initial conditions.

Let us illustrate in the case of zero initial velocity and given the initial deflection

$u(x,0) = f(x)$. By differentiating (4) we have $\dfrac{\partial u}{\partial t} = c\phi'(x + ct) - c\phi'(x - ck)$, \hfill (5)

where primes denote derivative with respect to the **entire** arguments x + ct and x - ct, respectively; Form (4) and (5) and the initial conditions, we obtain

$u(x,0) = \phi(x) + \psi(x) = f(x)$

$ut(x,0) = c\phi'(x) - c\psi'(x) = 0$

From the last equation, $\psi' = \phi'$. Hence $\Psi = \phi + k$ and from this and the first equation,

we have $2\phi + k =$ for or $\phi = \dfrac{1}{2}(f - k)$

With these functions ϕ and Ψ, the solution (4) becomes

$$u(x,t) = \frac{1}{2}\left[f(x+ct) + f(x-ct)]\right] \tag{6}$$

This is in agreement with the solution by separation of variables solution.

If the initial velocity g(x) is not identically zero, instead of (6), we similarly obtain

$$u(x,t) = \frac{1}{2}\left[f(x+ct) + f(x-ct) + \frac{1}{2c}\int_{x-ct}^{x+Ct} g(s)ds.\right] \tag{7}$$

Exercises

1 What is the solution of the boundary value problem
   $u_{xx} = u_{tt}, u(x,0) = x(1 - x), u_t(x,0) = 0, u(0,t) = u(1, t) = 0$ at the point x = 0.3 and t = 42.
   [Ans : u(x = 0.3, t = 4) = -0.21]

2 Show that the function u(x,t) = f(x+at) + g (x - at) satisfies the wave equation $a^2 u_{xx} = u_{tt}$

3 Solve the boundary value problem $u_{tt} = u_{xx}$ with the conditions $u(0,t) = u(1,t) = 0$, $u(x,0) =$
   $\dfrac{1}{2} x (1-x)$ and $u_t(x,0) = 0$, taking h = k =0.1 for $0 \le t \le 0.4$

4 Find the nodal values of the equation $u_{tt} = 16 uxx$ given that $u(0,t) = u(5.t) = 0$, $u(x,0) = x^2 (5-x)$ and $u_t(x,0) = 0$ taking h=1 upto one half of the periods of inebriation.

13.11.7 *Numerical Methods for Elliptic equations*

Two of the most important elliptic partial differential equations in mathematical physics and engineering are

Laplace equation, which has the following from in two variables:

$$\nabla^2 u = \frac{\partial^2 u}{\partial x^2} + \frac{\partial^2 u}{\partial y^2} = 0 \tag{1}$$

and the closely related to it is the Poisson's equation $\nabla^2 u = g(x, y)$ (2)

The solution of these equations is a function u(x,y) which is satisfied at every point of a region R subject to certain boundary conditions specified on the closed curve C surrounding the region R. To solve the Laplace equation on a region in the xy - plane, we subdivide the region with equispaced lines parallel to the x - and y- axes.

We cover the region R by mesh points $x_i = ih$, $y_j = jk$, $i, j \geq 0$, h and k are step lengths in the x and y directions. However it is common to take h as the step length in both directions, so that the discussion will be for square meshes in R.

∴ We consider $x_i = ih$ and $y_j = jh$, $i, j \geq 0$. It is convenient to introduce abbreviated notation.

$$u_{ij} = u(x_i, y_j)$$

$$\therefore (\Delta^2 u)_{ij} \frac{1}{h^2}\left(u_{i+1,j} + u_{i-1,j} + u_{i,j+1} + u_{i,j-1} - 4u_{ij}\right) = 0 \tag{3}$$

Note that five points are involved in the relationship of equation (3), i.e., points to the right left, above and below the central point (x_i, y_j), pictorially, the equation (3) becomes

$$\Delta^2 u_{ij} = \frac{1}{h^2}\begin{Bmatrix} & 1 & \\ 1 & -4 & 1 \\ & 1 & \end{Bmatrix} u_{i,j} = 0 \text{ and the error of approximation is } 0(h^2)$$

Hence $u_{i,j} = \frac{1}{4}\left(u_{i-1,j} + u_{i+1,j} + u_{i,j+1} + u_{i,j-1}\right)$ \hfill (4)

[Note : we can derive the nine point formula for the Laplace's equation and get

$$\Delta^2 u_{ij} = \frac{1}{6h^2}\begin{Bmatrix} 1 & 4 & 1 \\ 4 & -20 & 4 \\ 1 & 4 & 1 \end{Bmatrix} u_{i,j} = 0 \text{ With error of approximation as } 0(h^6)]$$

The equation (3) is written for computational purpose as a **standard 5- point formula**

$$u_{i,j} = \frac{1}{4}\ (u_i\text{ -1, }j+1+u_{i+1}\text{, }j\text{-1} + u_{i+1}\text{, }j\text{+1} + u_{i\text{ -1}}\text{, }j\text{-1}) \tag{5}$$

showing that the value of ui,j is the average of its value at the four neighboring diagonal mesh points. And is called the **diagonal 5-point formula:** Although (4) is less accurate than (4), it serves a reasonably good approximation for obtaining the starting values at the mesh points. Pictorially the above two formulas are represented as follows.

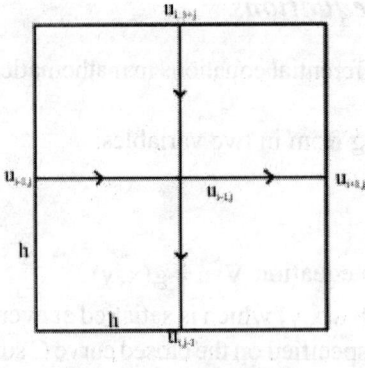

For formula (4)

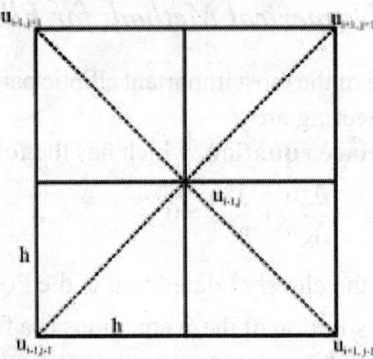

For formula (5)

Elliptic partial differential equations always occur as **boundary value problems**. The boundary conditions can be of the following three types.

    (i) **The Dirichlet or first boundary condition:** We have $u(x,y) = g(x,y)$, $(x,y) \in C$, where $g(x,y)$ is a prescribed function which is defined and continuous an C. This condition is called the **Direchlet Condition.** If $g(x,y) = 0$ on C then the boundary condition is called the homogeneous direchlet condition.

    (ii) **The Neumann on Second Boundary condition:** We have $\dfrac{\partial u}{\partial n} = g(x.y)$, $(x, y) \in C$,

where $g(x,y)$ is a prescribed function defined and continuous an C and n is the outwardly directed normal. This condition is called the **Neumann condition.** We may have homogeneous or inhomogeneous Neumann boundary condition.

    (iii) **The Third or Mixed Boundary Conditions:** We may have $\dfrac{\partial u}{\partial n} + \alpha(x,y)u$

$= g(x,y), (x, y) \in C$ where $\alpha(x,y) > 0$ and $g(x,y)$ are defined and continuous on C.

This condition is called the **mixed boundary condition.**

    (iv) In order to find the numerical solution of the Laplace equation $\Delta^2 u = \dfrac{\partial^2 u}{\partial x^2} + \dfrac{\partial^2 u}{\partial y^2} = 0$

along with the Dirichlet boundary conditions is called the Dirichlet problem. The difference equation for square mesh of step length h

is $u_{ij} = \dfrac{1}{4}(u_{i-1,j} + u_{i+1,j} + u_{i,j-1} + u_{i,j+1})$ for all i,j = 1,2,3 ....

In order to find the **initial values** of u at the **interior mesh** points, we first use the diagonal five point formula (5) and then use the standard five point formula (4).

Leibmann's Method

When applied to Laplace equation, the iterative technique is called **Leibmann's Method.** We illustrate the procedure for the problem.

$\Delta^2 u = 0$ for the square region R= $\{0 \leq x \leq 3, 0 \leq x \leq 3\}$ with u(x,0) = 0, u(x,3) = 1, u(0,y) = 0, u(3,y) = 0

SOLUTION :

We divide the region R into square meshes of size h=1 in both x and y directions using the standard 5-point formula we have.

$u_{ij} = \dfrac{1}{4}(u_{i+1,j} + u_{i+1,j} + u_{i,j+1} + u_{i,j-1})$ For all i,j = 1,2

$$i = 1, j = 1 \quad u_{11} = \frac{1}{4} (u_{21} + u_{01} + u_{12} + u_{10})$$

$$= \frac{1}{4} (u_{21} + u_{12})$$

$$i = 2, j = 1 \quad u_{21} = \frac{1}{4} (u_{31} + u_{11} + u_{22} + u_{20})$$

$$= \frac{1}{4} (u_{11} + u_{22})$$

$$i = 1, j = 2 \quad u_{12} = \frac{1}{4} (u_{22} + u_{02} + u_{13} + u_{11})$$

$$= \frac{1}{4} (u_{22} + 0 + 1 + u_{21})$$

$$= \frac{1}{4} (u_{11} + u_{22} + 1) = u_{21} + 0.25$$

$$i = 2, j = 2 \quad u_{22} = \frac{1}{4} (u_{32} + u_{12} + u_{23} + u_{21})$$

$$= \frac{1}{4} (0 + u_{12} + 1 + u_{21})$$

$$= \frac{1}{4} (u_{12} + u_{21} + 1) = u_{11} + 0.25$$

We have
$$u_{11} = \frac{1}{4} (u_{21} + u_{12})$$

$$u_{21} = \frac{1}{4} (u_{11} + u_{22})$$

$$u_{12} = \frac{1}{4} (u_{11} + u_{22} + 1)$$

$$u_{22} = \frac{1}{4} (u_{12} + u_{21} + 1)$$

Table showing successive iteration.

Points	Initial Value	Values at iteration number				
		1	2	3	4	5
u_{11}	0.1	0.075	0.10	0.125	0.1225	0.1225
u_{21}	0.1	0.075	0.10	0.125	0.1225	0.1225
u_{22}	0.2	0.325	0.35	0.365	0.3725	0.3750
u_{12}	0.2	0.325	0.35	0.365	0.3725	0.3750

The iteration is expected to converge

Jacobi's Method

Denoting the nth iterative values of $u_{i,j}$ by $u^{(n)}_{i,j}$, the iterative formula to solve the standard 5 point formula which is

$$u_{i,j} = \frac{1}{4}(u_{i-1,j} + u_{i+1,j} + u_{i,j+1} + u_{i,j-1})$$

is given by

$$= u^{(n+1)}_{i,j} = \frac{1}{4}\left(u^{(n)}_{i-1,j} + u^{(n)}_{i+1,j} + u^{(n)}_{i,j+1} + u^{(n)}_{i,j-1}\right)$$

It gives improved values of ui,j at interior mesh points and is called the **point - Jacobi's formula.**

Gauss Seidal method

In this method, the iteration formula is $u^{(n+1)}_{i,j} = \frac{1}{4}\left(u^{(n+1)}_{i-1,j} + u^{(n)}_{i+1,j} + u^{(n+1)}_{i,j+1} + u^{(n)}_{i,j-1}\right)$

It utilizes the latest iterative values available and scans the mesh points symmetrically from left to right along successive rows.

Solution of Poisson's equation

$$u_{xx} + u_{yy} = f(x,y)$$

The standard 5-point formula takes the form

$$u_{i-1,j} + u_{i+1,j} + u_{i,j+1} + u_{i,j-1} - 4u_{ij} = h^2 f(ij, jh)$$ for square mesh of size h in both x and y directions.

Applying the above formula at each interior mesh point, we will get a set of linear equations in the nodal values ui,j. These equations can solved by Gauss - Seidal method. The error of approximation is of the order $O(h^2)$ in solving positions equation for square mesh of step size h, in both x and y directions.

Example

For the boundary value problem
$\Delta^2 u = u_{xx} + u_{yy} = 4, -1 \le x \le 1, -1 \le y \le 1,$
$u = 0$ on the boundary.
Write a five point difference formulation with mesh spacing h in both direction. Solve the difference equation for $h = \dfrac{1}{2}$.

SOLUTION

Due to symmetry we consider the problem in $\dfrac{1}{8}$ th of the square namely the square

with vertices $(0,0), \left(\dfrac{1}{2}, 0\right), \left(0, \dfrac{1}{2}\right)\left(\dfrac{1}{2}, \dfrac{1}{2}\right)$. Let the solution values $u(0,0),\ u\left(\dfrac{1}{2}, 0\right), u\left(\dfrac{1}{2}, \dfrac{1}{2}\right)$

are denoted by a,b,c, respectively .

Applying the five point difference scheme at any nodal point (i,j) with mesh spacing h in both x and y directions, may be written as
$$u_{i+1,j} + u_{i-1,j} + u_{i,j+1} + u_{i,j-1} - 4_{ui,j} = h^2 \Delta^2 u_{i,j}$$
Applying this scheme at a,b and c and using the boundary conditions, we obtain a system of equation $4b - 4a = 1 : 2c + a - 4b = 1; 2b - 4c = 1$. Solving, we get the solution as
$$a = \dfrac{9}{8} -, b = -\dfrac{7}{8} \text{ and } c = -\dfrac{11}{16}.$$

4. Find the solution of $\Delta^2 u = 0$ in R, subject to the Dirichlet condition $u(x,y) = x-y$ an C, when R is the region inside the triangle with vertices (0,0), (14,0), (0,14) and C is its boundary. Assume step size $h = 4.0$, use the equation $u_{i-1,j} + u_{i+1,j} + u_{i,j+1} + u_{i,j-1} - 4u_{i,j} = 0$ at nodes (4,4) (8,4) and (4,8)

(a) Applying the difference formula and using the known boundary values obtain the system of equations

as
$$\begin{bmatrix} -1 & \dfrac{1}{4} & \dfrac{1}{4} \\ \dfrac{1}{3} & -2 & 0 \\ \dfrac{1}{3} & 0 & -2 \end{bmatrix} \begin{bmatrix} u_{11} \\ u_{21} \\ u_{12} \end{bmatrix} \begin{bmatrix} 0 \\ -4 \\ 4 \end{bmatrix}$$

(b) Show that the solution vector is $(0, 2, -2)^T$
(Note : these are exact values since the exact solution $u(x,y) = x-y$ is linear)

3. Find the approximate solution to Laplace's equation on a square
$\Delta^2 u = 0$, with boundary conditions $u(0,y) = 0, u(1,y) = 0$ for $0 \le y \le 1$

$u(x,0) = 0$, $u(x,1) = 1$ for $0 < x < 1$.

with step size $h = k = \dfrac{1}{3}$

13.11.8. *Exercises*

1. (a) Explain Leibmann's method to solve elliptic equation $u_{xx} + u_{yy} = 0$
2. Write down the finite difference analogue of the equation $u_{xx} + u_{yy} = 0$ and solve it for the region bounded by the square $0 \le x \le 3$ and $0 \le y \le 3$, the boundary conditions being $u = 0$ at $x = 0$ and $u = \dfrac{1}{2}(9 + y)$ at $x = 3$.

   $u = \dfrac{1}{2}x^2$ at $y = 0$ and $u = \dfrac{2}{3}x^2$ when $y = 3$, with $h = k = 1.0$. Use Gauss - Seidal method to compute the values of u at the internal mesh points.

17. (a) Explain how Poisson's equation $u_{xx} + u_{yy} = -f(x,y)$ is solved numerically, replacing it by differences, when u (x,y) is given on the sides of a square region.

    (b) Write the finite difference analogue of the equation $u_{xx} + u_{yy} = 0$. Compute the truncation error for square meshes of size h.

20. Using Liebmann's method, solve the equation $u_{xx} + u_{yy} = -10(x^2 + y^2 + 10)$ over the square with sides $x = 0 = y$, $x = 3 = y$; with $u = 0$ on the boundary and with square mesh of length $h = 1$.

21. The equation $u_{xx} + u_{yy} + u = 0$ is considered in a triangular domain bounded by the lines $x = 0$, $y = 0$ and $x + y = 1$. It is given that $u = 0$ on the boundary. The differential equation is replaced by a second order difference equation, which is considered in a square with mesh size h. Choosing first $h = \dfrac{1}{3}$ and then $h = \dfrac{1}{4}$, obtain approximations for the smallest eigen value.

18. Consider the boundary value problem $u_{xx} + u_{yy} = 4$, $|x| \le 1, |y| \le 1$, $u = 0$ on the boundary. Write a five point and nine point difference formulation with mesh spacing h in both directions.

    Solve the difference equations for $h = \dfrac{1}{2}$

4. Solve the Laplace equation $\dfrac{\partial^2 \phi}{\partial x^2} + \dfrac{\partial^2 \phi}{\partial y^2} = 0$ for a square region and having the boundary values $\{\phi(0,0) = 0;\ \phi(1,0) = 8.7;\ \phi(2,0) = 12.1,\ \phi(3,0) = 12.8;\ \phi(4,0) = 9.0\};\ \{\phi(0,1) = \phi(0,2) = \phi(0,3) = \phi(0,4) = 0\}\ \{\phi(4,1) = 17.0;\ \phi(4,2) = 21;\ \phi(4,3) = 21.9;\ \phi(4,4) = 18.6\}$ and $\{\phi(1,4) = 11.1;\ \phi(2,4) = 17;\ \phi(3,4) = 19.7\}$

5.

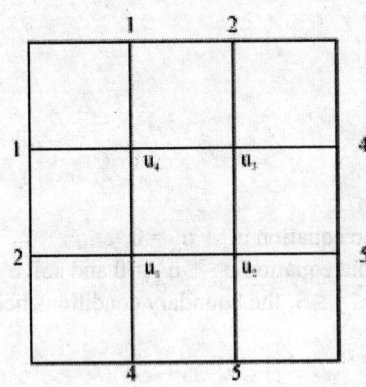

7.

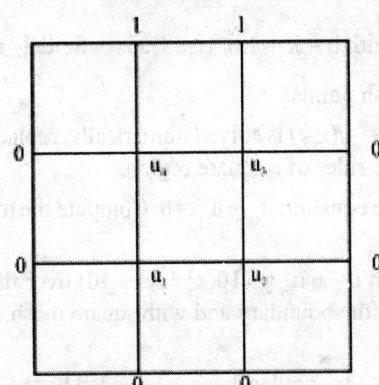

8.

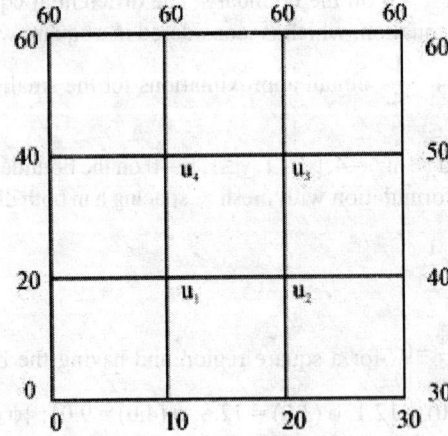

9.

[Ans : $u_1 = u_3$; $u_4 = u_6$; $u_7 = u_9$

i	$u_1 = u_3$	u_2	$u_4 = u_6$	u_5	$u_5 = u_9$	u_8
0	6.25	9.375	18.75	25	43.75	53.12
1	7.03	9.765	18.945	25.25	43.01	52.81
2	7.13	9.830	18.81	25.15	42.94	52.77

10. Applying Libmann's method of iteration find the solution of Laplace's equation $u_{xx}+u_{yy}=0$ in a square with vertices A(0,0), B(0,1),C(1,1), D(1,0). Use a spacing of h = $\frac{1}{3}$. The boundary conditions are: u|AB= 30y; u|BC = 30(1-x)|; u|CD=0=u|AD

11. Solve $u_{xx}+u_{yy}=0$ in a square region bounded by x=0, x=1, y=0, y=1 given that u =100 on the part of the boundary y =0 and that u=0 on the remaining part of the boundary. Take h=k = $\frac{1}{4}$

12. Write down the finite difference scheme for $u_{xx}+u_{yy}=0$ and solve it for the region bounded by the square $0 \le x \le 4$, the boundary conditions being u=0 at x=0 and u=8+2y at x=4; $\frac{1}{2}x^2$ u= at y=0 and u=x2 when y=4, with h=k=1. Use Gauss-Seidel method to compute the values of u at the internal mesh points.

13. Solve the mixed boundary value problem $\nabla^2 u = 0$, $\qquad 0 \le x, y \le 1$

$$\left.\begin{array}{l} u = 2x, \quad y = 0 \\ u = 2x - 1, \, y = 1 \end{array}\right\}, \qquad 0 \le x \le 1$$

$$\left.\begin{array}{l} u_x + u = 2 - y, \ x = 0 \\ u = 2 - y, \quad x = 1 \end{array}\right\}, 0 \le y \le 1 \text{ using the five points formula with h=k=}$$

[Ans : $u\left(0, \dfrac{1}{3}\right) = -\dfrac{1}{3}, u\left(\dfrac{1}{3}, \dfrac{1}{3}\right) = \dfrac{1}{3}, u\left(\dfrac{2}{3}, \dfrac{1}{3}\right) = 1$

$u\left(0, \dfrac{2}{3}\right) = -\dfrac{2}{3}; u\left(\dfrac{1}{3}, \dfrac{2}{3}\right) = 0, u\left(\dfrac{2}{3}, \dfrac{2}{3}\right) = \dfrac{2}{3}$

15. Solve the boundary value problem:

    (a) $\nabla^2 u = 0$, $0 \le x, y \le 2$, Using the five point difference scheme at any nodal

    point[i,j]with mesh spacing h=$\dfrac{1}{2}$ in both direction.

    With boundary conditions:

    $u = (x,0) = 0$; $u = (0, y) = 0$

    $u = (x,2) = 0$; $u = (2, y) = 0,$

    [Hint: Due to symmetry take the region R $\{1 \le x \le 2, 1 \le y \le 2\}$.

    Consider u(1,1)=a; $u\left(\dfrac{3}{2}, 1\right) = b$; $u\left(\dfrac{3}{2}, \dfrac{3}{2}\right) = c$ $u\left(1, \dfrac{3}{2}\right) = b.$

    Ans: $a = -\dfrac{9}{8}, b = -\dfrac{7}{8}, c = -\dfrac{11}{16}]$

    (b) $\nabla^2 u = 0 - 1 \le x, y \le 1$, $u(x - 1) = 0$ $-1 \le x \le 1$

    $u(x,1) = 0$, $-1 \le x \le 1, u(-1, y) = 0, -1 \le y \le 1$

    and $u(1, y) = 0,$ $-1 \le x \le 1$

    Consider u(0,0)=a, $u\left(\dfrac{1}{2}, 0\right) = b$, $u\left(\dfrac{1}{2}, \dfrac{1}{2}\right) = c$, $u\left(0, \dfrac{1}{2}\right) = b$

    Ans: a= $-\dfrac{9}{2}, b = -\dfrac{7}{2}$, $c \dfrac{-11}{4}]$

16. Solve the Laplace equation $\nabla^2 u = 0$ for the steady temperature distribution for a rectangular

    plate R = $\{0 \le x \le 20 \text{ cms}, 0 \le 9 \le 12 \text{ cms}\}$ with boundary conditions

    $\left.\begin{array}{l} u(x,0) \ = 50 \\ u(x,12) \ = 100 \end{array}\right\} 0 \le x \le 20;$

    $\left.\begin{array}{l} u(0, y) \ = 20 \\ u(20,4) \ = 20 \end{array}\right\} 0 \le y \le 12;$

19. (a) Solve $\nabla^2 u = xy$ we use h $\dfrac{1}{3}$. The values of u on the boundary are everywhere Zero. The

region is a square with corners at (0,0) and (2,2)

(b) If f (x,y)= (1-x) (1-y), find the solution

22. Solve the linear elliptic equation (x+1) uxx +(y+1)2 uyy = 1+u on $0 \le x \le 1$, $0 \le y \le 1$ with u(0,y) =y, u(1,y)=y² u(x,0)=0, u(x,1)=1. If h=k= $\frac{1}{3}$, then there are 16 points $\{u_{ij}|\ 0 \le i \le 3.\quad 0 \le j \le 3\}$ at which to determine an approximation to u(x,y). The points $\{u_{ij}|\ i = 0\ \text{or}\ i = 3\ \text{or}\ j = 0\ \text{or}\ j = 3\}$ are determined directly by the boundary conditions. Hence the only unknown that need to be determined are u_{11}, u_{12}, u_{21}, and u_{22}

(i) By discrediting the given diffential equation show that $(ih+1)\left[\dfrac{u_{i+1,j} - 2u_{i,j} + u_{i-1,j}}{h^2}\right]$

$$+(jk+1)^2 \frac{|u_{i,j+1} - 2u_{ij} + u_i, j-1|}{k^2} = 1 + u_{i,j}$$

(ii) Form the equations for the unknown $\{u_{ij}\}$ in matrix notation.

(iii) Show that the solution correct to four decimal places is
$$u_{11} \approx 0.0131; u_{12} \approx 0.3791,$$
$$u_{21} \approx -0.0265; u_{22} \approx 0.3419.$$

13.12 FINITE ELEMENT METHOD

Basic concept of the finite element method

13.12.1 Introduction

While studying physical phenomena, scientists and engineers exhibit interest in the **mathematical formulation** of the physical process and in **analyzing numerical** aspects of the mathematical model. The most distinctive feature of the finite element method that separates it from others is the **division of a given domain** into a set of simple subdomains, called **finite elements**. Any geometric shape that allows computation of the solution or its approximation or provides necessary relations among the values of the solution at selected points, called **nodes**, of the subdomain, qualifies as a **finite element**. There are also other features like **seeking continuous, polynomial approximations** of the solution over each element in terms of nodal values, and assembly of element equations by imposing the inter element continuity of the solution and **balance of inter element forces.**

The finite element analysis involve :

- Finite element discretization
- Element equations
- Assembly of element equations and solution
- Convergence and error estimate

In summary, the finite element method is characterized by the three features:

3. The domain of the problem is represented by a collection of finite elements known as the finite element mesh.

4. The physical process is approximation by functions of the desired type over each finite element and develop algebraic equations relating to physical quantities at the nodes.

5. The element equations are assembled using continuity and / or "balance" of physical quantities.

The finite element method is regarded as a technique of solving differential equations. A differential equation is said to describe a boundary value problem, if the dependent variable and possibly its derivations are required to take specified values on the boundary. An initial value problem is one in which the dependent variable and possibly its derivations are specified initially (i.e., at time $t = 0$).

13.12.2 *Definitions*

• Functionals

The term functional describes the functions defined by integrals whose arguments themselves are functions. The integral

$$I(u) = \int_a^b F(x, u, u')\, dx, \quad u = u(x), \quad u' = \frac{dx}{dx} \text{ is called a functional}$$

• Variational Symbol

The change αv in u, where v is a constant and v is a function is called the **variation** of u and is denoted by $\delta u = \alpha v$

The operator δ is called the **variational** symbol.

The first variation of F at u is defined by

$$\delta F = \frac{\delta F}{\delta u} \delta u + \frac{\delta F}{\delta u'} \delta u'$$

The total differential of F is defined as

$$d F = \frac{d F}{d x} dx + \frac{d F}{d u} du + \frac{d F}{d u'} du'$$

Since x is not varied during the variation of u to $u + \delta u$, $dx = 0$. "δ" acts as a differential operator with respect to dependent variables.

When the region R enclosed by the boundary is divided into a finite number of subdomains, called finite elements, we use of straight lines in the one dimensional case and triangular or rectangular elements in two dimensions. The curved boundaries are handled in a natural manner. At the boundaries of the elements called the interfaces, the inner element conditions are to be satisfied. The general requirement at the interface is that the approximating function, say w and its partial derivations upto order one less than the highest order derivative occurring in the differential equation or the variation principle must be continous. Elements satisfying this criterion are called compatible or conforming elements, otherwise they are called non - conforming elements. When the element size shrinks to zero, the function and its derivatives must have representation in the element, when the finite element satisfies this criterion, then it is called a complete element. When all the above conditions are satisfied then the solutions converge and the assembly of the individual element equations is meaningful in conformity with the interpolating functions remain unchanged.

13.12.3 *Weak formulation of Boundary value problems*

The motivation of integral formulations of boundary value problems comes from the fact that variational methods of approximation, e.g., the Ritz, Galerkin, leastsquares, collocation, or, ingeneral, weighted - residual methods, are based on weighted - integral statements of the governing equations. Since the finite element method is a technique for constructing approximation functions required in an element wise application of any variational method, it is necessary to study the weighted - integral formulation and the weak formulation of differential equations.

Consider the problem of solving the differential equation

$$-\frac{d}{dx}\left[a(x)\frac{du}{dx}\right] = q(x) \text{ for } o < x < L \tag{1}$$

for the solution $u(x)$, subject to the boundary conditions :

$$u(0) = u_0, \ a\frac{du}{dx}\Big|_{x=L} = Q_0 \tag{2}$$

A weak form of a weighted integral statement of a differential equation in which the differentiation is distributed among the dependent variable and the weight function and includes the natural boundary conditions of the problem.

We seek an approximate solution, over the entire domain $\Omega = [o, L]$ in the form

$$u \approx U_N = \sum_{j=1}^{N} c_j \ \phi_j(x) + \phi_0(x) \tag{3}$$

where the c_j are coefficients to be determined, and $\phi_j(x)$ and $\phi_0(x)$ are functions preselected such that the specified boundary conditions for the problem are satisfied by the N - parameter approximate solution U_N. This is accomplished by choosing N linearly independent weight functions in the integral statement.

Step 1

If w is the weight function, the weighted - integral or weighted - residual equivalent to the original differential equation is formed as follows :

$$0 = \int_0^L w \left[-\frac{d}{dx} \left(a \frac{du}{dx} \right) - q \right] dx \tag{4}$$

This allows to choose N linearly independent functions for w and obtain N equations for $c_1, c_2, \ldots\ldots\ldots c_n$ of (3)

Step 2

The resulting integral form will require weaker continuity conditions on ϕ, and hence the weighted integral statement is called the **weak form**. This requires weaker continuity of the dependent variable and often it results in symmetric set of algebraic equations in the co-efficients. Also, the **natural boundary conditions** of the problem are included in the weak form and hence the approximate solution UN is required to satisfy only by the **essential boundary conditions** of the problem. The coefficients of the weight function and its derivatives in the boundary expressions are termed the **secondary variables**. Specification of secondary variables at the boundary constitutes the **natural boundary conditions.**

The dependent variable of the problem is called the **primary variable** and its specification on the boundary constitutes the **essential boundary conditions**.

Step 3

We require the weight function w to vanish at boundary points where the essential boundary conditions are specified. The weight function w is required to satisfy w (o) = 0, because u (o) = u_0. Since w(o) = 0,

$$Q(L) = \left(a \frac{du}{dx} u_x \right)_{x=L} = \left(a \frac{du}{dx} \right)_{x=L} | = Q_0$$

The original problem reduces to the expression

$$0 = \int_0^L \left(a \frac{dw}{dx} \frac{du}{dx} - wq \right) dx - w(L)Q_0$$

These three steps constitute the weak or variation form of a differential equation.

13.12.4 (A) The Collocation Method

We start with the method of weighted residuals by considering the operator equation :

A(u) = f in Ω, (1) where A is a differential operator (linear or nonlinear), acting on the dependent variable u, and f is a known function of the independent variables. The operator A may be :

" $A(u) = -\dfrac{d}{dx}\left(a\dfrac{du}{dx}\right) + cu$

" $A(u) = -\dfrac{d^2}{dx^2}\left(b\dfrac{d^2u}{dx^2}\right)$

" $A(u) = -\left[\dfrac{\partial}{\partial x}\left(k_x\dfrac{\partial u}{\partial x}\right) + \dfrac{\partial}{\partial y}\left(k_y\dfrac{\partial u}{\partial y}\right)\right]$

" $A(u) = -\dfrac{d}{dx}\left(u\dfrac{du}{dx}\right)$

" $A(u,v) = u\dfrac{\partial u}{\partial x} + v\dfrac{\partial u}{\partial y} + \dfrac{\partial^2 u}{\partial x^2} + \dfrac{\partial}{\partial y}\left(\dfrac{\partial u}{\partial y} + \dfrac{\partial v}{\partial x}\right)$

If operator A is to be linear, it must satisfy the relation

$A(\alpha u + \beta v) = \alpha\, A(u) + \beta\, A(v)$

for any scalars and and and u and v being dependent variables.

In the collocation method, an approximate solution u_N A_o (1) is sought in the form

$$u_N = \sum_{j=1}^{N} c_j\, \phi_j + \phi_o \tag{2}$$

by requiring the residual in the equation to be identically zero at N selected points $x^i = (x^i, y^i)$, $i = 1,2, \ldots\ldots, N$ in the domain Ω :

$$R(x^i, y^i, c^j) = 0,\ i = 1,2, \ldots\ldots N \tag{3}$$

The collocation method will be a special case of

$$\int_{\Omega} \psi_i\,(x,y)\, R\,(x,y,cj)\, dx\, dy = 0,\quad i = 1,2,\ldots\ldots N \tag{4}$$

where Ω is a two - dimensional domain and Ψi are **weight functions**. The set $\{\Psi i\}$ must be a linearly independent set, when $\Psi i = \delta(x - x^i)$. Here $\delta(x)$ is the **Dirac delta function,** defined by

$$\int_{\Omega} f(x)\, \partial\,(x - ?)\, dx\, dy = f(\xi) \tag{5}$$

With this choice of weight functions, the weighted residual statement becomes

$$\int_{\Omega} \partial\,(x - x_i)\, R\,(x, c_j)\, dx\, dy = 0 \quad \text{or } R(xi, cj) = 0 \tag{6}$$

(B) The Petrov - Galerkin Method

The weighted - residual method is referred to as the Petrov - Galerking method, when $\Psi i = \phi i$. When the operator A is linear, the weighted residual equation

$$\int_\Omega \psi_i\,(x,y)\,R\,(x,y,c_j)\,dx\,dy = 0, \quad i = 1,2\ldots\ldots N$$

can be simplified to the form

$$\sum_{j=1}^{N}\left[\int_\Omega \psi_i\,A(\phi_j)\,dx\,dy\right]c_j = \int_\Omega \psi_i\,[f - A(\phi_o)]dx\,dy$$

or $\displaystyle\sum_{j=1}^{N}A_{ij}\,c_j = F_i$

Here the coefficient matrix [A] is not symmetric

$$A_{ij} = \int_\Omega \psi_i\,A(\phi_j)\,dx\,dy \neq A_{ji}$$

(C) The Galerkin Method

For the choice of weight function Ψi equal to the approximation function ϕi, the weighted - residual method is better known as the **Galerkin Method**. The algebraic equations of the Galerkin approximation are

$$\sum_{j=1}^{N}A_{ij}\,c_j = F_i$$

where $\displaystyle A_{ij} = \int_\Omega \phi_i\,A(\phi_j)dx\,dy, \quad F_i = \int_\Omega \phi_i\,[f - A(\phi_0)]\,dx\,dy$

(D) The Rayleigh - Ritz Method

Here the coefficients c_j of the approximation are determined using the weak form of the problem, and the choice is made as $w = \phi i$.

Consider the linear variational problem of finding the solution such that

$$B\,(w, u) = l\,(w) \tag{1}$$

for all sufficiently differentiable functions w that satisfy the homogenous form of any specified essential boundary conditions on u. When the functional B is bilinear and symmetric and l is linear, the problem (1) is equivalent to the minimization of the quadratic functional

$$I(u) = \frac{1}{2}B\,(u,u) - l(a) \tag{2}$$

We seek an approximate solution in this method, in the form of a finite series

$$U_N = \sum_{j=1}^{N} c_j \, \phi_j + \phi_0, \tag{3}$$

where the constants cj, called the

Ritz coefficients, are chosen such that (1) holds for $w = \phi i$, $i = 1,2, \ldots\ldots N$. It holds for N different choices of w, so that N independent algebraic equations in cj are obtained. The i - th algebraic equation is obtained by substituting ϕi for w :

$$B\left(\phi_i \, \sum_{j=1}^{N} c_j \phi_j + \phi_o \right) = 1(\phi_i), \quad i = 1,2,\ldots\ldots N \tag{4}$$

If B is bilinear, the summation and constants c_j can be taken outside the operator. We have

$$\sum_{j=1}^{N} B(\phi_i, \phi_j) c_j = 1(\phi_i) - B(\phi_i, \phi_o) \tag{5}$$

$$\text{or } \sum_{j=1}^{N} B_{ij} \, c_j = F_i, \; B_{ij} = B(\phi_i, \phi_j) \tag{6}$$

$F_i = 1(\phi i) - B(\phi i, \phi o)$ which represents the i th algebraic equation in a system of N linear algebraic equations in N constants c_j. The columns (and rows) of the matrix coefficients B_{ij} - B (ϕi, ϕj) must be linearly independent in order that the coefficient matrix in (5) can be inverted.

13.12.5 The Least - squares method

This method involves the determination of the parameters c_j by minimizing the integral of the square of the residual

$$R = A(U_N) - f = A\left(\sum_{j=1}^{N} c_j \, \phi_j + \phi_o \right) - f \neq o \tag{1}$$

$$\frac{\partial c}{\partial c_i} \int_{\Omega} R^2 \, (x,y,c) dx \; dy = 0 \text{ or } \int_{\Omega} \frac{\partial R}{\partial c_i} R \, dx \; dy = 0 \tag{2}$$

Comparison of (2) with the weighted integral so that the residual R to vanish, is given by

$$\int_{\Omega} \psi_i \, (x,y) \; R(x,y,cj) \, dx \; dy = 0, \qquad i = 1,2,\ldots\ldots N \tag{3}$$

where Ω is a two dimensional domain and Ψi are the weight functions. We get

$$\sum_{j=1}^{N}\left[\int_{\Omega} A(\phi_i) A(\phi_j)\, dx\ dy\right] c_j = \int_{\Omega} A(\phi_i)\left[f - A(\phi_i)\right] dx dy$$

or $\displaystyle\sum_{j=1}^{N} A_{ij}.c_j = F_i$, where $\qquad\qquad$ (4)

$$A_{ij} = \int_{\Omega} A(\phi_i)\, A(\phi_j)\, dx, \qquad F_i = \int_{\Omega} A(\phi_i)\left[f - A(\phi_0)\right] dx\,, \qquad (5)$$

Here the coefficient matrix Aij is symmetric

13.12.6 Rayleigh - Ritz method using calculus of variation

The calculus variation seeks to optimize (often minimize) a special class of functions called functional of the form

$$I\,[y] = \int_a^b F\left(x, y, \frac{dy}{dx}\right) dx$$

We seek to minimize I [y], by making the curves to pass through the end points (a, y(a)) and (b, y(b)). In addition for the optimal trajectory, the Euler - Lagrange equation must be satisfied :

$$\frac{d}{dx}\left[\frac{\partial}{\partial y} F(x, y, y')\right] = \frac{\partial}{\partial y} F(x, y, y'')$$

Consider a second order differential boundary value problem over [a,b].
$$y'' + Q(x) y = F(x), \quad y(a) = y_0, \ y(b) = y_n \qquad\qquad (1)$$
[Note: Here the end points are subject to Dirichlet conditions]
The functional that corresponds to equation (1) is

$$I\,[u] = \int_a^b\left[\left(\frac{du}{dx}\right)^2 - \phi u^2 + 2Fu\right] dx \qquad\qquad (2)$$

[Since $F(x, u, u') = (u')^2 - \phi u^2 + 2Fu$: $\dfrac{\partial}{\partial u'}$ [F (x,u,u')] = 2u' $\therefore \dfrac{\partial}{\partial x}\left(\dfrac{\partial F}{\partial u'}\right) = 2u''$

and $\dfrac{\partial F}{\partial u} = -2\partial u + 2F$

$$\therefore \frac{d}{dx}\left(\frac{\partial F}{\partial u'}\right) = \frac{\partial F}{\partial u} \text{ becomes}$$

$2u'' = -2Q u + 2F$

or $u'' + Q u - F = 0$, which is the same as the given differential equation.

$$\therefore I(u) = \int_a^b \left[\left(\frac{du}{dx}\right)^2 - Qu^2 + 2Fu\right] dx$$

The Raylist - Ritz method is based on the suitable choice for u (x). So we let u(x) which is the approximation to y(x) (the exact solution), be a sum :

$$u(x) = c_0\,\phi_0 + c_1\,\phi_1 + \ldots\ldots + c_n\phi_n = \sum_{i=0}^{n} c_i\,\phi_i \tag{3}$$

There are two conditions on the i's in equation (3) They must be chosen such that u (x) meets the boundary conditions and the individuals ϕ_i's must be linearly independent. We call the ϕ_i's **trial functions**; the c_i's and ϕ_i's are chosen to make u(x) a good approximation to the true solution of equation (1). We will use the functional of equation (2) to do this. If we substitute u (x) as defined by equation (3) into the functional (2) to do this. Hence the functional equational (2) becomes

$$I(c_0, c_1, \ldots\ldots\ldots, c_n) = \int_a^b \left[\left(\frac{d}{dx}\sum c_i\,\phi_i\right)^2 - \phi\left(\sum c_i\,\phi_i\right)^2 + 2F\sum c_i\,\phi_i\right] dx \tag{4}$$

To minimize, we take its partial derivations with respect to each unknown c_i and set to zero, resulting in a set of equations in the c_i's that we can solve. This will define u(x) in equation (3).

We now substitute then u(x) of equation (3) into the functional. If we partially differentiate with respect to, say c_i, where this is one of the unknown ci's, we will get

$$\frac{\partial I}{\partial C_i} \int_a^b 2\left(\frac{dy}{dx}\right)\frac{\partial}{\partial c_i}\left(\frac{dy}{dx}\right) dx - \int_a^b 2Qu\left(\frac{\partial u}{\partial c_i}\right) dx + 2\int_a^b F\left(\frac{\partial u}{\partial c_i}\right) dx \tag{5}$$

where we have broken the integral into three parts.

The procedure is clarified by the following example.

Examples

EXAM. 1:. Solve the equation $y'' + y = 3x^2$, with boundary conditions y(o) = 0 and y(2) = 3.5. Here Q = 1 and $F = 3x^2$. Use polynomial trial function upto degree 3. If we define u(x) as

$$u(x) = \frac{7}{4} x + c_2 (x) (x - 2) + c_3 x^2 (x - 2) \tag{6}$$

we have linearly independent ϕ_i's. The boundary conditions are met by the first term, since the other terms are zero at the boundaries, $u(x)$ also meets the boundary conditions. [It is customary to match the boundary conditions with the initial term (s) of $u(x)$ and then make the succeeding term equal zero at the boundaries].

Examination of equation (5), shows that we need these quantities.

$$\left. \begin{array}{l} \dfrac{du}{dx} = \dfrac{7}{4} + c_2 (2x - 2) + c_3 (3x^2 - 4x) \\[2mm] \dfrac{\partial}{\partial c_x} \left(\dfrac{du}{dx} \right) = 2x - 2; \quad \dfrac{\partial}{\partial c_3} \left(\dfrac{dy}{dx} \right) = 3x^2 - 4x \\[2mm] \dfrac{\partial u}{\partial c_2} = x(x - 2); \qquad\qquad \dfrac{\partial u}{\partial c_3} = x^2 (x - 2) \end{array} \right\} \tag{7}$$

Substituting from equations (7) into equations (5), we have two equations :

$$\frac{\partial I}{\partial c_2} : 0 = \int_0^2 2 \left[\frac{7}{4} + c_2 (2x - 2) + c_3 (3x^2 - 4x) \right] (2x - 2) dx$$

$$- \int_0^2 2 \, (1) \left[\frac{7}{4} x + c_2 \, (2x - 2) + c_3 \, (3x^2 - 4x) \right] (x^2 - 2x) dx$$

$$+ 2 \int_0^2 \left(3x^2 \right) \left(x^2 - 2x \right) dx , \tag{8}$$

and $\dfrac{\partial I}{\partial c_3} : 0 = \displaystyle\int_0^2 2 \left[\dfrac{7}{4} + c_2 \, (2x - 2) + c_3 \, (3x^2 - 4x) \right] (3x^2 - 4x) dx$

$$- 2 \int_0^2 2(1) \left[\frac{7}{4} x + c_2 \, (x^2 - 2x) + c_3 \left(x^3 - 2x^2 \right) \right] \left(x^3 - 2x^2 \right) dx$$

$$+ 2 \int_0^2 \left(3x^2 \right) \left(x^3 - 2x^2 \right) dx \tag{9}$$

On integration and simplifying (8) and (9) we get the pair of equations

$$\frac{16}{5} c_2 + \frac{16}{5} c_3 = \frac{74}{15}$$

and $\dfrac{16}{5} c_2 + \dfrac{128}{5} c_3 = \dfrac{36}{5}$ (10)

Solving for c_2 and c_3 from the above, the coefficients in u(x) are known. On expanding and simplifying, we find the

$$u(x) = \left(\frac{119}{152}\right) x^3 - \left(\frac{46}{57}\right) x^2 + \left(\frac{53}{228}\right) x$$

EXAM. 1: Solve the equation $y''+y=3x^2$, $y(0) = 0$, $y(2) = 3.5$ using collocation method.

SOLUTION:

We take u(x) as before to satisfy the boundary condition.

$$u(x) = \frac{7}{4} x + c_2 (x)(x-2) + c_3 (x^2)(x-2) \quad (1)$$

The residual is, after substituting u(x) for y(x), $R(x) = u'' + u - 3x^2$, which becomes, after differentiating u twice to get u'',

$$R(x) = c_2 (2) + c_3 (6x-4) + \frac{7}{4} x + c_2 (x^2-2x) + P3 (x^3-2x^2) - 3x^2 \quad (2)$$

Since $R(x)$ is forced to be zero and two points in [0,2]. We take arbitrarily as x=0.7 and x=1.3 (more or less equally spaced in [0,2].

From x = 0.7; $\dfrac{109c_2 - 437c_3 - 245}{1000} = 0$ $\Biggr\}$

From x =1.3; $\dfrac{1092c_2 + 2617c_3 - 2755}{1000} = 0$ (3)

Solving for c_2 and c_3 and substituting them in (2), we get

$$u(x) = \left(\frac{425}{500}\right) x^3 \left(\frac{61607}{55481}\right) - x2 + \left(\frac{140023}{221924}\right) x$$

EXAM. 2: Solve $y'' + y = 3x2$, $y(0) = 0$, $y(2) = 3.5$ by Galerkin method

SOLUTION :

Use the same u(x) as before,

$u(x) = \dfrac{7}{4} x + c_2(x)(x-2) + c3(x^2)(x-2),$

so that $\phi_2 = x(x-2)$ and $\phi_3 = x^2(x-2)$.

The residual is

$R(x) = y'' + y - 3x^2$, which become after substituting u' and u for y'' and y, respectively,

$R(x) = c_2(2) + c_3(6x-4) + \dfrac{7}{4} x + c_2(x)(x-2) + c_3 x^2(x-2) - 3x^2$ \hfill (1)

Since there are two unknown c_i's, we carry our two integrations:

Using $\phi_2 2$ as a W_i : $\displaystyle\int_0^2 [(x-2)] R(x)\, dx = 0$

Using ϕ_3 as a W_i : $\displaystyle\int_0^2 x^2 (x-2) R(x)\, dx = 0$

These two equations give in c_i's:

$\left. \begin{aligned} & -\dfrac{24c_2 + 24c_3 - 37}{15} = 0 \\[2mm] & -\dfrac{2(84c_2 + 160c_3 - 180)}{105} = 0 \end{aligned} \right\}$ \hfill (2)

Hence

$u(x) = \left(\dfrac{119}{152}\right) x^3 - \left(\dfrac{46}{57}\right) x^2 + \left(\dfrac{53}{228}\right) x$ \hfill (3)

EXAM. 3:Formulas for numerical integration are expressed in terms of the predetermined x-values at which the functional values are known. The weighting factors are applied to each of the functional values. The integral

$I = \displaystyle\int_a^b w(x) f(x)\, dx$, where $w(x) > 0$ in $[a,b]$ is the weight function, is approximated by

a finite linear combination of values of $f(x)$ in the form

$I = \displaystyle\int_a^b w(x) f(x)\, dx \approx \sum_{k=0}^n \lambda_k f_k$ where x_k, $k = 0(1)$ n are the abscissas or notes distributed

within $[a,b]$ and λ_k, $k = 0(1)$ n are called the weights of the integration rule. The coefficients (weight & factors) λ_k, $(n+1)$ in number are to be determined. A formula with $(n+1)$ parameters corresponds to a polynomial of degree less than or equal to n.

If the requirement that the function to be evaluated at predetermined n-values, then we have $2n+2$ unknown as ($(n+1)$ nodes x_k's and $n+1$ weights λ_k's) and the method can be made exact for polynomials of maximum degree $(2n+1)$. Formulas based on this principle are called **Gaussian quadrature formulas**.

The error of approximation is given as

$$R_n = \int_a^b w(x)f(x)dx - \sum_{k=0}^n \lambda_k f_k$$

EXAM. 4: Why the procedures of the Rayleigh-Ritz method and the Galerkin method are called variational methods?

The Rayleigh-Ritz and the Galerkin methods arise in solving boundary value problems related to finding the maximum or minimum of certain integrals involving an unknown function. These two methods are based on the branch of mathematics, the **calculus of variations.** Hence the procedures are called variational methods.

EXAM. 5: Explain briefly the Rayleigh - Ritz method.

Consider a second order linear boundary value problem are in [a,b]

$$y'' + Q(x)y = F(x), y(a);y0, y(b) = y_n \tag{1}$$

Since the boundary values at the end points are known, the end points are said to be subject to Dirichlet at conditions.

Let

$$I(y) = \int_a^b F\left(x, y\frac{dy}{dx}\right)dx$$

The curves yi (x) each pass through end points [a,y (a)] and [b, y(b)] and for the optimal trajectory, the Euler - Lagrange equation must be satisfied:

$$\frac{d}{dx}\left[\frac{\partial}{\partial y'}F(x,y,y')\right] = \frac{\partial}{\partial y}F(x,y,y')$$

The functional that corresponds to (1) is

$$I(u) = \int_a^b \left[\left(\frac{du}{dx}\right)^2 - Qu^2 + 2Fu\right]dx \tag{2}$$

[Note Q (x) ≥ 0 ; and u (x) is an approximation to y(x), given by u

$$(x) = \sum_{i=0}^n c_i\phi_i].$$

There are two conditions on the ϕ_i

- The ϕ_i's are chosen such that u (x) meets the boundary conditions and

- the individual ϕ_i's must be linearly independent]

  (2) becomes

$$I(c_o, c_2, ..., c_u) = \int_a^b \left\{ \left(\frac{d}{dx} \sum c_i \phi_i \right)^2 - \phi \left(\sum c_i \phi_i \right)^2 + 2F \left(\sum c_i \phi_i \right) \right\} dx \qquad (3)$$

To minimize, we must have

$$\frac{\partial I}{\partial ci} = \int_a^b 2 \left(\frac{dy}{dx} \right) \frac{\partial}{\partial ci} \left(\frac{du}{dx} \right) dx - \int_a^b 2\phi u \left(\frac{\partial u}{\partial c_i} \right) dx + 2 \int_c^b F \left(\frac{\partial u}{\partial ci} \right) dx = 0, \qquad (4)$$

$i = 0,1,2,...........n$

Solving these equations, we get the best values of the ci s.

EXAM. 6: Explain briefly the Galerkin method

The Galerkin is a " residual method" that uses

$R(x) = y'' + q y - F.$ $\qquad (1)$

We multiply R (x) by the weight functions $W_i(x)$, a good choice being equal to the trial

functions ϕ_i.

We compute the unknown coefficients by setting the integral over [a,b] of the weighted residual to zero:

$$\int_a^b W_i(x) R(x) dx = 0, i = 0,1,2,.....n \text{ , where } W_i(x) = \phi_i(x) \qquad (2)$$

EXAM. 7: Explain briefly the collocation method.

Collocation method is called a "residual method". The residual R(x) is defined as R

$(x) = y'' +$

We approximate y (x) again with u (x) equals to a sum of trial functions, usually chosen as linearly independent polynomials. We substitute u (x) into R (x) and attempt to make R (x)=0 by a suitable choice of coefficients in u(x). This is done by choosing selected points in the interval [a,b] at which we make R(x)=0.

EXAM. 8: Consider the differential equation

$$-\frac{d^2u}{dx^2} - u + x^2 = 0, u(0) = 0, u'(1) = 1$$

Find the solution by the collocation method, choosing the points $x = \frac{1}{3}$ and $x = \frac{2}{3}$ in

[0,1]

SOLUTION:

Let $\phi_0(x) = a + bx, \phi_0(0) = 0, \phi_0'(1) = 1$

$\therefore \phi_0(x) = x$ $\qquad (1)$

Let $\phi_1(x) = a + bx + cx^2$. Use $\phi_i(0) = 0, \phi'(1) = 0, i=1,2$

$\therefore \phi_1(x) = -cx(2-x)$. c can be set equal to 1, because it will be absorbed in the parameter c_1

$$\therefore \phi_1(x) = -cx(2-x)$$

(2)

For ϕ_2, we can take ϕ_2 a+ bx^2+ dx^3 with d≠ 0.

(Note : ϕ_2 does not contain all order terms in either case) For the second choice of ϕ_2,

we get $\phi_2(x) = x^2 \left(1 - \frac{2}{3}x\right)$

(3)

Let $U_N = \sum_{j=1}^{N} c_j \phi + \phi_0$

(4)

Now $R = -\frac{d^2}{dx^2} (U_N) - U_N + x^2$

$$R = \frac{d^2}{dx^2}\left[\sum_{j=1}^{n} c_j \phi_j + \phi_0\right] - \left(\sum_{j=1}^{n} c_j \phi_j + \phi_0\right) + x^2 = 0 \text{ for}_{j=1+j=2}$$

(5)

$$\left(-c_1\phi_1'' + c_2\phi_2'' + 0\right) - \left(c_1\phi_1 + c_2\phi_2 + \phi_0\right) + x^2 = 0$$

$$-c_2\left(x^2 - \frac{2}{3}x^3\right) - x + x^2 = 0 \text{ for x= } \frac{1}{3} \text{ and x= } \frac{2}{3}$$

$$\therefore R\left(\frac{1}{3}\right) = 0 \Rightarrow 117c_1 + 61c_2 = -18$$

(6)

$$R\left(\frac{2}{3}\right) = 0 \Rightarrow 90c_1 - 34c_2 = -18$$

(7)

Solving for c_1 and c_2. we get

$$c_1 = \left(\frac{1710}{9468}\right) \text{ and } c_2 = -\left(\frac{486}{9468}\right)$$

$$\therefore u_c = c_1\phi_1 + c_2\phi_2 + \phi_0$$

$$\frac{1710}{9468}\{-x(2-x)\} - \frac{486}{9468}\left(x^2 - \frac{2}{3}x\right)$$

$$= \frac{6048}{9468}x + \frac{1224}{9468}x^2 + \frac{324}{9468}x^3$$

$$= \frac{1}{9468}\left[6048x + 1224x^2 + 324x^3\right]$$

[Note: u(0)=0, u' (1)=1 are satisfied]

EXAM. 9: Solve the problem in Ex.4 by the least squares method.

SOLUTION:

Take $\Psi_i = \dfrac{\partial R}{\partial c_i}$. We have from

$$R = c_1\left(2 - 2x - x^2\right) + c_2\left(-2 + 4x - x^2 + \frac{2}{3}x^3\right) - x + x^2$$

$$\int_0^1 \left(2 - 2x - x^2\right)R\,dx = 0 \text{ and } -\int_0^1 \left(2 - 4x + x^2 - \frac{2}{3}x^3\right)R\,dx = 0$$

or $\dfrac{28}{15}c_1 - \dfrac{47}{90}c_2 - \dfrac{13}{60} = 0$

and $-\dfrac{47}{90}c_1 + \dfrac{253}{315}c_2 + \dfrac{1}{30} = 0$

Solving the above, we have $c_1 = \dfrac{1292}{9935}, c_2 = \dfrac{991}{19870}$.

The solution to the differential equation becomes

$$u_{LS} = 1.2601x - 0.08017x^2 - 0.03325x^3$$

EXAM. 10: Find the approximate solution of y"=3x+1, y (0) = 0 = y (1) by
(i). Rayleigh-Ritz method; (ii) Collocation method, setting the residual to Zero

$$\text{at } x = \frac{1}{3} \text{ and } x = \frac{2}{3}. \text{ (iii) Galerkin method.}$$

SOLUTION:

(i) Rayleigh-Ritz method.

Let $u(x) = cx\ (x-1)$ satisfying $u(0)=0=u(1)$ Using Euler-Lagrange

equation $\dfrac{d}{dx}\left[\dfrac{\partial F}{\partial u'}(x, y, u')\right] = \dfrac{\partial}{\partial u}F(x, y, u')$

Here $F(x, y, u') = \left(\dfrac{du}{dx}\right)^2 + 2Fu$

$\therefore \dfrac{\partial F}{\partial u'} = 2u' = \dfrac{d}{dx}\left(\dfrac{\partial F}{\partial u'}\right) = u''$

$$\frac{\partial F}{\partial u} = 2u = 2(3x+1)$$

$$\therefore I(c_1, c_2, \ldots c_n) = \int_a^b \left[\left(\frac{d}{dx} \sum c_i \phi_i \right)^2 + F\left(\sum c_i \phi_i \right)^2 \right] dx$$

As there is only one constant in u(x) = c x (x-1),

we have $\dfrac{du}{dx} = 2cx - c$ $\quad \dfrac{d}{dc}\left(\dfrac{du}{dx} \right) = 2x - 1$

$$\frac{d}{dc}[c(x)(x-1)] = x(x-1)$$

$$\therefore \frac{dI}{dc} = \int_0^1 2\left(\frac{du}{dx} \right) \frac{d}{dc}\left(\frac{du}{dx} \right) dx + 2\int_0^1 (3x+1)c(x)(x-1)dx$$

$$\int_0^1 2c\,(2x-1)\,(2x-1)\,dx + \int_0^1 2c\,(3x+1)\,(x^2-x)\,dx$$

$$\frac{dI}{dc} = 0 = 2c\int_0^1 (4x^2-4x+1)dx + 2\int_0^1 (3x^3-2x^2-x)dx$$

$$2c\left(\frac{4}{3} - \frac{4}{2} + 1 \right) + 2\left(\frac{3}{4} - \frac{2}{3} - \frac{1}{2} \right) = 0$$

$$\frac{2c}{3} - 2.\frac{5}{12} = 0 \;\; \text{or} \quad\quad c = \frac{5}{6} \times \frac{3}{2} = \frac{5}{4}$$

$$\therefore u(x) = \frac{5}{4} x (x-1)$$

(ii) Collocation method

R(x) = u''(x)-3x-1

Let u(x) = c_1 x(x-1) + c_2 x²(x-1); u(0)=0=u(1)

u'(x)=2c_1 x-c_1+3c_2x²-2c_2x

u''(x)=2c_1+6c_2x-2c_2

R(x) = u'' -3x-1

$\quad$ = 2c_1+6c_2x-2c_2-3x-1

$$R\left(\frac{1}{3}\right) = 2c_1 + 2c_2 - 2c_2 - 2 = 0 \text{ or } 2c_1 = 2; c_1 = 1$$

$$R\left(\frac{2}{3}\right) = 2c_1 + 4c_2 - 2c_2 - 2 - 1 = 0 \text{ or } 2c_1 + 2c_2 = 3, \text{ or } c_2 = \frac{1}{2}$$

$$\therefore u(x) = x(x-1) + \frac{1}{2}x^2(x-1)$$

$$= \frac{1}{2}x^3 + \frac{1}{2}x^2 - x = \frac{1}{2}(x^3 + x^2 - 2x)$$

Analytical Solution

$y'' = 3x+1$, $y(0)=0=y(1)$

A.E is $m^2=0$, $m=0$. C.F is $y=A+Bx$.

G.S: $y=A+Bx+P.I$

$$\text{P.I is} = \int (3x+1)dx = \frac{3x^2}{2} + x$$

$$\int \left(\frac{3x^2}{2} + x\right)dx = \frac{x^3}{2} + \frac{x^2}{2}$$

$$\therefore \text{ G.S is } y = Bx + \frac{x^3}{2} + \frac{x^2}{2}$$

$y(0) = 0$ $y(1) = 0 = B + 1 \therefore B = -1$

$$\therefore \text{ The G.S. or analytical solution becomes } y = -x + \frac{x^3}{2} + \frac{x^2}{2} = \frac{1}{2}(x^3 + x^2 - 2x)$$

$\therefore$ The Collocation method here gives the analytical solution.

(iii) Galerkin method

Let $R(x) = u'' - 3x - 1$ (1)

Let $u(x) = c_1 x(x-2) + c2\ x^2(x-2)$ (2)

$R(x) = (2c_1-2c_1-1) + (6c_2-c_1)\ x + (c_1-c_2)x^2 + c2x^3$ (3)

$$\phi_1(x) = x(x-1) \therefore \int_0^1 x\ (x-1)R(x)\ dx = 0 \qquad (4)$$

$$\phi_2(x) = x^2(x-1) \therefore \int_0^1 x^2(x-1)R(x)\ dx = 0 \qquad (5)$$

$(4) \Rightarrow 2c_1 + c_2 = 0$ (6)

$(5) \Rightarrow 49c_1 + 66c_2 = 35$ (7)

Solving (6) and (7), the approximate solutions is

$$u(x) = \frac{35}{73}\left(2x^3 - 3x^2 + x\right)$$

Exercises

1. Solve by collocation to find the approximate solution of y =3x+1, y(0)=0=y(1)using a quadratic approximation in x as (i) u(x) = cx(n-1), by setting the residual to o at (ii) u(x)=ax(x-1)+bx2(x-1) by setting the residuals at and to zero.
2. Use the Rayleigh-Ritz method to find the approximate the solution of y =3x+1, y(0)=0=y(1) using
  i. a quadratic approximation in x as u(x)=cx(x-1)
  ii. Using u(x) = ax(x-1)+bx2(x-1)
  iii. Using u(x) = ax(x-1)+bx(x-1)2
  iv. Using the boundary conditions as y (0)=1 and y(1)=3, and u(x) = ax(x-1)+ bx2(x-1).
3. Use the a Galerkin method to find the approximate solution of y =3x+1, y(0)=0=y)1) choosing
  (i). u (x)=ax(x-1)
  (ii). u (x)=ax(x-1)+bx2(x-1)
4. Solve the boundary value problem u +u=x, o?x?1, u (0) = 0, u(1)=0, with 1(x) = x(x-1) and 2(x) = x(x-1) and weight functions w1(x)=1 and w2(x)=x. Choose c1 and c2 so that the residual E(x) for u=c1 1+c2 2 is orthogonal to w1 and w2 on the interval o? x ?1.

      [Ans:

13.13 THE FINITE-ELEMENT METHOD APPLIED TO PARTIAL

DIFFERENTIAL EQUATIONS

Rayleigh-Ritz method minimizes the functional for the problem by setting partial derivatives to zero. Galerkin method sets integrals of a weighted residual to zero. The two methods are equivalent for most problems.

Integration by parts in two dimensions is made use of for representation of finite element methods. Some of the formulas are given below

$$\bullet \iint_R f\frac{\partial g}{\partial x}dA = \oint_c fg\ \mathbf{n.i}.ds - \iint_R g\frac{\partial f}{\partial x}dA \ ,$$ (1)

where f and g are continuously differentiable functions in R together with C, where curve C is a simple closed curve bounding region R and $\mathbf{n} = \left(\dfrac{dy}{ds}\right)\mathbf{i} - \left(\dfrac{dx}{ds}\right)\mathbf{j}$ is the outer normal.

If we apply Green's theorem $\oint (Pdx + Qdy)$ with P=0 and Q=fg and using

$$\iint_R f\frac{\partial h}{\partial y}dA = \oint_c fh(n.j)ds - \iint_R h\frac{\partial f}{\partial y}dA \tag{2}$$

and adding (1) and R, we get

$$\iint_R f\left(\frac{\partial g}{\partial x} + \frac{\partial h}{\partial y}\right)dA = \oint_c f(gi + hj).nds - \iint_R \left(h\frac{\partial f}{\partial x} - h\frac{\partial f}{\partial y}\right)dA$$

or with v = gi+hj

$$\iint_R f\,\mathrm{div}\,vdA = \oint_c fv.nds - \iint_R v.\nabla fdA \; , \tag{3}$$

since div fv=fdivv +v. ∇f. If we take v= ∇w in (3), we obtain the useful Green's formula

$$\iint_R f\nabla^2 wdA = \oint_c f\frac{\partial w}{\partial n}ds - \iint_R (\nabla f.\nabla w)dA \tag{4}$$

Interchanging f and w and subtracting, we obtain another Green's formula:

$$\iint_R \left(f\nabla^2 w - w\nabla^2 f\right)dA = \oint_c \left(f\frac{\partial w}{\partial n} - w\frac{\partial f}{\partial n}\right)ds \tag{5}$$

If we take here f= ∇^2u, we obtain another identify.

$$\iint_R \left(f\nabla^2 u\nabla^2 w - w\nabla^4 u\right)dA = \oint_c \left(\nabla^2 u\frac{\partial w}{\partial n} - w\frac{\partial^2 u}{\partial n}\right)ds \tag{6}$$

With the aid of the above formulas the Galerkin equations for two dimensional problems can be transformed in finite element method for partial differential equations.

We consider solving the elliptic equation

$$u_{xx} + u_{yy} + Q(x, y)u = F(x, y) \tag{7}$$

on region R bounded by curve L, with boundary conditions

$$u(x, y) = u_0 \text{ on } L_1 \text{ and } \frac{\partial u}{\partial n} = \alpha u + \beta \text{ on } L_2,$$

where $\dfrac{\partial u}{\partial n}$ is the outward normal gradient,

If is to be noted that we have Dirichlet conditions on some parts of the boundary and mixed boundary conditions on other parts.

We will use u(x, y) as the exact solution to (1) and v (x, y) as our approximation to u(x, y)

The procedure for applying finite-element method consists in following the steps:

1. Find the functional that corresponds to the partial-differential equation.
2. Subdivide the region into sub regions (elements)-(triangular elements) such that every node (vertices of the triangular elements) and every side of the triangles must be common with adjacent elements except for the sides on the boundaries.
3. Write an interpolating linear relation that gives values of the dependent variable within an element based on the values of at the nodes. We will write the interpolation function as the sum of three terms: each term involving a quantity ci. the value of v(x, y) at a node.
4. Substitute the interpolating relation into the functional and partial derivatives of the functional with respect to each c to Zero. This gives three equations, with the c's unknowns for each element.
5. Combine together the element equations of step 4 to get a system of equations, by adjusting these for the boundary conditions of the problem, and then solve, giving the values for the unknown nodal values, the c's that are approximations to u(x, y) at the notes.

The functional corresponding to equation (7) is

$$I(u) = \iint_R \left[\left(\frac{\partial u}{\partial x} \right)^2 + \left(\frac{\partial u}{\partial y} \right)^2 - Qu^2 + 2Fu \right] dxdy$$

$$- \int_{L2} \left(\alpha u^2 + 2\beta u \right) dl \tag{8}$$

Example

We consider the Galerkin formulations for solving two dimensional problems.

Solve $\nabla^2 u = x^2 y^2$ for $o<x<1$, $o<y<1$; $u=0$ for $x=0$ or $y=0$, $\dfrac{\partial u}{\partial n} = 0$ for $x=1$ or $y=1$.

We want u(x, y) to be such that the Boundary conditions hold and that $\iint_R \left(\nabla^2 u - x^2 y^2 \right)$ wdA=0 for an appropriate sequence of functions w satisfying the boundary conditions. Hence R is given as a square region, C is its boundary.

We transform the above equation using (4), with f replaced by w and w by u, obtaining

the equation. $\oint_C w \dfrac{\partial u}{\partial n} ds - \iint_R \left[(\nabla u . \nabla w + x^2 y^2 w) \right] dA = 0$

If w and u satisfy the boundary conditions, then the first term is 0, since w is 0 on part

of C and $\dfrac{\partial u}{\partial n} = 0$ on the rest of C. It is sufficient to make w=0 for x=0 or y=0 (essential

boundary conditions) and that we need not require $\dfrac{\partial w}{\partial x} = 0$ for x=1 or y=1 (natural boundary

conditions).

Using the equation $\iint_R (\nabla u . \nabla w + x^2 y^2 w) dA = 0$ and making it to hold for u=$c_1 \phi_1$(x,

y) + $c_2 \phi_2$(x, y) +........+ $c_n \phi_n$(x, y), where w is equal to any one of the ϕ_1(x, y),......., ϕ_n(x,
y). Here the functions ϕ_k(x, y) should satisfy the essential boundary conditions and the
infinite sequence {ϕ_n(x, y)} should be such that all functions u(x, y) satisfying the essential
boundary conditions can be arbitrarily well approximated, in terms of least-square error, by
an approximate linear combination of members of the sequence. We trey ϕ_1(x,y)=xy, ϕ_2(x,y)=
(x²y+xy² and are led to the two equations.

$\iint_R \left[c_1 (x^2 + y^2) + c_2 (x^3 + 2x^2 y + 2xy^2 + y^3) + x^3 y^3 \right] dA = 0,$

$\iint_R \left[c_1 (x^3 + 2x^2 y + 2xy^2 + y^3) + c_2 (x^4 + 4x^3 y + 8xy^2 + 4xy^3 + y^4) + (x^4 y^3 + x^3 y^4) \right] dA = 0$

These give $\dfrac{2}{3} c_1 + \dfrac{7}{6} c_2 + \dfrac{1}{16} = 0$ and $\dfrac{7}{6} c_1 + \dfrac{103}{45} c_2 + \dfrac{1}{10} = 0$ so that

$c_1 = -\dfrac{114}{712} = -0.1601, c_2 = \dfrac{27}{712} = 0.0379$ and we get the approximate solution as
u=-0.1601 xy+0.0379 (x²y+ xy²):

Also when x=1. $\dfrac{\partial u}{\partial n} = \dfrac{\partial u}{\partial x} = 0.1601y + 07.58y + 0.0379y^2$
=0.0843y+0.0379y².

13.13.1 *Dirichlet problem (Variational approach)*

Consider a Dirichlet problem for the Poisson equation $\nabla^2 u = f(x, y)$ in R, u=0 on C, (9).

The solution provides the unique minimum of the integral $J = \iint_R (|\nabla v|^2 + 2fv) dA$, (10)

among all functions v in R having the boundary values 0 on C.

Let $U(x, y) = v(x, y) - u(x, y)$.

Then $\iint\limits_{R} \left(|\nabla v|^2 + 2\,f.v \right) dA = \iint\limits_{R} \left(\nabla v.\nabla v + 2f.v \right) dA$

$= \iint\limits_{R} [(\nabla u + \nabla U)(\nabla u + \nabla U) + 2f(u + U)]dA$

$\iint\limits_{} |\nabla u|^2 + |\nabla U|^2 + 2\nabla u.\nabla U + 2f(u + U)dA \Big]$

Applying the relation (3) to the term $\nabla u. \nabla U$ we obtain

$= \iint\limits_{R} \nabla u.\nabla U dA = \oint\limits_{C} U \frac{\partial u}{\partial n}\,dx - \iint\limits_{R} U\nabla^2 u\,dA$

But $U=0$ on C and $\nabla^2 u = f$ in R. Hence $\iint \nabla u.\nabla U dA = \iint\limits_{R} fU dA$

Therefore $\iint\limits_{R} \left(|\nabla v|^2 + 2f.v \right)dA = \iint\limits_{R} \left(|\nabla u|^2 + 2f.u \right)dA + \iint\limits_{R} \left(|\nabla v|^2 + 2f.u \right)dA$

$= \iint\limits_{R} \left(|\nabla u|^2 + 2f.u \right)dA.$

Thus u does minimize the integral J. The minimum is attained only for $\nabla U=0$ Or U a constant Since $U=0$ on C, v must be identically 0, that is, the minimum is attained only for $v(x, y)$ $u(x,y)$.

We now select $\phi_1(x, y)$, $\phi_2(x, y)$,...., $\phi_n(x, y)$ to R with 0 boundary values on C and try to choose $c_1, c_2, ..., c_u$, to minimize the integral J among all functions

$v(x, y) = c_1 \phi_1 + c_2 \phi_2 + ... + c_n \phi_n$ (11)

If we replace v by this expression in (10) then J becomes a function of n real variables $c_1, c_2, ..., c_n$, say $F(c_1, c_2, ..., c_n)$. This function F attains its minimum at a point

where $\dfrac{\partial F}{\partial c_1} = 0$, ..., $\dfrac{\partial F}{\partial c_2} = 0.$ Now

$F(c_1, c_2, ..., c_n) = \iint\limits_{R} \left(\nabla v.\nabla v + 2fv \right)dA$

$= \iint\limits_{R} [(c_1\nabla\phi_1 + ... c_n\nabla\phi_n)]dA$

$+ 2f(c_1\nabla\phi_1 + ... c_n\nabla\phi_n\,dA)$

$$\iint_{R}\left[\sum_{i,j=1}^{n}\nabla\phi_{i}.\nabla\phi_{j}.c_{i}c_{j}+\ldots\ldots\ldots+2f\sum_{i=1}^{n}(.c_{i}\phi_{i})\right]dA \qquad (12)$$

$$\therefore\frac{\partial F}{\partial c_{i}}=2\iint_{R}\left(\sum_{j=1}^{n}\nabla\phi_{i}.\nabla\phi_{j}...+f\phi_{i}\right)dA, i=1.2\ldots\ldots u_{i}$$

The condition for minimum is $=\iint_{R}[(\nabla v.\nabla w+fw)]dA=0$

where v is an in equation (11) and w= $\phi_{1}, \ldots\ldots, \phi_{n}$

Thus we obtain the Galerkin equations. Then we follow the Rayleigh-Ritz procedure.

[Note: We can also show that for proper choice of $\phi_{1}, \ldots\ldots, \phi_{n}$ as linearly independent, over R, the function $F(c_{1}, c_{2}, \ldots\ldots, c_{n})$ has a unique minimum]

13.14.1 *Linear Element*

The linear element is also called the line segment element. The region R = [a, b] is divided into M non are lapping subintervals, denoted by $R_{i} = [x_{i}, x_{i+1}]$, $0 \leq i \leq M-1$, where x_{0}=a and x=b. we side the element $[x_{i}, x_{i}+1]$, where the approximating function is taken as

$u(x)=N_{i}+u_{i}+N_{i}u_{i}+1$ (1),

Where $N_{i+1} = \dfrac{x_{i+1}-x}{x_{i+1}-x_{i}} = \dfrac{1}{h_{i}}(x-x_{i})$ and

$N_{i} = \dfrac{x-x_{i}}{x_{i+1}-x_{i}} = \dfrac{1}{h_{i}}(x-x_{i})$ and also,

we have $u(x_{i})=u_{i}$ and $u(x_{i}+1)=u_{i+1}$, $h_{i}=x_{i}+1-x_{i}$.

If the nodal values u_{i} and u_{i+1} and the first order derivatives u_{i}' and u_{i}' +1 are used in constructing the piecewise polynomial in the interval $x_{i}\leq x\leq x_{i}+1$, then we have

$$u(x)= A_{i}(x)u_{i} + B_{i+1}(x)u_{i+1} + C_{i}(x)u_{i}' + D_{i+1}(x)u_{i+1}', \qquad (2)$$

where $A_{i}(x), B_{i+1}(x), C_{i}(x),$ and $D_{i+1}(x),$ are the Hermite cubic polynomials,

$$A_{i}(x)= 3N_{i+1}^{2} - 2N_{i+1}^{3}, B_{i+1}^{x} = 3N_{i}^{2} - 2N_{i}^{3}$$

$$C_{i}(x)= \left(N_{i+1}^{2} -N_{i+1}^{3}\right)(x_{i+1} - x_{i}), D_{i}(x)= \left(- N_{i}^{2} + N_{i}^{2}\right)x(x_{n+i} + x_{i})$$

(1) becomes

$$u(x)= \frac{1}{h_{i}}(x - x_{i+1})u_{i} + \frac{1}{h_{i}}(x - x_{i})u_{i+1},$$ which is a linear approximation function

(3)

The respective $A_i(x), B_{i+1}(x), C_i(x)$ and $D_{i+1}(x)$

become $A_i(x) = \dfrac{3}{h_i^2}(x - x_i)^2 + \dfrac{2}{h_i^3}(x - x_i)^3$

$$B_{i+1}(x) = \frac{3}{h_i^2}(x - x_i)^2 + \frac{2}{h_i^3}(x - x_i)^3$$

$$C_i(x) = \left[\frac{1}{h_i^2}(x - x_i)^2 + \frac{1}{h_i^3}(x - x_i)^3\right](h_i)$$

$$= \frac{1}{h_i}(x - x_i)^2 + \frac{1}{h_i^2}(x - x_i)^3$$

$$D_{i+1} = \left[-\frac{1}{h_i^2}(x - x_i)^2 + \frac{1}{h_i^3}(x - x_i)^3\right](h_i)$$

$$= -\frac{1}{h_i}(x - x_i)^2 + \frac{1}{h_i^2}(x - x_i)^3$$

$\therefore$ (2) becomes

$$u(x) = \left[\frac{3}{h_i^2}(x - x_i)^2 + \frac{2}{h_i^3}(x - x_i)^3\right](u_i)$$

$$+ \left[\frac{3}{h_i^2}(x - x_i)^2 - \frac{2}{h_i^3}(x - x_i)^3\right](u_{i+1})$$

$$+ \left[\frac{1}{h_i^2}(x - x_i)^2 + \frac{1}{h_i^3}(x - x_i)^3\right]u_i'$$

$$+ \left[-\frac{1}{h_i}(x - x_i)^2 + \frac{1}{h_i^2}(x - x_i)^3\right]u_{i+1}' \qquad (4)$$

The **shape functions** N_i's are also called **hat functions.**

The reason that the shape functions N's are also called "hat functions" is apparent when we look at the sketch of the N's for several adjacent elements as shown below

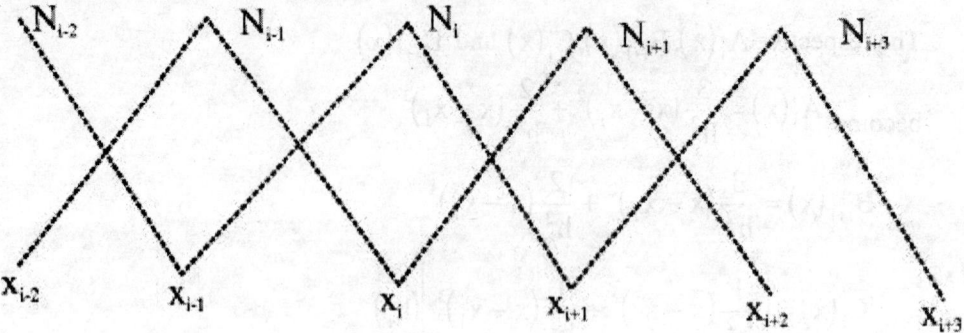

Interpolating relations for triangular elements

We will use a linear relation. The figure (a) drawn below is a sketch of typical element (i) whose nodes are numbered r, s, and t in the counter clock wise direction. The nodal values are cr, cs, and ct.

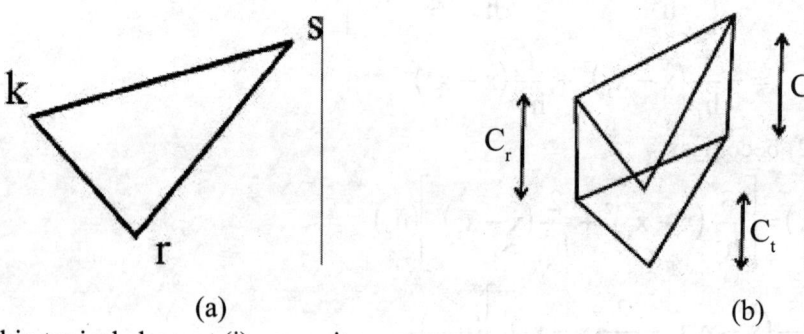

(a) (b)

Within typical element (i), we write

$$w(x, y) = N_r c_r + N_s c_s + N_t c_t = \sum_{j=r,s,.t} N_j c_j$$

$$\left(N_r, N_s, N_t\right) \begin{Bmatrix} c_r \\ c_s \\ c_t \end{Bmatrix} = (N)\{c_j\}, \tag{5}$$

where the N's (called **shape functions**) will be defined so that w(x, y) at an interior point is a linear interpolation from the nodal values, the c's. Equation (5) shows that w(x, y) can be expressed as the product of vectors (N) and {c}. We use parantheses to enclose a row vector and curly brackets to enclose a column a column vector. We indicate matrix M by [M]

Figure (b) suggests that w (x, y) lies on the plane above the element that passes through the nodal values. Equation (5) does not define w(x, y) outside the element (i); there will be similar expressions for the other elements, but their N's and c's will differ.

The entire region w(x, y) would be a surface composed of planar facets, each in a plane above an element. W(x, y) for the entire region is continuous, but w'(x, y) is not.

Another name for the N's of equation (5) is pyramid function. It is obvious that N's are independent of x and y, but the c's are independent of x and y. We are now in a position to develop expressions for the N's.

Since w(x, y) varies linearly with position within the element, an alternative way to write w(x, y) in a linear relation given by

$$w(x, y) = a_1 + a_2 + a_3 + y = (1, x, y)\{a\},$$ (6)

which must agree with the nodal values when (x, y) is (x_j, y_j), j=r, s, t. Hence

w at r: $c_i = a_1 + a_2 x_r + a_3 y_r$

w at s: $c_s = a_1 + a_2 x_s + a_3 y_s$

and w at t: $c_t = a_1 + a_2 x_t + a_3 y_t$

The above is a system of equations:

$$[M]\{a\} = \{c\}, \text{ where}$$ (7)

$$[M] = \begin{bmatrix} 1 & x_r & y_r \\ 1 & x_s & y_s \\ 1 & x_t & y_t \end{bmatrix}, \{a\} = \begin{Bmatrix} a_1 \\ a_2 \\ a_3 \end{Bmatrix}, \{c\} = \begin{Bmatrix} c_r \\ c_s \\ c_t \end{Bmatrix}$$

Solving for {a}, we get

$$\{a\} = [M^{-1}]\{c\}, \text{ where } M^{-1} \text{ is the inverse of } M, \text{ we have}$$

$$[M^{-1}] = \frac{1}{2(\text{Area})} \begin{bmatrix} (x_s y_t - x_t y_s) & (x_t y_r - x_r y_t) & (x_r y_s - x_s y_r) \\ (y_s - y_t) & (y_t - y_r) & (y_r - y_s) \\ (x_t - x_s) & (x_r - x_t) & (x_s - x_r) \end{bmatrix},$$ (8)

with 2 (Area)=det(M). The value of the determinant is the sum of the elements in row 1 of equation (8) within brackets. Here Area is the area of the triangular element.

We can write w(x, y) in terms of the shape function, for this purpose.

We have w(x, y)= a1+ a2x+ a3y =(1 x y) {a} $= (1 \quad x \quad y)[M^{-1}]\{c\}$ (9)

Also from (5), we have

W(x, y)=(N){c} (10)

Comparing (9) and (10), we have $(N) = (1 \quad x \quad y)[M^{-1}]$ (11)

Equation (11) indicates that each N is a linear function of x and y of the form

$$N_j = A_j + B_j x + C_j y \quad j = r, s, t$$ (12)

and that the coefficients are in columns j of $[M^{-1}]$.

Examples

For the triangular element shown in the adjoin figure with nots r, s, t, in counter clock wise order, find {a}. {N} and w(0.8,0.4)

Node	x	y	c
r	0	0	100
s	2	0	200
t	0	1	300

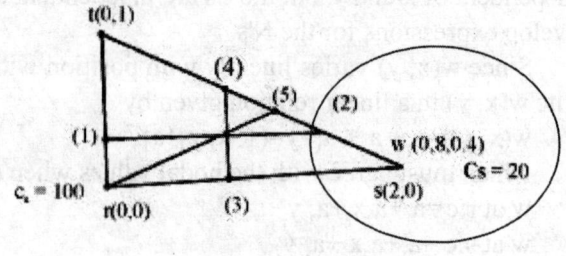

The point (1) is at (0, 0.4), so w there is $\dfrac{w-100}{0.4} = (300-100)\dfrac{200}{1}$ or by linear interpolation between nodes r and t. Similarly w at point (2) is 240. The point (0.8, 0.4) is

of distance from points (1) and (2), so $w(0.8, 0.4) = 180 + \dfrac{2}{3}(240-180) = 220$.

We get the same result by interpolating between points (3) and (4) and between node r and (5).

To get {a} we first compute $[M_{-1}]$.

$$[M] = \begin{bmatrix} 1 & 0 & 0 \\ 1 & 2 & 0 \\ 1 & 0 & 1 \end{bmatrix}, [M^{-1}] = \begin{bmatrix} 1 & 0 & 0 \\ -0.5 & 0.5 & 0 \\ -1 & 0 & 1 \end{bmatrix}$$

Then we compute $\{a\} = [M^{-1}]\{c\} = \begin{bmatrix} 1 & 0 & 0 \\ -0.5 & 0.5 & 0 \\ -1 & 0 & 1 \end{bmatrix}\begin{bmatrix} 100 \\ 200 \\ 300 \end{bmatrix} = \begin{bmatrix} 100 \\ 50 \\ 200 \end{bmatrix}$,

giving w(x, y)=100+50x+200y w(0.8, 0.4)=220.

Also $(N) = \begin{pmatrix} 1 & x & y \end{pmatrix}[M^{-1}] = (1 - 0.5x - y, 0.5x, y)$

or $N_r = 1 - 0.05x - y$
  $N_s = 0.5x$
and $N_t = y$

13.14.2 *Finite strip method*

Finite element method is well established as a powerful and reliable tour, and is frequently used as a a reference to calibrate other numerical, analytical and experimental results. However, FEM is a computationally intensive tool, to use because of the pre and post-processing required, and computer resources grow rapidly with the required precision level.

Cheung, Powell, Ogden and others introduced the Finite Strip Method (FSM), as a special case of the FEM. Finite Strip is a two dimensional strip for analysis of plates, based on simple polynomial functions in one direction, and continuously differentiable smooth series in the other direction. A less computer resource is needed without less of precision

A typical Finite Strip dirscretization of a ¢ rectangular plate is shown in the adjoining figure Strips of width b and length a are parallel to the y-axis and connected by nodal lines. The nodes, lead to the definition of the displacement function. As a first approach the strips are kept constant.

13.14.3 *Finite Boundary Element Method*

It is applicable most often to linear elliptic partial differential equations, often Laplace's equations. The problem of solving a partial differential equation within a given domain can be transformed into one of solving an equivalent **integral equation** on the boundary of the domain. The unknown in the integral equation will be the **"Charge density"**, on the boundary of the domain

The procedure is as follows.

We have the partial differential equation L [u(x)]=0, where L[.] is an elliptic differential

operator, with the Dirichlet or Neumann data: $u_s\Big| = f\left(x \text{ or } \dfrac{\partial u}{\partial n}\right)\Big|_s = g(x)$,

where S is the boundary of the domain.

Define $\psi(x, y)$ to be the Green's function of L[u(x)]=0 That is L[$\psi(x, y)$]=δ(x-y) where y is an arbitrary point inside the domain. Using Green's theorem the solution can be represented in anyone of the following forms.

i. $u(x) = \int_s \sigma(z)\Psi(x;z)dz$

ii. $u(x) = \int_s \mu(z)\dfrac{\partial\Psi}{\partial n}(x;z)dz$

iii. $u(x) = \int_s \left[\eta(z)\Psi(x;z) + \xi(z)\dfrac{\partial\Psi}{\partial n}(x;z)\right]dz$

In these equations $\alpha(z)$ and $\eta(z)$ represent the surface densities of the "single layer" potential, $\mu(z)$, $\zeta(z)$ represent the surface densities of the "double layer" potential, z represents a point on the boundary, and n represents the outward pointing normal. If $\alpha(z)$ or $\mu(z)$ or $\eta(z)$ and $\zeta(z)$ were known, then u(x) could be computed via one of the above three equations.

It turns out that the single layer potential is continuous across the boundary S, while double-layer potential has a jump of $\mu(y)$. This is because, as x tends to the boundary from inside of the domain.

$$u(y) = -\frac{1}{2}\mu(y) + \int_S \mu(y).\frac{\partial \Psi}{\partial n}(x;z)dy ,$$

while as x tends to the boundary from the outside of the domain,

$$u(y) = -\frac{1}{2}\mu(y) + \int_S \mu(y).\frac{\partial \Psi}{\partial n}(x;z)dy$$

Using representation (2), the solution of the Dirichlet problem, if we allow the point x to approach the boundary, we determine that

$$f(y) = -\frac{1}{2}\mu(y) + \int_S \mu(z).\frac{\partial \Psi}{\partial n}(z;y)dz$$

The Fredholm integral equation of the second kind can be solved for $\mu(y)$. After (y) is obtained, the value of u(x) may be computed.

When form (1) has been used to represent the solution of Neumann problem, then the following integral equation results $g(y) = -\frac{1}{2}\sigma(y) + \int_S \sigma(z).\frac{\partial \Psi}{\partial n}(x;z)dz$

After $\alpha(y)$ is in obtained by solving the above integral equation, the value of u(x) may be computed.

13.15 Qualitative Analysis of Ordinary Differential Equation- Stability of non linear systems

The solution to a differential equation is said to be stable if small perturbations in the initial conditions, foundary conditions or coefficients in the equation itself lead to "small" changes in the solutions. There are many different types of stability that are useful.

Stable

A solution y(x) of the system $\frac{dy}{dx} = f(y,x)$ that is defined for x>0 is said to be stable if, given any $\in > 0$, there exists a $\delta > 0$ such that any solution w(x) of the system satisfying $|w(0)-y(0)| < \delta$ also satisfies $|w(x)-y(x)| < E$.

Asymptotic stability

The solution u(x) is said to be asymptotically stable if, in addition to being stable, $|w(x)-u(x)|$ 0 as $x \to \infty$.

Relative stability

The solution u(x) is said to be relatively stable if|w(0)-u(0)|<δ implies that |w(x)-u(x)|< ∈ u(x).

The requirements of a numerical scheme are defined as **consistency, stability** and **convergence**. These three conditions cover different aspects of the relations between the discredited equation, the numerical solution and the exact analytical solution of he differential equation.

The **consistency** condition defines a relation between the differential equation and its discrete formulation. It is a condition on **structure** of numerical formulation.

The **stability** condition establishes a relation between the computed solution and the exact solution of the discretized equations. It is a condition on solution of the numerical scheme.

The **convergence** condition connects the computed **solution** to the exact solution of the differential equation. It is a condition as solution of the numerical scheme.

The conditions of **consistency stability and convergence** are related to each other and the precise relations is contained in the fundamental **Equivalence Theorem** of Lax.

While solving difference equation, we come across **single step methods** and **multistep methods**. While discretization of the unknown function u(x), the approximate values u_j will contain **errors**. It is necessary to be concerned with the effect of these errors on the solution when more and more grid points are taken to get a more accurate solution.

A method is **convergent** if, as more and more grid points are taken or step size is decreased, the numerical solution converges to the exact solution in the absence of round off errors.

A method is **stable**, if the effect of any single fixed round off error is bounded, independent of the number of mish points.

Von Neumann Test is applicable to finite difference shemes for partial differential equation. The test determines if the difference scheme is stable. For difference schemes with constant coefficients, the test consists of examining all exponential solutions to determine whether they grow exponentially in the time variable even when the initial values are bounded functions of the space variable. If any of them do increase without limit then the method is **unstable**. Otherwise, it is **stable**.

The test can also be applied to equations with variable coefficients by introducing new, constant coefficients equal to the frozen values of eh original ones at some specific points of interest.

The Von Neumann test for stability is applicable to parabolic partial differential equation.

Further parabolic scheme of $u_t=u_{xx}$, the step lengths h and k along the x and t directives must satisfies $k/h^2 \leq \frac{1}{2}$ for stability.

The courant - Friedrictis - Lewy consistency criterion if applicable for he stability test for hyperbolic partial differential equation.

Here the dependent variables with satisfy ordinary differential equation along the characteristics. These characteristics will propagate from the curves along which the initial date is given to every point in the domain. At any specific point as which the solution is desired, the characteristics through that point must be determined.

If a numerical scheme for a hyperbolic equation attempts to compute a numerical approximation to the solution at a appoint then all the relevant characteristics must be present or the method may not converge to the correct solution.

For the wave equation $u_{tt} = c^2 u_{xx}$ the step lengths handle along the x and y-directions, respective by should be chosen so that

(i) $e\dfrac{k}{h} > 1$, then the method cannot converge to the exact solution.

(ii) $e\dfrac{k}{h} < 1$, then the method may converge to the exact solution.

The term "stability" is used in a variety of ways in the description of differential equation and in particular numerical methods for solving differential equation. For some differential equation, any errors that occur in computation will be magnified regard less of the numerical methods. Such probles are called **ill conditioned**. Other differential equations require extremely small step sizes to achieve accurate results; these problems are called stiff.

We call a numerical method **stable** if errors incurred in one stage of the process do not tend to be magnified at later stages. The analysis of stability of a method often involves th investigation of he error for a simple problem, such as $\dfrac{dy}{dx} = \lambda y$. If the method is unstable for the model equation, it is likely to be have badly for other problems as well and the method is unstable. If $\lambda > 0$, the true solution grows exponentially and the error will increase as x increase. In the other hand, for $\lambda < 0$, the exact solution is decaying exponentially and the error also will tend to zero as x tends to infinity.

When we consider an m-step me for stability of the difference equation that defines a numerical method, the method is stable if all the roots of the corresponding characteristic polynomial

$$p(\lambda) = \lambda^m - a_1 \lambda^{m-1} + a_2 \lambda^{m-2} + \ldots \ldots a_m$$ satisfy $|\lambda_m| \leq 1$ and any root with $|\lambda_m| = 1$ is simple, It can be shown that for any method that is at least first-order accurate we must have $a_1 + a_2 + \ldots \ldots + a_m = 1$, so $\lambda_k = 1$ is a root. If the other m-1 roots satisfy $|\lambda_k| < 1$, the method is **strongly stable,** but not strongly stable, it is called **weakly stable.**

A method is called A-stable if any solution produced when the method is applied (with fixed step size h > 0) to the problem $\dfrac{dy}{dx} = \lambda y$ with ($\lambda = \alpha + \beta i$ and $\alpha < 0$) tends to zero as $n \to \infty$.

Since A-stability is difficulty to achieve, a some what less restrictive stability condition, known as **Stiff - stability**, is often sufficient. Methods for stiff ODE are **implicit** and often require iterative techniques for their solution.

BIBLIOGRAPHY

1. ANDERSON, T.W - An Introduction to Multivariate Statistical Analysis - Wiley and Sons. INC, Third Edition, 2003
2. ANDREWS, L.C and PHILIPS, R.L - Mathematical Techniques for Engineers and Scientists - Prentice Hall of India-2006
3. ANDREWS, L.C. and SHIVA MOGGI, B.K - Intergral Transforms for Engineers - Prentice Hall of India Private Limited, New Delhi, 2003
4. BALAGURUSAMY, E - Numerical Methods - Tata McGraw-Hill Publishing Co- New Delhi-1999
5. BLCK LEY, W.G. and THOMPSON, R.H.S – Matrices - Their meaning and Manipulation - B.I. Publications, Delhi-1964
6. BONDY, J.A and MURTHY, V.S.R - Gruph Theory with Applications - Macmillan, 1977
7. BOYEE AND DIPRIMA - Elementary Differential Equations and Boundary Value Problems with ordinary Differential Equation – Architect CD, Eight th Edition, 2005.
8. CHURCHIL RUEL, V and JAMES WARD BROWN - Complex variables and Applications - MCGRAWHILL BOOK, Company, Fourth Edition, 1986
9. CURTES F. GERALD, PATRIC, O, WHEAL BY–Applied Finite Element Method – Harwood Publishing Co, Chichester.
10. CURTIS F. GERALD and PATRICE WHEATLEY – Applied Numerical Analysis – Addison - Wesley Publication Company, Third Edition – 1984.
11. DALLAS E. JOHNSON ETAL–Applied Multivariate Methods for Data Analysis – Thomson and Duxbbury Press, Singapore, 1998.
12. DANIEL D. MCCRACKEN and WILLAM S. DORN – Numerical Methods and Fortran Programming - Wiley series INC–Wiley International Edition , 1964.

13. DANIEL ZWILLINGER – Hand Book of Differential Equations–Academic Press INC, Newyork, 1989.
14. DONALD GROSS and CARL M. HARRIS – Fundamentals of Queueling Theory – JOHN WILEY AND SONS Newyork, Second Edition, 1985
15. DOUGLAS, C.MONTGOMERY, GRORGE C. RUNGER – Applied Statistics and Probability for Engineers, - JOHN WILEY AND SONS, Third Edition, Reprint – 2008.
16. ELGOLTS, L – Differential Equations and the Calculus of Variations – MIR publishers, Moscow, 1973.
17. FINALYSON, A – The Method of weighted Residuals and Variational Principles – Academic Press – 1972.
18. FREUND, J.E – Mathematical Statistics – John Wiley and Sons 1NC, Asia – Fifth Edition – 2005.
19. FROBERG, C.E–Numerical Mathematics – The Bensjamin / Cummings Publishing Co. INC–1985.
20. GASS, S.L.–Linear Programming MCGRAW HILL BOOK Company.
21. GEE, S.G and RAGHAVENDRA, V- Ordinary Differential Equations and stability Theory – TATA MCGRAW HILL Pupblishing co., 1980.
22. GEOFFREY GORDEN–System Simulation – Prentice Hall of India Private Limited, New Delhi, Second Edition – 2006
23. GEORGE J. KLIR and YUAN, B-Fuzzy sets and Fuzzy Logic – Theory and Application – Prentice Hall of India Private Limited – 1997.
24. GHILDYAL, C.D., RAO, K.V and BALA CHAN DRAN, S – Numerical Analysis – An Integrated Approach – Macmillan India Limited – 1985.
25. GOEL, B.S, MITTAL, S.K – Operations Research – Pragate Prakashan, Meerut – Fourth Enlarged and Research ed Edition – 1980.
26. GOULT, R.T, HOSKIN, R.F MILNER, J.A and PRATT, M.J- Applicable Mathematics – A course for Scientists and Engineers – Macmillan – 1972.
27. GREMALDI – R.P – Discrete and Combinational Mathematics – Pearson Educational, INC – 1999.
28. GREWAL, B.S – Higher Engineering Mathematics – Khanna Publications, 40th Edition – 2007.
29. GRIWAL, B.S – Numerical Methods in Engineering and Science, Khanna Publications, 40th Edition, 2007.
30. GROSS, D, HARRIS, C.M – Fundamental of Queuing Theory, John Wiley and sons, Third Edition – 2002
31. GUE THOMAS – Mahtematical Methods of Operations Research
32. GUETHOM GUPTA, A.S Calculus of Variations with Applications – Prentice Hall of India Private Limited, New Delhi – 1997.

33. GUPTA, P.K and HIRA, D.S – Operations Research – An Introduction, S. Chand & Company limited, New Delhi f- Second Reprinted Edition – 1986.

34. GUPTA, P.K and MAN MOHAN – Linear Programming and Theory of Games – Sultan Chand and Sons. New Delhi – Second Extensively Revised Edition 1978

35. GUPTA, P.P, MALLIK, G.S – Calculus of Finite Differences and Numerical Analysis – Krishna Pra Kashan Mandir, Meerut, U.P – Sixth Rivised Edition – 1984.

36. GUPTA, S.C and KAPOOR, V.K – Fundamentals of Mathematical Statistics – Sultan Chand – Extensively revised edition 1982.

37. GUPTA, S.C, and KAPOOR, V.K – Fundamentals of Mathematical Statistics – Sultan Chand and sons, New Delhi – 2001.

38. GUPTA, S.K – Numerical Methods for Engineering – New Age Publications – 1995.

39. IRWIN MILLER and MARYLEES MILLER – John E. Freund's Mathematical statistics – Prentice Hall of India Private Limited, New Delhi – Sixth Edition – 2002.

40. JAIN, M.K, IYENGAR, S.R.K and JAIN, R.K – Computational Mathematics for Partial Deferential Equations – New Age International Private Limited, 2003.

41. JAIN, M.K, IYENGAR, S.R.K and JAIN, R.K – Numerical Methods for Scientific and Engineering Computation – Wiley Eastern Limited – 1987.

42. JAMES, G – Advanced Modern Engineering Mathematics – Pearson Education, Third Edition – 2004.

43. JAY L. DEVORE – Probability and statistics for Engineers – CEN GAGE Learning, India Edition, Singapore – 2008.

44. JOHNSON, R.A and GUPTA, C.B. Millen and Freund's Probability and statistics for Engineering – Pearson Education – Asia, Seventh Edition – 2007.

45. JOHNSON, R.A, WICHERMAN, DEAN D.W – Applied Multivariate Statistical Analysis – Pearson Education – Asia – Fifth Edition – 2002.

46. KAMBO, N.S – Mathematical Programming Techniques – Affiliated East West Press Private Limited – New Delhi – 1984.

47. KANTISWRUP; GUPTA, P.K and MAN MOHAN – Operations Research – Sultan Chand and Sons Publishers, New Delhi – Second Revised Edition – 1986.

48. KAPUR, J.N and SEYHANA, H E. Mathematical Statistics – S. Chand and company limited – Twentieth Edition – 2001.

49. KLIR, G – Fuzzy Sets and Fuzzy Logic – Theory and Applications – Prentice Hall – New Jersy – 1995.

50. KREYSZIG, E-Advanced Engineering Mathematics – John Wiley and Sons – Eigth Edition – 2004.

51. KRISHNA RAJU, N and MUTHU, K.U – Numerical Methods for Engineering Problems – Macmillan India Limited – Reprint & Second Edition – 1992.

52. KWAK, N.K – Mathematical Programming with Business Applications – MCGRAW HILL - 1973

53. MANN, P.S – Introductory Statistics John wiley and sons INC – Fifth Edition – 2004.
54. MATHEWS.J.H and HOWELL, R.W – Complex Analysis for Mathematics and Engineering – Narosa Publishing House, New Delhi – 1997.
55. MEDHI, J – Statistical Methods – Introductory Text – New Age International Private Limited, New Delhi – 2000.
56. MITCHELL, A.R and GRIFFITH, D.F – The Finite Difference Method in Partial Differential Equation – John Wiley and Sons, New Yark – 1980.
57. MONTGOMERY, D.C, RUNGER, G.C – Applied Statistics and Probability for Engineers, John Wiley and Sons – Third Edition – 2007.
58. MOON, T.K, STERLING, W.C – Mathematical Methods and Algorithms for Signal Processing – Parason Education, 2000.
59. MORTON, K.W and MEYERS, D.F – Numerical Solutions of Partial Differential Equation – Cam bridge University Press, Cambridge – 2002.
60. NARAYANAN, S; MANICAVASAGAM PILLAI, J.K and RAMANIAH, G – Advanced Mathematics for Engineering Students – Volume III – S Viswanathan Publishers Private Limited, Chennai – 1990 (a) NARASINGH DEO – system simulation with digital computer – Prentice
61. O' NEIL, P.V – Advanced Engineering Mathematics – Thomson, Asia Private Limited, Singapore – 2003.
62. PREM S.MANN – Introductory Statistics, John Wiley and Sons, INC, Asia – Fifth Edition – 2005.
63. PUSHPAVANAM, S – Mathematical Methods in Chemical Engineering – Prentice Hall of India.
64. RAJASEKARAN, S – Numerical Methods in Science and Engineering – A Practical Approach – A.H Wheeler and Company Private Limited – 1986.
65. RAMANIAH, G – Tensor Analysis – S. Viswanathan Publishers Private Limited, 1990.
66. REDDY, J.N – An Introduction of the Finite Element Method – Tata Mc-graw hill Book Company Limited, New Delhi – Second Edition – 2003.
67. RICHARD JOHNSON and GUPTA, C.B – Miller and Freund's Probability and Statistics for Engineers – Prentice Hall of India Private Limited, New Delhi – Seventh Edition – 2007.
68. RICHARD, A JOHNSON and DEAN W. WICHEM – Applied Multivariate Statistical Analysis Pearson Education, Asia – Fifth Edition – 2003.
69. ROBERT – V, HOGS; ELLIOT, A; Probability and Statistical Inference – Pearson Education, Indian Seven Edition, 2006.
70. ROBERT, Azz1, T.G – Computer Networks and Systems – Queueting Theory and Performance Evaluation – Springer Verlag – Third Edition – Reprint – 2002.
71. SPIEGEL, M.R – Theory and Problems of Complex Variable and its Applications – Schaum's Outline Series – McGraw Hill Book Company, Singapore – 1981.

72. STEPHENSON, G; RADMORE, P.M – Advanced Mathematical Methods for Engineering and Science Students Cambridge University Press – 1999.

73. TAHA, H.A – Operations Research An Introduction – Pearson Education, Asia Edition, New Delhi – Seventh Edition – 2003.

74. TIMOTHY, J. ROSS – FUZZY set and Fuzzy Logic - Engineering Applications – Wiley India Second Edition, Third Reprint – 2008.

75. VOHRA, N.D – Quantitative Techniques in Management – Tata McGraw Hill Company Limited, 2007.

76. WAGNER, M. HARVEY – Principle of Operation Research Prentice Hall

77. WALPOLE, R.E , MEYEARS, R.H , MEYERS, S.L and YE, K-Probability and Statistics for Engineers and Scientists, Pearson Education, New Delhi – Asia Eight Edition – 2007.

78. WARD CHENEY, DAVID KINCAID – Numerical Mathematics and Computing – Thompson – Books / COLL – Vicas Publishing House, Fourth Edition – 1999.

79. WILEY SERIES: Numerical Computations – Fundamentals of Numerical Discretiztion of Internal and External Flows Volume I.

80. WINSTON, W.L – Operations Research – Thomson – Books – Coll – Fourth Edition – 2003.

81. ZEIGLER, BERNARD, P – Theory of Modeling and Simulations- John Wiley and Sons, INC – 1976.

74см